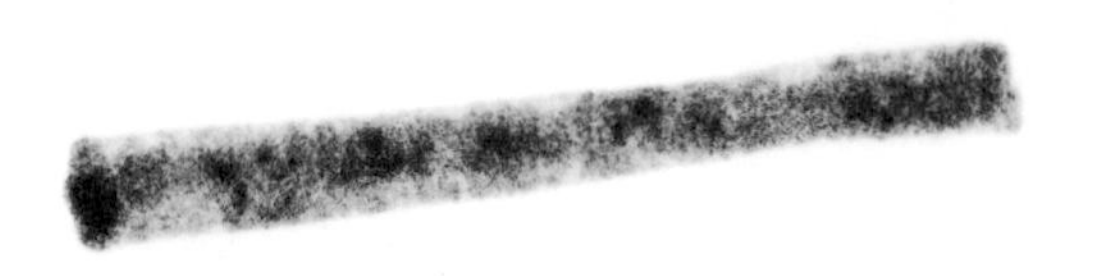

Optical Fiber Sensors Volume Four

Applications, Analysis, and Future Trends

For a complete listing of the *Artech House Optoelectronics Library*,
turn to the back of this book.

Optical Fiber Sensors Volume Four

Applications, Analysis, and Future Trends

John Dakin
Brian Culshaw

Artech House
Boston • London

Library of Congress Cataloging-in-Publication Data

Optical fiber sensors.

(The Artech House Telecommunication library)
Includes bibliographies and indexes.
Contents: v.1. Principles and components-v.2. Systems and applications-v.3 Components and subsystems-v4. Applications, Analysis, and Future Trends.
1. Optical fibers. 2. Optical detectors. 3. I. Dakin, John, 1947–
II. Culshaw, B. III. Series.
TA1800.0666 1988 621.36'92 88-7611
ISBN 0-89006-317-6 (v.1)
ISBN 0-89006-376-1 (v.2)
ISBN 0-89006-932-8 (v.3)
ISBN 0-89006-940-9 (v.4)

British Library Cataloguing in Publication Data
Culshaw, B. (Brian)
Optical fiber sensors
Vol. 4 : Applications, Analysis, and Future Trends
1. Optical fiber detectors
I. Title II. Dakin, John, 1947–
621.3'692

ISBN 0-89006-940-9

Cover design by Jennifer Makower

685 Canton Street
Norwood, MA 02062

International Standard Book Number: 0-89006-940-9
Library of Congress Catalog Card Number: 88-7611

10 9 8 7 6 5 4 3 2

Contents

Preface

In this volume we again visit the applications area without which none of the intriguing science and technology would be relevant, and would, by now, have all slipped into ignominy. In Volumes One and Two we also devoted all of the second book to applications concepts. In this volume, there are some recurrent themes and some slight changes of emphasis, but the degree of communality between the topics covered in Volumes Two and Four provides encouraging evidence that optical fiber sensors are beginning the find their place.

The conventional wisdom of optical fiber sensors points toward medical and biomedical engineering, the electrical power industry, structural monitoring, and specialized industrialized sectors as the principal users of OFS technology. The gyroscope sits squarely in its own idiosyncratic slot and services the specific needs of some sections of the rotation measurements community. The same conventional wisdom also points out that distributed measurements are critical and multiplexing is highly desirable.

All of these topics were covered, to some extent, in Volume Two. Here in Volume Four we see a significant consolidation in these sectors and energetic exploitation within a select range of applications.

The first four chapters consider chemical, biochemical, and biomedical applications. Spectroscopy is a key concept and the principles of spectrometers for optical fiber measurements have already been covered in some detail in Volume Three. There are two basic approaches involving either direct spectroscopy, where the light interacts with the measurand of interest (discussed by John Dakin and his colleagues), or involving intermediate indicators, where the light interacts with an independent measurand-sensitive dye (discussed by Otto Wolfbeis). Light scattering is a very important technology and Bob Carr discusses some of its multifaceted characteristics in Chapter 9. The medical and biomedical sector has emerged as a natural applications area for optical fiber sensors, and Francesco Baldini and Anna Mignani conclude the subsection on this particular topic with a description of some of the ongoing work in in-vivo biomedical sensors. Many of these are now emerging as commercial products.

In Chapter 11, Kazuo Hotate focuses on gyroscope features and discusses the evolution of the low cost and the expensive in gyroscope technology. The many potential applications ranging from missiles to guided lawnmowers are particularly intriguing. The gyroscope is covered in great detail in Hervé LeFèvre's book in this Optoeletronic Series and is now a relatively mature technology with numerous corporations worldwide at the preproduction phase.

Julian Jones and John Berthold in Chapters 12 and 13 cover some of the very wide-ranging industrial application of optical fiber sensors. The interesting feature of this sector is the specialized nature of the vast majority of the applications and the need, perhaps more than elsewhere, for complementary techniques to augment the guided wave optics. Many of the applications are very specific and therefore far more cost-tolerant, since in many cases the optical fiber offers a unique solution that cannot be attained using alternative technologies.

Multiplexed and distributed sensing techniques are very important benefits of optical fiber sensor technology, and again, both were introduced in Volume Two. Distributed sensing has moved on considerably. The Raman temperature sensing system is a commercial entity and some forms of microbend strain sensing systems have also come to fruition. However, distributed sensing, while undoubtedly very attractive as a concept has achieved relatively little practical presence.

In Chapter 14 we look at some of the emerging techniques in distributed sensing that promise to substantially broaden the market address by combining unique solutions with cost-effective technology. Tsuneo Horiguchi has achieved much with Brillouin backscatter systems for distributed sensing. These systems measure mechanical strain and/or temperature in single-mode fibbers with high precision (a few tens of miscrostrains) over ranges up to 100 km. They can most definitely address problems well outside the remit of any conventional technology and are already commercially available but at a very high price. Alan Rogers looks at the possibilities for the Kerr effect in distributed sensing whereby strain or temperature fields can be turned into a difference frequency that can be accurately demodulated. While these systems are currently some way from commercial reality, the potential they offer is significant. Craig Michie and George Stewart, in contrast, examine some of the systems for distributed chemical measurement, including evanescent wave spectroscopy and some very simple chemical:mechanical systems using microbend transducers now in engineering prototype phase. This scratches the surface of a most important technology. There are many other distributed measurement systems using Rayleigh backscatter, evanescent spectroscopy in sol-gel coats, and distributed fluorescence, but our selection here will give a flavor of some exciting developments.

Alan Kersey brings us up to date on the extremely important topic of sensor multiplexing. It is here that the Bragg grating described in detail in Volume Three will probably make its biggest contribution as a sensor system element, and much of this chapter is devoted to multiplexing systems for gratings. There are other things as well: Multiplexed interferometers feature strongly and successful trials of very large arrays have been completed. Paradoxically, the simplest of optical fiber sensors using intensity modulation or straightforward color modulation have not been extensively incorporated into systems, though perhaps their relatively high intrinsic losses preclude all but the simplest networks. It is interesting to note that in Volume Two the discussion on multiplexing was largely theoretical. A few years on we see very significant changes with a large number of successful multiplexed systems implemented and running in realistic environments.

Smart structures have a mixed press but the mixed press agrees on their importance in structural engineering in the coming decades. Early repair is undoubtedly preferable to longer term replacement. There is also considerable agreement that optical fibers will be an important-but not exclusive-sensing technology in such structural instrumentation systems. Bill Spillman who contributes Chapter 16 of this volume has a uniquely balanced, even pragmatic, view of the subject, and presents his views on how optical fiber sensors will complement all the other new techniques that can be constructively brought to bear on a most important problem.

Some sort of market insight is an essential feature for a text of this nature, and Sam Crossley, who has undertaken market surveys in the area and interacted strongly with manufacturers, completes the main body of this volume with an account of ongoing current commercial activity in OFS. Again, by definition, any discussion of this nature is bound to omit some important activity but this chapter paints a picture of a relatively buoyant, if small, industry supporting viable activity and addressing specialized market areas. In common with all other instrumentation techniques, optical fiber sensors are inevitably niche oriented and generally the domain of the small and specialized rather than the large and omnipresent corporation. The gyroscope, with its need for extensive test facilities and specialized marketing, is probably the only example of OFS technology that has taken a significant part in large corporate portfolios.

We have missed out current and voltage sensing, not because it is unimportant but because we feel that relatively little has changed at the conceptual or engineering level since Volume Two was published, and intrinsic (Faraday rotation) and extrinsic (magneto-optic crystals) sensors were described in principle. Certainly the materials have been improved; certainly the engineering of real-life systems has turned concept into reality, but the principle remains that all these sensors rely on electric or magnetic fields inducing a change in birefringence. We have also not mentioned the important area of sensing in telecommunications, where the use of the OTDR coupled with advanced data analysis techniques has revolutionized network testing and where a range of new components have emerged to, for example, detect the presence of moisture in remote cable ducts and manholes.

So, much has changed since the publication of the first two volumes. Some of the consolidation we anticipated at the time has happened and dominant applications areas have emerged. These sectors have been identified for some time and are, with the benefit of hindsight, naturals for optical instrumentation. Are there any future niches? If we look at needs for advanced instrumentation then environmental assessment driven by both legislative and social pressures must loom large in the future. Optics will have its place principally through chemical and turbidity measurements. There will be new technologies too: integrated optics chips will become more intricate, tunable sources more readily available, and the fiber Bragg grating will inevitably appear as a (hopefully) low-cost production item with properties primarily geared initially toward the telecommunications industry. OFS has most certainly gained from telecommunications research and development in the past. There has, however, been an important philosophical change since Volume Two. In most major countries the telecommunications industry has shifted its interest in optoelectronics hardware to network software and components, so the device research that gave us the fiber amplifier and Bragg grating is much more subdued than it was a decade ago. However, in other disciplines, including environmental chemistry and structural engineering, new philosophies are also evolving, and the need for elegant and effective instrumentation is emerging as a buoyant interest from which fiber sensors will reap the benefits.

In many ways the crystal ball is less clear than it was at the end of the 1980s. While the dominant applications sectors for fiber optic sensing have emerged, the ones which could emerge are far less self-evident, but emerge they will. We would happily respond to the best suggestion for these new emerging techniques and applications we receive before the end of 1997 with a bottle of champagne or (preferably) a good malt whisky. If you have read this far and are inclined to throw in your views, we look forward to your thoughts.

We both greatly enjoyed bringing these volumes together and again we would like to express our sincere thanks to our families and our many friends and colleagues, particularly to Julie and Susanna at Artech and to the ever-tolerant Aileen at Strathclyde and Janet at Southhampton, who helped to bring it all about.

Brian Culshaw
Glasgow, Scotland

John Dakin
Southampton, England

March 1997

Chapter 7

Optical Fiber-Chemical Sensing Using Direct Spectroscopy

John P. Dakin, Steven J. Mackenzie,
and Jane Hodgkinson
University of Southampton, England

7.1 INTRODUCTION

It is the quantized nature of our universe, most evident at the atomic and molecular level, which allows so much information about the constituents of matter to be deduced from optical spectra. Because molecules and atoms can only emit or absorb photons (particles of light) with energies that correspond to certain allowed transitions between quantum energy states [1], optical spectroscopy is one of the most valuable tools of the analytical chemist. It can provide a rapid nondestructive analysis of many important compounds and radicals, and optical fibers permit remote online monitoring.

Two basic approaches are possible: either direct optical interaction with the analyte or indirect analysis using chemical indicators (i.e., compounds that change their optical properties by reaction with the analyte). This chapter concentrates on the first method, that of direct spectroscopy. As stated above, the advantages are that the method is nondestructive to the sample under test and is usually very rapid. The disadvantage is that it is often not as selective as indicator chemistry, because many families of compounds exhibit similar optical properties when monitored directly.

We shall summarize the principal methods of optical spectroscopy, describe how a sensing head can be "remoted" using optical-fiber probes, and discuss the advantages and disadvantages of several techniques. In Chapter 10, by O. Wolfbeis, the complementary indirect methods based on optical indicators will be considered.

7.2 BASIC TECHNIQUES OF OPTICAL SPECTROSCOPY

In this section, we shall not explain the features present in any spectrum, but will simply describe how one or more features of a spectrum may be measured to determine the presence of certain chemical species. Several case studies of optical-fiber remote measurements are presented in Section 7.3 of this chapter, but first we will present the general techniques of optical spectrometry.

7.2.1 Transmission Spectroscopy

Light is absorbed due to rotational and vibrational transitions of molecules, and electronic transitions of molecules and atoms [1]. The region of the electromagnetic spectrum that is efficiently transmitted (with attenuation below 10 dB/km) by silica optical fibers extends from 600 to 1,900 nm [2,3]. This makes the electronic transitions and the overtones of molecular vibrational transitions accessible to remote investigation over standard optical fibers. Over much shorter lengths, near-UV spectra can be measured. For IR analysis, fluoride, silver halide, and chalcogenide glasses extend the (short range) possibilities to about 8,000 nm, allowing the fiber-remoted study of vibrational transitions, although such fibers are both expensive and fragile.

7.2.1.1 Basic Concepts of Transmission Spectroscopy

This is the method most commonly used by the chemist, using an instrument called a spectrophotometer, which measures the variation of transmission of light through a sample as a function of optical wavelength.

At each particular wavelength λ and over a small (resolution) wavelength interval $\delta\lambda$, the transmitted power $P(\lambda)$ is measured. This is given by Lambert's Law, which defines $P(\lambda)$ as

$$P(\lambda) = P_0(\lambda) \cdot \exp[-\alpha(\lambda) \cdot \ell] \tag{7.1}$$

where $P_0(\lambda)$ is the input power over the same wavelength interval, ℓ is the optical path length through the sample, and $\alpha(\lambda)$ is the attenuation coefficient at the wavelength λ. The sample characteristics are described by either a transmission factor,

$$T(\lambda) = \exp[-\alpha(\lambda) \cdot \ell] \tag{7.2}$$

or an absorbance factor,

$$A(\lambda) = \log_{10}(P_0(\lambda)/P(\lambda)) \tag{7.3}$$

For dilute solutions of absorbing compounds, the attenuation coefficient a, and

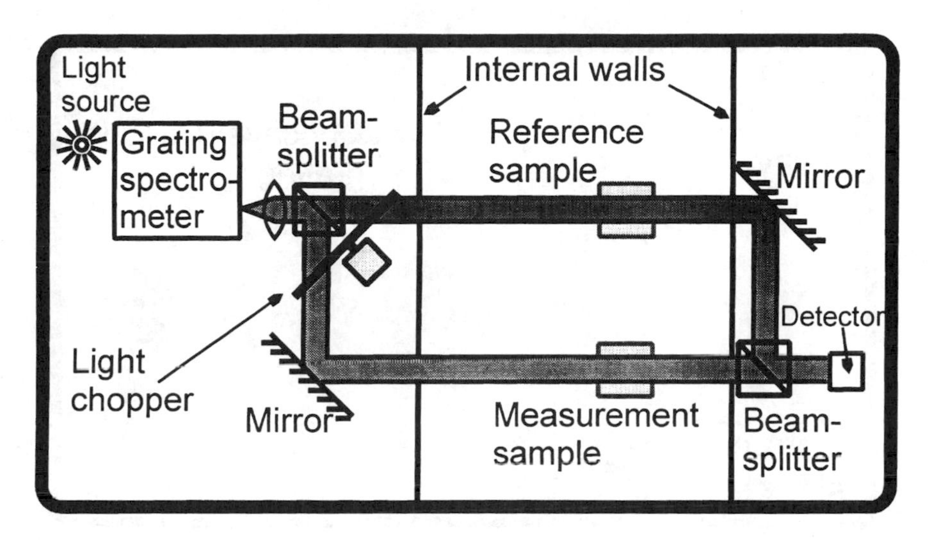

Figure 7.1 Schematic of transmission spectrophotometer.

the absorbance A are usually both proportional to the solute concentration C. This is known as Beer's Law. The formula for the absorbance can be expressed mathematically as $A = \epsilon \cdot c \cdot l$, where ϵ is the (decadic) molar absorption coefficient and c is the molar concentration.

The basic principle of the spectrophotometer is shown in Figure 7.1. A few years ago, most instruments used a grating spectrometer as a tunable filter. In early instruments, the output beam from this was directed alternately via a sample (or measurement) channel and a separate reference channel. Most modern instruments are single-channel, taking advantage of the greater stability of modern optoelectronics and computer processing. With these latter types, the measurement sequence involves a first measurement using a reference cell, storage of the transmitted signal, then measurement with the sample in the cell, and, finally, division of the new signal by the stored reference.

The latest generation of instruments (Figure 7.2), use a modified optical system, with a grating spectrometer system (with a focal plane detector array) placed after the sample. The detector array is now usually based on charged-coupled device (CCD) technology. This arrangement effectively allows simultaneous parallel detection of transmitted signals at each wavelength. This has the attraction of greatly improving the signal to noise ratio (or reducing the signal averaging time). The rapid progress in CCD detector arrays suitable for the UV/visible/near-IR region (i.e., 0.2 to 1.05 μm) has led to dramatic improvement in several aspects, in particular the detection sensitivity, the number of detectors (pixels) possible, and in the uniformity of response across the array. For this spectral region, the method has largely overtaken the alternative approach of Fourier transform spectroscopy (FTS, see Section 7.2.8).

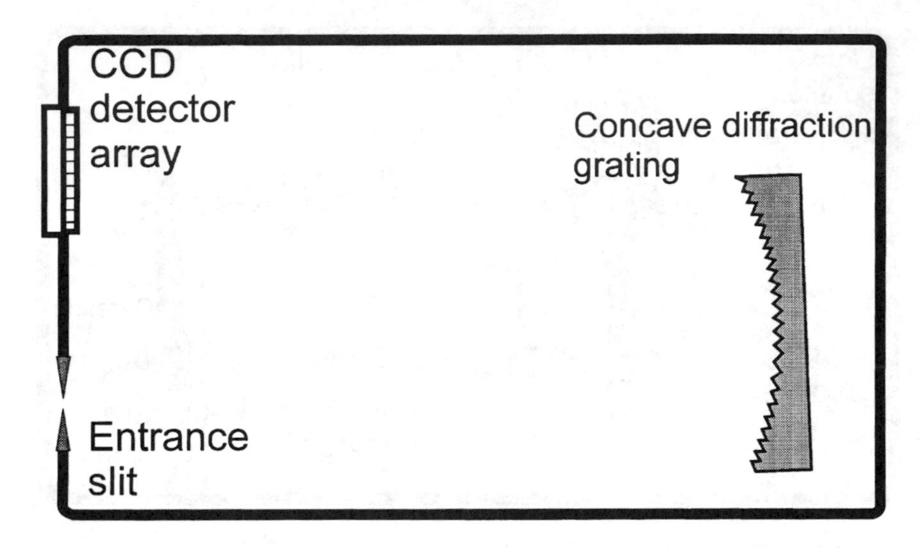

Figure 7.2 The latest generation of spectrophotometer, with a grating and focal plane detector.

However, when it is desired to perform ultrahigh spectral resolution, or if measurements are required in the mid- and near-IR region of the spectrum, (regions for which detector array technology is less advanced), FTS still has a strong position.

Working in the IR is made difficult by the presence of water, which has very strong IR absorption bands, which might interfere with measurements made. Working in the UV avoids this problem, although scattering from small particles may necessitate filtering of analytes.

7.2.1.2 Fiber Probes for Use With Spectrophotometers

In order to modify the spectrophotometer for optical-fiber use, it is convenient to focus the light source into the outgoing fiber and couple the return fiber into the input slit of the spectrometer, but unfortunately this usually involves losses due to the small core diameter and acceptance angle typical of optical fiber. At the remote measurement head (Figure 7.3), it is usually necessary to collimate the light from the outgoing fiber through the sample cell (or region) and refocus the transmitted light into the return fiber.

However, many design variations of the sensor heads are possible, for example, ones having "folded" light paths to allow the fibers to enter and leave from the top end of a "dip-stick" type probe. For very highly absorptive samples, no optics are needed, as the measurement sample can be simply allowed to enter a narrow gap between aligned fiber ends, an arrangement that has low losses if the beam diameter of light from the transmitting fiber has not diverged to greatly exceed the core diameter of the receiving fiber.

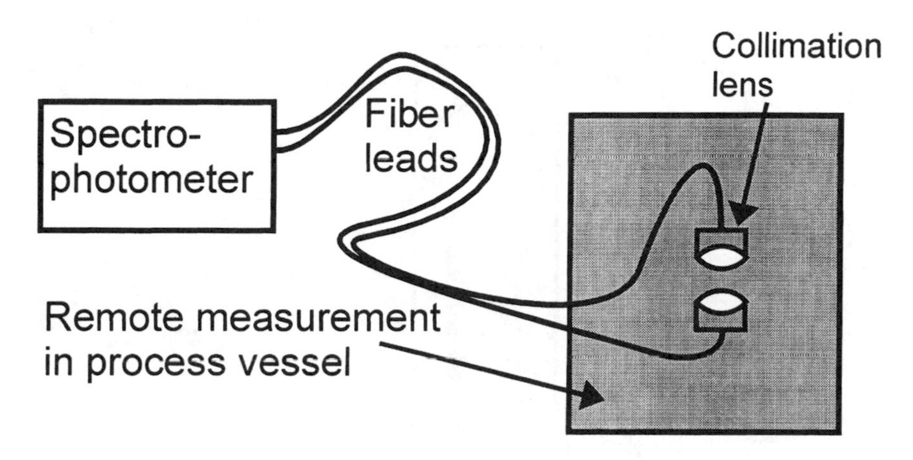

Figure 7.3 Remote measurement of sample absorption through optical fibers.

This arrangement was used in an early process control investigation, reported by Freeman [4], shown in Figure 7.4. The ends of two fibers are separated by 2 mm and the analyte (in this case, copper sulphate solution in an electroplating bath) is allowed to flow between them. The design of the probe is such that air bubbles (used to agitate the solution) do not pass between the two fibers.

Fiber-optic probes can easily be used to monitor samples that have strong spectral absorption peaks, as the transmission loss in short lengths (few tens of meters) of fiber is low. The low-loss fiber transmission windows for silica-based fibers are in three main areas. The first window covers the region 700 to 900 nm (typical losses of 3 to 5 dB/km), the second 1,050 to 1,350 nm (typical losses of 0.5 to 2 dB/km), and the third 1,450 to 1,750 nm (typical losses of 0.2 to 3 dB/km). In low OH^- (dry) fibers, the first two windows effectively merge into one, as in such types the OH^- absorption at 950 nm is extremely low, whereas the peak at 1,400 nm remains significant for transmission paths over 1 km.

If the OH^- content could be kept even lower than at present, there would be a reasonably low loss window from 700 to 1,750 nm. For short-distance transmission (few tens of meters), however, losses for most silica fibers are moderate over the entire range from 400 to 2,100 nm. However, if the sample has a flat, featureless absorption spectrum, or if a quantitative measure of the absorption is required, it is necessary to provide a reference signal in order to determine the sample properties more precisely. One method is to provide a fiber switch to alternately direct the guided light through the sample probe and through a reference fiber path before recombining the signals. The main technical problem is to avoid relative variations in transmission between these paths, such as may occur if there are any interchannel variations arising from the switch, fiber-bending losses, connectors, or fiber couplers.

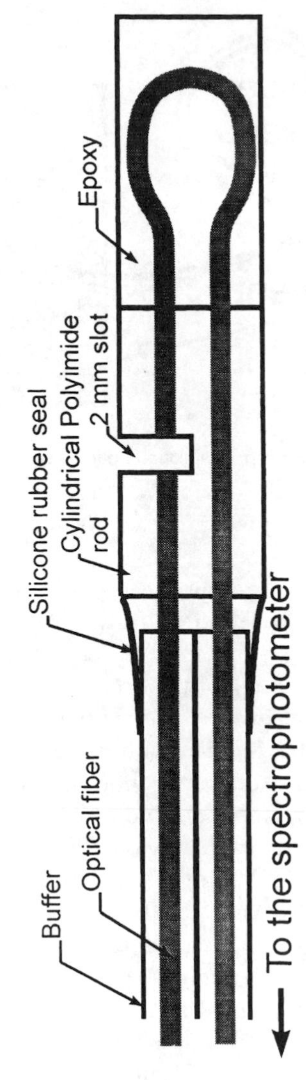

Figure 7.4 An early transmission probe as described by [4].

The other problem, mentioned above, is the small acceptance aperture of the fiber. Although the small size of the fiber can be effectively enlarged using a focusing lens, this results in an effective narrowing of the acceptance angle. The effective throughput *TP* of a step-index fiber, as defined by the approximate relation below, remains unchanged:

$$TP = A \cdot \pi \, (NA)^2 \tag{7.4}$$

where A is the area of the fiber core and NA is the numerical aperture. At the launch end of the system, the throughput *TP* represents the ratio of the power launched into fully guided modes of the fiber from a large area source, when the latter is butted against the launch end of the fiber, to the on-axis radiance of the source. The throughput has half the above value for graded-index fibers. The throughput also determines the light intensity collected by the fiber from diffusely scattering systems such as Raman scatterers. Clearly, large-core, high NA fibers are the most efficient if all the light transmitted through the fiber can be usefully employed. However, the useful numerical aperture of an optical fiber may be constrained by the acceptance NA of the spectrophotometer.

The total power launched into the fibers from high-radiance LEDs is rarely above a few hundred microwatts, and the spectral radiance of incandescent filament lamps is usually at least an order of magnitude less. Thus, well-designed light detection systems are required to produce practical sensors. Use of expensive high-intensity arc lamps can improve matters, but laser sources are far better if ones of suitable wavelength are available.

With laser sources there is no problem in achieving launch efficiencies of over 80%, particularly into multimode fibers, and the detection system constraints are eased substantially. Obtaining absorption spectra is much more difficult, except when using expensive tunable types (low-cost semiconductor lasers are, however, tunable over narrow ranges).

7.2.2 Absorption Measurements by Attenuated Total Reflection/Evanescent Field Absorption

The technique of attenuated total reflection (ATR) spectroscopy is a commonly used tool for the study of the IR absorption spectra (from 2 to 10 μm) of, among other things, highly absorbing or turbid liquids [5]. Recently, the technique has been extended through the use of optical fibers, where it is more usually referred to as evanescent field absorption (EFA) spectroscopy or evanescent wave spectroscopy (EWS) [6].

When light incident from a medium of refractive index n_1 to one of lower index n_2 is reflected at an interface by total internal reflection, some of the energy of the reflected wave penetrates the lower index medium as an evanescent field [7].The evanescent field is not a traveling wave, and does not propagate energy away from the boundary. As the angle of incidence α approaches the critical angle, the penetration depth can become very large, with a proportionally larger fraction of the reflected energy present in the evanescent wave. The penetration depth also increases with increasing wavelength. If there are any absorbing components in the lower index medium, then energy is lost from the reflected wave. In this way the absorption spectra of highly opaque materials may be measured simply, without the need for ultra-thin transmission cells.

A typical bulk-optics ATR attachment (infra-red element (IRE)) for an IR spectrometer is a parallel sided glass block [5] with a partially collimated beam of light propagating in a sawtooth path down the long axis undergoing total internal reflection (TIR) at its top and bottom surfaces. When the IRE is immersed in, for example, an absorbing liquid and the light passing through the IRE analyzed by a spectrometer, the resulting spectrum will contain the same information (in a slightly distorted form) as a conventional thin-cell IR transmission spectrum.

The evanescent field of a length of optical fiber (with a suitably thin cladding) may be used to perform a similar measurement. If the fiber is multimoded, then the resulting spectrum will be further distorted from the conventional transmission spectroscopy measurement due to the uneven removal of power from each guided mode of the fiber [6]. Evanescent field sensors are particularly prone to surface contamination.

To increase the sensitivity and remove potential masking absorptions, EFA sensors for contaminants in water often consist of polymer-coated tapers. The tapered shape increases the evanescent field penetration into the polymer coating, and the latter selectively absorbs (preconcentrates) many organic compounds, while excluding water. However, this polymer renders the sensor thermally sensitive and there are delays in response due to diffusion times.

The topic of evanescent field sensors is dealt with more fully in Chapter 3.

7.2.3 Absorption Measurement by Photoacoustic Spectroscopy

There has been much interest in photoacoustic spectroscopy (PAS) recently, and a number of textbooks have been devoted to the subject [8–10]. Light energy absorbed by an analyte is converted to heat, which causes the analyte and surrounding matrix (gas, liquid, or solid) to expand. If the light source is pulsed or chopped, the resulting series of pressure waves may be detected using a microphone. A typical photoacoustic system is shown schematically in Figure 7.5. Normalized absorbance detection levels (per joule of laser pulse energy) down to 10^{-9} cm^{-1}/J have been measured [11].

There are a number of advantages to using PAS:

1. In conventional spectrophotometry, scattered light would markedly reduce the amount of light reaching the detector thus giving a spurious absorption measurement. It has less effect on the photoacoustic signal level, and so the method

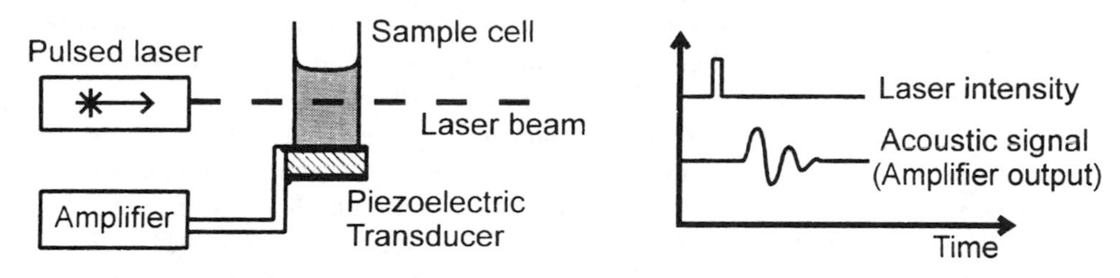

Figure 7.5 A typical arrangement for photoacoustic spectroscopy.

is less sensitive to turbidity. Light that is deflected from its path, but still absorbed, will give rise to a signal, although the effective optical path length through the sample will change by a small amount.

2. The applicable concentration range of the technique can be large: low levels of absorption can be detected with the same equipment as high levels of absorption.
3. The photoacoustic signal is directly proportional to the intensity of light absorbed, so there is no need to measure small differences between large light intensities, as in absorption spectroscopy. The signal to noise ratio may be much greater, and in particular the effects of source flicker noise and shot noise are greatly reduced.

The magnitude of any photoacoustic or photothermal effect will be proportional to the quantity $\beta/C_p\rho$ of the substance under analysis, where:

β = volume thermal expansion coefficient of the sample;

C_p = specific heat capacity of the sample (constant pressure);

ρ = density of the sample.

This figure of merit is generally larger for gases and solids than for liquids. In particular, water has one of the smallest thermal expansion coefficients, vanishing to zero at 4°C, making it one of the most difficult solvents to work with.

7.2.4 Fluorescence (Luminescence) Spectroscopy

7.2.4.1 Basic Principles of Fluorescence Spectroscopy

The process of fluorescence in a compound involves the photon-induced excitation of electrons to higher energy levels (i.e., an absorption process), followed by their spontaneous return to a lower energy level, with consequential re-emission of a photon.

The re-emitted photon usually has a lower energy than the incident photon, as energy is often lost by phonon-excitation processes (loss of energy to molecular vibrations).The lower energy emission band is called a Stokes band. A band with higher energy than the incident beam is called an anti-Stokes band.

Many aromatic compounds exhibit fluorescence when excited by UV light, but there are also some compounds and materials (e.g., certain organic dyes) having high fluorescence efficiency of optical re-emission when excited by less energetic photons (at other, longer wavelengths). The conversion efficiency is often expressed as a percentage, known as the quantum efficiency η, which is the percentage of the number of absorbed incident photons that result in re-emitted fluorescent photons:

$$\eta = \frac{\text{No. of fluorescent photons}}{\text{No. of absorbed photons}} \times 100\% \tag{7.5}$$

It is essential to effectively separate the desired fluorescent light from the scattered incident light. Fortunately, this problem is assisted by the difference in wavelength arising from the inelastic nature of the fluorescence process. Therefore, it is only necessary to provide effective optical filters to remove the incident light from the detected fluorescence signal. If several fluorescent compounds are present, each having different fluorescent wavelengths, they may be detected independently using wavelength-selective bandpass filters or a grating spectrometer with a focal-plane detector array. Simple forms of probe for the detection of fluorescence (or Raman scattered light) in liquids are described in Section 7.2.7.

7.2.4.2 Time-Resolved Fluorescence Spectroscopy

This technique takes advantage of the statistical nature of the fluorescence processes. If a large number of molecules are excited by a short pulse of light to the same excited state, and then begin to fall back to their ground state, then

$$I = I_0 \exp(-t/\tau) \tag{7.6}$$

where the fluorescent light intensity I decays exponentially with the time t after optical excitation. The value I_0 is the peak intensity and τ the fluorescent lifetime.

In order to measure τ, two basic methods are used. The first (time-domain analysis) involves measuring the decay function, following short-pulse optical excitation, and computing the value of τ from this function. The second (frequency-domain analysis) uses a source with a sinusoidally modulated incident light intensity. Either the frequency variation of the fluorescent light intensity, as the modulation frequency is varied, or the phase delay between fluorescence and excitation signals (both of which are related to the value of τ) can be monitored. Because of the weak received signal, the frequency-domain methods usually use a coherent electronic detector based on a mixer circuit. This recovers the desired frequency component in the detected signals corresponding to the original sinusoidal modulation signal. If there are compounds in the analyte having distinctly different fluorescent lifetimes, they may be separated using either of the above time-resolved techniques. Also, the time-resolved techniques are complementary to any method of separation of signals in the wavelength domain.

A more typical fluorescent process consists of the excitation of an electron, its nonradiative decay to an intermediate level, and subsequent radiative transition back to the ground state. In this case, the nonradiative decay is also described by an exponential decay, with its own characteristic time constant. As long as the nonradiative time constant is much shorter than the radiative time constant (as it often is), (7.6) remains valid.

7.2.4.3 Limitations and Problems of Fluorescence Spectrometry

One of the primary problems of fluorescence spectroscopy is the nonlinear variation with concentration at high levels of analyte (i.e., when absorption of the incident light becomes large). This causes a reduction in the fluorescent signal for two reasons. Firstly, it reduces the mean optical excitation level in the sample and secondly, at high absorption levels causes all the absorption to take place close to the point of entry of light into the sample. (In the latter case, efficiency of light collection may be more or less due to geometric effects of the measurement apparatus).

Other problems can occur due to a strong dependence of the fluorescent signals on a variety of environmental parameters. Oxygen usually quenches (i.e., reduces) fluorescence. The pH of a solution, its temperature, and any impurities can all influence the fluorescence lifetime and intensity. In addition, many fluorescent materials can become "bleached" during light absorption. This "photobleaching" can be reversible if it is merely due to a long fluorescent lifetime (saturation behavior), or it may, if it is due to a nonreversible photochemical reaction, gradually cause permanent depletion of the fluorophore. As expected, photobleaching is most serious at high illumination levels, but, unfortunately, intense sources are often used for trace analysis.

7.2.5 Light Scattering

There are many varieties of light-scattering photometers that can be constructed using optical fibers. However, this subject is covered in detail in Chapter 9.

7.2.6 Raman Spectroscopy

7.2.6.1 Basic Concepts

Raman scattering is observed when photons are *inelastically* scattered (i.e., frequency-shifted) by vibrating molecules. The Raman spectra of solids, liquids, or gases may be observed when a monochromatic light source (typically a laser) is used to excite the sample under investigation. Light shifted to both higher and lower frequencies can be seen, the magnitude of the shifts being equal to the characteristic vibrational frequencies of the molecule. The intensity of the light shifted to higher frequencies is, under normal circumstances, much lower than the down-shifted light.

As in fluorescence spectroscopy, photons that are inelastically scattered to lower frequencies are termed the Stokes lines in the spectrum, and those photons scattered to higher energies are anti-Stokes lines. Unlike fluorescence, any wavelength of light may be used to excite the characteristic Raman spectrum of a compound, and Raman lines are often very narrow ($<20\ cm^{-1}$). Raman-scattered light is typically several orders of magnitude weaker than fluorescent light.

The total differential scattering cross-section $(d\sigma/d\Omega)_{90}$, or simply σ_{90}, is usually used to describe the Raman activity of a molecular vibration. It represents the probability of a single incident photon being scattered into a particular Raman line, in a solid angle $d\Omega$, perpendicular to the polarization vector of the incident light. It has units of $cm^2/molecule^{-1}/sr^{-1}$. The flux of light scattered into the collection optics of a Raman system is given by (7.7), where P_D is the incident laser power, D_a is the analyte number density, A_D is the area of the analyte under observation, and Ω is the solid angle of collection.

$$\Phi = P_D\,(d\sigma/d\Omega)_{90}\,D_a A_D\,\Omega \tag{7.7}$$

The magnitude of the scattering cross-section increases very rapidly (like Rayleigh-scattered light): it is proportional to the fourth power of the frequency of the scattered light. However, many compounds have absorption bands in the visible and UV regions of the spectrum. If the excitation light is close to one of these bands, then the efficiency of the Raman-scattering process can be enhanced by up to several orders of magnitude. However, under these conditions the Raman light may be masked by fluorescent emission, or subsequently absorbed by the material, reducing the size of the effect.

The polar distribution pattern of Raman light scattered by a molecule is similar to that of an oscillating electric dipole (i.e., it is zero in the direction of oscillation of the electric field vector of the incident light and is maximum at all directions in the plane at 90 degrees to this direction). In a liquid, where all the molecules are orientated at random to one another, the radiation pattern is simply the sum of the scattering from each molecule, illustrated (for polarized light excitation) in Figure 7.6.

Only those vibrations resulting in a change in the polarizability of a molecule will scatter light inelastically, and are said to be Raman-active. Diatomic molecules always exhibit Raman-active vibrations, and, in general, if a vibration preserves all of the symmetry elements of the molecule, then the vibration will be Raman-active. Vibrations that result in particularly large changes in polarizability produce more intense Raman signals. These are often vibrations of atoms bonded by π-bonds, or large resonance bonds (e.g., benzene). Stronger Raman bands are normally expected from compounds of elements in the second and subsequent rows of the periodic table

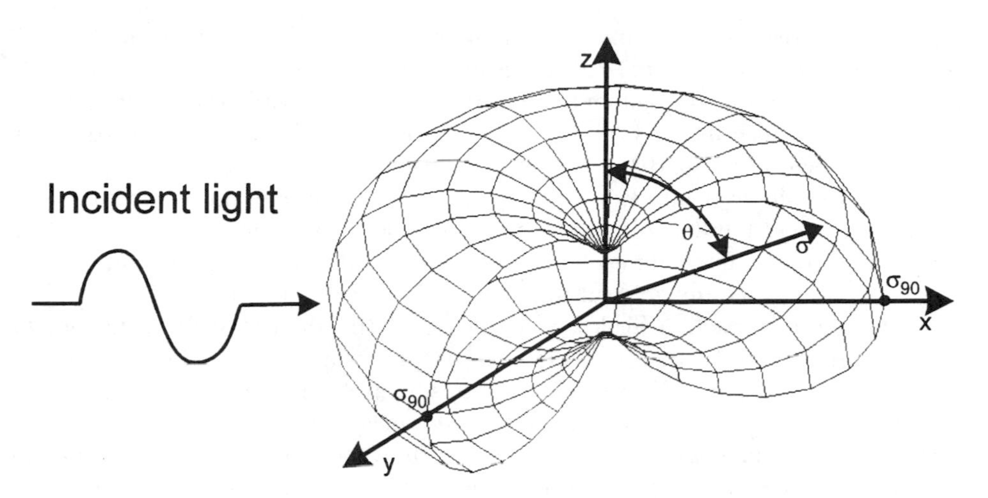

Figure 7.6 The average distribution of scattered light from a single molecule is the mean average of the randomly distributed scatterers throughout the scattering medium, illustrated here for the case of water.

(as they have more electrons), from cyclical molecules (such as benzene), and from hydrogenic molecules (those containing hydrogen). Liquids with large intermolecular interactions (such as those due to hydrogen bonding, in the case of water) will have broad bands, as the motion of each molecule is affected by that of its neighbors.

Typically, in a transparent homogeneous material, a proportion on the order of 10^{-2} of the incident photons will be scattered elastically, and a fraction of only 10^{-4} of these will be shifted in wavelength. Use of higher frequency exciting radiation increases both Rayleigh and Raman scattering. However, a compromise arises when choosing the optimum wavelength for Raman analysis, as eventually the photon energies will correspond to electronic transitions within the sample. A sufficiently high-energy photon may be absorbed by a molecule (rather than scattered) and then re-emitted as fluorescence (usually after a characteristic time much greater than the 10^{-13} seconds of a Raman-scattering event). Fluorescence bands are spectrally broad and can often be four to six orders of magnitude stronger than the weak Raman lines [6]. Fluorescence is not normally a problem with IR excitation, but becomes significant with incident radiation in the visible to UV region. Except for rare two-photon absorption events, fluorescence usually results in photons of a lower energy than the radiation that excites it, and so is seldom a problem in the study of anti-Stokes bands, despite the fact these are even weaker than the Stokes lines.

7.2.6.2 Surface-Enhanced Raman Scattering (SERS)

Enhancement factors as high as 10^6 have been observed in the scattering cross-sections of molecules adsorbed onto suitably prepared metal surfaces. The experimental results are remarkable with regard to both the magnitude of the observed enhancement and the wide range of analytes amenable to the technique. Silver, copper, and gold are commonly used substrates, the most frequently used being silver. Pyridine, adsorbed on to electrochemically roughened silver substrates displays an enhancement of up to 10^6, over a wide wavelength range extending from 500 to 700 nm. The enhancement factor varies for different adsorbates, for each vibrational band of a molecule, and in addition, is a function of the excitation wavelength. Silver substrates display a maximum enhancement in the visible, between 500 and 700 nm [12], whereas copper and gold substrates are most efficient in the near-IR [13].

Despite its attraction for low-level measurement, the use of SERS for *in situ* analysis has the practical disadvantage that it requires a carefully prepared surface, which must, by the nature of the technique, be susceptible to fouling. (One method to avoid fouling is to use a second surface layer, permeable to the analyte, e.g., a polymer on top of the metallic layer. However, the sensor will then be thermally sensitive and the response will be slower due to the diffusion-time delay.)

7.2.7 Optical-Fiber Probes for Raman Scattering and Fluorescence Measurements

Fluorescent light is radiated in all directions, so the coupling efficiency into the receiving optical fibers is poor at distances more than a few core diameters from the site of the emission. In addition, if the absorption coefficient is low, relatively little energy will be absorbed from the incident beam to be re-emitted. Both these aspects can be improved by using more sophisticated optical cell designs, to provide multiple passes of the incident beam through the sample and to more efficiently direct the fluorescent light into the receiving fiber.

7.2.7.1 Single-Fiber Probes

Only a single optical fiber is necessary to perform extrinsic measurements of wavelength-shifted (i.e., *inelastically*) scattered light (Figure 7.7).

Probes of this type are small, cheap, and efficient collectors of scattered light. They are commonly used for fiber-remoted measurements of fluorescence and absorption in the UV/VIS/NIR regions of the electromagnetic spectrum.

The sensitivity of measurements made with a single-fiber probe can be limited by backreflections (Fresnel reflection) from the fiber tip. Polishing the tip of the fiber at an angle can eliminate backreflections from the probe tip, but distributed backscatter from the whole length of the fiber is unavoidable. Although a spectrometer with suitably high stray light rejection may be able to remove scattered and reflected light at the incident wavelength, Raman-scattered light and fluorescent light originating within the fiber core will interfere with weak signals from the analyte when a single-fiber probe is used.

The unwanted fiber fluorescence, which is normally particularly troublesome with UV or short- wavelength visible excitation, is less intense when using high OH^-

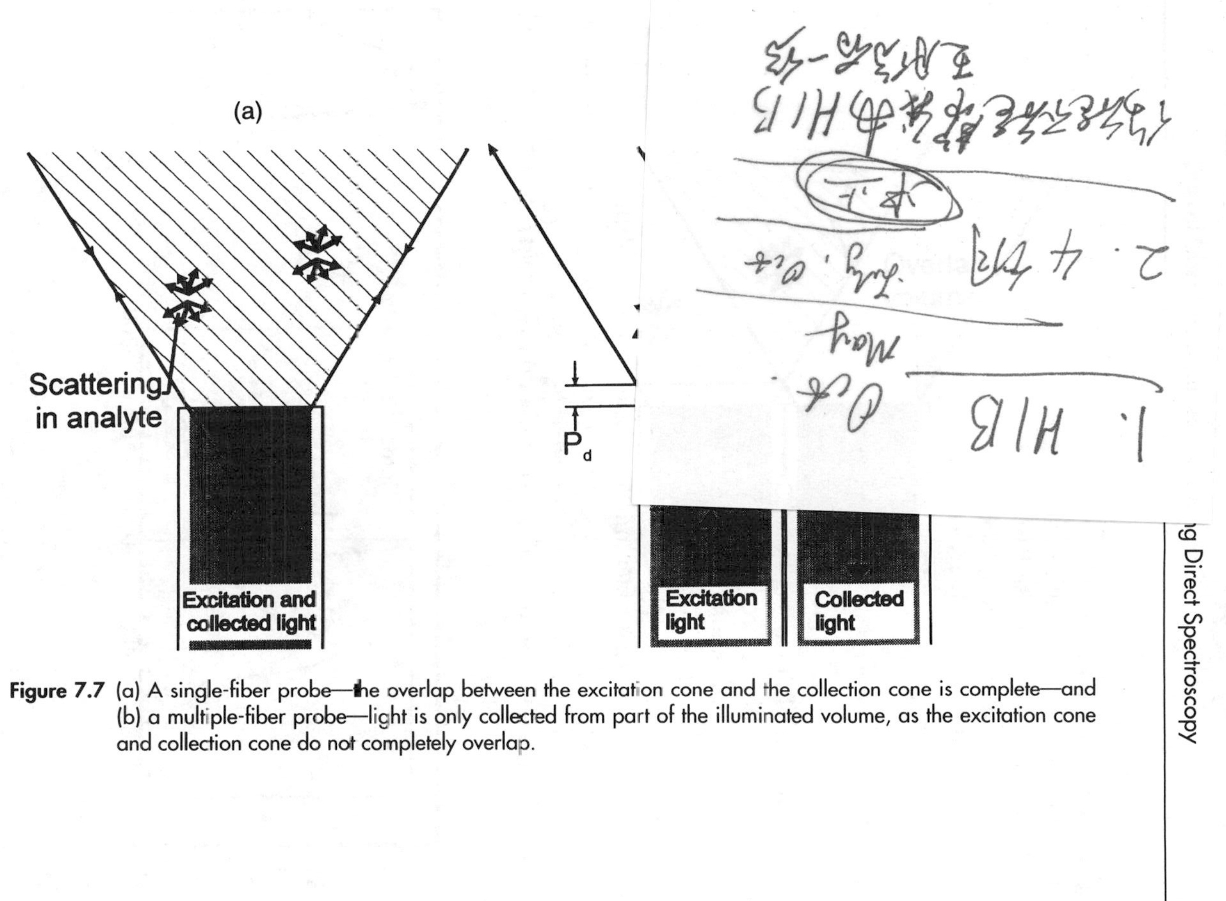

Figure 7.7 (a) A single-fiber probe—the overlap between the excitation cone and the collection cone is complete—and (b) a multiple-fiber probe—light is only collected from part of the illuminated volume, as the excitation cone and collection cone do not completely overlap.

content (wet) silica, UV-grade fiber. Generation of Raman light within the silica cannot be similarly avoided, as this broadband scattering results from the vibrations in the glass itself, although these vibrations are all below 1,400 cm^{-1}. A measured Raman spectrum of pure silica glass is shown in Figure 7.8. These interfering components can be almost entirely rejected from measurements made through multiple optical-fiber probes [14,15].

7.2.7.2 Multiple-Fiber Probes

By using separate optical fibers to carry light to and from the probe, the optimum placement of optical filters can be used to reject any potentially interfering light. A narrow bandpass filter after the delivery fibers ensures only light at the excitation wavelength reaches the sample, and is essential for probes to be used in Raman measurements. A notch filter (or a long or shortpass filter) before the collection fibers removes any unshifted light that could generate fiber Raman or fluorescence between sample and spectrometer. In fact, much more unshifted light than shifted is usually collected in a Raman measurement, so removing this component can dramatically reduce the stray light within the spectrometer, too.

The designs of the probes themselves fall into two broad categories: those where

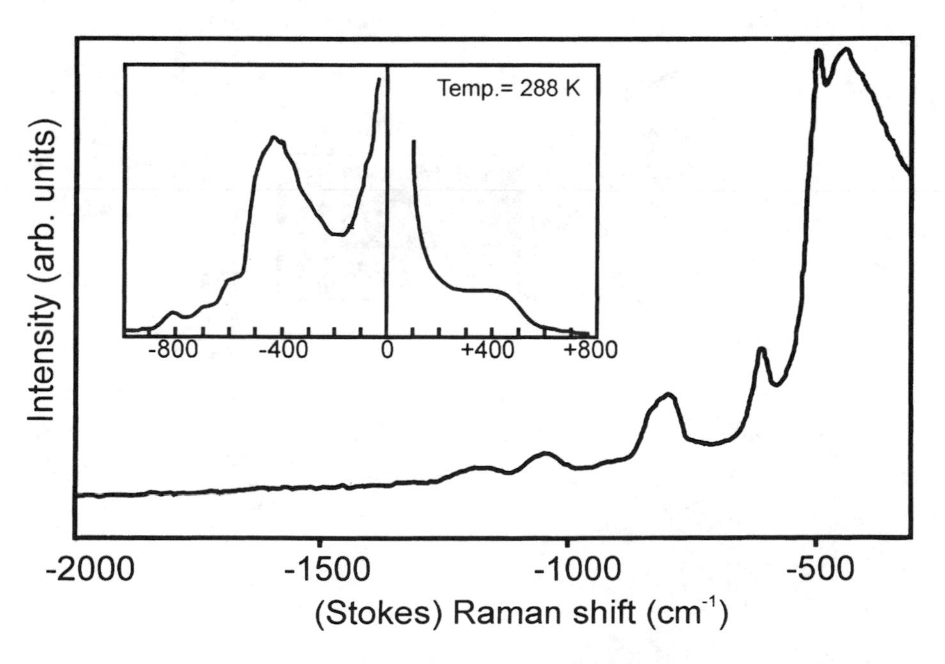

Figure 7.8 The (Stokes) Raman spectrum of a fused silica fiber, recorded by Ma [15], with the low wavenumber and highly temperature-dependent anti-Stokes spectrum shown inset.

separate delivery and collection fibers are used in the probe, and those where the light from the delivery and collection fibers is multiplexed into a (short) length of a single fiber. Probes in the first category are relatively inefficient due to the incomplete overlap of their emission and collection cones. In fact, no light at all is collected from points closer than P_d (see Figure 7.7). The distance P_d is linearly related to the separation between the fiber cores D:

$$P_d = \frac{2 \cdot r_c - D}{2 \cdot \tan(\alpha)} \tag{7.8}$$

where r_c is the core radius of the fibers and α is the collection half-cone angle. It has been proposed that probes formed of concentric rings of collection fibers around a central delivery fiber (a fairly common arrangement) could be used to perform depth-profiling measurements [16]. While this idea is sound in principle, it is hard to think of possible practical applications.

An example of the second category of probe is shown in Figure 7.9. Light is coupled in and out of the short sensor stub via a color separation filter, which can efficiently reflect light in one band of wavelengths while transmitting light outside that band. The excitation and collection cones then overlap as in a simple single-fiber probe, with a similar high collection efficiency. Unfortunately, filter transmission is typically below 80%, or lower if a sharp transition between transmitting and blocking (or reflecting) regions is required.

7.2.7.3 Improving the Efficiency of Optical-Fiber Probes

An immediate advantage of using bundles of optical fibers to collect light scattered from an analyte is that a circular light-collection area is easily transformed to a linear

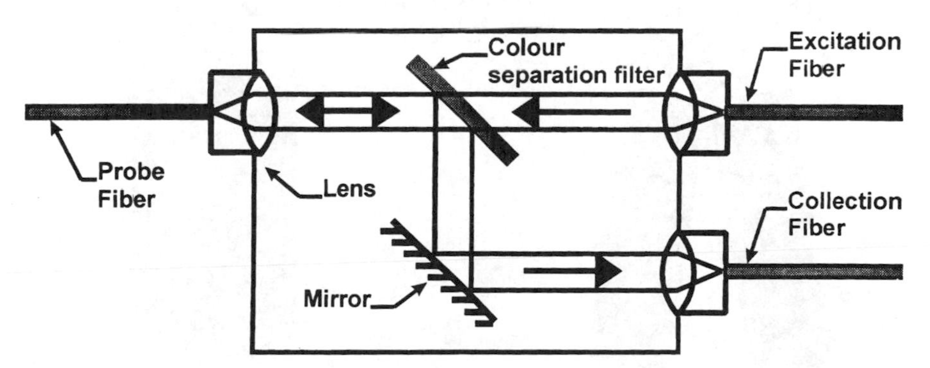

Figure 7.9 Use of a color separation filter allows excitation and collected light to be combined in a short length of fiber at the probe tip. Short-wavelength light is transmitted by the color separation filter, whereas longer wavelength Raman-scattered light is reflected.

slit at the spectrograph input. Light is usually most efficiently collected from a circular region of the analyte (for instance, around a focused spot of excitation light). The resolution of a dispersive spectrometer (i.e., one in which the different wavelengths of light are separated by a diffraction grating or prism) is usually limited by the width of the input slit. By rearranging the collection fibers to be a linear array at the spectrometer input, the maximum possible throughput is achieved at a given resolution.

The amount of light collected by multiple-fiber probes may be increased either by arranging the collection fibers at an angle to the excitation fiber or by angle-polishing the tips of the emission and collection fibers with respect to one another (Figure 7.10). An angle of 9 degrees between 0.22 NA fibers will increase the collected light intensity by about 70% with respect to parallel fibers in a clear colorless analyte. The optimum angle increases as the fiber NA increases [17].

The amount of light collected by an optical-fiber bundle can be increased by using a lens to focus the excitation light to a spot in the analyte, and the scattered light back onto the bundle. This may be economically achieved using a sapphire ball lens. Use of lenses also facilitates noncontact measurements.

For liquid analytes, light may be confined within the numerical aperture of the probe fibers over even longer distances by enclosing the analyte in a reflective tube of

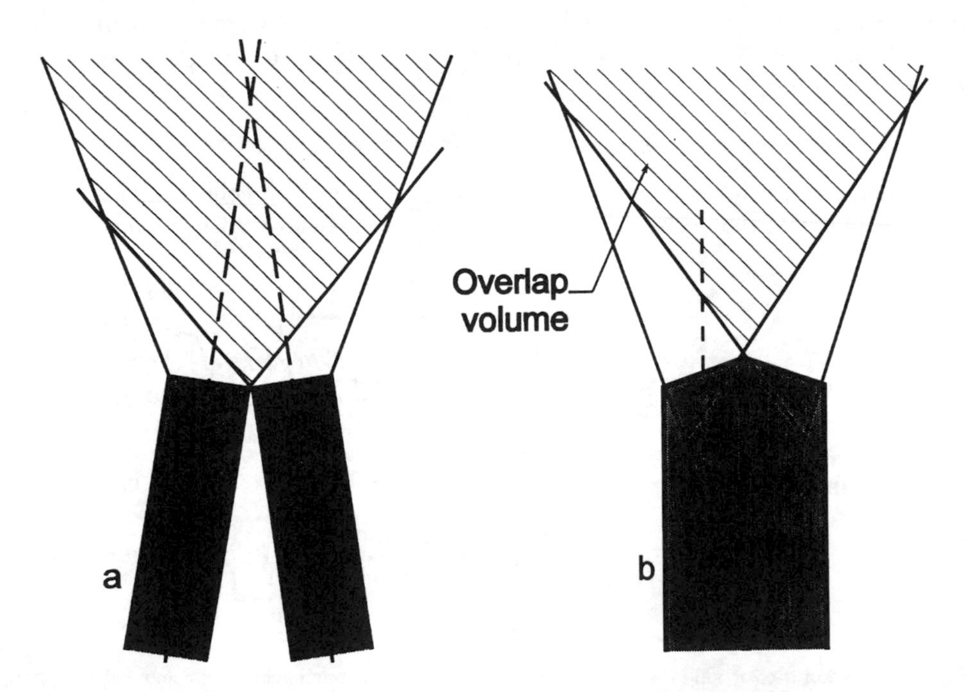

Figure 7.10 The overlap between the emission and collection cones may be increased by (a) angling the two fibers with respect to one another and (b) angle-polishing the tips of the fibers.

a diameter approximately equal to the diameter of the fiber probe. In this way, a long measurement length may be used, while maintaining a confined beam (with consequent increase in collected signal intensity). Although the sampling length is increased, the actual volume is very low when using small-diameter tubes.

7.2.7.4 Commercially Available Optical-Fiber Probes

Convenient and rugged examples of all the probe arrangements described above are available commercially. Polytec manufactures a range of probes for optical-fiber-remoted measurements of absorption (both transmission and ATR), reflectance, and scattering, mainly intended for use with their X-dap diode array spectrograph [18]. They also sell a fiber-coupled cuvette holder and a flow-through cell.

Dilor manufacture probes such as the Dilor Super-Head distributed through Instruments SA, illustrated in Figure 7.9. This is marketed specifically as a probe for Raman spectroscopy, and may be arranged to multiplex measurements at several different points. Hellma manufactures immersion probes for transmission measurements, designed to work with any standard photometer. The Savannah River Technology Company manufactures a probe with angle-polished tips for increased collection efficiency and microfilters at the probe tips. This probe has been used for diffuse reflectance, fluorescence, and Raman measurements [19].

Probes that can be interfaced via a standard cuvette holder are available as add-ons, for example, from Hellma, for any transmission spectrophotometer [20]. The probes can be made rugged and chemically resistant, and due to the optical nature of the measurements, the method is intrinsically safe in flammable areas and immune from electromagnetic interference. Efficient probes for collecting fluorescent or Raman light are available from FCI [21,22].

7.2.8 Fourier Transform Spectroscopy

This is not so much a spectroscopic arrangement, rather a generic means of processing the optical signals without the traditional monochromator to analyze the light spectrum. The method involves detection of the optical signal, via an interferometer having a scanned path-length difference. A two-path interferometer acts as an optical filter, having a periodic transmission $T(\nu)$, with a function of optical frequency ν where

$$T(\nu) = 1 + \sin(k(\nu)) \tag{7.9}$$

The parameter k depends on the optical path differences in the interferometer. The detected signal $S(\nu)$ when a complex light spectrum $I(\nu)$ is, after passage through the interferometer then incident on a detector with a spectral response $D(\nu)$. $S(\nu)$ is given by

$$S\,(\nu) = \int_{\nu = \nu_{\max}}^{\nu = \nu_{\min}} I(\nu) \cdot T(\nu) \cdot D(\nu) \cdot d\nu \tag{7.10}$$

The function S represents the correlation between the sinusoidal transmission function of the interferometer and the combined spectral responses of the input light spectrum and the spectral response of the detector. Clearly, the correlation is best with spectral variations having the same (optical frequency) periodicity as the interferometer. In the simplest case, where $I(\nu)$ is a narrow-line spectrum of constant frequency, the signal S represents a single (sample) point on the sinusoidal response. If the path length of the interferometer is now scanned, its periodic transmission-response function will translate across the frequency band (and also change its period versus frequency). The detected signal will therefore vary as the correlation between the spectral features and the periodic interferometer response varies. In the simplest case of a single frequency line spectrum, the detected signal will vary sinusoidally as the interferometer is swept. More complex spectra can be considered to consist of a linear addition of a series of such narrow-line spectra, each having an appropriate amplitude. For a complex spectral response, the variation of the detected signal with time, as the interferometer is swept, represents the inverse Fourier transform of the original spectrum. The spectrum can therefore be recovered by Fourier transformation of the temporal variations in detected signals, resulting when the interferometer is scanned. This is again illustrated by the simple example of the line spectrum, which results in a sinusoidal temporal response. The Fourier transform of a sinusoidal signal has a single value, corresponding to the single frequency of the line spectrum.

When compared to conventional spectrometers, the Fourier transform system has several advantages. The first is that it is relatively easy to obtain high spectral resolution by using a long "path" difference interferometer (10-cm path difference at 1-μm wavelength, results in a fractional resolution of around 1 part in 10,000). The second is that a significant fraction of the source light is incident on the detector at all times, because the mean transmission of the interferometer is much higher than that of a narrowband grating monochromator, thereby improving the optical efficiency. The third is that the interferometer can be designed with a large optical aperture, giving much higher optical throughput from radiance-limited optical sources. The advantages are less significant when only moderate resolution is needed and low-noise (e.g., CCD) detector arrays can be used, as these also allow parallel detection of each spectral component.

The main disadvantage is normally the need for a precisely aligned interferometer, with its necessary thermal and mechanical stability. Fourier transform spectroscopy is applicable to a wide variety of spectroscopic techniques, including transmission, reflectance, fluorescence, and Raman spectroscopy, as in all of these cases it can be used for spectral analysis before the detection system.

7.2.8.1 All-Fiber Fourier Transform Spectrometer

This ingenious method [23] involves interference between unequal-length, guided-wave light paths in monomode optical fiber. Light from a broadband source is passed through an absorbing medium, and then through a variable-path all-fiber Michelson interferometer. To produce the spectrum of the transmitted light, the Fourier transform is taken of the time-varying detected signal as the path difference is mechanically scanned. The fiber Michelson interferometer has one piezoelectrically stretched fiber arm and one unstretched arm. The system has yet to be tried for low-contrast applications, such as optical gas detection, but has been successfully used to measure the spectrum of a semiconductor laser source at various levels of injection current. For applications such as gas detection, the usual signal/noise advantage when operating at high resolution in the Fourier transform domain should, however, be achievable. The restriction arising from single-mode operation will, unfortunately, cause a severe loss of light compared with the multimode approach generally used in gas-sensing systems.

One feature of the fiber-based FT spectrometer not associated with normal types is that the optical delay will be wavelength-dependent due to fiber dispersion.

7.3 CASE STUDIES OF FIBER-OPTIC SENSORS USING DIRECT SPECTROSCOPY

In this section, we shall describe a number of areas where fiber optics have been used for direct spectroscopy.

7.3.1 Liquid-Phase Sensing

Many liquid-sensing applications may be carried out using extension leads from a commercial spectrophotometer. Some manufacturers now sell such attachments as optional extras for their commercial instruments. High-concentration solutions often exhibit strong absorption spectra, so there is seldom a need to develop the specialist instrumentation of the type required for low-contrast gas detection (see the next section). The measured spectra can be used for the simultaneous determination of several absorbing solutes by using multivariate analysis methods [24,25]. These are essentially computerized methods, capable of adding the various spectral features characteristic of each *expected* analyte in appropriate proportions and finding the best match to the experimentally observed spectrum. Clearly multivariate analysis becomes difficult, if not impossible, if several analytes with very similar absorption spectra are present. The problem becomes insoluble if unexpected contaminants with strong unknown spectral characteristics are present.

Better results are claimed in these circumstances for techniques incorporating artificial neural networks [26,27], which are easily implemented as computer software. Neural networks can be fast and tolerant of imperfect data, but care must be taken to use them within their limitations [28]. The network must be optimally "trained" by presenting it with a suitable number of reference spectra that should encompass data in every form in which it might be subsequently encountered (e.g., buried in noise or in the presence of contaminants).

In addition to simply measuring the transmission of a collimated beam of light through a sample, there are a number of other ways in which the qualitative absorption spectrum can be deduced. Indeed, the first recorded demonstration of an optical-fiber absorption measurement (by M. Polani) [29], used 50-μm core glass fibers to measure the diffuse reflectance of blood at 805 and 660 nm. From these measurements, the absorption spectrum of the blood around 660 nm, and hence its oxygen saturation, could be determined. By using the nonabsorbed wavelength of 805 nm as a reference, fluctuations in parameters such as fiber-bending losses could be corrected for. The small size of optical-fiber probe heads (and chemical resistance of silica components) mean that very little interference is caused to any process under observation.

Many recent applications of spectroscopy to industrial process control (many using optical fibers) are presented in recent SPIE volumes (e.g., [30,31]) and in journals such as *Process Control and Quality*. We shall now examine examples of spectroscopic methods in more detail.

7.3.1.1 Transmission Spectroscopy

This technique has found wide application in chemical, biological, and environmental monitoring and process control due to its generic nature, intrinsic safety, and ease of application. In 1988, Boisdé reported that over a kilometer of optical fiber had been installed at French Atomic Energy Commission sites for the purpose of online process monitoring [32]. The samples monitored ranged from measurements of a single species in a restricted analytical medium, through simultaneous determination of several species (possibly with mutual interference), to trace measurements in a complex medium. In some cases, single-wavelength measurements were appropriate; in others, full spectral measurements were necessary. It was claimed that the first (unpublished) work was done as early as 1974, with differential measurement at wavelengths of 477 nm (absorption peak) and 520 nm (low-absorption reference wavelength) monitoring Pu (IV) concentration in an aqueous solution.

The simple fiber-fiber probe of Freeman [4], which was described in Section 7.2.1.2, was used to determine copper-sulphate concentration in an electroplating bath. Light from a 820-nm LED was coupled into one of the fibers, and a fraction of the light transmitted across the gap was collected by the other fiber. The intensity of this light was then measured by a photomultiplier (although a photodiode could probably have been used). Freeman found excellent correlation between the Cu^{2+} concentration and the absorbance of the light transmitted between the fibers over the concentration range 0.2 to 0.4M.

Of the other constituents of the plating bath, sulfuric acid was found to influence the transmission the most, with increased sulfuric-acid concentration lowering the apparent absorbance of the solution.

For a simple fiber-fiber probe, changes in the refractive index of the analyte can modify the output-light cone angle to a different extent at the reference wavelength than at the measurement wavelength (optical dispersion). This problem can be reduced by lensing the fibers so that the light is collimated before it enters the sample. Researchers at the Westinghouse Savannah River Company (Aiken, USA) have been developing absorbance probes for inline monitoring since the 1970s [21]. A lens assembly from one of their probes is shown in Figure 7.11.

These probes have been used in pairs, for instance across a process stream, with light transmitted through the analyte from one assembly to another. Because the light is collimated when it leaves the flat glass-liquid interface, changes in the refractive index of the analyte have less effect on the intensity of collected light.

An alternative arrangement is to position a mirror facing the end of a single optical fiber. A medical application of a single-fiber probe using a fixed mirror is described by Coleman, who used an optical fiber threaded through a hypodermic needle with an aluminum reflector close to its tip [33], shown in Figure 7.12. Coleman described applications for *in vivo* analysis in regions previously too small to sample, and presented *in vitro* measurements of bilirubin in human blood. By using a 25-mW argon ion laser source at 457.8 nm, Coleman measured a minimum detectable absorbance of 0.005m^{-1} corresponding to bilirubin concentrations between 0.05M and 1.3M.

7.3.1.2 Evanescent Wave Spectroscopy (EWS)

This topic is covered extensively in Chapter 3 of this volume, but we shall briefly discuss a few applications here. The arrangement is similar in each case: light is transmitted through a section of optical fiber, either unclad or with a very thin cladding, which is immersed in to the analyte. As mentioned earlier, a polymer buffer layer is often applied to the fiber in the sensing region. This can serve the dual purpose of excluding solvent from the evanescent wave region, and preconcentration of the sensed species: a hydrophobic species may dissolve in very low concentrations in

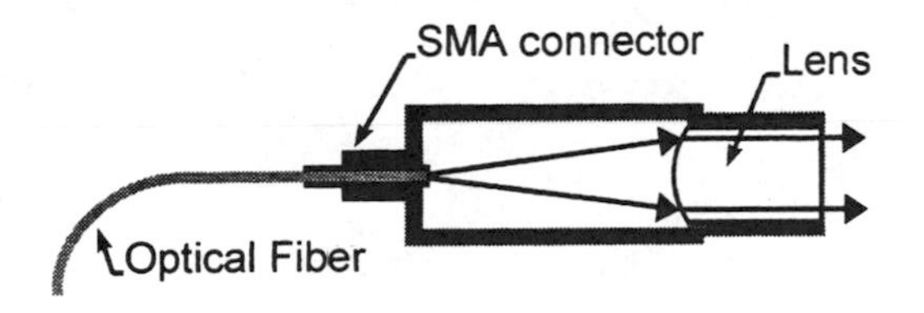

Figure 7.11 Savannah probe lens assembly [21].

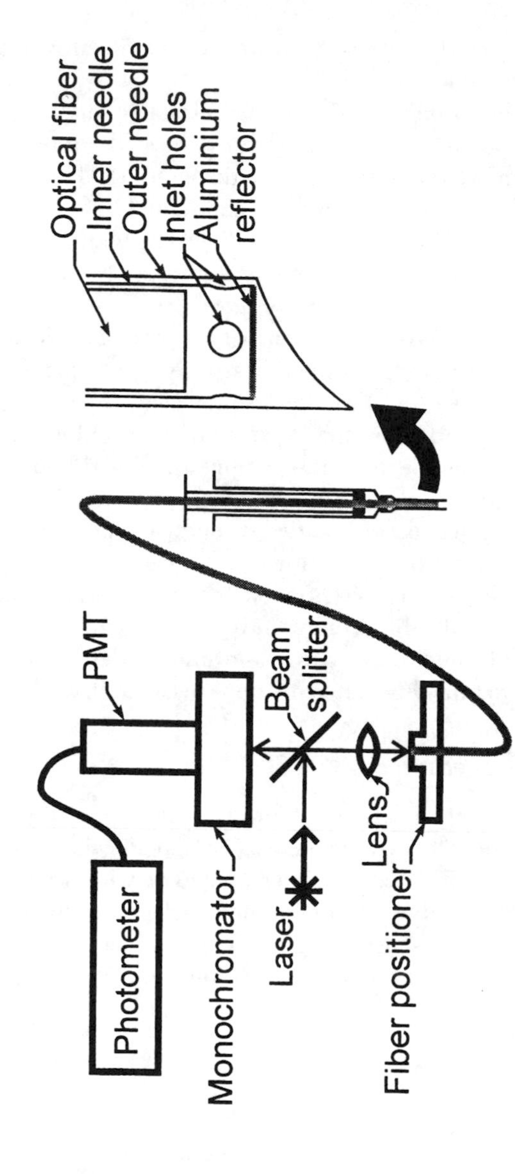

Figure 7.12 In vivo arrangement described by [33]. The optical path length is twice the distance between the fiber tip and the reflector.

water, but will preferentially move into a polymer layer. Sensitivity is increased by using longer and smaller diameter sensing fibers, and is also enhanced by coiling the fiber with a small bend radius.

Mid-IR fibers have been used for EWS, but only as short detection elements, limiting the range of remote measurements to a few tens of meters from the interrogation system [34–37]. Krska has used 10-cm lengths of silver-halide fibers, coated with low-density polyethylene (LDPE) to detect 50 mg/l of trichloroethylene in water, using a blackbody source and FTIR spectrometer. By using a 0.5-mW lead-salt laser diode, temperature-tuned over the range 847 to 1,099 cm^{-1} (approximately 12 to 9 μm), and coupled to a simple photodiode detector, a concentration of 0.1 mg/l could be detected in a measurement time of 55 sec (albeit after allowing a 10-minute interval for enrichment by diffusion into the polymer).

Degrendpre and Burgess have used 1.5m of 400-μm core polysiloxane clad fiber coiled on a 3-cm diameter Teflon support [38] to study the absorption (in the range 1 to 2.2 μm), of neat organic solvents, and mixtures of chloroform in carbon tetrachloride and of carbon tetrachloride in toluene. Bürck has reported detection limits of between 80 mg/l (dichloromethane) and 0.1 mg/l (1,2,4-trichlorobenzene) for various chlorinated hydrocarbons. Using a 10.58m long, 400-μm core sensing fiber, clad with polysiloxane [39], the 90% response (diffusion) times ranged from 0.5 min for dichloromethane and 71 min for 1,2,4-trichlorobenzene [40]. Initial experiments by Schwotzer with polysiloxane or Teflon-AF (5 to 20-μm thickness) coated fiber of 140-μm core diameter have shown detection limits of 10 mg/l for toluene and 0.05 mg/l for napthalene (both in water) [41]. These results were obtained using 0.5m of fiber coiled with a diameter of 1 cm.

7.3.1.3 Photoacoustic Spectroscopy

As previously mentioned, a photoacoustic signal results from the absorption of radiation, followed by dynamic pressure changes due to thermal expansion. The following case studies illustrate the variety of methods used to detect a photoacoustic response. The technique is applicable to gas or liquid-phase sensing (although the detection schemes may vary), and the examples in Section 7.3.2.6 may also be of interest.

Detection of Photoacoustic Surface Movement

In condensed phase samples, a bulk pressure wave may cause the displacement of a free surface, which can be sensitively detected in a number of optical ways [42]. These include measurement of the curvature, gradient, or displacement of the surface.

Hand and others have used a fiber-optic Michelson interferometer to measure the displacement of the surface of liquids [43,44]. A pulsed Nd:YAG laser was used as the pump source, directed onto one focus of an elliptical cell. Acoustic waves generated in this region were reflected by the cell walls to form a second focus at the liquid surface of the cell, thus amplifying the displacement. The resulting transient surface deflection was detected with the interferometer.

Two different types of cells were used; the first with an open liquid surface at the top of the cell, and the second using a thin, more highly reflective membrane in contact with the liquid surface. Using the latter method (Figure 7.13), the interferometer had a noise floor of 2×10^{-14} m/Hz$^{-1/2}$, and a pressure sensitivity of 0.1 Pa/Hz$^{-1/2}$ was demonstrated.

Using pulsed optical excitation offers a number of advantages, at the expense of an increase in system cost and complexity:

- The acoustic pulses produced have a large high-frequency component, enabling the short-wavelength pressure waves to be focused by a small cell (giving a fourfold signal enhancement, as reported by the authors).
- Time-gating the transient response removes any spurious signals arising from absorption at the cell windows, which can otherwise be a significant limiting factor in photoacoustic spectroscopy [45].

Detection of Photothermal Refractive Index Changes

A photoacoustic volume change may cause a corresponding change in the refractive index of the sample medium, which can be detected using thermal lensing techniques, photothermal deflection, or measurement of the optical path length through a sample.

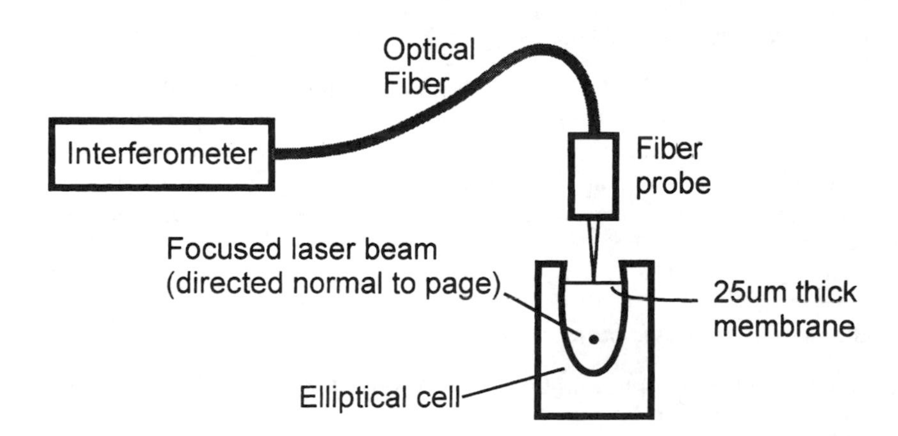

Figure 7.13 Apparatus used to detect surface movement caused by photoacoustic waves in an elliptical cell (*modified from* [43]).

Thermal lensing and photothermal deflection spectroscopy (PDS) have been comprehensively reviewed by Tam [46], and are particularly suited to the analysis of liquid-phase samples. Thermal lensing occurs when a beam of light (usually from a laser) is partially absorbed by an analyte to cause a thermal gradient in the fluid. The resulting refractive index changes create a transient lens in the sample, which usually defocuses the beam. This may be detected using a photodetector having a small acceptance aperture.

In photothermal deflection spectroscopy, higher sensitivity can be gained using a second CW beam, not absorbed by the analyte, to probe the refractive index changes. A detector may be split into two halves such that a deflected beam has an increased intensity on one half and a decreased intensity on the other. The ratio of the two intensities gives a measurement that is unaffected by changes in the source intensity. Different possible configurations are the following:

- So-called "collinear PDS," which has the following characteristics:
  - Probe beam at a small angle to the pump beam, to give a large interaction length;
  - Probe beam parallel with the pump beam, but displaced to the point of maximum response (maximum $\partial n/\partial r$).
- "Transverse PDS," characterized by the following:
  - Probe beam orthogonal to the pump beam.

Bohnert and coworkers have used both collinear and transverse PDS for the trace detection of pesticides in water [47]. A schematic diagram of the apparatus used for transverse operation is shown in Figure 7.14.

Transverse PDS was compared with a state-of-the art spectrophotometer (Cary 2400), which had a low noise level corresponding to an absorbance of 0.0002 AU. Detection limits for a number of pesticides were better for the photothermal method, by factors ranging between 2 and 20 for various compounds. Collinear PDS improved the detection limits by a factor of four, so that, for example, 2,4-dinitrophenol would be detected at the 0.5-ppb level. The fact that different pesticides responded differently might be explained by differing photoacoustic generation efficiencies for each chemical. Clearly, a strong photoacoustic signal is only possible if the energy absorbed by a molecule is converted to heat rapidly (i.e., within the integration time of the experiment). Other nonthermal relaxation pathways are possible for molecules, including fluorescence.

Refractive index changes have been observed in the liquid phase by Vegetti and others [48], who placed their sample (water containing trace quantities of a dye) in one arm of an optical interferometer, and detected the resulting phase change in the probe beam (see Figure 7.19). They employed an argon ion laser (modulated at 4 kHz) as their excitation source, arranged to be close and parallel to one arm of a

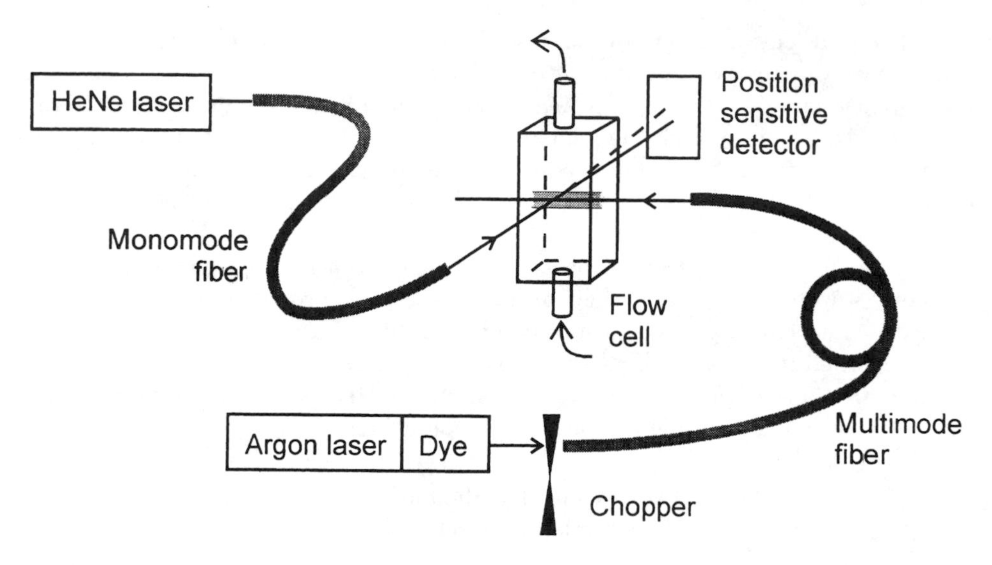

Figure 7.14 Schematic diagram of sample cell used for transverse photothermal spectroscopy of pesticides in water. Picture *modified from* [47].

Michelson interferometer. With this equipment, detection of phase differences as small as 5×10^{-7} rad were demonstrated, corresponding to a detection limit of 1.5×10^{-4} absorbance units for a 1-cm path length.

7.3.1.4 Fluorescence Spectroscopy

Remote monitoring of sample fluorescence via optical fibers is a longstanding *in situ* monitoring technique. Measurements made by fiber probes can provide qualitative and quantitative information, with an additional degree of selectivity provided by the choice of excitation wavelength and the fluorescent lifetime of a particular analyte. Although broadband sources can be used to excite fluorescence, lasers are usually preferred for use with optical fibers, as their output is more easily focused to a small spot. The acronym LIF is often used for laser-induced fluorescence.

Ground-water monitoring of Uranium was carried out by workers at the Lawrence Berkeley and Lawrence Livermore laboratories in the United States [49]. Their attractive method of uranium detection was based on its fluorescence, when excited at 488 nm by an argon ion laser source. There are strong fluorescence lines at 520, 527, and 530 nm and the technique has now been used in the laboratory down to levels of 10^{-14} molar.

Because fluorescence bands are quite broad, the additional information provided by fluorescent lifetime measurements is often necessary to distinguish compounds.

For instance, Chudyk found that phenol, toluene, and xylene (all single-ring aromatic compounds with peak emission around 295 nm) were initially indistinguishable in their vapor phase [50]. However, by measuring the fluorescent decay time τ with a pulsed 266-nm source, phenol (τ = 2.1 ns in cyclohexane [51]) could be easily distinguished from toluene (τ = 34 ns) and xylene (τ = 31 ns).

The electronics required to perform time-resolved fluorescence spectroscopy (TRFS) has been simplified by Bublitz [52]. By counting photons from the fluorescence in two adjacent arrival-time windows and ratioing the counts, engine oil in water was detected at concentrations below 1 mg/l, and polyaromatic hydrocarbons (PAH) concentration at the μg/l level. The fluorescence was excited by a 2-ns pulse from a multigas UV excimer laser, using 248 nm to excite the BTEX compounds (benzene, toluene, ethylbenzene, and xylene) and 337 nm for PAHs. Light was emitted between 260 and 370 nm for BTEX and between 370 and 500 nm for PAHs. The initial laser pulse triggered two-gated photon counters, first from 0 to 100 ns, and then from 120 ns to 220 ns after the initial laser pulse.

The light emission from impurities in natural water had a short decay time, and the photons from this source were counted in the first time window. The signal count in the second window was thus due to the longer decay-time contaminants. By taking the ratio of the two signals, the concentration of impurities could be deduced from a lookup table. The PAH emission wavelength coincided with the emission of the natural humic matter, and a further parameter had to be introduced to account for this.

Better detection limits for detecting PAHs *in situ* [53] were achieved by Panne and Neissner. Full TRFS was used, and detection limits down to the ng/l range were achieved for benzo-fluoranthenes and benzo-pyrenes. The probe consisted simply of two 600-μm core diameter silica fibers, angled at 11 deg to each other, with their end faces close together. The collected light was transmitted to the input of an f/3.8 spectrometer via a circular-to-linear fiber-bundle converter (for increased resolution), and a photodiode array at the focal plane was used to resolve the wavelength spectrum. The laser was pulsed (0.6 ns FWHM) and the detected signal was gated to integrate for 5 ns at any time after the pulse. After deconvolution of the system response, a time resolution of 0.5 ns was claimed.

One problem encountered with high optical power at wavelengths below 350 nm is photodegradation of the optical fiber. Interaction of intense UV light with silica fibers leads to fluorescence and a decrease of fiber transmission through the formation of color centers. Although this damage is minimized by using so-called "UV-enhanced," high OH^- fiber, Hillrichs proposed placing a frequency-doubling arrangement at the probe end of the transmission fibers [54]. The excitation light is then efficiently transmitted through the fiber at 532 nm, with a relatively low attenuation coefficient (0.16×10^{-4} cm^{-1}, compared with 11.5×10^{-4} cm^{-1} at 266 nm) and without significant photodegradation of the fiber. Despite the fact that the energy is converted (in a small BBO nonlinear optical crystal [55]) to 266 nm with only 1% efficiency, Hillrichs calculated that for fiber lengths over 30m, this is the most efficient way to transmit light to the sample, and, in any case, avoids cumulative damage to the fiber.

7.3.1.5 Raman Spectroscopy

Fiber remote Raman spectroscopy has found many applications in a wide range of process control and remote monitoring applications. The interest in the technique stems from its generality and its compatibility with cheaply available optical-fiber components. Approximately as wide a group of compounds as for IR are amenable to the technique, and qualitative and quantitative information can be deduced from the Raman spectrum without any need for indicator chemistry. Any wavelength of light may be Raman scattered by a molecule, and the optimum for many analytes lies conveniently within the transmission window of silica optical fibers, which efficiently transmit visible and near-IR light (but not the wavelengths used in most IR absorption work). The major weakness of the technique is the low intensity (generally four to six orders of magnitude lower than typical fluorescence) of Raman-scattered light, and much work has been done to efficiently collect this light and separate it from any interference.

A common problem in measuring Raman spectra of (in particular) organic molecules is sample fluorescence. By using time-resolved methods as described in Section 7.2.4.2, longer lived fluorescence can be eliminated from (effectively instantaneous) Raman scattering by pulsing the excitation light source and gating the detector so that only "early" Raman light is received. For samples containing only one fluorescent component, Raman light can be separated by phase resolved methods, such as that demonstrated by Demas [56].

Demas demonstrated the nulling of unwanted fluorescence signals from Raman spectra by modulating the excitation light. The fluorescence lifetime of a fluorescent compound is usually significantly greater than that for the Raman scattering process, so the frequency of modulation can be varied until the fluorescence signal is 90 degrees out of phase with the Raman signal. Lock-in detection can then be used to reject the fluorescent light. Demas resolved the Raman spectrum of water from a solution of rhodamine 6G laser dye (excited at 514.5 nm). Although results were good, the phase-resolved technique is essentially analog, and so it is most suitable for use with scanning monochromators (rather than spectrographs using multielement output arrays), which makes for long measurement times. A time constant of 1 sec was used to resolve Raman peaks barely visible in the nonphase-resolved measurement, much more successfully than by subtraction of a normalized background measurement of rhodamine fluorescence. Demas postulated that slight errors in the system's cancellation of the fluorescent background may be due to variations of the fluorescent lifetime of two overlapping bands of the single fluorescent molecule, limiting the techniques ultimate sensitivity.

An obvious means of maximizing the amount of Raman-scattered light generated within a sample is to pass the excitation beam through the cell multiple times, or to use an arrangement in which the excitation light has a large interaction length with the sample and subsequently scattered light is guided to the collection optics, such as a capillary cell. If the walls of the capillary are sufficiently thin, light launched coaxially travels mostly within the analyte. Even better, if the walls of the capillary are made of a material with a lower refractive index than that of the analyte, a waveguide is formed that confines the excitation and scattered light emitted within the collection optics of the spectrograph to the capillary. Walrafen reported enhancements of up to 1,000 times by using such capillaries of up to 25m in length for the Raman spectrum of benzene. Unfortunately, large numbers of potential analytes have refractive indices lower than that of silica (1.47) (e.g., most aqueous solutions). The fabrication of capillaries with a refractive index lower than water (1.33) is very difficult; however, new results using members of the Teflon-AF family of fluoropolymers ($n < 32$) have demonstrated water-cored waveguides with potential for wide application for analysis of water-based analytes [57].

The cure process of epoxy resins has been studied using Raman spectroscopy via optical fiber, for instance by Chike [58]. Fully utilizing the information available in the Raman-scattered light from the material, both the extent of the cure process and the temperature of the system were measured. The temperature could be measured by comparing the intensity of the anti-Stokes Raman-shifted light (light shifted to a shorter wavelength), which changes exponentially with temperature. The variation of the Stokes lines is only slight and hence the ratio of the two measurements can be used to deduce the temperature at the sampling point. The degree of cure was calculated by taking the ratio of the epoxide ring stretch at 1,240 cm^{-1}, which is linearly dependent on the progress of the cure and the 1,186-cm^{-1} vibration of a component not affected by the cure. These results were compared with infrared absorbance measurements (through 1 mm of sample) and found to be in excellent agreement, and although the infrared measurements were made at close to optimal conditions, the authors felt that there was room to improve the apparatus for the *in situ* Raman measurement.

The arrangement used to gather the Raman light from within the epoxy composite was simply a pair of parallel 200-μm core optical fibers bonded into an SMA connector and polished down to a 3-μm finish. This "pencil probe" arrangement was simply dipped into liquid samples prepared with a suitable epoxide concentration to simulate the glue at various stages of cure. In practical applications such parallel fibers could be left within the material after cure, possibly for use as sensors for chemical degradation or to measure the temperature of the material in use.

An application in the nuclear industry for which the technique has been investigated is the detection of water in sodium nitrate slurry [59]. Although Raman is not the most obvious technique for this application (water has much strong IR absorption bands and would be easily identified by these), the Raman information would be obtained as a byproduct of other measurements made. Two approaches were investigated: first the direct detection of the water bending vibration around 1,630 cm^{-1}, and more successfully by taking the ratio of the intensities of the solid sodium nitrate Raman line and the intensity of the dissolved nitrate line (which is shifted by 17 cm^{-1}).

Direct measurements of the 1,630-cm^{-1} line of water (chosen in preference to the more intense lines around 3,300 cm^{-1} because of its much smaller variation in the presence of electrolytes and with temperature) could be detected down to a concentration of 2.5%. Referencing the intensities of the two phases of the nitrate peak (in conjunction with measurements of the temperature of the system to correct for variations of the solubility of the nitrate), limits of detection below 1% were achieved. Again, only two fibers were used to make these measurements (parallel 400-μm core fibers), so improvements in the collection optics and hence detection efficiencies are possible.

To reject the elastically scattered light from a sample, larger Raman spectrometers usually prefilter the input light using a monochromator. Cooney and his coworkers [60] have described the use of holmium-doped sections of optical fiber as an effective means of filtering out the elastically scattered light from a Raman spectrum. By using the 488-nm line of an argon laser, the elastically scattered light lay within an absorption band of the doped glass and was removed, whereas the Raman light was not significantly affected. Clearly, however, one must avoid subsequent detection of undesirable fluorescent light from such a filter.

7.3.2 Gas Sensing

The most common method of optical gas sensing is spectral transmission analysis, which is employed in two principal regions of the optical spectrum. Absorption or emission lines in the shorter region, from 250 to 500 nm (the UV to visible blue region) arise from electronic transitions. This is a very useful region of the spectrum for sensing the energetic changes that can occur within atoms or molecules of a large number of gaseous species.

The longer wavelength region, 1 to 8 μm, covers the near and mid-IR bands of the spectrum, a region where the vibrational absorptions of materials are more significant.

Typically, a vibrational absorption "line" will have a degree of fine structure superimposed on it, corresponding to the (usually) lower energy transitions associated with the rotational energy steps. (All these levels are, of course, quantized into discrete allowed steps according to the usual laws of quantum mechanics.)

Unfortunately, none of these transmission windows corresponds to any region of the spectrum where gas absorption is high. Electronic absorptions usually occur in the UV and violet/blue regions of the spectrum, where fiber loss is very high. Most of the strong fundamental vibrational absorptions occur in the mid-IR at 2,700 nm or longer, where silica is almost opaque. Thus, if it is desired to use conventional silica-based fiber, then the NIR absorption lines must be used. The use of long-path or multipass absorption cells is then necessary, to achieve even a moderate contrast in the measurement, and a well-designed optoelectronic system is necessary to reliably detect low levels of gases. As mentioned already, there are IR fibers based on more exotic materials (e.g., fluoride, silver halide, or chalcogenide compounds) but, as stated, these fibers are expensive and fragile compared to silica fibers.

Examples of the sensing of gases by conventional optical methods are too numerous to list in full and can only be briefly mentioned here. A sample list of recent references in this area is given [61–71]. This list includes direct absorption methods [61,64–69] and a photothermal method where the resulting refractive index change is monitored by interferometry [62]. More recent papers use compact diode laser sources [63-68] or special long-wavelength LEDs [64]. Because of the narrow absorption lines of many gases, the measurement contrast is generally much higher with narrowband laser sources than with LEDs. A significant number [61,63,65,66] have used the infrared region of the spectrum, beyond the reach of silica-based fiber systems, where gaseous absorption is generally greater. A number of more sophisticated methods have also been reported, for example correlation spectroscopy [69] and the use of frequency-modulated laser sources that are rapidly swept through absorption lines [64].

In view of the extensive literature on gas spectrometry over many years, it is somewhat surprising that the first recognition of the potential for fiber-optic-remoted gas sensing should not be until 1979 [71]. The researchers from a group at Tohoku University, Japan, pointed out the large number of (albeit rather weak) spectral absorption lines, which lie within the transmission window of a typical silica-based optical fiber. They pointed out the attraction of performing long-distance remote measurement over such links. In addition, the possibility of using liquid-core fibers (silica tubes, filled with carbon tetrachloride) for the mid-IR region was suggested, as such guides offer long-wavelength transmission well beyond the cutoff of silica fibers, with a loss of only 56 dB/km as far out as 3.39 μm (This is in spite of using a silica cladding, which contributes to the losses by evanescent-field absorption.)

7.3.2.1 Nitrogen Dioxide Sensing

The first workers to demonstrate the technique in practice were from the same group at Tohoku University [72]. The gas chosen to demonstrate their method was nitrogen dioxide, a common impurity in vehicle exhaust gases, with a relatively long-wavelength electronic absorption line at 496.5 nm. The method involved a single-channel fiber-remoted spectrometer with two-wavelength referencing: one wavelength on the absorption line, the other displaced from the line of interest. The source, providing both the measurement line and a nearby reference line was a multiline argon ion laser. With a measurement-cell optical path length of 20m (using a multi-pass cell design to reduce size), and a response time constant of 1 sec, a noise-limited sensitivity of approximately 17 ppm was obtained. Measurements of nitrogen dioxide in exhaust gases from a motorcycle were taken, over a concentration range from 0 to 100 ppm.

7.3.2.2 Methane Gas Detection

The first demonstration of methane gas detection over optical-fiber paths (Figure 7.15) [73] was at the Electronics Research Laboratory of the Norwegian Institute of Technology, Trondheim. Their laboratory system used a broadband white light source, rotating chopper, and interference-filter arrangement to sequentially interrogate the transmission of the sample cell via a fiber-optic cable link, and that of a more direct (free-space) reference path. The measurement was broadband in nature, as the interference filter covered all the fine rotational line structure in the methane absorption band, (i.e., a 70-nm wide region, centered on 1.665 μm). The system effectively averaged the detected signal level over this wavelength range, giving rise to a very low contrast. Even on the peak of the individual lines, there is a relatively weak absorption at the low concentrations (usually <5% maximum) required to be measured. However, despite this, it was the first reported demonstration of methane detection over optical fibers and had a respectable noise-limited sensitivity of 0.5% of the lower explosive limit (LEL) of methane (the LEL is approximately 5%

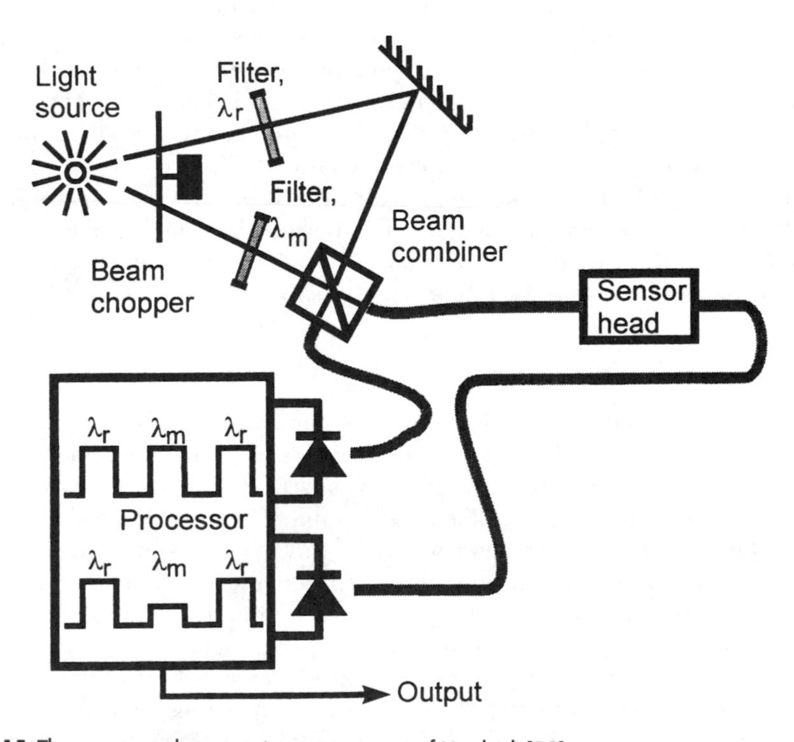

Figure 7.15 The remote methane sensing arrangement of Hordvick [73].

methane in air). However, neither the long-term drift characteristics nor the cross-sensitivity to other gases were critically examined in this early paper.

The first use of semiconductor LED sources for fiber-remoted gas detection, in this case in conjunction with a fiber-remoted methanometer, was again reported by the Inaba group from Tohoku University [74]. A quaternary InGaAsP/InP device having a laser structure was operated under lasing threshold level to provide an ELED source having a center wavelength of 1.61 μm and an 80-nm linewidth. The system was operated as a single-beam absorption system (i.e., no reference path). The ELED was square-wave modulated and light was launched directly into the transmission fiber (Figure 7.16). This fiber-guided light over a 1-km path to the single-pass sensing cell (0.5m long) and a similar fiber collected the transmitted light and guided it to a cooled germanium detector, followed by a lock-in amplifier. The noise-limited resolution was equivalent to ≈0.07% of methane. In this simple laboratory demonstrator, however, there was no provision of a reference path to guard against long-term drift effects.

The first fiber-remoted methane detection scheme to be field-tested was reported by Stueflotten [75] of A/S Elektrik Bureau, Norway. This system had much in common with the one just described (i.e., it used a compact chopped LED source and synchronous detection). However, steps were taken to enhance the long-term stability of the system by using a dual-LED system, with one LED source centered on the absorption band and the other at an adjacent (nonabsorbing) region of the spectrum (Figure 7.17). These sources were alternately pulsed and the outputs combined into

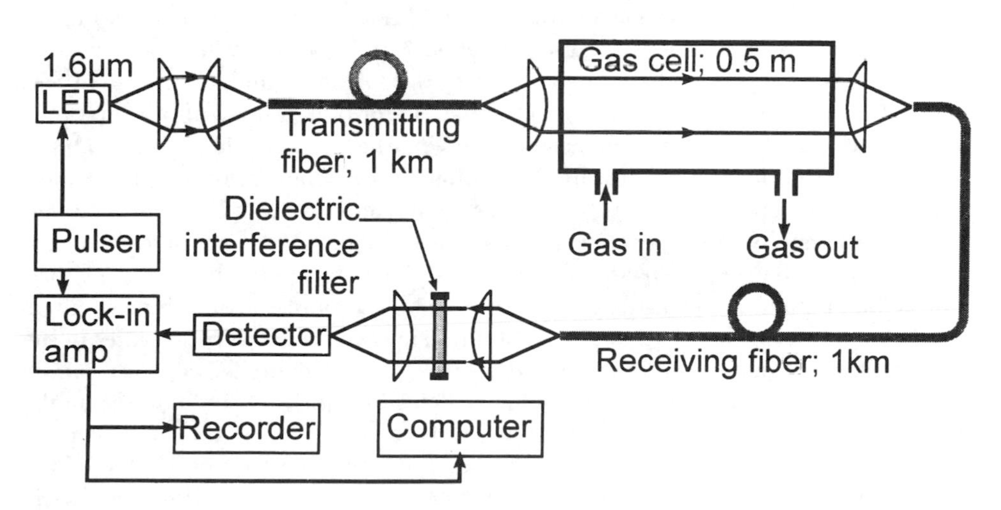

Figure 7.16 The remote methane sensing arrangement of Chan [74].

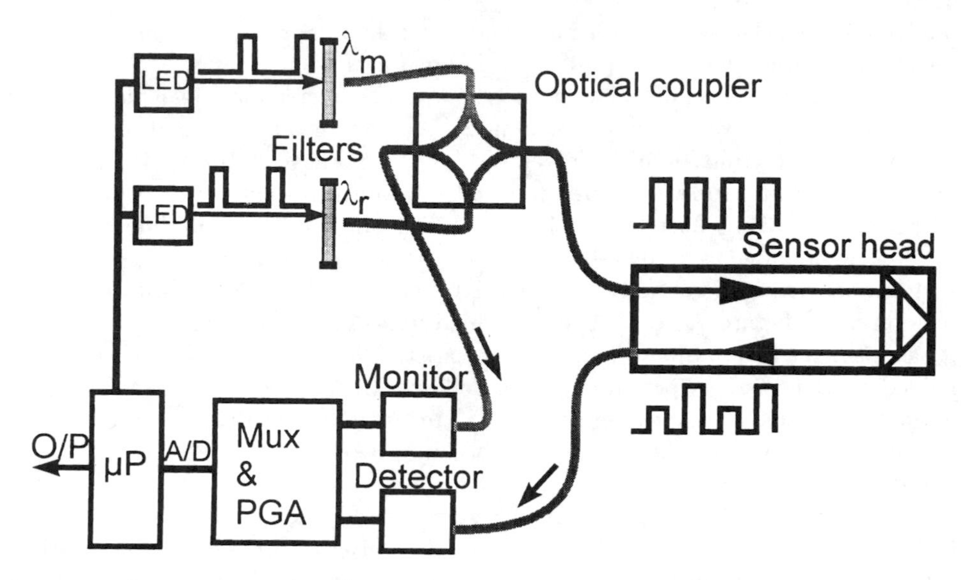

Figure 7.17 The arrangement of Steuflotten [24].

the input fiber using a passive coupler. On their return to the detector, after passage through a two-pass cell and the return fiber, the detected signal amplitudes from each source were electronically compared with a directly derived sample of the transmitted light signals. This allowed estimation of the degree of absorption that had taken place in the sample cell. The rms noise-limited sensitivity of the system, with a 1-sec time constant, was ±1.5% LEL (equivalent to 0.075% methane). The system was reported to have been tested on a North Sea gas rig for 6 months. With this first prototype, it was indicated that problems in achieving the necessary long-term stability were experienced due to temperature fluctuations in LEDs and wavelength-selective variations in the optical couplers, connectors, and cables. It was claimed, however, that a new design had been developed that was expected to overcome these problems.

All the methane gas detection methods described above were based on relatively broadband illumination of the gas, using either LED sources or interference filters having linewidths of the order of 20 to 100 nm. However, the absorption lines in the gas have natural linewidths much less than this, and a far higher contrast can be achieved using narrowband sources or filtered-detection systems. (The contrast in the measurement is an extremely important aspect, as a large fractional change helps to dominate problems of drift in signal level due to undesirable systematic effects.) However, the use of a narrow-line lasers as source can give rise to severe interferometric noise problems due to speckle effects (i.e., the modal noise phenomena familiar in fiber-optics communications systems). A method to help avoid this dilemma is to use a "comb" filter, having a regularly spaced series of narrow passbands, with wavelength spacing corresponding to that between the rotational absorption lines of the gas [76,77], as shown in Figure 7.18(a).

This increases the useful power that can be extracted from a broadband incandescent or LED source, as the combined effect of several absorption lines may be monitored and effectively decreases the coherence of the source. A suitable filter is a Fabry-Perot cavity, arranged to have the correct spacing to achieve the desired free spectral range. A further attraction of the approach is that the Fabry-Perot filter may be mechanically scanned to produce an ac measurement and allow referencing of the absorption signal (i.e., the "dip" in the signal when the filter coincides with the gas absorption lines) to the peak transmitted signal, (the flat peak level when the filtered bands lie between the gas absorption lines), as shown in Figure 7.18(b).

A further advantage is the strong selectivity of the method, because it tends to "fingerprint" the gas absorption spectrum. The reported methods used, typically, six rotational lines for the measurement, making the detected signal dependent on a matching of the Fabry-Perot spacing to the characteristic line spacing, in addition to the absorption band location.

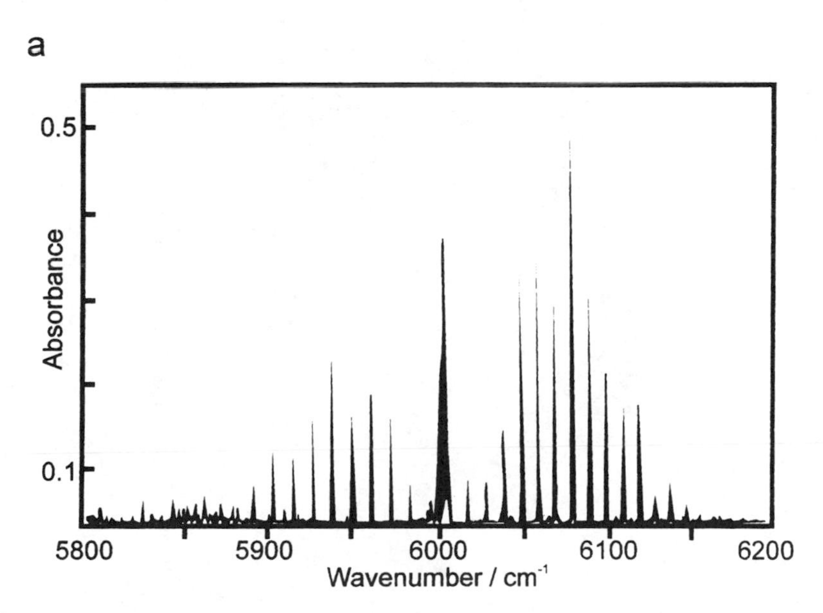

Figure 7.18 (a) Methane detection using a scanned Fabry-Perot comb filter, (b) probe layout using scanning filter, and (c) response to methane gas.

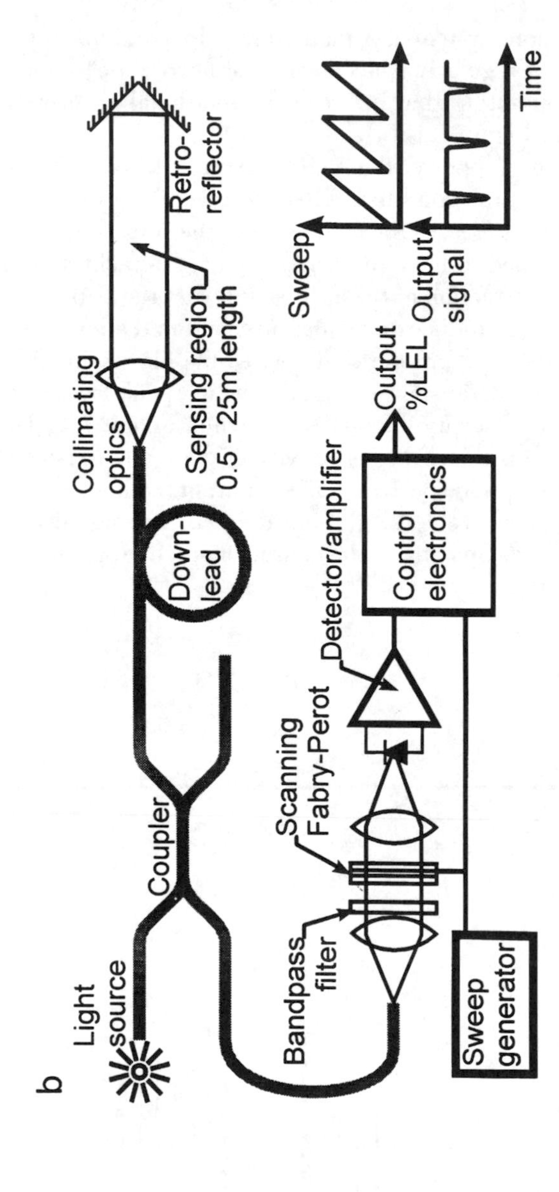

Figure 7.18 (continued)

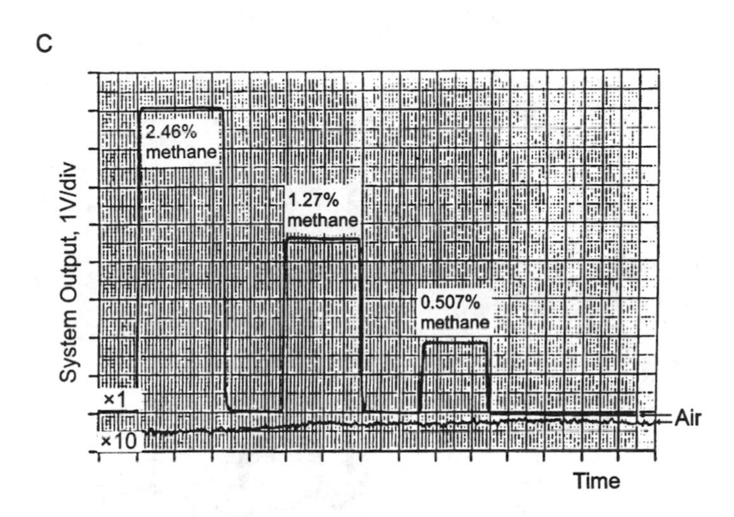

Figure 7.18 (continued)

The response to methane gas, using a twin-pass (2 × 0.5m) absorption cell, is shown in Figure 7.18(c) [78]. With a 1-sec measurement time constant, the noise-limited detection level of this method corresponded to ±0.003% rms methane.

Two other narrow linewidth systems for the detection of methane have recently been reported. The first system, shown in Figure 7.19, is based on the use of a twin-longitudinal mode 1.33-μm GaInAsP (Fabry-Perot type) semiconductor laser diode [79]. The sensor also contains a scanning Fabry-Perot etalon to alternately select one or the other longitudinal mode of the laser. The laser was thermally tuned until one mode corresponded with an absorption line of the methane gas. In spite of the very low absorption of the 1.33-μm band, the sensitivity of the results was good, mainly because of the high source power available in a very narrow line. (A noise-limited sensitivity of the order of ±0.05% methane was reported in a 1m path length cell). The higher power available from the source should also enable multipass cells to be used, (although longer path lengths can give rise to nonlinearity in the response to the methane level).

The second laser-based method [80] used a Tm^{3+} fiber laser, tuned through a 1.685-μm methane line by stretching a wavelength-selective grating mirror. No particular laboratory problems with either of these methods were reported, although it might be anticipated that "speckle" effects (i.e., the well-known modal noise phenomena) and, in the first case, laser-mode partition effects could be a possible source of error, particularly under more severe operational environments.

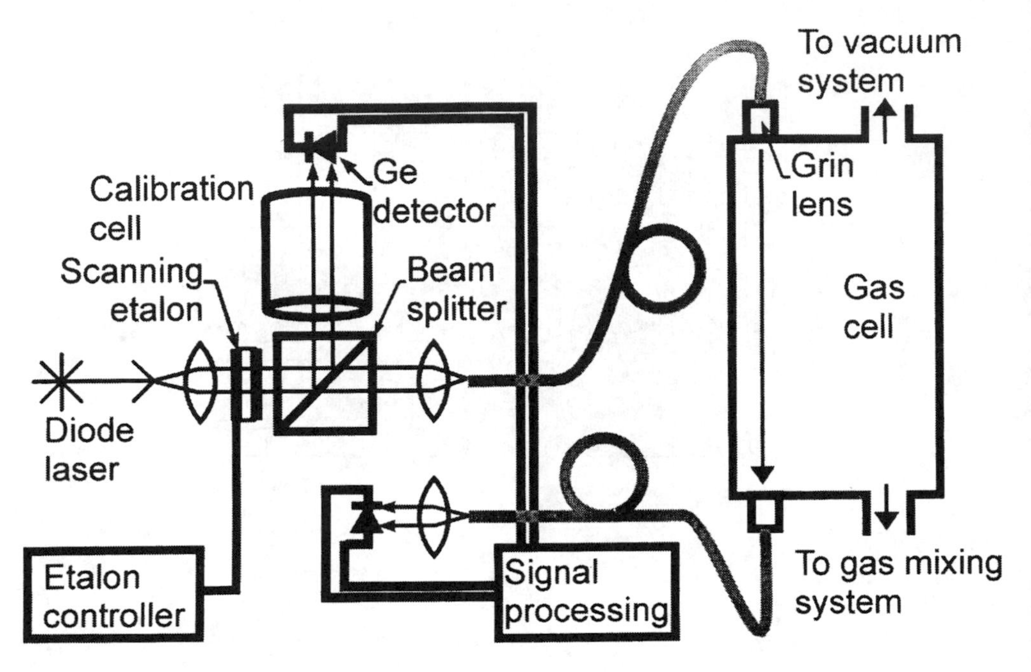

Figure 7.19 Gas sensor using diode laser source and etalon to select the appropriate laser mode.

7.3.2.3 Hydrogen Gas Sensing

A novel system for hydrogen gas detection [81] is based on the dimensional expansion experienced by palladium metal when it adsorbs hydrogen gas. This occurs by an interesting process in which the gas is occluded at interstitial sites within the atomic lattice. The metal, in the form of thin wire, was bonded to one fiber arm of an all-fiber Michelson interferometer (see Figure 7.20). The resulting linear dimensional change in the palladium, which is proportional to the square root of the hydrogen partial pressure, is transferred to the fiber, resulting in changes in the output state of the interferometer. A hydrogen partial-pressure resolution of ±2 Pascals was observed, equivalent to a hydrogen concentration of 20 ppm. A problem with this initial laboratory system was that the attachment of the palladium wire resulted in a differential expansion between the two dissimilar arms. Hence a temperature change of only 0.3K could cause an effect equivalent to the 20-ppm detection limit. However, in a practical system, this source of error could be prevented, either by careful temperature control or by expansion balancing of the arms of the fiber interferometer.

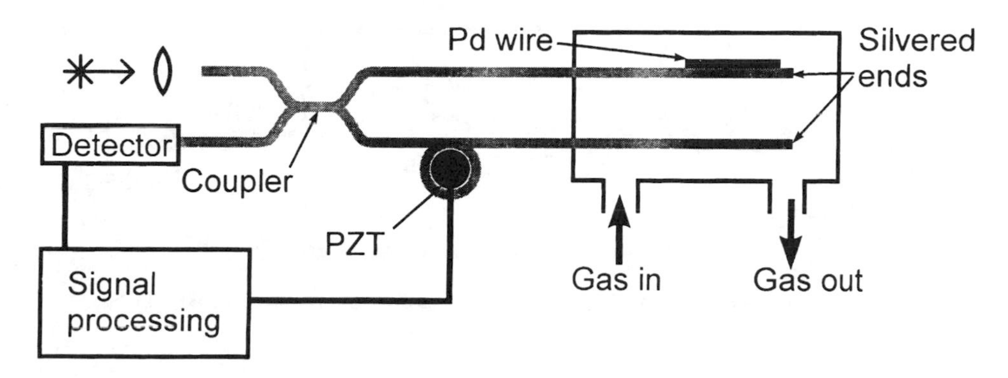

Figure 7.20 Experimental arrangement of H_2 sensor using occlusion in Pd [77].

7.3.2.4 Evanescent Wave Gas Sensing

All the above methods for gas sensing, with the exception of the palladium-based hydrogen sensor, have been of the extrinsic type (i.e., one using the fiber merely as a light guide to and from the sensing region). It is now appropriate to describe intrinsic-type sensors that use the evanescent field extending beyond the fiber core region to detect the gas in the vicinity of the fiber. The first method of this type was reported by Tanaka [82] (see Figure 7.21). This paper reported an intrinsic sensor using a 50 μm/125 μm fiber, which was fused and stretched so that it tapered down to form a short, ultrathin sensing region. Here, the evanescent field could leak into the air surrounding the fiber. The sensor was used to detect the strong 3.39-μm methane absorption line. However, in this region the high losses in the silica fiber limited the total length of fiber to less than 3m. In addition, the noise-limited resolution limit was relatively poor (of the order of 2% methane). Therefore, although of technical interest, it is less obvious how this technique might ultimately be applied unless lower loss infrared fibers were to be used. However, even when the fiber loss problem is solved, there could be severe sensitivity to surface contamination.

An alternative to using tapers, which require sections of specially treated fiber, is to use noncylindrical fibers that have the core region close to the outside surface. One example of this is the D-fiber, so called because of its D-shaped cross-section, which brings the core close to the flat edge of the fiber [83]. Culshaw has used this method for gas detection [84]. The attraction is that because of the constant cross-section of the fiber, construction of a long sensing section is possible, with the further possibility of distributed sensing with an OTDR system. Jin has attempted to compensate for surface contamination [85], but this is likely to be difficult in real

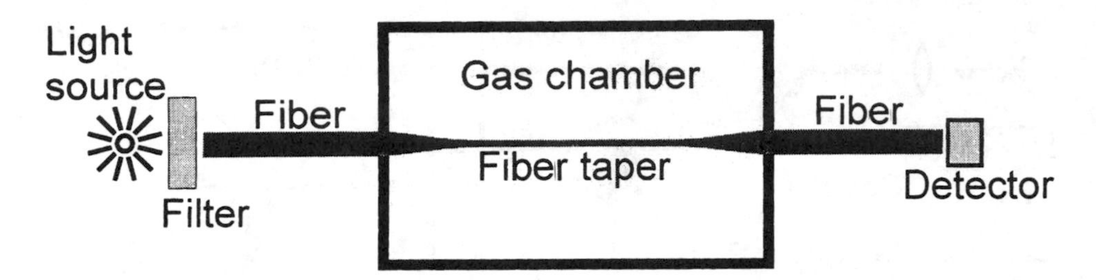

Figure 7.21 Measuring the absorption of a gas by fiber evanescent wave spectroscopy (FEWS).

environments with variable temperature, nonuniform contamination, and bend sensitivity.

At the UV end of the spectrum, Potyrailo has used 1.8m of coiled polysiloxane-clad fiber to measure ozone absorption at 254 nm in concentrations between 0.06 to 0.35% by volume [86]. The polysiloxane cladding was permeable to gaseous ozone, with a typical diffusion time of the order of a minute, and showed no degradation over the two-month period of the investigation. The limit of detection was calculated as 0.02%, and applications in water treatment as well as ozone manufacturing were foreseen.

7.3.2.5 Porous Glass Sensors

A final intrinsic sensor uses the ingenious idea of a vitreous gas-permeable fiber, constructed from a glass that has been designed to undergo phase separation. This glass composition permits chemical leaching out of an alkali-rich phase to leave a porous fiber structure of lower alkali content [87]. The microscopic dimensions of the pore structure (typically ≈1,000Å diameter) in such fibers permits a strong optical interaction with the gaseous species and the method therefore shows promise for high-sensitivity detection of many gases. The first demonstration of the porous fiber method was for the detection of water vapor over a 0 to 50% relative humidity range. A response time of the order of 1 min was achieved, but no long-term measurements were taken. Problems with the method could possibly occur due to surface adsorption or due to the ingress of liquids into the pores by capillary action. As stated above, the great attraction of such intrinsic sensors are their potential for truly distributed sensors using OTDR. In order to be successful, the fiber attenuation and the degree of coupling between the incident optical field and the gas must be well characterized. In addition, good reversibility will normally be required. Although the early methods reported [82–90] show varying degrees of promise for such application, much more work must be done before they can be seriously used in systems. So far, none have yet been used in conjunction with OTDR systems, although reports of distributed measurements are expected to appear in the literature.

7.3.2.6 Photoacoustic Gas Sensing

Modulating or pulsing the photoacoustic excitation source causes a varying thermal expansion in the sample, which generates a series of acoustic waves propagating away from the optical path. These are detected using a microphone, which usually consists of a piezoelectric transducer made of a ceramic such as lead zirconate titanate (PZT), or a polymer such as poly [vinylidene difluoride] (PVDF). Condenser microphones have been used for gas-phase detection. However, as an interesting variation, an optical-fiber sensor has been used with the eventual aim of enhancing sensitivity.

Optical Detection of Photoacoustic Pressure Waves

Two types of optical-fiber interferometers for acoustic detection have been compared by Bregeut and coworkers [91]. A resonant azimuthal cell was used, with an optical microphone situated at a pressure antinode and the optical excitation beam situated at a pressure node (chopped electrically at the frequency of cell resonance, 4 kHz). The microphone consisted of 2m of monomode optical fiber coiled and bonded onto an aluminum plate.

Photoacoustic waves in the cell caused the microphone plate to be deflected, and the fiber to be periodically stretched, resulting in optical phase differences between the sensing beam and the reference beam. A Michelson interferometer and a Sagnac interferometer were compared, as shown in the schematic diagrams in Figure 7.22.

The Sagnac interferometer had the advantages of greater stability (low-frequency environmental disturbances affect both interfering beams equally) and greater simplicity (there was no need to equalize the lengths of the sensing arm and the reference arm). Unfortunately, its performance was worse than an electrical microphone, which had a signal-to-noise ratio (SNR) seven times greater. The authors suggested that this may have been due to greater shielding of the electrical microphone from environmental acoustic noise. The sensitivity of the optical microphone could have been increased by lengthening the sensing fiber and by better acoustic shielding, and therefore had promise of superior ultimate performance.

The work illustrates the advantage of using a resonant photoacoustic cell to provide a mechanical "amplification" of the photoacoustic pressure waves by an amount depending on the quality factor Q of the resonance. Since the precise frequency of resonance will vary with temperature and pressure, active locking of the source modulation frequency to the resonant peak is usually needed to ensure large Q values, and so the system can become rather complex. It has been calculated that without compensation, a change in temperature of 1K, in a resonant cell with a Q of 500, can reduce the photoacoustic signal by a factor of 2 [92].

7.3.2.7 Raman Spectroscopy for Gas Sensing

Raman scattering, which has excellent potential for general chemical analysis, has only recently been applied to optical-fiber gas sensing [93] (Figure 7.23). As

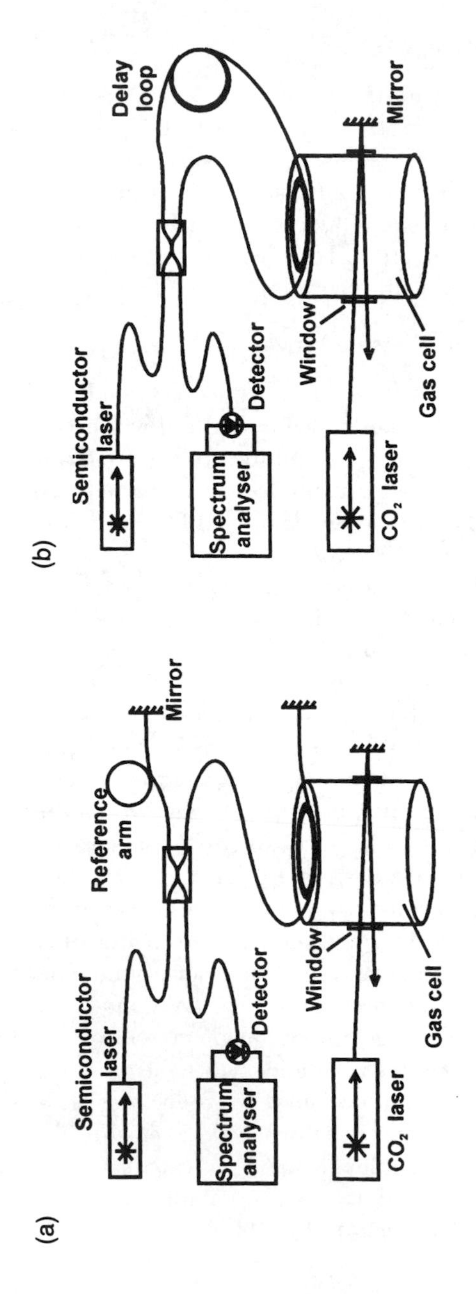

Figure 7.22 Interferometric detection of standing pressure waves in a resonant azimuthal photoacoustic cell with (a) a Michelson interferometer and (b) a Sagnac interferometer. Picture modified from Bregeut [91].

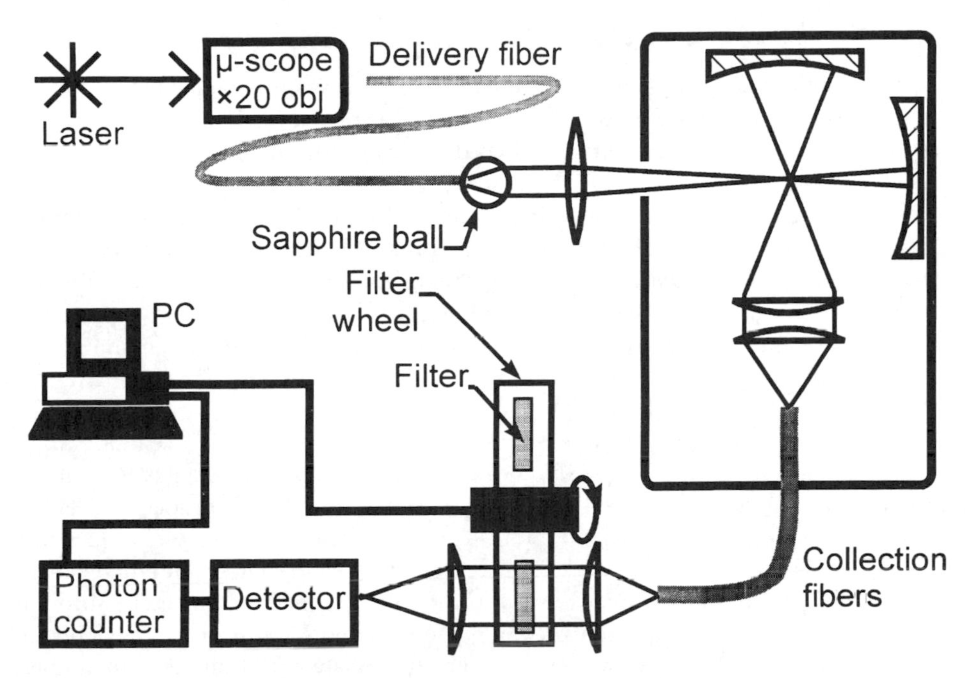

Figure 7.23 Gas sensing by Raman spectroscopy.

described earlier in this chapter, the advantage of the method is its capability to interrogate energy levels normally associated with the mid and far infrared, while using visible or near-infrared light in both the excitation and scattered beams. In addition, gases such as nitrogen, having no directly measurable IR absorption bands, can be measured due to the different selection rules associated with Raman transitions. However, Raman-scattered light from gas samples is extremely weak. It was found necessary to use a photomultiplier, in photon counting mode, and to average for tens of seconds in order to detect the weak Raman light from a relatively concentrated gas sample [93], although little was done to maximize the signal by design of the Raman cell (e.g., by arranging multiple reflections). An internally gold-coated capillary cell has been used (as described for liquid phase analysis in Section 7.2.7.3) by Dyer to demonstrate the possibility of using FT Raman instrumentation for the study of gases [94]. Using the capillary cell and 2.5W excitation at 1064 nm, the nitrogen in air could be seen at a 2:1 signal to noise ratio. Although the recorded spectra were of poor quality due to the long-wavelength excitation and poor detector responsivity, this technique is suited to remoting via optical fibers using shorter excitation wavelengths and dispersive instrumentation.

7.3.2.8 Correlation Spectroscopy for Gas Sensing

Early fiber-optic-based gas-sensing methods measured transmission using broadband LED sources [73] and thus had poor selectivity. Other recent methods have used laser sources [95], but these can present problems due to the long coherence length (in particular, modal noise effects in multimode fiber). Even when single-mode fiber is used to prevent this, effects such as Fabry-Perot etalons at connectors, and more complex interference patterns within the measurement cell and launch optics, can potentially cause severe practical limitations. Correlation spectroscopy [96–100] allows the use of a broadband source, yet still monitors the fine spectral features of the gas spectrum. It has the further advantage of employing all of the spectral information contained in the selected gas-absorption band rather than just a single line, thereby enhancing the cognitive nature of the technique.

The basic methods of real-time correlation spectrometry all involve modulation of the absorption spectrum of the reference gas sample relative to the gas to be measured. Modulation can be achieved directly by varying the absorption spectrum of the reference gas. In an early paper [99], progress with three modulation techniques was reviewed, each using the correlation spectroscopy method. These included Stark, pressure, and phase modulation. The apparatus comprised two gas cells, through which light from a broadband source was passed sequentially prior to detection. In the fiber-remoted version, light was conveyed to the cells via multimode optical fiber, and a bandpass filter was included before the detector in order to attenuate light outside the absorption band of the relevant gas. The gas to be detected was directed into the measurement cell, whereas the reference cell was filled with a known concentration of the gas to be detected and then sealed.

Modulation of the absorption spectrum of either gas cell resulted in a change in the correlation between the spectrum of the measurement gas sample and that of the reference gas, thereby causing a variation in the transmission of the system and hence in the detected signal. The absorption spectrum of the gas contained in the reference cell was modulated, either directly or indirectly, in order to produce the desired signal. The synchronously detected output signal then depended on the concentration of the correlating gas in the measurement cell.

Stark modulation of gases occurs only with polar molecules and results from the splitting (or, at atmospheric pressure, the broadening) of individual absorption lines when a large electric field is applied. The pressure-modulation technique is more generally applicable. This involves periodically pressurizing the gas within the reference cell, causing a variation in both the strength and width of the absorption lines.

A third modulation approach is indirect and involves a redistribution of the optical spectrum of the light as it passes from the measurement to the reference cell. Angle modulation of the light (using phase or frequency modulation) is a means of achieving this, which can be achieved using electro-optic (Pockel's cell) or acousto-optic (Bragg cell), respectively. Two more recent papers have considered improvements to correlation spectroscopy systems [101,102].

7.4 CONCLUSIONS

The direct detection of chemicals using direct spectroscopy has been reviewed at some length. The treatment has been extremely detailed in view of the generic nature and industrial importance of many of the techniques. This is an area where many real commercial systems have already been developed (albeit many being simple extensions for laboratory spectrophotometers) and is likely to be an extremely important future area for fibers.

ACKNOWLEDGMENT

The authors would like to thank Dr. Richard Harris for proofreading the final manuscript.

References

[1] Thorne, A. P., *Spectrophysics*, (2nd Ed.), Chapman and Hall Ltd. (London), 1988.

[2] Heitmann, W., "Attenuation analysis of silica-based single-mode fibers," *Journal of Optical Communications*, 11, 1990, pp. 122–129.

[3] Senior, J. M., *Optical Fiber Communications Principles and Practice*, (2nd Ed.), Prentice Hall International (UK) Ltd., Hemel Hempstead. 1992.

[4] Freeman, J. E., et al., "A fiber-optic based sensor for bioanalytical absorbance measurements," *Analytical Chemistry*, 56, 1985, pp. 2246–2249

[5] Willis, H. A., van der Maas, J. H., and Miller, R.G.J. (Eds.), *Laboratory Methods in Vibrational Spectroscopy*, (3rd Ed.), John Wiley and Sons, 1987.

[6] Wolfbeis, O. S. (Ed.), *Fiber Optic Chemical Sensors and Biosensors*, Vol. I, CRC Press, 1991.

[7] Hecht, E., *Optics*, (2nd Ed.), Addison-Wesley Publishing Company, Reading MA, 1987.

[8] Rosencwaig, A., *Photoacoustics and photoacoustic spectroscopy*, John Wiley & Sons, ISBN 0-89464-450-5, 1980.

[9] Pao, Yoh-Han (Ed.), *Optoacoustic spectroscopy and detection*, Academic Press, 1977.

[10] Hess, P. (Ed.), "Photoacoustic, photothermal and photochemical processes in gases," *Springer-Verlag Topics in Current Physics*, 46, 1989.

[11] Jackson, W. B., et al., "Photothermal deflection spectroscopy and detection," *Applied Optics*, 20, 1981, pp. 1223–1345.

[12] Allen, C. S., et al, "Tunable laser excitation profile of surface enhanced Raman scattering from pyridine adsorbed on a copper electrode," *Chemical Physics Letters*, 75, 1980, p. 201.

[13] Chang, R. K., and Furtak, T. E. (Eds.), *Surface Enhanced Raman Scattering*, Plenum Press, 1982.

[14] Myrick, M. L., and Angel, S. M., "Elimination of background in fiber-optic Raman measurements," *Applied Spectroscopy*, 44, 1990, p. 565.

[15] Ma, J., and Ying-Sing, L., "Fiber Raman background study and its application in setting up optical fiber Raman probes," *Applied Optics*, 35, 1996, pp. 2527–2533.

[16] Zhu, Z. Y., and Yappert, M. C., "Determination of the effective depth for double-fiber fluorometric sensors," *Applied Spectroscopy*, 46, 1992, pp. 919–924.

[17] Plaza, P., et al., "Simulation et optimisation des capteurs á fibers optiques adjacents," *Applied Optics*, 25, 1986, pp. 3448–3454.

[18] Polytec X-dap product literature, Lambda Photometrics Ltd., Harpenden, UK.

[19] Nave, S. E., et al., "Fiber-optic Raman spectroscopy at the Savannah river site: uses and techniques," *Process Control and Quality*, 3, 1992, pp. 43–48.

[20] House of Hellma product literature, Hellma (England) Ltd., 23 Station Road, Westcliffe-on-Sea, Essex, SS0 7RA, UK.

[21] Nave, S. E., O'Rourke, P. E., and Toole, W. R., "Sampling probes enhance remote chemical analyses," *Laser Focus World*, Dec. 1995, pp. 83–87.

[22] FCI Environmental, Inc., 1181 Grier Drive, Las Vegas, Nevada. (Geotechnical Instruments, in UK).

[23] Kersey, A. D., et al., "Single-mode fiber Fourier transform spectrometer," *Electronic Letters*, 21, 1985, pp. 463–464.

[24] Jie, L., and Brown, C. W., "Near-IR fiber optic probe for electrolytes in aqueous-solution," *Analytical Chemistry*, 65, 1993, pp. 287–292.

[25] O'Rourke, P. E., "Chemometrics/on-line measurements," *JNMM, Journal of the Institute of Nuclear Materials Management*, 18, 1988, pp. 85–94.

[26] Mittermayr, C. R., et al., "Neural networks for library search of ultraviolet-spectra," *Analytica Chimica Acta*, 294, 1994, pp. 227–45.

[27] Long, J. R., Vasilis, G. G., and Gemperline, P. J., "Spectroscopic calibration and quantitation using artificial neural networks," *Analytical Chemistry*, 62, 1990, pp. 1791–1797.

[28] Zupan, J., and Gasteiger, J., "Neural networks: A new method for solving chemical problems or just a passing phase?," *Analytica Chimica Acta*, 248, 1991, pp. 1–30.

[29] Polani, M. L., "In Vivo Oximeter with Fast Dynamic Response," *Review of Scientific Instruments*, 22, 1962, p. 1050.

[30] "Optically Based Methods for Process Analysis," *Proc. of SPIE - The International Society for Optical Engineering*, 1681, Somerset, NJ, USA, (Conf. code 17498), Mar. 23–26, 1992.

[31] "Optical Sensing for Environmental and Process Monitoring," *Proc. of SPIE - The International Society for Optical Engineering*, 2365, McLean, VA, USA, (Conf. code 22254), Nov. 7–10, 1994.

[32] Boisde, G., Blanc, F., and Perez, J-J., "Chemical measurements with optical fibers for process control," *Talanta*, 35, 1988, pp. 75–82.

[33] Coleman, J. T., Eastham, J. F., and Sepaniak, M. J., "Fiber-optic based sensor for bioanalytical absorbance measurements," *Analytical Chemistry*, 56, 1984, pp. 2246–2249.

[34] Krska, R., et al., "Fiber optic sensor for chlorinated hydrocarbons in water based on infrared fibers and tunable diode lasers," *Applied Physics Letters*, 63, 1993, pp. 1868–1870.

[35] Simhony, S., and Katzir, A., "Fourier transform infrared spectra of organic compounds in solution and as thin layers obtained by using an attenuated total internal reflectance fiber-optic cell," *Analytical Chemistry*, 60, 1988, pp. 1908–1910.

[36] Krska, R., et al., "Polymer coated silver halide infrared fibers as sensing devices for chlorinated hydrocarbons in water," *Applied Physics Letters*, 61, 1992, pp. 1778–1780.

[37] Ertan-Lamontagne, M. C., et al., "Polymer-coated, tapered cylindrical ATR elements for sensitive detection of organic solutes in water," Applied Spectroscopy, 49, 1995, pp. 1170–1173.

[38] Degrendpre, M. D., and Burgess, L. W., "Fiber-optic FT-NIR evanescent field absorbance sensor," *Applied Spectroscopy*, 44, 1990, pp. 273–279.

[39] Bürck, J., Conzen, J. P., and Ache, H. J., "A fiber optic evanescent field absorption sensor for monitoring organic contaminants in water," *Fresenius Journal of Analytical Chemistry*, 342, 1992, pp. 394–400.

[40] Bürck, J., et al., "Fiber-optic evanescent wave sensor for in situ determination of non-polar organic compounds in water," *Sensors and Actuators B*, 18-19, 1994, pp. 291–295.

[41] Schwotzer, G., et al., "Fiber optic evanescent field sensor for hydrocarbon monitoring in air and water applying UV absorption," *OFS 11*, Sapporo, Japan, May 1996, pp. 21–24.

[42] Sontag, H., and Tam, A. C., "Optical detection of nanosecond acoustic pulses," *IEE Transactions on Ultrasonics, Ferroelectrics and Frequency Control*, UFFC-33, 1986, pp. 500–506.

[43] Hand, D. P., et al., "Detection of photoacoustic waves in liquids by fiber-optic interferometer," *Optics Communications*, 104, 1993, pp. 1–6.
[44] Hand, D. P., et al., "Optical fiber interferometry for photoacoustic spectroscopy in liquids," *Optics Letters*, 20, 1995, pp. 213–215.
[45] Tam, A. C., "Applications of photoacoustic sensing techniques," *Reviews of Modern Physics*, 58, 1986, pp. R83–R121.
[46] Tam, A. C., "Applications of photoacoustic sensing techniques," *Review of Modern Physics*, 58, 1986, pp. 381–431.
[47] Bohnert, B., Faubel, W., and Ache, H. J., "Comparison of collinear and transverse photothermal deflection spectroscopy for trace analysis of pesticides in water," *Fresenius Journal of Analytical Chemistry*, 343, 1992, pp. 513–517.
[48] Vegetti, G., et al., "Photothermal detection of trace chemicals by fiber-optic interferometric probe," *Springer series in optical sciences vol 69: Photoacoustic and photothermal phenomena III*, Bicanic, D. (Ed.).
[49] Klainer S., et al., "A monitor for detecting nuclear waste leakage in a sub-surface repository," Report, Livermore-Berkeley Laboratory, 1981.
[50] Chudyk, W. A., Carrabba, M. M., and Kenny, J. E., "Remote detection of Ground water contaminants using far-ultraviolet laser induced fluorescence," *Analytical Chemistry*, 57, 1985, p. 1237.
[51] Berlman, I. B., *Handbook of Fluorescent Spectra of Aromatic Molecules*, (2nd Ed.), Academic Press, New York, 1974.
[52] Bublitz, J., et al., "Fiber-optic laser-induced fluorescence probe for the detection of environmental pollutants," *Applied Optics*, 34, 1995, p. 3223.
[53] Panne, U., and Neissner, R., "A fiber-optical sensor for polynuclear aromatic hydrocarbons, based on multidimensional fluorescence," *Sensors and Actuators B*, 288, 1993, pp. 13–14.
[54] Hillrichs, G., Karlitschek, P., and Neu, W., "Fiber optic aspects of UV laser spectroscopic in situ detection of water pollutants," *SPIE Proceedings Vol. 2293 Chemical, Biochemical, and Environmental Fiber Sensors VI*, 07/24–07/29/94, San Diego, CA, USA.
[55] Dmitriev, V. G., Gurzadyn, G. G., and Nikogosyan, D. N., *Handbook of Nonlinear Optical Crystals*, Springer Verlag, 1991.
[56] Demas, J. N., and Keller, R. A., "Enhancement of luminescence and Raman spectroscopy by phase-resolved background suppression," *Analytical Chemistry*, 57, 1985, p. 539.
[57] Mackenzie, S. J., and Dakin, J. P., "Internally Teflon-AF coated capillary cell for optical fiber remote spectroscopy," *CLEO/Europe-EQEC*, Hamburg, Sept. 8–13, 1996.
[58] Chike, K. E., et al., "Raman and near-infrared studies of an epoxy resin," *Applied Spectroscopy*, 47, 1993, p. 1631.
[59] Lombardi, D. R., Mann, C. K., and Vickers, T. J., "Determination of water in slurries by fiber-optic Raman spectroscopy," *Applied Spectroscopy*, 49, 1995, pp. 220–224.
[60] Cooney, T. F., et al., "Rare-earth doped glass fiber for background rejection in remote fiber-optic Raman probes: Theory and analysis of Holmium-bearing glass," *Applied Spectroscopy*, 47, 1993, p. 1683.
[61] Edwards, D. K., "Absorption by infrared bands of carbon dioxide at elevated pressures and temperatures," *Journal of the Optical Society of America*, 50, 1960, pp. 617–626.
[62] Davis, C. C., and Petuchowski, S. J., "Phase fluctuation optical heterodyne spectroscopy of gases," *Applied Optics*, 20, 1981, pp. 2539–2554.
[63] Koga, R., Kosaka, M., and Sano, H., "Field methane tracking with a portable and real-time open-gas monitor, based on a cw-driven Pb-salt laser," *Optics and Laser Technology*, 17, 1985, pp. 139–144.
[64] Aagard, R. T., et al., "Development of a selective natural gas detector," *Proc. of International Gas Research Conference*, Toronto, Sept. 1986.

[65] Forrest, G. T., "Simple diode laser spectroscopy from 6 to 10 microns," *Spectrochimica Acta 42a*, 1986, pp. 281–284.
[66] Silver, J., and Stanton, A. C., "Airborne measurements of humidity using a single-mode Pb-salt diode laser," *Applied Optics*, 26, 1987, pp. 2558–2566.
[67] Milton, M. J., and Woods, P. T., "Pulse averaging methods for a laser remote monitoring system using atmospheric backscatter," *Applied Optics*, 26, 1987, pp. 2598–2603.
[68] Cassidy, D. T., "Trace Gas detection using 1.3 μm InGaAsP diode laser transmitter modules," *Applied Optics*, 27, 1988, pp. 6120–6140.
[69] Margolis, J. S., McCleese, D. J., and Martonchik, J. V., "Remote Detection of Gases by Gas Correlation Spectro-radiometry, Optical and Radar Remote Sensing," edited by Killinger, D. K., and Mooradian, A., *Springer Series on Optical Sciences*, 39, 1983, pp. 114–117.
[70] Wang, L., et al., "Comparison of approaches to modulation spectroscopy with GaAlAs semiconductor lasers: water vapour," *Applied Optics*, 27, 1988, pp. 2071–2077.
[71] Inaba, H., et al., "Optical fiber network system for air-pollution monitoring over a wide area by optical absorption method," *Electronics Letters*, 15, 1979, pp. 749–751.
[72] Kobayasi, T., Hirama, M., and Inaba, H., "Remote monitoring of nitrogen dioxide molecules by differential absorption using optical fiber link," *Applied Optics*, 20, 1981, pp. 3279–3280.
[73] Hordvik, A., Berg, A., and Thingbo, D., "A fiber optic gas detection system," *Proc. of 9th European Conference on Optical Communications*, Geneva, Switzerland, 1983, p. 317.
[74] Chan, K., Ito, H., and Inaba, H., "An Optical fiber-based gas sensor for remote absorption measurements of low-level methane gas in the near-infrared region," *Journal of Lightwave Technology*, 2, 1984, pp. 234–237.
[75] Stueflotten, S., et al., "An infrared fiber optic gas detection system," *Proc. of OFS'84 International Conference*, Stuttgart, 1984, pp. 87–90.
[76] Dakin, J. P., et al., A novel optical fiber methane sensor, *Proc. of SPIE International Conference on Fiber Optics '87*, London, 33, 1987, p. 734.
[77] Dakin, J. P., Croydon, W. F., and Hedges, N. K., "A fiber optic methane sensor having improved performance," *Proc. of SPIE International Conference on Fiber Optics '88*, London, 30, 1988, p. 949.
[78] Dakin, J. P., and Croydon, W. F., "Applications of fiber optics in gas sensing," *Review at ECOC/LAN*, Amsterdam, June 1988. (No full written version of this review paper was given in proceedings - results described in the current text were shown during the presentation).
[79] Mohebati, A., and King, T. A., "Remote detection of gases by diode laser spectroscopy," *Journal of Modern Optics*, 35, 1988, pp. 310–324.
[80] Barnes, W. L., et al., "Tunable fiber laser sources for methane detection at 1.68μm," *Proc. of SPIE International Conference on Chemical, Biochemical and Environmental Sensors IV*, Boston, Sept. 8–11, 1992, p. 1796.
[81] Farahi, F., et al., "Fiber Optic interferometric hydrogen sensor," *Proc. of OFS'86 International Conference*, Tokyo, 1986, pp. 127–130.
[82] Tanaka, et al., "Fiber-optic evanescent wave gas spectroscopy," *Proc. of OFS'86 International Conference*, Tokyo, Post Deadline Paper 2-1, 1986.
[83] Stewart, G., and Culshaw, B., "Optical wave-guide modeling and design for evanescent field chemical sensors," *Optical and Quantum Electronics*, 26, 1994, pp. S249–S259.
[84] Muhammad, F. A., Stewart, G., and Jin, W., "Sensitivity enhancement of D-fiber methane gas sensor using high index overlay," *IEE Proc.-Journal of Optoelectronics*, 140, 1993, pp. 115–118.
[85] Jin, W., et al., "Compensation for surface contamination in a D-fiber evanescent-wave methane sensor," *Journal of Lightwave Technology*, 13, 1995, pp. 1177–1183.
[86] Potyrailo, R. A., Hobbs, S. E., and Hieftje, G. M., "Near ultraviolet evanescent-wave determination of ozone with fiber optics," prepared for submission to *Analytical Chemistry*,
[87] Shahriari, M. R., et al., "Porous Fiber optics for a high sensitivity humidity sensor," *Proc. of OFS'88 International Conference*, New Orleans, 1988, pp. 373–381.

[88] Surgi, M. R., "The design and evaluation of a reversible fiber optic sensor for determination of oxygen," *Proc. of OFS'88 International Conference*, New Orleans, 1988, pp. 349–352.
[89] Beyler, L. L., Ferrara, J. A., and MacChesney, J. B., "A plastic-clad silica fiber chemical sensor for ammonia," *Proc. of OFS'88 International Conference*, New Orleans, 1988, pp. 369–372.
[90] Liebermann, R. A., Beyler, L. L., and Cohen, L. G., "Distributed fluorescence oxygen sensor," *Proc. of OFS'88 International Conference*, New Orleans, 1988, pp. 346–348.
[91] Breguet, J., Pellaux, J. P., and Gisin, N., "Photoacoustic detection of trace gases with an optical microphone," *Sensors and Actuators A*, 48, 1995, pp. 29–35.
[92] Tilden, S. B., and Denton, M. B., "Theory and evaluation of a windowless non-resonant optoacoustic cell," *Applied Spectroscopy*, 39, 1985, pp. 1022–1029.
[93] Samson, P. J., "Fiber optic gas sensing using Raman spectrometry," *Proc. of the 14th Australian Conference on Optical Fiber Technology*, Brisbane, Dec. 1989, pp. 145–148.
[94] Dyer, C. D., and Hendra, P. J., "Near-infrared Fourier transform Raman spectroscopy of gases," *Analyst*, 117, 1992, pp. 1393–1399.
[95] Kobayasi T., Hirana M., and Inaba, H., "Remote monitoring of NO_2 molecules by differential absorption using optical fiber link," *Applied Optics*, 20, 1981, p. 3279.
[96] Goody, R., "Cross-correlating spectrometer," *Journal of the Optical Society of America*, 58, 1986, p. 900.
[97] Taylor, F. W., et al., Radiometer for remote sounding of the upper atmosphere, *Applied Optics*, 1, 1972, p. 135.
[98] Edwards, H. O., and Dakin, J. P., "A novel optical fiber gas sensor employing pressure-modulation spectrometry," *Proc. of 7th Optical Sensors Conference*, Sydney, Australia, 55, 1990.
[99] Dakin, J. P., and Edwards, H. O., "Gas sensors using correlation spectroscopy, compatible with fiber optic operation," *Sensors and Actuators B*, 11, 1993, pp. 9–19.
[100] Dakin, J. P., and Edwards, H. O., "Progress in fiber-remoted gas correlation spectrometry," *Optical Engineering*, 31, 1992, pp. 1616–1620.
[101] Dakin, J. P., Edwards, and H. O., Weigl, B. H., "Progress with optical gas sensors using correlation spectroscopy," *Sensors and Actuators B*, B29, 1995, pp. 87–93.
[102] Dakin, J. P., "Latest developments in gas sensing using correlation spectroscopy," *Proc. of SPIE, Optics of Environmental and Public Safety*, 2508, Munich, June 19–23, 1995.

Chapter 8

Chemical Sensing Using Indicator Dyes

Otto S. Wolfbeis
University of Regensburg, Germamy

This chapter describes the design, fabrication, and properties of materials that respond to the presence of a chemical species by a change in their optical properties. Ideally, such effects are reversible. They may be detected by conventional methods of absorption, reflection, or luminescence spectrometry, and applied in various formats such as test strips and disposable tests, but preferably by making use of optical waveguides including optical fibers, integrated optics, capillary type devices, and the like. Specifically, we describe the design and use of appropriate indicator dyes, polymers, and additives, with a particular focus on materials for sensing pH, oxygen, carbon dioxide, ammonia, and certain ions. These materials (or their solutions in an appropriate solvent) may be deposited on various supports including simple plastic strips (e.g., by spin-coating), on fibers (e.g., by dip coating), inside porous materials (e.g., by soaking), on integrated waveguides or walls of disposable cuvettes (e.g., by spreading the solutions as thin films), inside capillaries (e.g., by passing the solution through the capillary), on the bottom of microtitre plates, or inside disposable vials.

8.1 INDICATORS

Indicators (*probes*) are synthetic dyes that undergo color changes on interaction with chemical species. The purpose of using a so-called *indicator chemistry* (i.e., a dye in or on a polymeric support) in optical sensing is to convert the concentration of a chemical analyte into a measurable optical signal. In other words, the indicator acts as a transducer for a chemical species that frequently cannot be determined directly by optical means. This has an important implication in that it is the concentration of the *indicator* species that is measured rather than that of the *analyte* itself.

The chemistry of indicators is fairly established [1–4], but not optimized in many cases for sensing purposes. In fact, many indicators cannot be used in fiber-optic chemical sensors because of unfavorable analytical wavelengths, poor photostability, low molar absorbance, the need for additional reagents (such as strong acid or alkali, which are frequently used in conventional spectrometry in order to adjust for optimal conditions), or simply because they are not available in a purity required for sensing applications. The spectral range for the kind of dyes treated in this chapter extends from 350 to 900 nm. However, optical sensor systems are preferably operated between 450 and 800 nm. It is noted at this point that indicator dyes are available for numerous ions (including practically all metals ions and the proton), but not for most organic species of clinical or environmental significance.

On interaction with the target analyte, most indicators undergo a change in color or fluorescence (with one band appearing as the other disappears) rather than a change in intensity of one single band, which is only the case for certain (quenchable) fluorophores. Usually, both the complexed and uncomplexed indicator species have absorptions (but much less so emissions) of comparable intensity. Such indicators are referred to as two-wavelength indicators. They are advantageous over other indicators in that they lend themselves to two-wavelength internal referencing methods.

Fluorescent indicators, in contrast, are frequently of the yes/no type in that only one of the species (i.e., the complexed or the uncomplexed form) is fluorescent. In such cases, fluorescence intensity can be measured with no background resulting from the presence of a second species. Obviously, however, two-wavelength internal referencing is impossible. Measurements of decay time or polarization are then preferred over other internally referenced methods. Another disadvantage of fluorescent indicators results from the fact that they are prone to quenching by species other than the analyte. Finally, many fluorophores display low molar absorbance (when compared to color indicators), and are not excitable by green, yellow, or red LEDs, or by semiconductor lasers. On the other side, fluorescent indicators provide distinctly improved sensitivity (which is important in case of minute sensor size) because of the unsurpassed sensitivity of luminescence. Finally, luminescence offers a broad variety of techniques including measurement of intensity, lifetime, polarization, energy transfer, and combinations thereof, since processes occurring in both the ground and the excited state can be monitored.

There is a general trend visible now toward the use of longwave absorbing indicator dyes for the following reasons:

1. Shortwave emitting light sources are expensive and often require high power, while LEDs as well as diode lasers are inexpensive, easy to drive, and require low power.
2. Photodiodes are inexpensive photodetectors that—unlike PMTs—do not require

high voltage and display best sensitivity in the 600- to 900-nm range (with exceptions).

3. Many dyes suffer from photobleaching if exposed to blue or UV light.
4. Optical waveguides display measurable intrinsic absorption at below < 450 nm and this is particularly true for plastic waveguides; simultaneously, background luminescence increases.
5. Most biological matter has good permeability for light at > 600 nm and < 900 nm only, so this is the window at which *in vivo* sensors are preferably operated at.
6. Scattering of light usually decreases with λ^4.

Consequently, present day fiber-optic chemical sensors preferably are based on LED light sources, photodiode detectors, glass waveguides, and indicator dyes absorbing in the 450- to 800-nm range.

8.1.1 pH Indicators

These are mostly weak acids (less often, weak bases) whose color or fluorescence is different in the dissociated and the associated (protonated) form, respectively [1,5]. Figure 8.1 shows the pH dependence of the excitation and emission spectra of the widely used fluorescent pH probe HPTS. It has two bands in the excitation spectrum, a fact that allows for two-wavelength excitation and hence internal referencing because the ratio of the fluorescence intensities obtained at an excitation of, for example, 405 and 460 nm, is independent of dye concentration, the intensity of the light source, and the sensitivity of the photodetector (unless they vary within the time required for making the two measurements).

An important parameter for characterization of a pH indicator is its pK_a value (i.e., the pH at which the dye is present in the undissociated and the dissociated form at 50% each). The pK_a is the negative log of the binding constant (which in turn is the inverse of the stability constant K_s):

$$pK_a = -\log ([\mathrm{Ind}^-][\mathrm{H}^+]/[\mathrm{H} - \mathrm{Ind}]) \tag{8.1}$$

where [H-Ind] represents the concentration of the undissociated indicator molecule while $[\mathrm{Ind}^-]$ denotes the concentration of the anion (the dissociated form which, in case of phenolic dyes, is more intensely colored), and $[\mathrm{H}^+]$ is the concentration of protons (i.e., the negative antilog of the pH). At the transition point of the titration curve, $\mathrm{pH} = pK_a$.

A typical titration plot as obtained from pH-dependent fluorescence emission spectra is shown in Figure 8.2, from which it is obvious that 1) pH indicators are most

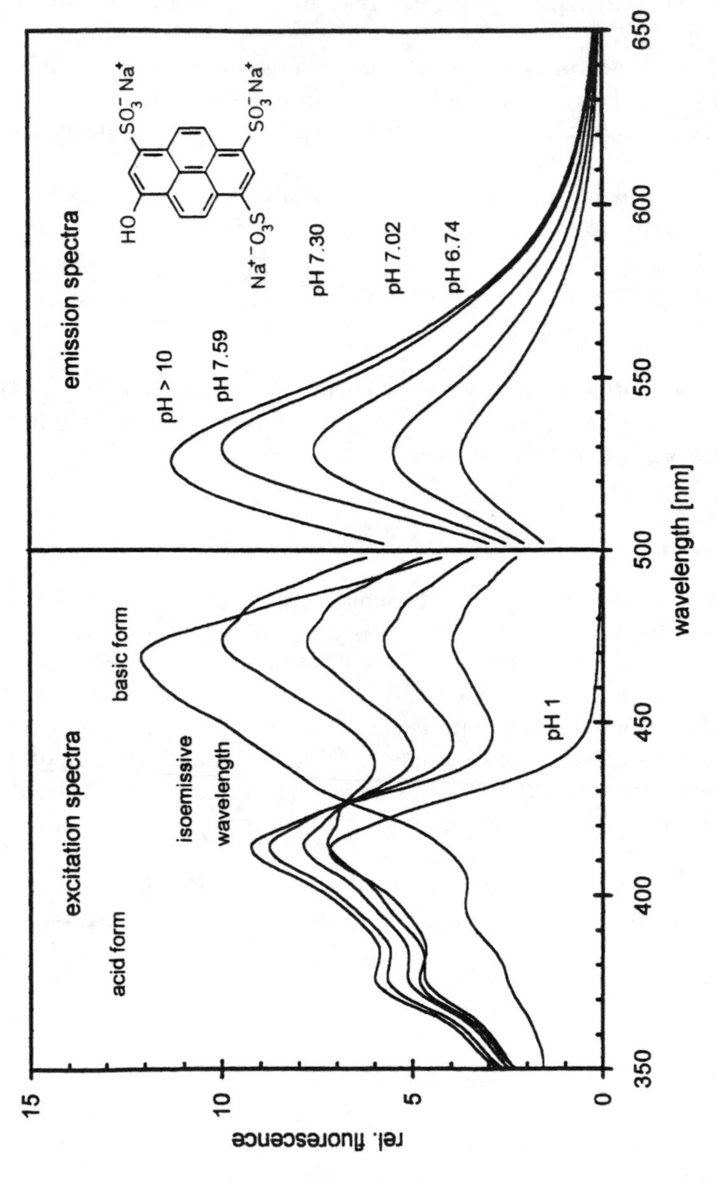

Figure 8.1 The pH-dependent fluorescence excitation and emission spectra of the pH indicator HPTS covalently immobilized on cellulose.

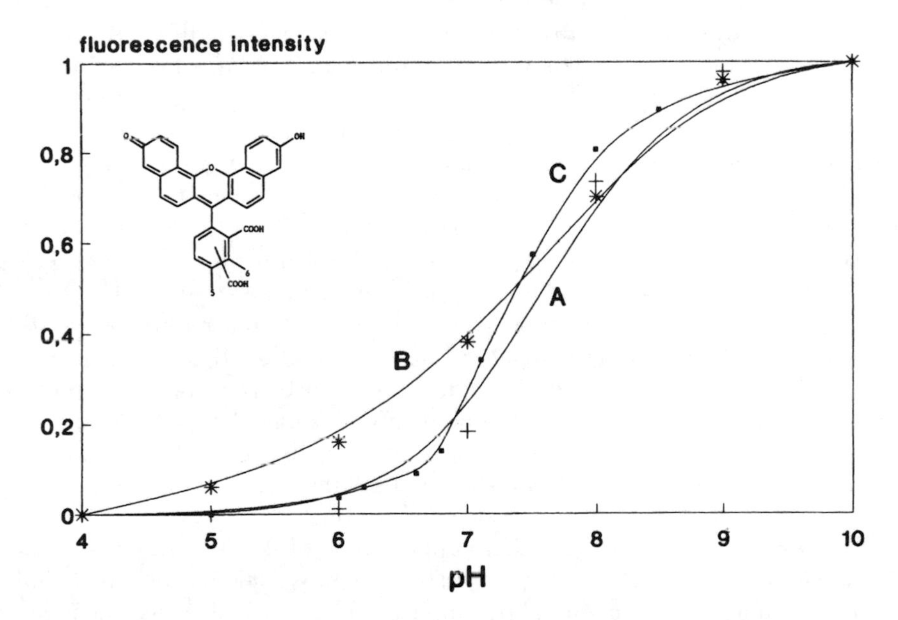

Figure 8.2 pH-Dependent fluorescence intensities ("titration plots") of the naphthofluorescein dye CNF (a) in dissolved form, (b) immobilized on cellulose, and (c) in a sol-gel.

sensitive at pHs near the pK_a, 2) their dynamic range covers a pH range of approximately (pK_a +/− 1.5) units, and 3) the shape of the curve is different for the dissolved and immobilized forms of the dye.

The relation between pH, pK_a and absorbance (or fluorescence intensity) of the two species is given [5,6] by

$$pH = pK_a + \log [Ind^-]/[H - Ind] \tag{8.2}$$

pK_a's may be determined by spectrophotometry or fluorimetry from a titration plot using (8.3):

$$pK_a = pH - \log(E_x - E_A)/(E_B - E_x) \tag{8.3}$$

where E_x *is* the absorbance (or luminescence intensity) at a given wavelength and a certain pH, and E_A and E_B are the absorbances at this wavelength for the pure acid and base forms, respectively. The values E_A and E_B are obtained by acquiring the spectra of the indicator at pHs of < ($pK_a - 2$) and >($pK_a + 2$), respectively.

The fact that optical pH sensors measure over a limited range of pH is disadvantageous but inevitable in view of the mass action law that governs response (see (8.2)). No single indicators are available that allow measurements to be performed over the pH 1 to 13 range (as can electrodes). Rather, different indicators have to be employed. The most important range is the one in the near neutral (physiological) pH range. However, few indicators only meet the requirements for use in pH sensors for physiological samples. Desirable properties include 1) an appropriate pKa (7–8); 2) absorption/excitation maxima at or above 450 nm to allow the use of inexpensive waveguide optics and light sources; 3) high molar absorbance; 4) photostability and chemical stability; and 5), ease of immobilization. Tables 8.1 and 8.2 summarize some of the more common absorption indicators which, however, if immobilized on a solid support, may undergo significant shifts in both their pK_a values (and, hence, pH transition ranges) and-less so-their absorption maxima.

Fluorescent indicators have been applied more often than absorbance-based indicators in optical sensors ("optrodes"). The 7-Hydroxy-coumarins are pH indicators for cell studies, but have found little application in sensors because their spectral maxima are in the UV (or the blue) part of the spectrum. Fluoresceins, in contrast, form a widely used class of pH probes. Their popularity results from the close match

Table 8.1
Spectral Maxima (in nm) and pKa Values of the Common Absorbance-Based Indicators for Physiological pHs

Indicator	*Λmax of Acid/Base Form*	pK_a
Bromothymol blue	430/617	6.8
o-chlorophenol-indophenol	555/625	7.1
Chlorophenol red	460/530	6.3
Dibromo-xylenol blue	ca. 420/614	7.6
a-naphthol-phthalein	428/661	6.7 and 7.9
Neutral red	527/453	7.4 and 5.9
Nitrazine yellow	460/590	6.5
Palatine chrome black	520/643	7.4
Phenol red	432/576	7.6
Phenoltetrachloro-sulfonaphthalein	435/575	7.0
Solochrome violet RS	515/562	4.35, 7.4, 9.35
Styryl acridine	455/684	7.50 *

* in plasticized pvc.

Table 8.2
Selected Longwave Absorbing pH Indicators

Indicator	*Color of Acid/Base Form*	*pKa Value or pH Range*
Methyl violet	Yellow/blue	0.0–1.6
Malachite green	Yellow/blue-green	0.2–1.8
Cresol red	Red/yellow	1.0–2.0
m-cresol purple	Red/yellow	1.2–2.8
Bromophenol blue	Yellow/blue	2.8–4.8
Congo red	Blue/red	3.0–5.0
Bromocresol green	Yellow/blue	4.6
4-phenylazo-1-naphthylamine	Red/yellow	4.0–5.6
Bromocresol purple	Yellow/purple	6.3
Meta-cresol purple	Yellow/purple	7.4–9.0
4,4′-bis(4-amino-1-naphthylazo)-2,2′-stilbenedisulfonate	Blue/red	8.0–9.0
Naphtholbenzein	Orange/blue	8.2–10.0
Ethyl bis(2,4-dinitrophenyl)-acetate	Blue/yellow	10.5
Alizarin yellow R	Yellow/red	10.0–12.0
Alizarin	Red/purple	11.0–12.4
Indigocarmine	Blue/yellow	11.4–13.0
Tetraethyl anilinesulfophthalein	Blue/yellow	13.2

of their absorption with the emission of the blue LED and the 488-nm line of the argon laser and their availability in activated form (e.g., FITC), which facilitates covalent immobilization. Notwithstanding their popularity, many fluoresceins are poor pH probes in having small Stokes' shifts, overlapping pK_as, and limited photostability. The spectral properties of fluoresceins are similar to bilirubin and flavins, which therefore may interfere in blood and serum measurements. More longwave emitting fluoresceins therefore are preferred. Table 8.3 lists the most common fluoresceins along with their properties. The naphthofluoresceins and the SNARF and SNAFL dyes have dual emissions, which enables dual-wavelength measurements.

Most pH sensors have been obtained by immobilization of pH indicators on hydrophilic supports such as cellulose, where shifts in pK_a due to immobilization remain small. More recently, pH sensors have been developed based on polymers like plasticized PVC or polyurethane. Classical indicators are insoluble in such polymers, but have good solubility in water. In order to make pH probes soluble in hydrophobic polymers, they have been made lipophilic by either eliminating charged functions such as sulfo groups or by introducing long alkyl side chains to render them more lipophilic. It is noted, however, that lipophilic pH indicators undergo massive shifts in their apparent pK_a when incorporated in lipophilic polymers, as shown in

Table 8.3
Absorption and Emission Maxima (in nm) as Well as pKa Values of Various Fluoresceins

Indicator	$pK_a(s)$	*Excitation/Emission Maxima at pH 10*	*at pH 3*
Fluorescein	2.2, 4.4, 6.7	490/520	a)
Eosin	3.25, 3.80	518/550	460/536
2′,7′-dichlorofluorescein	0.5, 3.5, 5.0	502/526	460/520
Dimethylrhodole	ca. 6.0	510/545	b)
5(6)-carboxy-fluorescein	ca. 6.4	505/530	a)
5(6)-carboxy-eosin	ca. 3.6	525/560	465/540
Carboxy naphthofluorescein	ca. 7.0	590/665	510/565
SNARF[c)]	ca. 7.6	560/625	530/575
SNAFL[c)]	ca. 7.6, 7.3	550/620	515/540
Vita blue	ca. 7.5	610/665	524/570

a) No fluorescence.
b) A hybrid between fluorescein and rhodamine.
c) Registered trade name of Molecular Probes, Inc. (Eugene, OR).

Figure 8.3 for a pH-sensitive membrane made from plasticized polyurethane hydrogel. Such sensors are particularly easy to fabricate because they can be deposited as thin films from respective solutions ("cocktails") by conventional techniques.

8.1.2 Effects of Ionic Strength on pK_as

The ionic strength dependence of pH optrodes represents the major limitation for precise optical determination of pH [5–9]. The effect of ionic strength on a typical titration plot is given in Figure 8.4. The main sources of error are the effects of ionic strength and dissolved polyelectrolytes (viz. proteins), of added solvent, and of surface structural effects of optrodes. It has been concluded [7] that for thermodynamic reasons neither optical nor electrochemical sensors can measure pH precisely, but that, on grounds of error minimization in electrodes, the electrochemical measurements of ion activities are superior to the optical.

8.1.3 Metal Chelators

There are many types of dyes that form colored complexes (*chelates*) with metal ions [1–3] and therefore may be employed as indicators in optical sensors. However, the color reaction must be sufficiently selective and the value of the stability constant of the complex formed should be such as to make the reaction reversible in order to make the device a sensor rather than a single-shot probe. This appears to be a problem with most sensors for heavy metals [10].

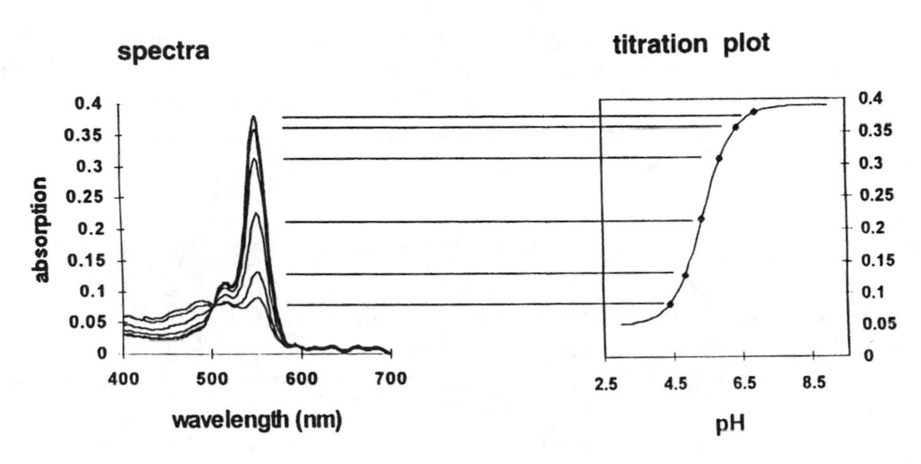

Figure 8.3 Titration plot di-iodofluorescein octadecylester incorporated into a membrane of polyurethane (plasticized with 66% NPOE). The apparent pK_a value is 7.3 (as opposed to ca. 4.2 in water).

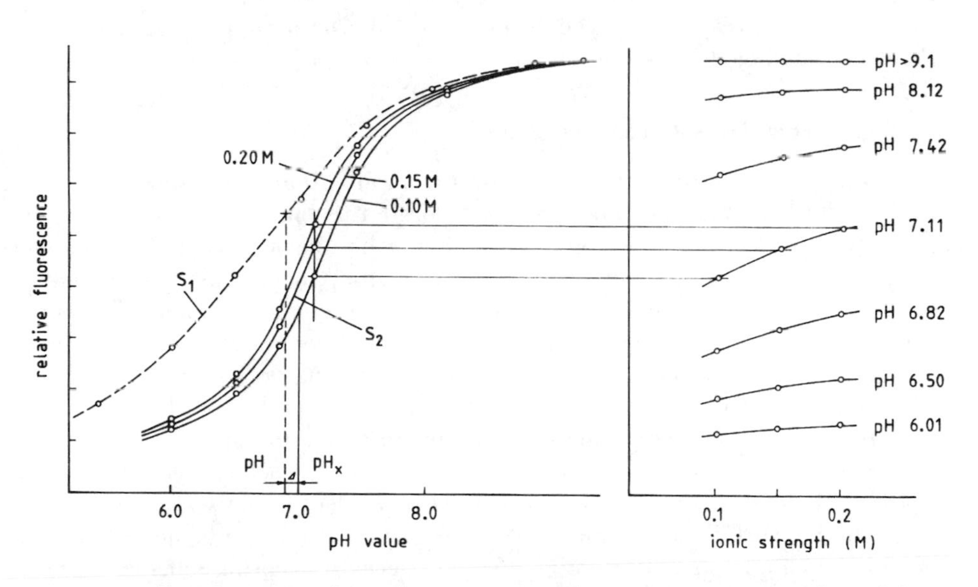

Figure 8.4 (a) Effect of ionic strength (NaCl in concentrations from 0.10 to 0.20 mM) on the work function of a pH sensor (S_2) and (b) effect on the accuracy of the measurement.

Metal indicators are usually salts of polybasic acids, which change color when the acidity of the solution is varied. It is therefore mandatory to buffer the pH of the sample solution when an indicator of this type is used. The theoretical basis of the use of metal indicators can be discussed in terms of the so-called conditional constant K_s (8.4). When a metal ion M reacts with an indicator in a molar ratio of 1:1,

$$K_S = [M - \text{Ind}]/([M'][\text{Ind}']) \tag{8.4}$$

where [Ind′] denotes the concentration of the indicator, which is not bound in the complex M-Ind, and [M'] the concentration of the metal ion that is not bound to the indicator as (M-Ind). In view of (8.2), K_s can be highly pH-dependent. A more extensive theoretical treatment can be found in [1].

Excellent textbooks and reviews on metal chelators are available [1–3], so there is no need to go into detail. The state of the art in optical probing of heavy metals has been reviewed [10]. Several manufacturers offer optical strip tests with reflectometric readout, but all act irreversibly. There is an obvious lack of optical indicators for alkali and earth alkali ions to work at near neutral pH. In view of the tremendous interest in sensing these species in clinical samples, alternative approaches have been made, which shall be discussed next.

8.1.4 Crown Ether Dyes (Chromoionophors)

This class of indicators dyes has attracted particular attention with respect to sensing alkali ions [11–15]. Chromoionophores incorporate two functions in one molecule, namely 1) that of a crown ether (or a more complex binding site) capable of binding alkali or alkaline earth ions (but also certain main group metal ions), and 2) that of a chromophore that is designed to bring about specific color changes. The chromophoric groups can bear one or more dissociable protons or can be nonionic (Figure 8.5). In the former, the ion exchange between the proton and appropriate metal cations causes the color to change, while in the latter the coordination of the metal ion to the chromophoric donor of the dye molecule induces a change of the charge transfer (CT) band of the dye. If complexation is associated with the release of a proton, the sensor obviously will have a pH-dependent response.

In the above CT-based systems, charges are shifted along the conjugated □-bonds of a chromophore or fluorophore. An alternative sensing scheme has been described that relies on an effect called photoinduced electron transfer (PET) [16–20]. In such systems, an electron is shifted from a tertiary nitrogen atom to a fluorophore through space (not along a chemical bond) as shown in Figure 8.6 (with an anthracene moiety acting as the fluorophore). This photoinduced process-which can only be observed in fluorescence-is suppressed if the free electrons of the nitrogen atom are blocked by binding to an ion such as potassium, or by a proton. This is a most promising and widely applicable sensing scheme.

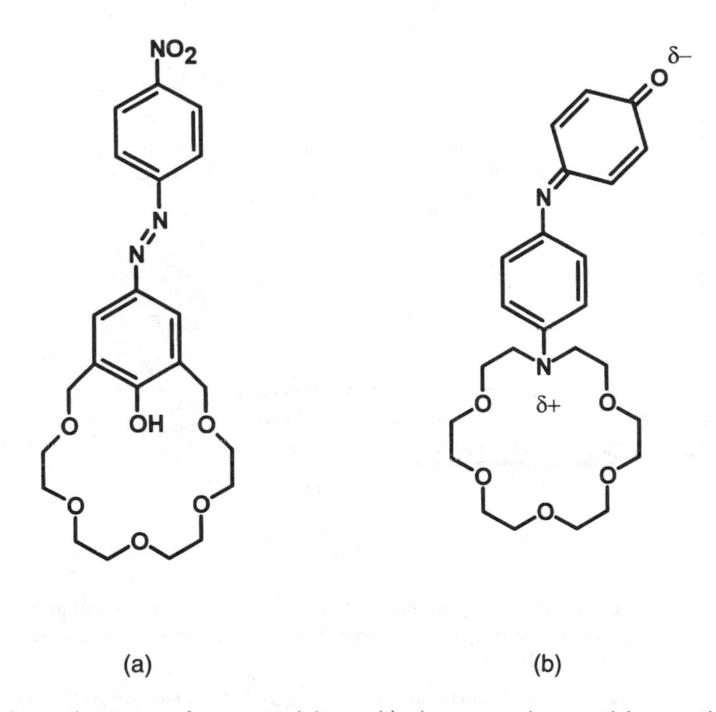

Figure 8.5 Chemical structures of (a) a typical dissociable chromoionophore, and (b) a nondissociable chromoionophore, both containing a cyclic polyether ring (the "crown") and a chromophore.

8.1.5 Chelators for Calcium and Magnesium

In recent years, indicator dyes have been developed for the clinically important bivalent cations calcium and magnesium [4,21], which are different from previous probes in that they chelate at physiological pH and in the concentration encountered in practice (see (8.4)). Hence, they meet the need for monitoring calcium or magnesium at physiological pH and over a wide range of concentrations.

8.1.6 Potential-Sensitive Dyes

These comprise a quite different class of dyes that respond to transport processes occurring at a sensor/sample interface [22]. Thus, they do not directly report the concentration or an activity of a chemical species. They provide an interesting alternative to sensors based on conventional chromogenic chelators. Potential-sensitive dyes (PSDs) (also referred to as polarity-sensitive dyes) are usually placed directly at the site where a "potential" is created by chemical means, usually via an ion carrier (such as valinomycin) incorporated into a lipid membrane.

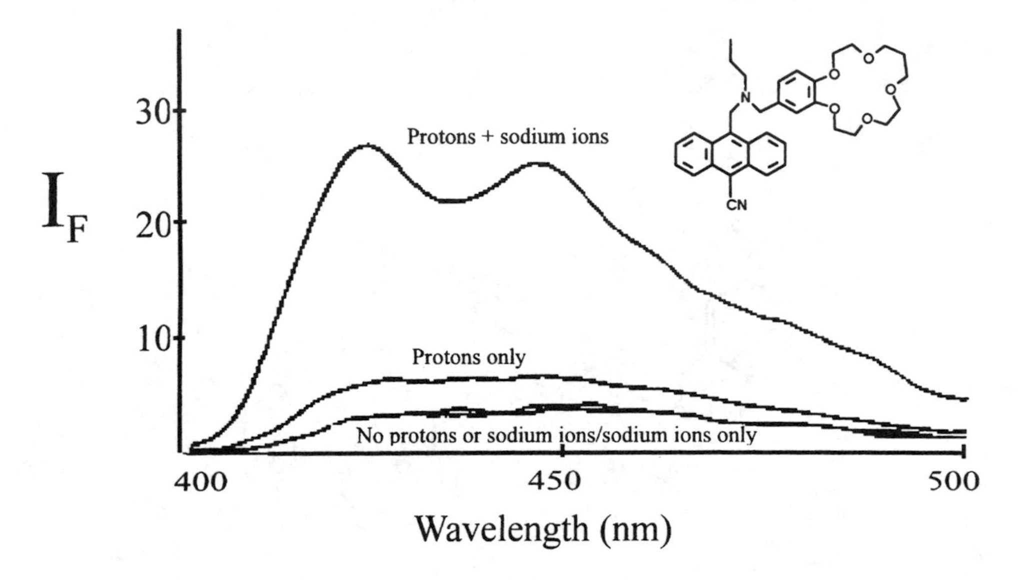

Figure 8.6 Chemical structure of a crowned fluorophore which, dissolved in methanol, on addition of 10 mM of sodium acetate gives a sixfold increase in fluorescence intensity. (*Source:* [20].)

The response of PSDs to electrolytes [22–26] is based on one or more of the following effects: 1) a field-dependent distribution of the dye between regions of different polarity within the lipid membrane, resulting in a solvatochromic effect; 2) changes in the otherwise homogeneous distribution of the fluorophore within the membrane when an electric field is created, leading, for example, to aggregation and self-quenching; 3) the *Stark* effect (i.e., a change in the absorption and emission spectrum of a fluorophore when an external field is applied); and 4) potential-induced changes in the solvatation of the dye. The exact mechanisms of PSD-based sensors (and the relative contributions of the above effects) are not clear yet.

Figure 8.7 gives a schematic representation of the mechanisms occurring at the sensor/sample interface of a sensor membrane that fully reversibly responds to potassium ion. The dye is redistributed and undergoes a change in its microenvironment as a result of the transport of a cation into the membrane. Obviously, the kind of charge of the ion plays an important role in this process. It is assumed that the major effect results from the displacement of the PSD from an environment where it is strongly fluorescent to an environment where it is less fluorescent (or vice versa).

Another parameter to be considered is the hydrophilicity/lipophilicity balance (HLB) of the sensor membrane and the dye contained in it. If both the PSD and the polymer are highly lipophilic, the PSD will not be displaced into the direction of the aqueous sample and hence will not undergo a significant spectral change. The same is true if both are highly hydrophilic. It follows that the choice of the appropriate HLB of PSD and polymer dictates the relative signal change of such sensors.

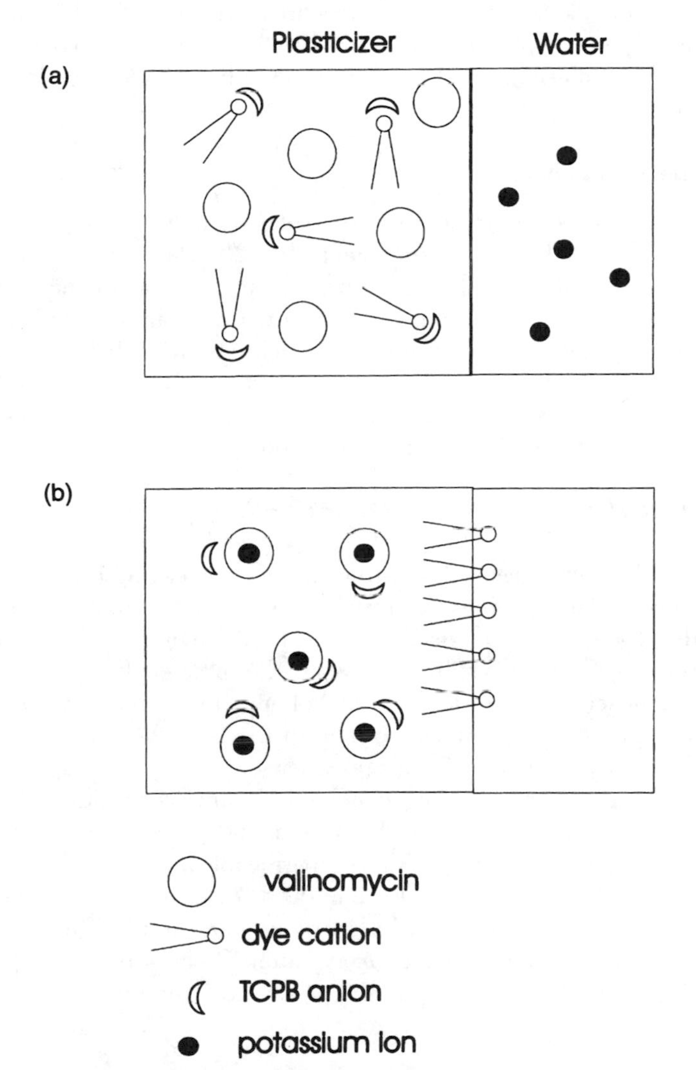

Figure 8.7 Schematic of the dye distribution in a sensor based on the use of a PSD (a) before and (b) after potassium ion has been carried from the water phase into the membrane phase.

The rhodamine dyes comprise an important class of cationic PSDs and have been used in sensors for alkali ions [22,25]. They are nontoxic and highly fluorescent. Acridine Orange [23] and certain merocyanines [25,26] have been used as well. The preferred polymers are plasticized PVC, certain PVC-PVA-PVAc copolymers, and polyurethane hydrogels.

8.1.7 Quenchable Fluorophores

Both the fluorescence intensity and the decay time of certain fluorophores are reduced in the presence of so-called dynamic quenchers [27,28]. The process of dynamic quenching is fully reversible (i.e., the dye is not consumed in a chemical reaction). Hence, quenchable fluorophores comprise an important class of indicators for reversible sensing. In the case of dynamic quenching, the interaction between quencher (analyte) and fluorophore is in the excited state only. The relation between luminescence intensity (I) and decay time (τ) on one side, and analyte concentration on the other is described by the Stern-Volmer equation (8.5)

$$(I_O/I - 1) = (\tau_O/\tau - 1) = K_{SV}[Q] = K_q \cdot \tau_O \cdot [Q] \quad (8.5)$$

where I_O and I are the luminescence intensities in the absence and presence, respectively, of the quencher Q present in concentration $[Q]$, τ_O and τ are the luminescence decay times in the absence and presence, respectively, of quencher Q, K_{SV} is the overall (Stern-Volmer) quenching constant, and K_q is the bimolecular quenching constant. At higher quencher concentrations, Stern-Volmer plots tend to deviate from linearity. Figure 8.8 gives typical Stern-Volmer profiles of the quenching of the luminescence of a dye by oxygen in various polymers.

Oxygen is known to be a notorious quencher of luminescence, and this is widely exploited for sensing purposes [28–35]. Interferences by ionic species can be eliminated by immobilizing the fluorophore in ion-impermeable materials such as silicone or polystyrene. This is discussed in more detail in the section on oxygen sensors. Other dynamic quenchers of luminescence include bromide and iodide [36], halothane (which quenches by virtue of the so-called heavy atom effect of bromine) [37], and the transition metals (which quench due to the presence of unpaired spins) [10,38].

8.2 POLYMERIC SUPPORTS AND COATINGS

Polymer chemistry forms an integral part of sensor technology since all "chemistries" rely on the use of one of the many polymers and related supports. In indicator-based sensing schemes, polymers are expected to be optically inert. Their function is that of 1) a solid support onto which indicator dyes are being immobilized and 2) a material possessing a certain permeation selectivity for the species of interest while rejecting others. The choice of polymer is mainly dictated by the above considerations, but also

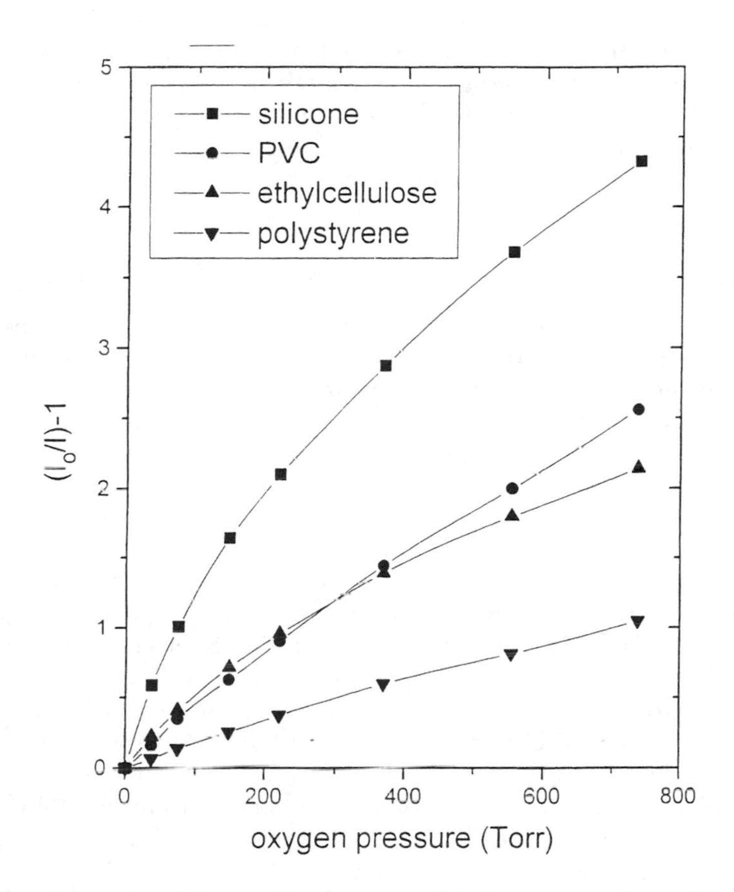

Figure 8.8 Stern-Volmer plots of the quenching by oxygen of the luminescence of ruthenium-tris(dpp) perchlorate dissolved in various polymers [29].

by the polymer's compatibility with the sample (e.g., blood). Polymer properties are compiled in various books and reviews [39–45] to which reference is made.

The choice of polymer material has a pronounced effect on the performance of the sensor. The response time, for example, will be governed by the diffusion coefficients of gases or ions, and the quenching efficiency by both the diffusion and solubility of the analyte in the polymer. Solubility and diffusion coefficients for various gas/polymer combinations have been compiled [46–48]. However, numerous new materials are available for which data are scarce. It is also known that copolymers and polymer mixtures do not necessarily display the properties that may be expected from averaging the data of the pure components.

Certain polymers such as polystyrenes and polyesters display intrinsic fluorescence under UV excitation, while poly(vinyl chlorides), poly(vinyl alcohols), and polysiloxanes are fairly "clean." Most organic polymers have added plasticizers to make them softer and more permeable. Among these, esters of phthalic acid are fluorescent under UV excitation and can give rise to a considerable background signal. NPOE, in turn, is a plasticizer widely used in electrodes but acts as a dynamic quencher of the luminescence of many luminescent indicators.

8.2.1 Silicones

Silicones have unique properties [46,49–51] in possessing a higher permeability for most gases (including water vapor!) than any other polymer, but being impermeable to ions, including the proton. The selectivity of sensors for carbon dioxide, for example, results from the fact that interfering protons do not pass hydrophobic membranes and therefore cannot interact with a dissolved pH indicator.

Silicones also have excellent optical and mechanical properties, and unique gas solubility. In case of oxygen, it exceeds all other polymers. Numerous silicone prepolymers are commercially available and allow easy manufacturing of membranes, emulsions, suspensions, or other kinds of sensing chemistries. One can differentiate between one-component and two-component silicone prepolymers. The former cure in the presence of moisture (e.g., in air) by splitting off acetic acid, methanol, or amines (which are bases!). In two-component prepolymers, a catalyst is added to one component in order to cause an addition reaction of component A to component B to give a long chain polymer. The catalyst is usually contained in one of the prepolymers. Some catalysts have been found to act as quenchers of the fluorescences of charged indicators. Many silicones are of the room-temperature vulcanizing (RTV) type, and the respective prepolymers may be dissolved in aprotic solvents such as toluene or chloroform. This greatly facilitates handling.

Notwithstanding their advantages in terms of permeability and permselectivity, silicones, once formed, do not easily lend themselves to surface modification. Hence, covalent immobilization of indicators on cured silicone rubber is extremely difficult. Moreover, silicones have limited compatibility with other polymers and are difficult to glue onto many other materials (with the notable exception of glass, which has excellent adhesion to RTV silicones). As a matter of fact, certain sensor types described in the literature and based on silicone rubber materials in combination with other materials have extremely poor long-term stability because of material incompatibility, particularly if stored in buffer. Finally, silicones are very good solvents for most gases including oxygen. This may lead to a depletion of gas when the sample volume comes to lie below the hundredfold volume of the sensing layer.

The main application of silicone materials is in sensors for oxygen and other uncharged quenchers such as sulfur dioxide and chlorine, and as gas-permeable covers in sensors for carbon dioxide or ammonia. Silicones cannot be easily plasticized by conventional plasticizers, but form copolymers that may be used instead [52].

Blackened silicone is a most useful material for optically isolating gas sensors in order to make them insensitive to the optical properties of the sample [30].

8.2.2 Other Hydrophobic Polymers

Poly(vinyl chloride) (PVC), polyethylene, poly(tetrafluoroethylene) (PTFE), polystyrene, and ethylcellulose comprise another group of hydrophobic materials that efficiently reject ionic species. Except for polystyrene, they are difficult to chemically modify so that their function is confined to that of a "solvent" for indicators, or as a gas-permeable cover. However, the diffusion of analytes through, and the solubility of gases in such membranes is quite different from silicones and results in drastically limited quenching constants.

Plasticized PVC is the preferred matrix for ion sensors (including pH sensors) if provided with a carrier (such as a crown ether). Unless plasticized, PVC is not suitable for ion-sensing purposes. Useful plasticizers include DOS (dioctyl-sebacate), TOP (trioctyl-phosphate), DOP (dioctyl-phthalate), NPOE (nitro-phenyl-octyl-ether), and related long chain esters and ethers. Plasticizers are added in fractions up to 66% [22–25], and this can completely modify quenching constants and binding constants. Since NPOE is a notorious quencher of luminescence, trifluoromethyl-POE and cyano-POE (both of which do not quench due to the lack of nitro groups) have been suggested as alternatives [53], the former being commercially available. Water-equilibrated thin films of plasticized PVC, in fact, are not a homogeneous medium but rather may be imagined as a inhomogeneous system resembling a microemulsion as shown in Figure 8.9.

PVC is soluble in THF solvent only, and this represents a major disadvantage in view of the toxicity and flammability of THF. Modified PVC (PVC-CP, a copolymer of poly(vinyl chloride), poly(vinyl alcohol) and poly(vinyl acetate) is a useful alternative to PVC since it is soluble in the much less toxic solvent ethyl acetate, but otherwise displays very similar properties. Finally, carboxy-PVC (PVC-COOH) is a commercially available PVC copolymer containing free -COOH groups and has been used for immobilizing amines such as proteins [54].

Polystyrene (PS) has been used in sensors for oxygen [35] because the quenching constants are much smaller in PS than in silicone, which is advantageous in the case of luminescent probes with very long decay times (which makes them extremely sensitive to oxygen). PS is soluble in various organic solvents, including ethyl acetate and toluene. Polystyrene may be plasticized by the same materials as are PVC and PVC-CP.

8.2.3 Silica Materials

Glass is widely used for manufacturing optical fibers. It is unique in terms of mechanical stability, optical transparency, and complete impermeability to any analyte. Aside

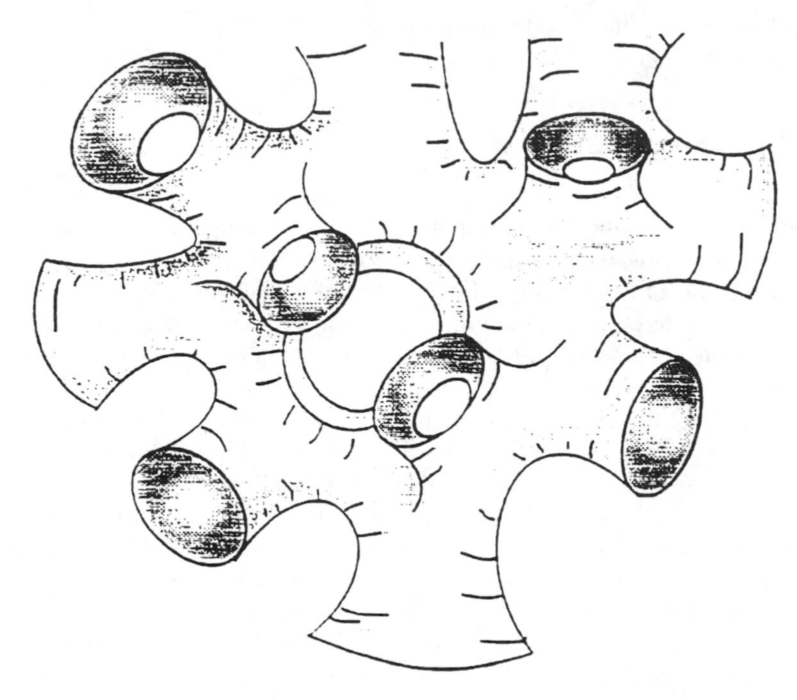

Figure 8.9 Schematic representation of the microinhomogeneity of thin films of water-saturated plasticized PVC. (*Source:* [22].)

from their function as a waveguide, glass fibers also have served as mechanical supports. Their surface may be made either hydrophilic or hydrophobic by treatment with a proper surface modification reagent [55,56]. Surface derivatization is usually performed with reagents such as amino-propyl-triethoxysilane, which introduces free amino groups onto the surface of glass to which dyes or proteins may be covalently attached. Glass does not measurably swell, but is difficult to handle in view of its brittleness. Many polymers have poor long-term adhesion to glass, which should be kept in mind when designing integrated optical chemical sensors.

Sol-gels form an attractive alternative to conventional glass [33,57]. They are obtained by hydrolytic polycondensation of tetraethoxysilane (TEOS) or related materials to give a fairly inert inorganic glassy matrix whose porosity and size of pore network can be varied to a wide degree by polymerization conditions, including time, pH, temperature, and silane:water ratios. Numerous organic dyes have been incorporated into sol-gel glasses at room-temperature conditions. Sol-gels support the transport of small molecules. Because they have no absorption in the near UV and visible, sol-gels are well-suited for fabrication of dyed materials in the form of films, fibers, or monoliths.

8.2.4 Hydrophilic Supports

Hydrophilic supports are characterized by a large number of hydrogen-bridging functions such as hydroxy, amino, or carboxamide groups, or by anionic groups (mainly carboxy and sulfo) linked to the polymer backbone. Typical examples are the polysaccharides (celluloses), polyacrylates, polyacrylamides, polyimines, polyglycols, and the variety of so-called hydrogels. Depending on the degree of polymerization and crosslinking, they are water soluble or water insoluble. All swell in water. Throughout, they are easily penetrated by aqueous solutions and display poor compatibility with hydrophobic polymers such as silicone and polystyrene. Most hydrophilic polymer membranes are easily penetrated by both charged and uncharged low molecular weight analytes, but not by large proteins.

Cellulose in either the bead or membrane form has found widespread application as a support for indicators [4,39,58–64]. The ease of penetration by water results in short response times. Cellulose membranes as thin as 6 μm are commercially available, but require careful handling [62,63]. Beads are easier to handle, and after dyeing can be immobilized in a hydrogel matrix [64]. Aside from plain membranes, cellulose bound to polyester also is commercialized and has found application to pH sensing [62,63]. In addition to cellulose, other polysaccharides including dextrans and agarose have been used for dye immobilization to produce sensing chemistries for water-soluble analytes, but with no obvious advantages over cellulose. All saccharides are readily populated by bacteriae and algae.

Chemical modification of cellulose by introducing either hydrophilic or lipophilic groups results in entirely different but extremely useful materials. Celluloses also may be rendered with charged groups so to make them ion-exchangers. Such materials are offered by various manufacturers, albeit optimized for chromatography purposes. When cellulose membranes dry out, they become very brittle and are difficult to handle. Once dry, cellulose requires a considerable time to completely rehydrate and thereby undergoes considerable swelling, resulting in signal drift. Both the swelling rate and the hydration number are pH-dependent.

Polyacrylamides (PAAs), poly(hydroxyethyl acrylate) (poly-HEMA), poly(vinyl alcohols), poly(vinyl pyrrolidones), polyurethanes, and polyglycols [39,40,43,65] are good polymeric solvents for a number of indicators, but are water-soluble unless crosslinked. They can be retained on a support by cellulosic membranes, but dissolve quite an amount of water when in contact with aqueous samples. Crosslinked PAAs form mechanically stable and water-insoluble supports that are easily handled and chemically modified, but lack the good permeability of cellulose. PAAs are also available in bead form, and their surface can easily be modified by functional groups such as carboxy or primary amine. However, an excess of these functions may 1) introduce a considerable buffer capacity, resulting in very long response time at the respective pH range and 2) establish an undesired Donnan potential.

Hydrogels are crosslinked macromolecular networks swollen in water or biological fluids and possess excellent biocompatibility, probably due to their high water content and special surface properties [65,66]. They are well-suited for pH and ion sensing, but covalent immobilization of indicator dyes is more tedious.

8.2.5 Diffusion and Permeation of Gases Through Polymers

The most important parameters for characterization of diffusion and permeation of gases through polymers are the diffusion coefficient D, the gas solubility S, and the permeation coefficient P. The permeation of small molecules through flawless and pinhole-free polymers occurs through consecutive steps of solution of a permeant in the polymer, and diffusion of the dissolved permeant through the inner free volume of the polymer, so that

$$P = D \cdot S \tag{8.6}$$

The temperature coefficients of P, D, and S can be represented in Arrhenius-type equations:

$$P = P_O \cdot \exp(-E_P/RT) \tag{8.7}$$

$$D = D_O \cdot \exp(-E_D/RT) \tag{8.8}$$

$$S = S_O \cdot \exp(-\Delta H_S/RT) \tag{8.9}$$

where E_P and E_D are the respective activation energies, and ΔH_S is the solution enthalpy. Permeability P generally decreases with increasing density of the polymer, its crystallinity, and orientation. Crosslinking a polymer reduces P, as do added fillers (such as silicagel), while adding plasticizers can increase it. Humidity increases the P of some hydrophilic polymers. Permeation coefficients for numerous gases can be found in a useful compilation [46], and typical solubility data are compiled in Table 8.4.

It should be stated at this point that polymers are not ideal solvents, and that indicator dyes incorporated into polymers in almost any case have thermodynamic

Table 8.4
Solubility S (in 10^{-6} mol · N^{-1} · m^{-1}) of Simple Gases in Polymers at Room Temperature

Polymer	*Hydrogen*	*Methane*	*Nitrogen*	*Oxygen*	*CO_2*	*Water Vapor*
Poly(vinyl chloride)	12.0	76	11.0	13	214	39,200
Polycarbonate	0.3	—	—	2	58	758
Polystyrene	—	—	—	25	290	—
Poly(ethyl methacrylate)	—	—	34.0	38	500	110,000
Polyethylene (d 0.914)	—	49	10.0	21	116	—
Polyethylene (d 0.946)	—	23	5.4	8	9	—
PTFE*	—	—	0.7	1.3	5	—
Natural rubber	17.0	—	25.0	50	401	—
Silicone rubber**	33.0	—	89.0	138	310	High

* Teflon.
** Containing 10% (w/w) filler.

properties (such as pKa values, lifetimes, or quenching constants) that are different from the respective data in solution. This represents a serious challenge in the design of materials for use in optical chemical sensors.

8.3 IMMOBILIZATION TECHNIQUES

Following the choice of indicator and polymeric support, the next step in sensor design involves immobilization of the dye in-or on-a support to result in the so-called sensing chemistry or working chemistry. Three methods are important for the preparation of sensing chemistries, viz.. mechanical, electrostatic, and covalent immobilization. Several reviews cover all aspects of the chemistry and physics of immobilized reagents and dyes, proteins, and even whole cells [60,61,67,68]. Immobilization of dyes is not confined to reactions occurring in aqueous solutions, and may involve several steps. However, the immobilization chemistry should be kept as limited as possible, and procedures giving high yields at mild reaction conditions are highly preferred. Immobilization of most dyes results in a change of their spectral characteristics, pK_a values, binding constants, and-in particular-dynamic quenching constants. The changes reflect the various interactions that occur between neighboring dye molecules in, or on, the polymer, interactions between dyes and polymer, and electronic effects of covalent bonds on the chromophor.

8.3.1 Mechanical and Physical Immobilization

Methods for mechanical (physical) immobilization include 1) adsorption, 2) inclusion of dyes into spheres that they cannot leave (for example, into the void volume of polymers [69]) inside microspheres or the inner domains of sol-gels [33,57,70] or zeolites [71]; and 3) dissolution of indicators in a polymeric "solvent." Adsorption is the most simple technique, but of limited practicability. While many proteins, lipophilic dyes, plasticizers and detergents adsorb very well on moderately polar surfaces such as polystyrene, they also are slowly washed out into samples or buffers and tend to diffuse into other materials.

Mechanical immobilization is more attractive. A good example is provided by the incorporation of cationic oxygen probes into silicone rubber, where they do not dissolve because of their positive charge. To overcome this problem, they were first absorbed onto silica gel particles, which then were dispersed into silicone prepolymer, which in turn was cured in air [72]. Alternatively, they may be deposited on fillers contained in silicones [73]. In another example, it has been shown [69] that copolymerization of acrylamide with methylene-bis(acrylamide) in the presence of phenol red leads to microspheres with the dye firmly bound to the polymer. Such nondiffusible forms of pH indicating dyes are obtained by emulsion copolymerization of phenol red with aqueous acrylamide in the presence of emulsifier and toluene under nitrogen to give microspheres that are useful for optical pH sensing.

Another method of immobilization involves the use of indicators that are highly lipophilic and hence dissolve in lipophilic polymers from which they are not readily washed out because of their much better solubility in the lipophilic phase. Typical examples include oxygen-sensitive polymers incorporating lipophilic nonionic dyes [34,35], lipophilic pH indicators dissolved in plasticized PVC [74–76] and lipophilic ion pairs (i.e., a pair of positively and a negatively charged organic species, one of which is an indicator) [30,77,78]. Such coatings are particularly easy to make and highly reproducible because fabrication only involves dissolution of the dye in a polymer solution, and casting this "cocktail" onto the surface of a waveguide. However, dyes may slowly diffuse into other materials wherein their solubility is better.

8.3.2 Electrostatic Immobilization

If a surface of a rigid support contains charged functions (such as sulfo groups or quaternized ammonium groups), it is capable of binding ions of opposite charge. Sulfonated polystyrene, for instance, binds cations with varying affinity. This effect is widely used for separation of anions or cations from a solution, and for enrichment of traces of ions. Cations subsequently may be displaced from the solid phase by strong acid, and anions by strong base. Many indicators are either cations or anions, and consequently may be immobilized this way.

Ion exchangers are commercially available and may be classified into "strong" and "weak" forms. This refers to the affinity of the material for the respective cations or anions. Both membrane- and bead-type ion exchangers are available. In order to firmly bind organic ions, the use of strong ion exchangers is preferred in order to prevent washout over time. Typical examples of indicators immobilized this way include bromothymol blue (on anionic polystyrene) [79,80] and hydroxypyrene-trisulfonate on cationic styrene [81].

The major advantages of electrostatic immobilization are the ease of the procedure and its reproducibility. Dye loading can be easily governed by the time of immobilization. The fabrication is very simple in that the charged polymer is immersed, for a defined period of time, into a solution of the dye. Because the indicator molecules are situated at sites on the surface of the polymer that are easily accessible to protons, but often not easily accessible to proteins, the corresponding pH sensors are said to display no protein error [81]. Many ion exchange materials show pH-dependent swelling and this may cause slow drifts in intensity and, even worse, the work function.

8.3.3 Chemical (Covalent) Immobilization

Covalent immobilization is the preferred method because it results in dyes that are firmly bound, via a covalent bond, to the polymer backbone and hence cannot be washed out by a sample. On the other side, the methods are more tedious than previous ones in that they require the presence of reactive groups on both the dye and the polymer, and at least one must be activated to freely undergo a chemical reaction with the partner. Numerous methods of surface modification (and activation) of polymers exist and can yield materials capable of covalently binding indicators via their reactive groups. With respect to reproducibility, it is preferred, though, to make use of preactivated commercial materials. Excellent reviews have been given on the immobilization of metal chelators on cellulose [60] and of pH indicators on various materials including celluloses for use in optical-fiber sensors [61]. Among those, we find the Remazol procedure (which is the preferred method for making commercial pH paper strips) to give best results in case of celluloses [62] and related hydroxy polymers. The respective chemical bonds are shown in Figure 8.10.

Covalent surface modification of quartz, sol-gel, silica gel, conventional glass, and even metals such as iron and platinum, and elemental carbon, is almost exclusively performed with reagents of the type $(RO)_3Si\text{-}R'$, with R being ethyl or methyl, and R′ being 3-aminopropyl, 3-chloropropyl, 3-glycidyloxy, vinyl, or a long chain amine [55,56,82]. An alternative reaction sequence that introduces amino groups involves the use of epichlorhydrine (which reacts with hydroxy groups) and then ammonia. The resulting materials then are easily reacted with the indicator or peptide to be immobilized. Porous glass with various types of organofunctional extension arms is commercially available and has been widely used for the design of waveguide biosensors.

From our experience, the recommended procedure for immobilizing dyes possessing -COOH groups onto amino-modified surfaces is via the N-hydroxysuccinimidoyl (NHS) esters of carboxylic acids, which is highly reproducible and

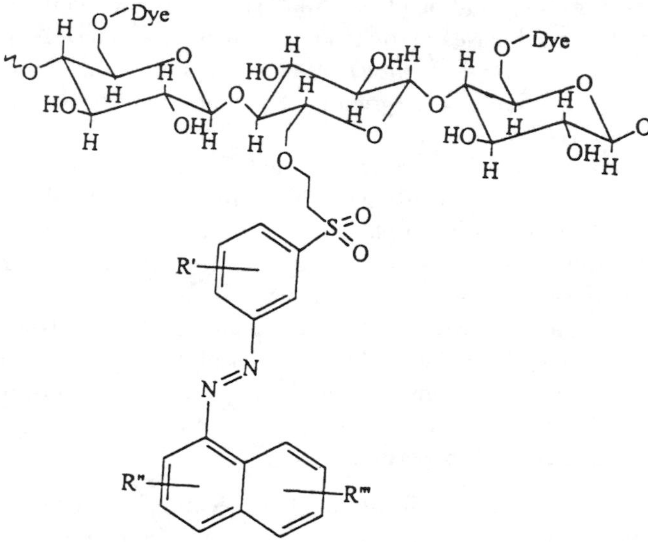

Figure 8.10 Schematic of a covalent link (the -SO_2-CH_2-CH_2-O- group) between a dye and a cellulose backbone. Substituents R′, R″ and R‴ can be varied to govern color and pK_a of the indicator dye.

proceeds under controlled and moderate conditions at room temperature. It is also recommended to use spacer groups (of a typical length of 6 carbon atoms) when immobilizing dyes or prteins so as to minimize undesired interactions between dye and support. A final method of immobilization of dyes is based on photopolymerization of dye-doped monomers, or by copolymerizing dyes possessing polymerizable groups with a monomer, typically acrylamide [69,83,84].

8.4 pH SENSORS

The kind of pH optrodes covered in this section is based on pH-dependent changes of the optical properties of an indicator-dyed layer attached to the tip or surface of an optical lightguide through which these changes are detected. The dye reversibly interacts with the protons of the sample to result in a pH-dependent absorption, reflection, or fluorescence. A selection of suitable dyes is given in Section 8.1.1., while suitable polymers are discussed in Section 8.2. Because the indicator dye and the sample are in different phases, there is necessarily a mass transfer step required before a constant signal is obtained. This leads to relatively long response times. Photobleaching and leaching, interferences by ambient light, nonideal optoelectronic equipment, the lack of violet LEDs, and inexpensive blue lasers are further problems encountered in development of fiber-optic pH sensors.

Numerous optical sensors for pH have been reported [5,64]. They differ mainly in the kind of chemical transducer and the optical sensing scheme employed. Most work so far was on sensors for physiological pHs (i.e., from 5 to 8). In the past years, however, the working range has been extended to other pHs as needed in certain industrial applications because it has been recognized that pH optrodes have the potential of becoming useful in special fields of application where potentiometric methods fail or because they can offer considerable economic and sampling advantages.

One of the limitations of optical sensors is their sensitivity to changes in ionic strength (IS; a parameter for total ion concentration) at constant pH (see Section 8.1.2). The error in pH measurement caused by the IS of a sample also depends on the charge of the dye and is largest if the IS of the calibrant is highly different from that of the sample. The theory of the IS dependence of optical pH sensors has been described [5–9,64] and has resulted in a sensor for measurement of IS [9].

The preferred polymers for use in optical sensors are cellulose and related hydrophilic supports (see Section 8.2.4 and below). More recently, alternative solid supports for indicator dyes have been found. Sol-gels, for example, have excellent compatibility with glass fibers and may be deposited on both the distal end of a fiber, or may even replace the cladding of a waveguide [57,85–87]. Other materials that have been used more recently include rather hydrophobic ones such as plasticized PVC into which a fully lipophilic dyes were incorporated to give sensors with pK_a around 7.5 [54,88]. However, such sensors have pK_a values that strongly depend on the charge and quantity of additives, and on the fraction of plasticizer added [89]. Lipophilic pH probes (such as certain eosins) may be incorporated into Langmuir-Blodgett films to give pH-sensitive lipid bilayer membranes with pK_a's quite different from the respective data in aqueous solution [90]. Recently, it was discovered that films of polypyrrol display pH-dependent absorptions between 600 and 1,000 nm [91]. Finally, it was shown that certain (nonsilicious) optical pH sensor materials are much more resistant to pH than are glass electrodes [63].

8.4.1 Absorbance and Reflectivity-Based Sensors

In the case of absorbance-based pH sensors, the Lambert-Beer law can be applied that relates absorption with the concentration ($[D]$) of the dye species

$$A = \log(I_0/I) = \epsilon \cdot [D] \cdot l \tag{8.10}$$

where I_0 and I, respectively, denote the intensity of transmitted light in the absence and presence of the dye at the analytical wavelength, l is the effective path length, and ϵ the molar absorption coefficient ($cm^{-1} \cdot mol^{-1}$) at the given wavelength.

In case of reflectance-based pH sensors and only absorption from the alkaline (longwave absorbing) species occurring at the analytical wavelength, the absorbance can be described by

$$A = \log(k \cdot I_O^{ref}/I) = (A_{max}/10(-\Delta + 1)) \qquad (8.11)$$

where $k = I_O/I_O^{ref}$, Δ = pH − pK, and $A_{max} = T.\epsilon.l$, A is the absorbance at a given pH, and A_{max} is the absorbance of the completely dissociated dye. I is the transmitted light intensity at the analytical wavelength, and I_O^{ref} is the transmitted reference light intensity. I_O^{ref} can be measured at any wavelength where the intensity of multiple reflected light is independent of pH. Typically, it is measured at the isosbestic point or at a wavelength at which neither form absorbs. The reference measurement is frequently needed to compensate for optical and instrument variations. A_{max}, k, and pK are intrinsic constants of the sensor. Other theories for reflectometric sensors do exist as well [92].

One of the first absorption-based pH-sensitive "chemistries" that had been developed made use of phenol red, which was incorporated into polyacrylamide beads [69,93] to give a sensor material with a pK_a of 7.92 +/−0.02 at zero ionic strength, and a pK_a of 7.78 +/− 0.02 at 0.25M ionic strength. The temperature coefficient of the system, expressed as the change in pH indication per °C was 0.017 between 20 and 40°C, and a change of 0.01 pH units was observed per 11% change in ionic strength over the range of 0.05 to 0.3M. The response time for the signal to drop to 63% of its initial value is 0.7 min [93]. A schematic of the sensor is shown in Figure 8.11.

Fiber-optic pH sensors for sea water monitoring were obtained [94] by immobilizing phenol red on XAD-type ion exchangers. The dyes were adsorbed onto the polymer by placing the dry beads in a 0.1% indicator solution in methanol for four hours. While easy to fabricate, this material tends to undergo a pH-dependent swelling, which causes long-term drifts and to release the dye at high ionic strength.

Sensors for process control and physiological studies are based on thymol blue and bromophenol blue [95], cresol red [96], bromocresolgreen, and bromothymol blue adsorbed on cellulose strips [97], or on related absorber dyes [79–82,98–102]. The temperature coefficients of the XAD-immobilized bromothymol blue and thymolphthaleine, respectively, between 25 and 45°C are 0.013 +/− 0.003 and 0.015 +/− 0.003 per °C. The response time for 63% of the total signal change to occur is 1 min.

The preferred method for making pH-sensitive optical materials is clearly via covalent immobilization. Mohr & Wolfbeis [24] have designed a general logic for making pH sensors for various pH ranges, starting from a single precursor that was reacted with various components to give azo dyes with widely varying pK_a's. These were covalently immobilized on cellulose-coated polyester films to result in sensors with pK_a's ranging from 0.5 to 11.3. Because of the stability of the dyes and of the covalent bond, the sensors are stable over years, have long operational lifetimes, and achieve response times in the order of 30 to 60 sec because the active layer (which is

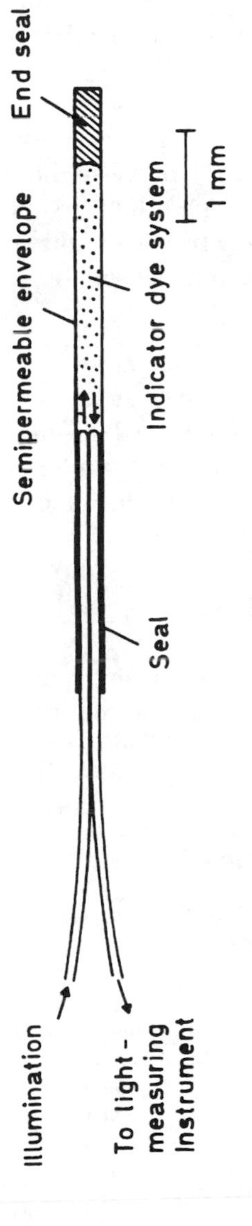

Figure 8.11 Fiber-optic pH sensor with reflective pH indicator chemistry contained in a semipermeable envelope at the tip.

a film of cellulose acetate on polyester, [103] is ca. 1 μm thick only. Table 8.5 summarizes the various types and pK_a's of the such membranes.

Polyphthalate esters (like Mylar) are the preferred materials for depositing sensor chemistries to obtain planar sensors or sensor spots. Such sensors are now being made in large quantities for use in blood gas analysis, both continuously and single shot. Sensors are made by coating the polyester films with the respective materials by methods such as spin coating, or spreading (frequently using dissolved materials) as they are known in the film industry (see Figure 8.12), and sensor spots are then punched out to be used as either planar sensors (e.g., in disposable cassettes) or at the tip of an optical fiber.

8.4.2 Fluorescence-Based Sensors

Fluorescence is particularly well-suited for optical sensing owing to its sensitivity. For weakly absorbing species (i.e., when $A < 0.05$), the intensity I_f of fluorescent light returning from the sensor tip is proportional to the intensity of the exciting radiation, I_O, and the concentration ($[D]$) of the fluorescent dye in the sensor:

$$I_f = k' \cdot I_O \cdot \phi \cdot \epsilon \cdot l \cdot [D] \tag{8.12}$$

where l is the length of the light path in the sensing layer, ϵ is the molar absorptivity, ϕ the quantum efficiency of fluorescence, and k' the fraction of total emission being measured. At constant I_O (8.12) can be simplified to give

Table 8.5
Absorption Maxima of Acid and Conjugate Base Forms and pKa Values of Dyed Cellulose Sensor Membranes (at 21°C) for Various pH Ranges

Membrane	*λ_{max} (nm) (Base Form)*	*λ_{max} (nm) (Acid Form)*	*pK_a*
M-1	553	460	9.37
M-2	535	501	9.26
M-3	541	473	7.55
M-4	517	491	7.83
M-5	518	455	7.34
M-6	476	487	11.28
M-7	474	479	10.68
M-8	481	488	10.64
M-9	507	509	3.68
M-10	492	518	~0.5
M-11	486	503	2.24

Source: [24].

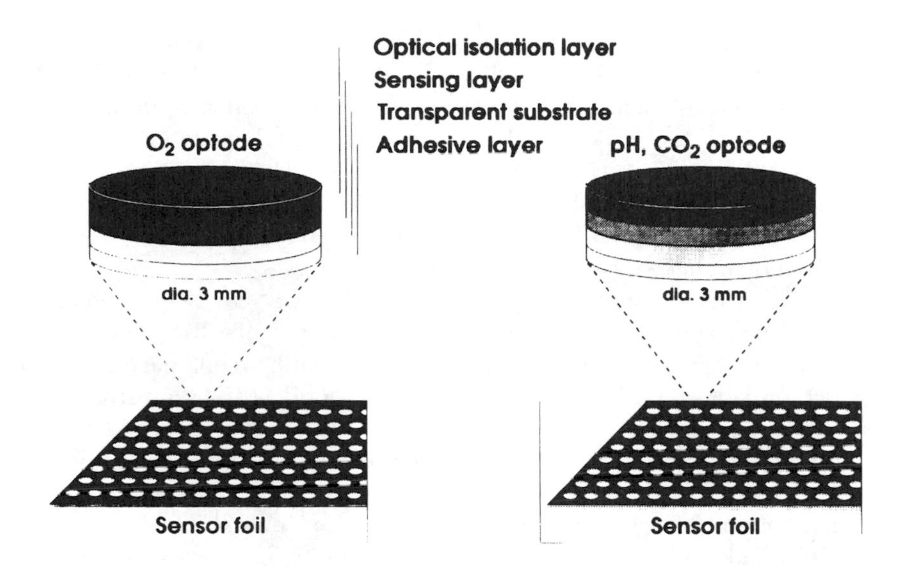

Figure 8.12 Cross-sections through sensor foils for mass fabrication of optrodes for oxygen and pH/CO_2.

$$I_f = k \cdot [D] \tag{8.13}$$

where $k = k' \cdot I_0 \cdot \phi \cdot \epsilon \cdot l$. Seitz [61], as well as Leiner & Hartmann [64], derived equations to relate fluorescence intensity with actual pH, and the former also has addressed calibration issues.

A variety of fluorescent indicators is known [85], but only a few meet the requirements of an excitation maximum beyond 450 nm, to allow the use of inexpensive and flexible plastic fiber optics as light guides, and of light-emitting diodes as excitation sources (see Table 8.3). Large Stokes' shifts are also desirable in order to conveniently separate scattered excitation light from fluorescence using inexpensive optical filters. Further desirable properties include photostability, the presence of functional chemical groups suitable for covalent immobilization, and the lack of toxicity.

One of the first pH fluorosensors was obtained [81] by electrostatic immobilization of hydroxypyrene trisulfonate (HPTS) on an anion-exchange membrane. It allowed for measurement of pH in the range 6 to 8 by relating the ratio of fluorescence intensities emitted at 510 nm and excited at 405 nm (specific for the acid form) and 470 nm (specific for base form). The ratio was not affected by source fluctuations and slow loss of reagent, all of which can affect a single intensity measurement. The sensor showed an approximately 10% loss in intensity after four hours of continuous illumination. An important observation is the effect of indicator loading on the response curve of a sensor. With increasing indicator amounts being immobilized, the relative signal change becomes smaller and the pK_a is shifted to lower values [61].

Two related optical sensing materials for measurement of near-neutral pH values were also described in some detail [8,9]. HPTS and the pH probe 7-hydroxycoumarin-3-carboxylic acid (HCC), respectively, were covalently immobilized on surface-modified controlled porous glass (CPG). Analytical excitation and emission wavelengths were, respectively, 410 and 455 nm for the HCC-based sensor, and 465 and 520 nm for the HPTS based sensor. On CPG, the pK_a values were distinctly lower than those determined in solution, and the HPTS-based sensor was more sensitive to ionic strength than then sensor with the coumarin dye.

In an alternative approach [64,104], HPTS was covalently bound to a hydrophilic cellulose matrix, which then was deposited at the distal end of a single optical fiber. The cellulose matrix was further covered with an opaque cellulose overcoat, which provides both mechanical integrity and optical isolation from environmental optical interferences. Fluorescence intensity again was measured at 520 nm under 460-nm and 410-nm excitation (see Figure 8.1). The ratio of the two signals can be related to pH, and is relatively insensitive to optical throughput.

Other fluorescence-based pH sensors use immobilized fluoresceins [59,105,106] and naphthofluoresceins [85] on cellulose films or in sol-gels. Rather than attaching preformed sensing materials onto single fibers, the sensor chemistry has been deposited directly on the end face of a fiber via thermal or photopolymerization [84,107]. In a typical experiment, a fluoresceinamine was copolymerized with acrylamide and N,N,-methylenebis(acrylamide) onto the distal end of surface-modified single glass/glass optical fibers of 100/140-μm diameter. The polymer-modified single fiber sensors had a response time of < 10 sec. One disadvantage is the poor sensor reproducibility due to the difficulty in controlling the polymerization process, the need for an argon ion laser, and a rather sophisticated detection system.

One of the trends in the design of optical-fiber probes is miniaturization. A near-field fluorescence optical technique was applied to design submicron-sized pH sensors [108]. Multimode fibers were drawn into submicron fiber tips and coated with aluminum to form minute optical-fiber light sources with a distal-end pH chemistry. Such sensors have excellent spatial resolution and very short response times. Long-wave absorbing fluorophors (such as the SNARF dyes) immobilized on dextranes were employed for measurement of physiological pH. Similarly, oxygen-sensitive coatings have been deposited at the tip of fiber only 15- to 40-μm thick, and the resulting sensors were used to measure oxygen profiles in marine sediments. Figure 8.13 shows the tip of such a fiber with and without the black optical isolation.

8.4.3 Energy Transfer-Based Sensors

Jordan et al. [109] developed a single fiber-optic pH sensor based on energy transfer from a pH-insensitive fluorophore, eosin (the donor), to a pH-sensitive absorber, phenol red (the acceptor). The dyes were co-immobilized with acrylamide on the distal end of a surface-silanized single optical fiber. The pH-sensitive layer had a thickness of approximately 10 μm. The emission spectrum of eosin overlaps with the absorption

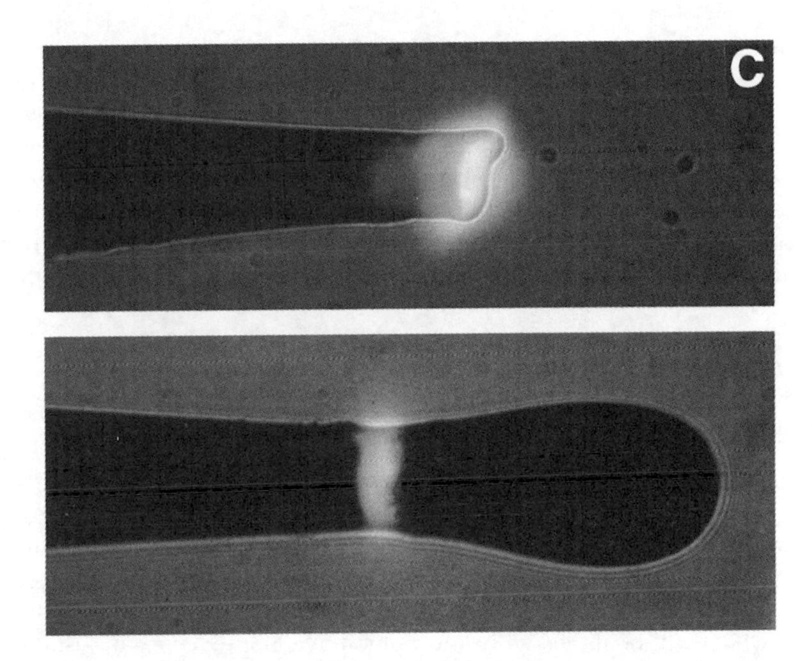

Figure 8.13 Oxygen micro-optrode (tip diameter ca. 15 µm) showing the red luminescence of the oxygen-sensitive chemistry without optical isolation (top) and with an optical isolation (a black silicone; bottom). (*From:* I. Klimant, V. Meyer and M Kühle, Limnol. Oceanogr. 40 (1995) 1159.)

spectrum of the basic form of phenol red. As the pH increases, the concentration of the base form of phenol red increases, resulting in an increased energy transfer from eosin to phenol red and in a diminished fluorescence intensity of eosin. Thus, changes in the absorption of phenol red, as a function of pH, are detected as changes in the fluorescence signal of eosin. The intensity of the fluorescence signal was observed with a photon-counting detector at the emission maximum of eosin (546 nm), after passing a dichroic mirror, a longpass filter, and a monochromator. The precision was +/− 0.008 pH units, when measured with standard buffer solutions. The time required for 63% response is 4 to 5 sec for a pH change from 7.1 to 6.5. Some photobleaching of the base form of phenol red has been observed when continuously exciting the sensor over a 10-min period. Because the efficiency of energy transfer depends on the sixth power of the average distance of the two dyes, such sensors are highly sensitive to leaching and bleaching, particularly if this occurs at different rates.

8.4.4 pH Sensors Based on Measurement of Decay Time

In intensity-based pH sensors, it is usually the ratio of two signals (obtained at two wavelengths) that is related to pH. An alternative self-referenced method is based on measurement of fluorescence decay time [110–112]. Unlike in oxygen sensors (where the population of the excited state of a single indicator is reduced due to collisional quenching by oxygen), decay-time pH sensors are based on the measurement of the relative contributions of the acid and base form of an indicator to the total decay time. Such sensors are likely to have excellent operational lifetime, but-like all 2-wavelength methods-cannot compensate for the most serious problem of indicator-based sensing (i.e., the temporal drift of the work functions). Like in all optical sensors, long-lived probes are desired for reasons of instrument simplicity and resolution.

8.5 OXYGEN SENSORS

8.5.1 Indicators and Polymeric Supports

Oxygen has almost exclusively been sensed via quenching of luminescence [28], while photometry plays a minor role. Chemiluminescent methods are irreversible and require addition of a reagent. The variety of indicators known for oxygen include polycyclic aromatic hydrocarbons such as decacyclene and perylene dibutyrate, and longwave absorbing metalorganic complexes of ruthenium, osmium, palladium, and platinum, which will be discussed below. Some have long-lived excited states (up to 1 ms), which makes them useful for lifetime-based oxygen sensors. Certain dyes also undergo quenching of their *phosphorescence* when absorbed on solid supports. These include trypaflavine, benzoflavine, chlorophyll, and hematoporphyrin [113–115].

In order to obtain an oxygen-sensitive material for use in optical sensing, it is necessary to immobilize the oxygen-sensitive dye in a polymer subsequent to being deposited on a waveguide. However, most indicators do not have appropriate functional groups suitable for covalent immobilization, so that physical immobilization or immobilization on ion-binding membranes usually is preferred. In addition, it has been found that the Stern-Vomer quenching constants (K_{SI} in (8.5)) are reduced by 30 to 50% on covalent binding of a dye onto a rigid surface.

A simple way for immobilizing lipophilic oxygen indicators is to dissolve them in hydrophobic polymers such as poly(vinyl chloride) (PVC) or silicone. Silicone has excellent oxygen permeability and solubility, but is a poor solvent for most dyes. PVC, on the other hand, is a good solvent for most polycyclic aromatic hydrocarbons (PAHs), but has slow oxygen diffusion. In order to make PAHs more lipid-soluble and less water-soluble, they may be fitted with tertiary butyl groups, which results in a five- to twentyfold improved solubility in silicones and other materials [37]. Cross-linked poly(hydroxyethyl methacrylate) is another solvent that retains polycyclic aromatics such as diphenylanthracene [116] or ruthenium-tris(dipyridyl) [117].

The widely used oxygen indicator ruthenium-tris(bipyridyl) may be immobilized electrostatically on cation exchange membranes or physically adsorbed on particles entrapped in silicone [72]. The incorporation of decacyclene into a Langmuir-Blodgett quadruple layer has been reported as well [90], and an excellent sensitivity and fast response to oxygen was observed. However, LB films are not stable on contact with many samples encountered in practice.

Oxygen-sensitive luminescent coatings also were obtained by 1) dissolving pyrene in dimethylformamide solvent [118], polyethylene [119] or poly(dimethylsiloxane) copolymers [52]; 2) by incorporating osmium-organic complexes on silica in silicone rubber [32]; 3) by embedding phosphorescent metal complexes of ferrone in silicone rubber films or binding them to anion exchanger beads [120]; 4) by dissolving camphor-quinone in PMMA, PVC, or polystyrene [121]; 5) by incorporating luminescent platinum porphyrins in silicone rubber [122]; and 6) by either dissolving porphyrins in polystyrene/toluene and spreading the cocktail as a film [35] or depositing porphyrins on various solid supports [123]. Certain (histidinato)cobalt complexes undergo reversible changes in absorption on exposure to oxygen, but are nonluminescent [124].

The quenching by oxygen of the luminescence of a ruthenium complex dissolved in various types of silicones was investigated by various groups. Since the work function was found to vary over time for quite a while, it was concluded that such sensor materials need to be recalibrated after prolonged storage. The type of matrix has a pronounced effect on the overall performance of such sensor materials, and decay kinetics are more complex in being multi-exponential, the various species being differently susceptible to quenching by oxygen [29,122,125,126]. Figure 8.14 shows the effect of oxygen on each of the three lifetimes of an oxygen sensor membrane composed of the dye Ru(dpp) in plasticized PVC. Physical models have been established to interpret the observed quenching processes [29,122,125 127].

8.5.2 Oxygen Sensors Based on Measurement of Luminescence Intensity

The first oxygen fluorosensors were based on the use of PAHs [119,128–134]. Fluoranthene, pyrene, benzoperylene, and decacyclene are typical indicators, but all suffer from interferences by halothane, sulfur dioxide, chlorine, and nitrous/nitric oxide. Typical polymeric solvents are silicone rubber, plasticized PVC, polystyrene, poly(hydroxyethyl methacrylate) (p-HEMA), or porous glass. Detection limits are in the order of 1 torr (as with most types of fluorescence-based oxygen optrodes). Not unexpectedly, the diffusion of oxygen through poly(dimethylsiloxane) (PDMS) and p-HEMA is quite different, the diffusion constants being 3.56 3 10^{25} cm^2/sec^{21} in PDMS (at 25°C), but 1.2 3 10^{26} only in PHEMA (at 20°C).

Peterson and others [133] found perylene dibutyrate adsorbed on polystyrene beads to be a viable sensing material. It has excitation and emission maxima of, respectively, 468 and 514 nm. It is stable, and is efficiently quenched by oxygen, thereby allowing a resolution of +/−1 torr up to 150 torr. The sensor measures the ratio of scattered blue light and green fluorescence. Others have used decacyclene dissolved in PDMS [37,129]. Interferences by the inhalation narcotic halothane were eliminated by covering the sensor with an 8-μm layer of teflon, which also acts as an

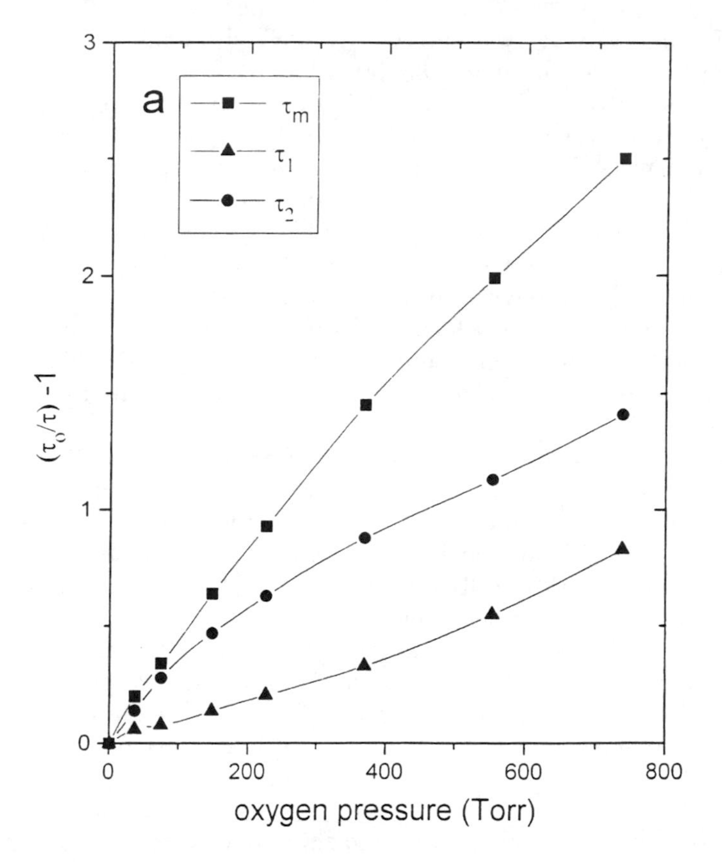

Figure 8.14 Effect of oxygen partial pressure on each of the three components of the decay of the luminescence of an oxygen sensor. (*Source:* [29].)

optical isolation [30,37]. An interesting application of this sensor is for simultaneous determination of oxygen and an interfering quencher such as halothane [37].

Semiconductor lasers have unique advantages over other types of conventional and laser light sources: they have a fairly large output (sufficient for most fiber applications), a narrow bandwidth, and a low price. Unfortunately, their wavelengths do not match the absorption spectra of PAHs. Okazaki and others [134] therefore have frequency-doubled the 780-nm emission of a semiconductor laser to obtain a 390-nm line with 50-μW intensity, which is suitable to excite benzo(g,h,i)perylene. The beam was launched into a fiber that guides light to the sensing material (the indicator dissolved in silicone grease) at its end. A second fiber was used to collect fluorescence at 430 nm. Oxygen was determined in the 0 to 30% range at atmospheric pressure.

In recent years, metalorganic complexes have become the preferred type of probes for oxygen. They have long lifetimes (0.2-1,000 μs) and therefore are efficiently quenched by oxygen. Their excitation wavelengths range from 450 to 600 nm, which makes them excitable by LEDs, and some display extraordinarily large Stokes' shifts. For example, ruthenium-tris(bipyridyl) [31,72] and the respective tris(phenanthroline) complex [135] were adsorbed onto silicagel particles and then entrapped in silicone polymer. The fluorescence of the resulting material is strongly quenched by oxygen, but a Stern-Volmer plot is nonlinear. Bacon & Demas [136] found ruthenium(II)tris (4,7-diphenyl-1,10-phenanthroline) (Ru(dpp)) to be a most viable oxygen probe that can be incorporated into a silica-filled silicone polymer. The resulting membranes, ca. 100 μm in diameter, were found to be useful for optical oxygen sensing and were investigated with respect to the effect of oxygen on fluorescence intensity and lifetime. The respective Stern-Volmer plots are practically superimposable, but not linear. Ru(dpp) is one of the most widely used oxygen probes at present.

Table 8.6 summarizes figures of merit for longwave oxygen probes. Among those, the metallated ketoporphyrins [35] look to be almost ideal indicators by virtue of their longwave absorptions and emissions, a 150 to 200 nm Stokes' shift, full compatibility with diode light sources, excellent photostability that by far exceeds all other probes of Table 8.6, solubility in various polymers, long decay time, compatibility with LEDs and diode lasers, 15% quantum yield, and the fact that they are uncharged. A generalized chemical structure, with Me representing the metal ion, is given in Figure 8.15.

Most of these indicators are highly specific if applied as a solution in a lipophilic organic polymer. No, or only small, interferences are found for water vapor, nitrogen, noble gases, carbon monoxide, carbon dioxide, methane, and higher alkanes at realistic pressures, although these usually can pass the polymer membranes. Potential interferents are sulfur dioxide (a notorious quencher), halothane (an inhalation narcotic), chlorine, and nitrogen oxides (except N_2O). Transition metal ions and heavy atoms also quench, but usually cannot enter the polymer membrane and therefore remain inert.

Table 8.6
Figures of Merit for Longwave Luminescent Oxygen Probes

Probe	*Excitation/Emission Maxima**	*Decay Time***	*Ref.*
Ru(dipy)	460/605	ca. 0.8 μs	[31]
Ru(dpp)	460/610	ca. 4 μs	[136]
Os-bis(2,2′,2″-terpy)	650/710	270 ns	[32]
Pd-octaethylporphyrin	545/670	990 μs	[137]
Pd-keto-porphyrin	600/780	480 μs	[35]
Pt-keto-porphyrin	590/760	64 μs	[35]
Al-Ferrone	390/600	unknown	[120]
Camphorquinone	470/560	ca. 1 ms	[121]

* In nanometers
** Under nitrogen.

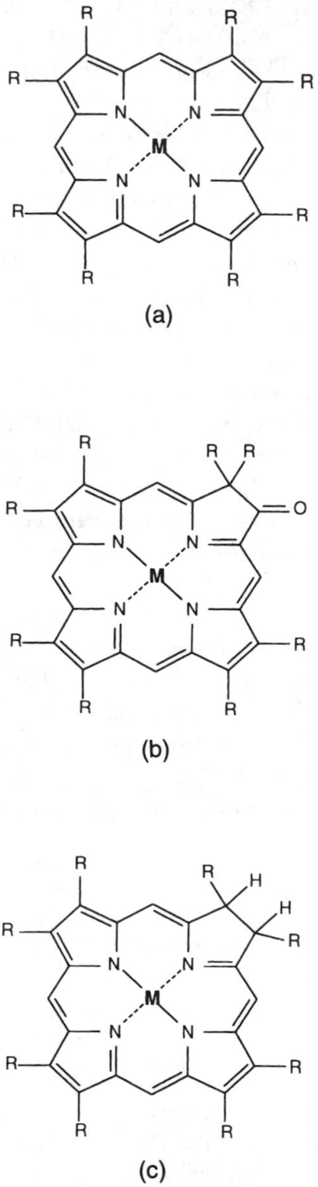

(a)

(b)

(c)

Figure 8.15 Chemical structures of luminescent oxygen probes: (a) porphyrins; (b) ketoporphyrins; and (c) chlorins. The central metal atom preferably is Pt or Pd.

8.5.3 Oxygen Sensors Based on Measurement of Decay Time

Equation (8.5) predicts that oxygen not only affects the intensity of the luminescence of a luminophore, but also its decay time τ(see Figure 8.14). The long decay of metalorganic oxygen indicators renders them particularly suitable for sensors based on measurement of decay time. In the first sensor of that kind [31], phase fluorimetry was applied to measure the oxygen-dependent phase shifts of a sinusoidally excited luminophor and a frequency-modulated blue LED served as a light source. Compared to former sensor types, decay time-based sensors display decisive advantages: They have negligible signal drift arising from leaching and bleaching because the decay time is independent of fluorophore concentration (in a first approximation). Secondly, they display excellent long-term stability because the system is internally referenced. Finally, no drift arising from light source intensity and photodetector sensitivity fluctuations is to be expected because it is not the absolute intensity that is measured, but rather the phase shift between excitation and fluorescence.

Numerous other reports on oxygen sensors based on measurement of decay time exist [31–35,138–140]. However, the aim of this chapter is on materials rather than methods and hence a discussion of decay time-based sensors is beyond its scope. An interesting application of decay time-based sensors is in transcutaneous oxygen sensing [140] where intensity-based systems are clearly inferior due to the strongly varying fluorescence background and light permeability of skin.

8.5.4 Oxygen Sensors Based on Measurement of Energy Transfer

An interesting type of oxygen sensor has been described [132] that is based on electronic energy transfer from a donor (whose fluorescence is quenched by molecular oxygen) to an acceptor (whose fluorescence is less affected by oxygen). Pyrene was employed as the donor, and perylene as the acceptor. The fluorescence emission band of the donor shows good overlap with the absorption band of the acceptor. When excited at 320 nm, the two-fluorophore system showed strong fluorescence at 476 nm, where pyrene itself is nonfluorescent. Although perylene is not efficiently quenched by oxygen, the system strongly responds to oxygen because fluorescence is quenched with an efficiency that by far exceeds the quenching efficiency for pyrene or perylene alone. There is an almost fourfold increase in the quenching constants of the energy transfer system (as compared to the conventional system). A fiber-optic oxygen sensor was developed by incorporating the two dyes into a silicone polymer matrix that had been attached to the end of an optical fiber. Oxygen was detected in the 0 to 150-torr range with a 0.5 to 3-torr resolution.

8.6 SENSORS FOR CARBON DIOXIDE

Traditionally, *gaseous* carbon dioxide (CD) has been assayed via infrared absorptiometry. The preferred method for measuring *dissolved* CD is via electrochemistry by measuring changes in the pH of a buffer solution retained in front of a pH electrode by a CD-permeable but proton-impermeable polymer. Two important indicator-based methods exist. The first (named the *Severinghaus* method) works by analogy to the electrochemical approach (i.e., via a change of pH of an immobilized buffer). The other (so-called plastic type) sensors are based on a hydrated pH indicator anion entrapped (without any buffer!) in a proton-impermeable polymer.

8.6.1 Severinghaus-Type Sensors

Such sensors are obtained by entrapping a buffer (such as a 10-mM hydrogencarbonate solution) along with a dissolved indicator (such as bromophenol blue or HPTS) in a gas-permeable polymer (such as silicone). The response of this sensor to CD occurs as a direct result of the proton concentration in the buffer in the sensitive material, which is related to the concentration of CD through the following series of chemical equilibria:

1. CO_2 (aq) + H_2O <=======> H_2CO_3 (hydration)
2. H_2CO_3 <=======> $H^+ + HCO_3^-$ (dissociation, step 1)
3. HCO_3^- <=======> $H^+ + CO_3^{2-}$ (dissociation, step 2)

These are governed by the following equilibrium constants:

$$K_h = [H_2CO_3]/[CO_2](aqu) = 0.0026 \tag{8.14}$$

$$K_1 = [H^+][HCO_3^-]/[H_2CO_3] = 1.72.10^{-4} \tag{8.15}$$

$$K_2 = [H^+][CO_3^{2-}]/[HCO_3^-] = 5.59.10^{-11} \tag{8.16}$$

In most studies on CD sensors, the total analytical concentration of carbon dioxide (i.e., ($[CO_2]aq + [H_2CO_3]$)), has been related to the response.

In the first optrode for CD, Lübbers and Opitz [118] followed the changes in fluorescence intensity of a membrane-covered solution of 4-methyl-umbelliferone in a carbonate internal buffer as a function of the partial pressure of CD. The internal buffer (also containing the indicator) was covered with a 6-µm PTFE membrane, which is permeable to CD but impermeable to protons and other ionic species. The ratio of fluorescence intensity at 445 nm measured under excitation at 318 nm and 357 nm was related to pressure. The same scheme was applied to construct a compact instrument (5 by 6 by 14 cm in size) for CD. It consisted of a blue LED as a light source, the longwave-absorbing indicator hydroxypyrene-trisulfonate (HPTS) dissolved in bicarbonate and placed behind a PTFE layer, two optical filters, and two photodiodes for detection of light.

Zhujun and Seitz [141] used HPTS in bicarbonate, covered with a silicone membrane, to sense CD. Fluorescence was measured with a bifurcated fiber system. Complete response occurs within a few minutes. Both sulfite and sulfide were found to interfere (probably as their membrane-diffusible forms SO_2 and H_2S). The equation used to relate the CD partial pressure to hydrogen ion concentration is

$$[H^+]^3 + N[H^+]^2 - (K_1 \cdot C + K_w)[H^+] = K_1 \cdot K_2 \cdot C \tag{8.17}$$

where N is the internal bicarbonate concentration, K_1, K_2, and K_w are the dissociation constants of carbonic acid ((8.15) and (8.16)) and water, respectively. The value C is the analytical concentration of CD (in both the hydrated and unhydrated form). It was shown that, within a limited range, there is linearity between CD pressure and $[H^+]$ according to

$$(H^+) = K_1/C \cdot N \tag{8.18}$$

In practice, the internal HCO_3^- concentration should be such that the CD concentrations of interest yield pH changes in the range of the pKa of the indicator. In the case of the widely used HPTS with its pKa of 7.3, the external pCO_2 should adjust a pH between 6.5 and 8.0 in the internal buffer.

Heitzmann & Kroneis [142] prepared CD-sensitive fluorescent membranes by soaking cross-linked polyacrylamide beads with a solution of HPTS in bicarbonate, and embedding them in silicone rubber matrix deposited at the tip of a fiber (Figure 8.16). The response to CD was varied by adding different quantities of bicarbonate, carbonate, and HPTS, all of which act as buffers. The poly(acrylamide) beads may be omitted so that an emulsion of the HPTS/carbonate solution in silicone rubber is obtained. These sensors have excellent long-term performance, but when stored in media of low pCD they tend to become destabilized. It takes several hours to obtain a stable baseline again after having been exposed to higher pCD levels. This is probably due to some dehydration and also a shortage of water molecules available for reaction (1) which, in turn, results in some contraction and expansion of the droplets and beads because of osmolarity effects.

Typically, a 20-mM bicarbonate solution would act as an internal buffer. With a given indicator, it is the choice of buffer that primarily determines the slope of the response curve. The slopes of the response curves also are governed by the pK_a of the indicator and the total ionic strength of the internal buffer. Unfortunately, high buffer concentrations result in very long response times. These are further prolonged with increasing thickness of the sensor coating, additional covers, slow kinetics of the hydration, and-in particular-slow dehydration of CD. For the system described above, the response for a change from 0 to 5% CD was 15 sec for 90% of the final

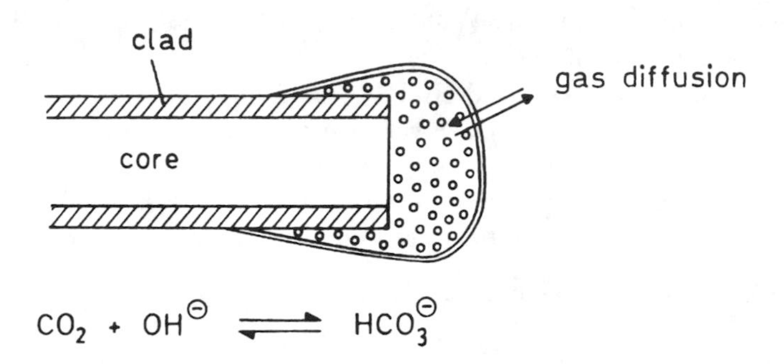

Figure 8.16 Optical sensor for carbon dioxide with a buffer solution emulsified into a drop of silicone placed at the tip of a fiber and covered with a gas-permeable black overcoat.

value. The reverse change required approximately 30 sec. It should be noted that most other optical CD sensors described so far have a much slower response. The addition of carbonic anhydrase can accelerate response time [143].

A CD sensor with an indicator directly immobilized on the fiber has also been described [144]. A fluorescent dye was covalently bonded onto the glass fiber tip, so the miniature size of the sensor was preserved. The fiber was then coated with a silicone-carbonate copolymer that rejects protons, whereas CD can pass. Direct casting of the polymer onto the sensor chemistry imposed major problems, and it was decided to use a preformed cover membrane. When two sensing layers with different spectral properties, and being selective for oxygen and CD, respectively, are attached to the end of a fiber, a single fluorosensor for both species can be obtained [145]. Blue excitation results in two fluorescences: The green fluorescence reports CD, while the red fluorescence reports oxygen.

Vurek and others [146] devised an absorbance-based CD sensor that relies on the principle of a previously designed pH sensor. An isotonic solution of salt, bicarbonate, and phenol red was covered with a CD-permeable silicone rubber membrane. The device uses two fibers: one carrying the input light, the other reflected light. The sensor responds over the physiological range and its performance was demonstrated *in vivo*. Similarly, a fluorescein-based CD-sensitive system was reported by Hirschfeld and others [147].

Leiner and others [148] have developed planar CD-sensitive chemistries for mass fabrication (Figure 8.12). Polyester membranes served as planar supports onto which a pH-sensitive material was fixed using a proton-permeable glue. After soaking with a bicarbonate buffer, the cellulosic material was covered with hydrogel and, finally, with a CD-permeable silicone-polycarbonate copolymer. CD, permeating into the internal buffer, changes the internal pH and hence causes fluorescence intensity to change. The sensing membranes can be prepared in large sheets that later can be punched into small (i.e., 1-5 mm) spots and placed at the distal end of a fiber. They also can serve as sensor sheets in measurement of surface pCD of the skin. Given the low costs for the fabrication of such sensor spots, another important application is in disposable kits for determination of CD along with oxygen and pH in blood. Figure 8.17 shows the design of such a kit.

A mechanically highly resistant and sterilizable sensor for CD was obtained by covalently immobilizing a commercial pH probe on a thin cellulose film, soaking it with buffer and covering it with a proton-impermeable film of silicone rubber [149]. The top layer also contained a highly reflective material that acts as an optical isolation. CD was measured over the 0 to 760-torr range, and effects of buffer capacity and buffer pH studied in detail. Response times for dissolved CD are in the order of several minutes, and no cross-sensitivity to pH was observed at all. Sensor sterilization with hydrogen peroxide did not affect the calibration graph. An interesting sensing scheme for CD that also is applicable to decay time-based sensing makes use of tris(pyrazinyl)thiazole complexes of ruthenium(II) whose luminescence is quenched by the protons formed by reaction of CD with water [143,150]. The probe was immobilized onto anionic dextrane gel soaked with phosphate buffer (8.5), and the resulting fiber sensor measured CD between 0 and 760 torr.

8.6.2 Plastic-Type Sensors

This type of sensor for carbon dioxide is being constructed without using an internal aqueous buffer solution. Rather, the pH-sensitive dye is placed directly in the organic polymer in the form of its anion, the cation usually being a quaternary ammonium ion. Mills and coworkers have described various types of plastic-type film sensors for CD. They are made by casting a cocktail composed of a pH indicator anion (D^-), an organic quaternary cation (C^+), and a polymer such as ethyl cellulose, all dissolved in

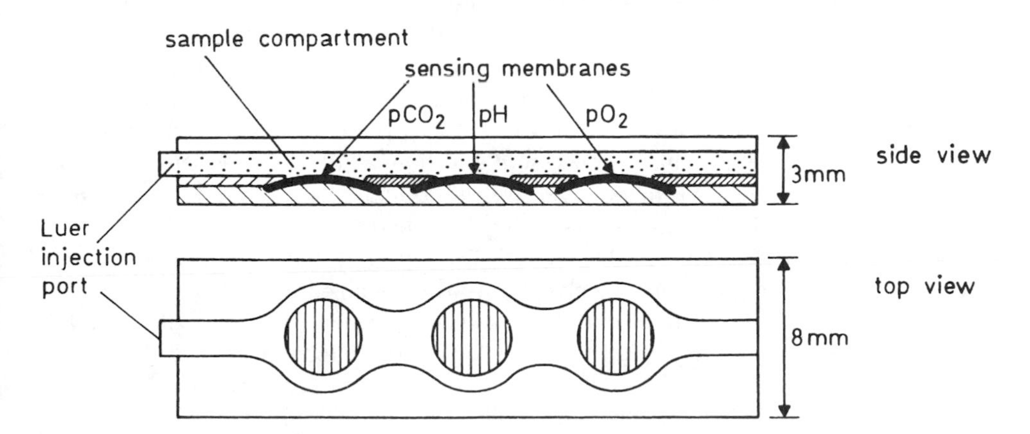

Figure 8.17 Top and side view of a disposable kit with three sensor spots (pH, oxygen, carbon dioxide) integrated into a compartment accommodating a calibrant or a blood sample.

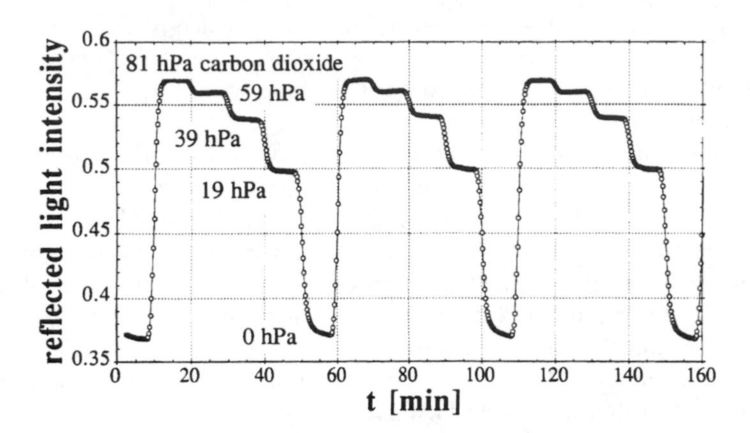

Figure 8.18 Changes in the reflectivity of a sensor for carbon dioxide with the partial pressure of carbon dioxide in water solution. (*Source:* [156].)

an organic solvent such as toluene onto a solid support and evaporating the solvent. Such films undergo very fast color changes on exposure to even small concentrations of CD [151–153], as can be seen from Figure 8.18. The response can be fine-tuned by variation of the components, in particular the quaternary ammonium base [154,155], and silicones have been suggested as an alternative to other polymers when measuring dissolved CD [156].

The scheme was extended to an energy-transfer fluorosensor composed of an absorber dye (m-cresol purple) and on inert fluorophore (sulforhodamine 101) entrapped in an ethyl cellulose film. CD modulates the decay time of the fluorescence of the rhodamine dye, and this serves as the analytical information. The sensor has an excellent long-term stability, is compatible with the 635-nm laser diode, and has response times in the order of seconds. A major problem may arise, though, if the two dyes leach/bleach at different rates in view of the extremely strong distance-dependence of the energy transfer process. In addition, all plastic-type sensors require storage in the complete absence of even traces of acidic gases, which tend to irreversibly deactivate plastic film sensors with their inherently low buffer capacity.

Another bufferless CD sensor material was reported that was prepared by dispersing fluorescein in poly(ethylene glycol) and then depositing it at the distal end of an optical fiber [157]. Evaporation of the solvent is reported to be negligible. The dynamic range is from 0 to 28% (v/v) for CD, with a detection limit of 0.1%. Full response is achieved within 10 to 20 sec. The outer membrane, ca. 10-µm thick, is composed of poly(ethylene glycol)s with molecular weights of 200 and 1540 Dalton, respectively, in a 20:80 (w/w) ratio.

8.7 AMMONIA SENSORS

Three major optical sensing schemes are known for ammonia. In the first, the absorption of light in the NIR by ammonia is exploited in plain fiber sensing. This approach is not very sensitive and response depends on humidity, but sensors are simple in design and display good stability. Absorption can be measured in both the transmission mode (in a gas cell) or by the evanescent wave technique. The latter, however, even more strongly depends on the relative humidity of the gaseous sample due to adsorption of water on the sensor/sample interface. The method cannot be applied to aqueous samples. In the second approach, ammonia is reacted with a dye such as ninhydrin to yield a purple coloration. This is an irreversible reaction, so that the "sensor" actually is a single-shot probe. In the third approach, the basic properties of ammonia are exploited: It is capable of changing the color of pH indicators immobilized on a waveguide. Only the third sensing scheme is both indicator-based and reversible and will be discussed here.

A reversible optical waveguide sensor for ammonia vapors was reported [158] that consists of a small capillary glass tube fitted with an LED and a phototransistor detector to form a multiple reflecting optical device. When the capillary was coated with a thin solid film composed of a pH-sensitive oxazine dye, a color change occurred on contact with ammonia. The instrument was capable of reversibly sensing ammonia and other amines. Vapor concentration from 100 to below 60 ppm ammonia were easily and reproducibly detected. A preliminary qualitative kinetic model was proposed to describe the vapor-film interactions. The method was applied to design a distributed sensor for ammonia [159].

Ammonia sensors based on the same principle as electrochemical ammonia sensors (viz., the change in the pH of an alkaline buffer solution) have been reported by various groups: Arnold and Ostler [160] followed the changes in the absorption of an internal buffer solution to which p-nitrophenol was added. Ammonia passes by and gives rise to an increase in pH, which causes a color change of the indicator to occur. Wolfbeis and Posch [161] entrapped a fine emulsion of an aqueous solution of a fluorescent pH indicator, which simultaneously may act as a buffer, in silicone rubber. Alternatively, 0.001M aqueous ammonium chloride may be used as internal buffer. The buffer strength strongly determines both response time and slope of the response curve. Detection limits are in the order of 5 to 20 μM, and equilibration is very slow, particularly in the back direction and with aqueous sample solutions. Another type of fluorescent sensor for ammonia was obtained by entrapping a 50-μM solution of a carboxyfluorescein in an ammonium chloride buffer in front of a fiber optic. The device was extremely sensitive and used for measurement of extracellular ammonia [162].

Hydrophilic ammonia sensor films were obtained by immobilizing bromothymol blue in a hydrophilic polymer and measuring the changes in reflectance induced by ammonia in the gas phase [163]. The working range was from 1.5 to 30 mM, and possible interferents were investigated. Similar films have been used in a portable photometric ammonia gas analyzer [164].

Shahriari and others [165] developed a new porous glass for ammonia detection whose structure imparts a high surface area to the fiber core. Ammonia vapors penetrating into the porous zone pretreated with a reversible pH indicator produce a spectral change in transmission. The resultant pH change is measured by in-line optical absorbance and is said to be more sensitive than sensors based on evanescent wave coupling into a surrounding medium. The signal can be related to the ambient ammonia concentration down to levels of 0.7 ppm. In order to speed up response time, a porous plastic material, exhibiting very high gas permeability and liquid impermeability, was used in another type of ammonia sensor [166]. The porous plastic fibers were prepared by copolymerization of a mixture of monomers (methyl methacrylate and triethylene glycol dimethyl acrylate), which can be cross-linked in the presence of an inert solvent (such as octane) in a glass capillary. After thermal polymerization, the plastic fibers were pulled out of the capillaries and used in the sensor.

Ammonia, being a basic gas, causes the color of appropriate pH indicators to change. This is exploited in a plastic-type ammonia sensor that works by analogy to the respective sensors for carbon dioxide, except that pH-changes go in the other direction. Again, silicone is the preferred material, but usual dyes are insoluble therein. As a result, they have to be chemically modified, for example, by making a lipophilic (silicone-soluble) ion pair composed of the dye (usually on anion) and an organic cation (such as cetyl-trimethyl-ammonium ion. The silicone matrix acts as a perfect barrier for hydrogen ions ("pH"), which would interfere, and the resulting sensors display very low limits of detection (LODs) that range from 20 to 100 ppb [167]. Figure 8.19 shows a typical response. The method has been extended to fluorescence where LODs are even lower [168], and the coating was applied in an integrated optical disposable [169].

A fluorescent type of sensor for monitoring ammonia in air was obtained by impregnating porous cellulose tape with a solution of eosine bluish, p-toluenesulfonic acid, and glycerol [170]. On exposure to ammonia, the fluorescence of the dye at 550 nm increases and is proportional to the concentration of ammonia gas at constant sampling time and flow rate. One hundred ppb of ammonia were detectable and interference studies revealed a remarkable selectivity, although acidic gases are likely to reverse the response of ammonia. Obviously, the sensor is inadequate for detection of ammonia in water.

Generally, all types of ammonia sensors based on pH effects also respond to other uncharged amines such as methylamine, pyridine, or hydrazine because they are strong bases, too, and can pass almost all polymers used in ammonia-sensitive materials. Secondly, all acidic gases including CO_2, SO_2, and HCl, but also vapors of organic acids such as acetic acid, will interfere once the sensor is loaded with ammonia. Hence, the specificity of such sensors is limited. One way to overcome

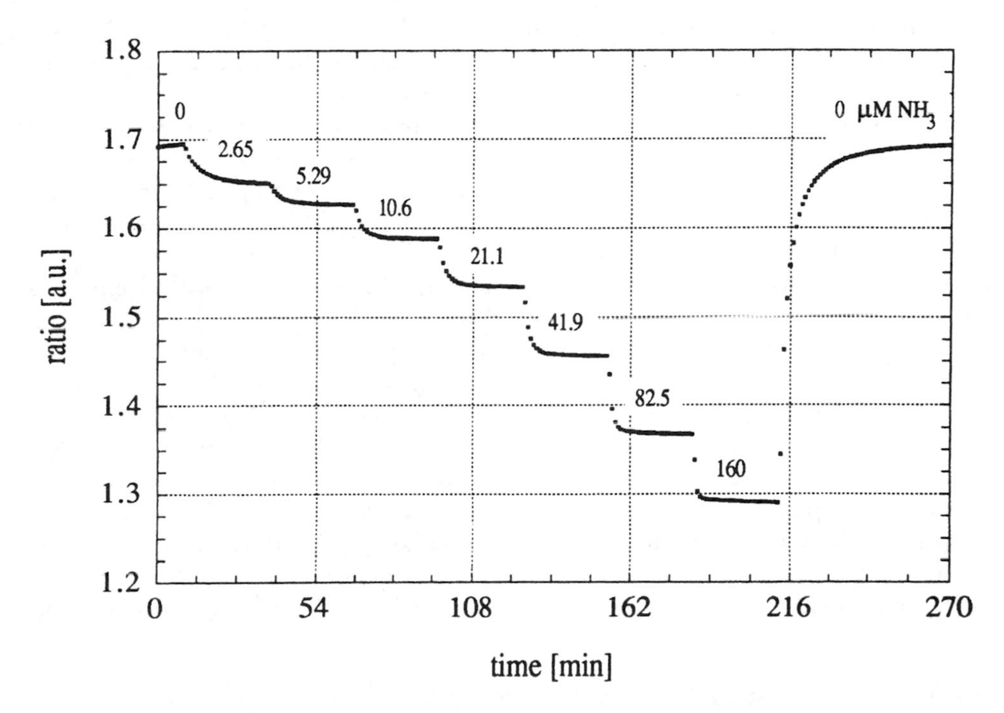

Figure 8.19 Response of an optical sensor to micromolar levels of dissolved ammonia. (*Source:* [167].)

interferences by acidic species is to make the sample strongly alkaline (if possible), which converts the acids into their nondiffusible salt forms.

8.8 ION SENSORS

Several schemes exist for sensing ions. They are based on either the use of 1) so-called chelators (i.e., dyes that bind a metal ion and thereby undergo a change in color; 2) ions carriers (i.e., uncolored, frequently cyclic ethers or esters that are capable of specifically binding (alkali) ions and to transport them into lipid sensor films); 3) chromoionophores (which, in essence, are a combination of a dye with an ion carrier, both contained in the same molecule); or 4) enzymes that undergo metal-induced change in their optical properties or activity.

8.8.1 Chelator-Based Ion Sensors

In this scheme, ions are determined, making use of so-called indicator dyes that undergo a binding reaction with ions, preferably of multiple charge. This reaction is accompanied by a change in the absorption or fluorescence of such "chelators." Numerous chelators exist [1–4,10], but most bind irreversibly or with a high or low pH so that they cannot be used for continuous sensing at near-neutral pH or at pHs, which are strongly different from the sample to be monitored, but rather act as single-shot probes. The respective sensor materials are obtained by immobilizing an indicator dye in an ion-permeable matrix such as cellulose or a hydrogel. A major disadvantage is based on the fact that for practically each ion, a different dye, and hence a different analytical wavelength, has to be applied.

Oehme & Wolfbeis have reviewed the state of the art in optical probing (as opposed to continuous sensing) of heavy metals (HMs) [10]. Aside from reporting on existing probes for the main group HMs (mainly copper, zinc, cadmium, mercury, silver) and for the transition HMs (Fe, Cr, Mn, Co, Ni), they also discuss unspecific probes (i.e., sensors for *total* HMs). Unfortunately, practically all existing sensors for HMs are different in terms of dye (i.e., analytical wavelengths), method of immobilization, and polymeric support, a fact that is highly disadvantageous and does not allow simple optoelectronic sensor systems to be designed that can be applied to all sensor chemistries. A uniform protocol would be highly desirable, but is unlikely to exist. A detailed discussion of all these sensor materials-which have found their most widespread application in the form of tests strips-is, however, beyond the scope of this chapter.

8.8.2 Sensors Based on the Use of Ion Carriers and Chromoionophores

No chelators are available for the clinically important alkali ions including potassium, sodium, and lithium, to work at pH 5–8 and to cover the clinical ranges, which are 110–180 mM for sodium, 1–10 mM for potassium, 0.4 to 2 mM for calcium, and 50 to 170 mM for chloride. While certain probes have become available in recent years from commercial sources, these are designed for cytological studies where much lower ion concentrations are encountered. Hence, these probes cannot be used to measure ions in extrastitial fluids. Hence, other sensing schemes have to be applied. The most general approaches make use of organic "hosts," capable of binding an ion (the "guest") inside its cavity or cyclic structure. If incorporated into a polymeric matrix (such as plasticized pvc), the host may even extract the guest from an aqueous sample phase. Typical hosts for use in ion sensing include the natural antibiotic valinomycin (which binds potassium ion) and numerous synthetic carriers (such as crown ethers, podands, and coronands) that organic chemists have synthesized and are known to bind alkali and earth alkali ions [16,171,172].

8.8.2.1 Sensors Based on Ion Exchange and on Coextraction

While a cation can be extracted from an aqueous into a lipid phase by a guest carrier, the counterion (the anion) usually cannot, and the process therefore would come to a quick end for reasons of electroneutrality. If, however, at the same time a proton can be released from the membrane (in exchange for the cation), then a complete ion exchange may take place ("cation in, proton out"; see Figure 8.20). An indicator dye contained in a nonpolar sensor membrane acts as the donor for the proton. On deprotonation, it undergoes a change in color that is related to the concentration of the ion.

An alternative scheme is referred to as coextraction. Here, a lipophilic anion such as chloride, salicylate, or the erythrosine anion is extracted into the lipid phase along with a cation, usually the proton. However, other cations may be extracted as well (via ionophors). The scheme has found its widest application for sensing anions such as chloride, bromide, and iodide, as well as nitrate. A schematic is shown in Figure 8.21.

Several approaches have been described for both the coextraction and the ion exchange process. Charlton and others have introduced two detection schemes. In the first [173,174], a plasticized PVC film containing valinomycin as the ion carrier is contacted with a sample to which a lipophilic and highly colored anion (such as erythrosin B) was added. When extracted into the PVC phase, potassium coextracts the a counterion, which in this case is the one of highest lipophilicity (i.e. erythrosin). As a result, the membrane turns pink. A linear relation exists between reflectivity at 550 nm and the potassium ion concentration over the 2 to 10-mM concentration range. This scheme forms the basis for the Ames Seralyzer solid-state potassium sensor strip.

In Charlton's second approach ([175]; also see [176, 177]), the basis for the assay is an ion exchange mechanism rather than a coextraction mechanism. The sensitive material is composed of PVC, a plasticizer, an ion carrier (such valinomycin), and a deprotonable dye. When valinomycin carries a potassium ion into the membrane, a proton is simultaneously released from a protonated dye (such as MEDPIN) contained in the membrane. On deprotonation, the dye undergoes a spectral change. This sensing scheme turned out to be extremely successful and has led to a number of commercial applications, including the Reflotron test and others. Fluorescent

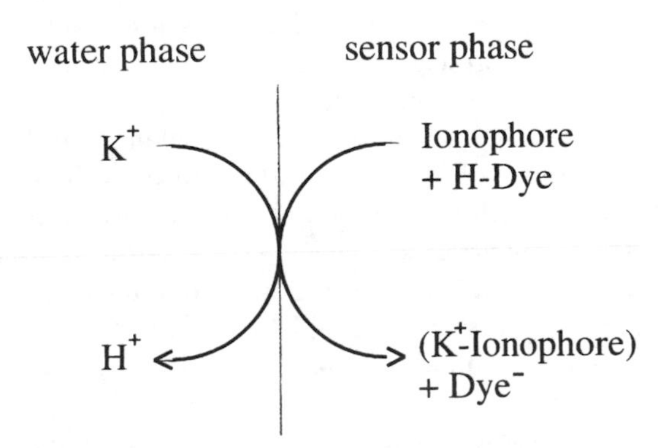

Figure 8.20 Schematic of the mechanisms leading to the exchange of a proton for a potassium ion inside a thin sensor membrane containing an ionophore and a protonated dye.

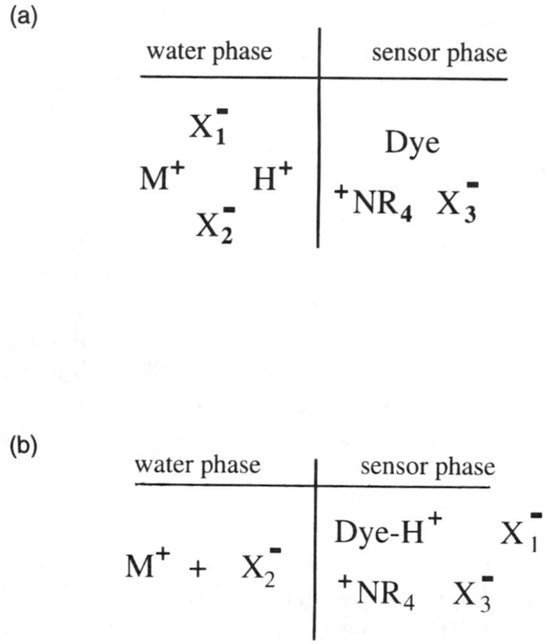

(a)

water phase	sensor phase
X_1^- M^+ H^+ X_2^-	Dye $^+NR_4$ X_3^-

(b)

water phase	sensor phase
M^+ + X_2^-	Dye-H^+ X_1^- $^+NR_4$ X_3^-

Figure 8.21 Distribution of charges in the water phase and the sensor phase (a) before and (b) after a coextraction process has occurred. The value X_1^- (unlike X_2^-) is highly lipophilic and, hence, extractable. For reasons of electroneutrality, a proton is coextracted, which protonates the indicator dye, which in turn results in a change in the optical properties.

modifications of the reflectometric methods for potassium using valinomycin as the carrier and based on either fluorescence energy transfer [178] or the inner filter effect using fluorescent beads [179] have been described.

The scope of both sensing schemes (i.e., ion exchange and coextraction) have been widely enlarged by the work of Simon and coworkers who discovered the beneficial effect of added lipophilic anions (such as the tetraphenylborates), which improve selectivity and response time [74,180–184], but also affect the work function. Numerous ions, gases, and neutral organic analytes can be analyzed now if appropriate carriers or receptors are found [185,180]. In fact, the sensing scheme is limited only by the selectivity and affinity of the molecular receptor.

More recently, synthetic crown ethers were found to represent an attractive alternative to natural ion carriers such as valinomycin and nonactin [183], and disposable ion-sensing probes for determination of blood electrolytes were obtained by incorporating respective ionophores ("carriers") along with a blue anionic dye into thin films of plasticized PVC. Again, the principle is based on ion exchange (in the case of cations) and coextraction (in the case of chloride), respectively. At constant pH, the color given by the sensor films can be related to the concentration of ions by equations derived from the Lambert-Beer and mass action laws [184].

By combining a dye moiety with an ionophor, so-called chromoionophors (or fluoroionophors) can be obtained. Both chromophores [11,186] and fluorophores [13,187] have been covalently linked to crown ethers, such that an ion-binding oxygen atom or nitrogen atom is part of the π-electron system of the dye. (In this context, the term crown ether refers to podands, coronands, and cryptands.) The event of binding the ion brings about a change in color, and-more strongly-in fluorescence, because the free electrons of the crown are now occupied by the alkali ion and cannot fully participate in the π-electron system of the chromophore or fluorophore. In most crown ethers, the binding constant (i.e., the turning point of the titration curve) strongly depends on the fraction of water in the solvent, because water acts as a competitive solvating agent for the crown.

The coextraction and ion-exchange sensing schemes are most promising because the selectivity to certain ions is exclusively governed by the carrier and the additives, while the dye (and hence the optical system) can be the same throughout. Their major disadvantage is the strong pH-dependence, which makes them not applicable to samples of unknown pH.

8.8.2.2 Sensors for Anions

Anions are more difficult to detect with high specificity than are cation because of the lack of adequate optical probes and the lack of selective carriers for anions. Usual carriers display a selectivity that follows the so-called Hofmeister pattern according to which lipophilic anions like perchlorate, salicylate, and iodide are more readily extracted than “hard” anions like fluoride and bicarbonate. Frequently used anion carriers include tetraalkylammonium salts, tri-alkyltin chloride, and certain porphyrins. They have been used both in coextraction-based sensors and in PSD-based sensors. Halides are known to quench the fluorescence of certain heterocyclic fluorophors and this has resulted in the design of respective sensors that cover the 1 to 100-mM concentration range (Figure 8.22) and hence are not suitable for trace analysis.

8.8.2.3 Sensors Based on Potential-Sensitive Dyes

An entirely different approach was introduced by Wolfbeis & Schaffar [188,189] in that the binding of ions, and their transport into a membrane, which results in the formation of an interface potential, was monitored by optical means using a potential-sensitive dye (PSD). The scheme has been extended [190]. Initially, Langmuir-Blodgett (LB) films were used as supports, because the potential (V/cm-1) is very high only if sensing films are very thin, typically a few nanometers in the case of LB films. Because of the poor stability of LB films, plasticized PVC was used in more

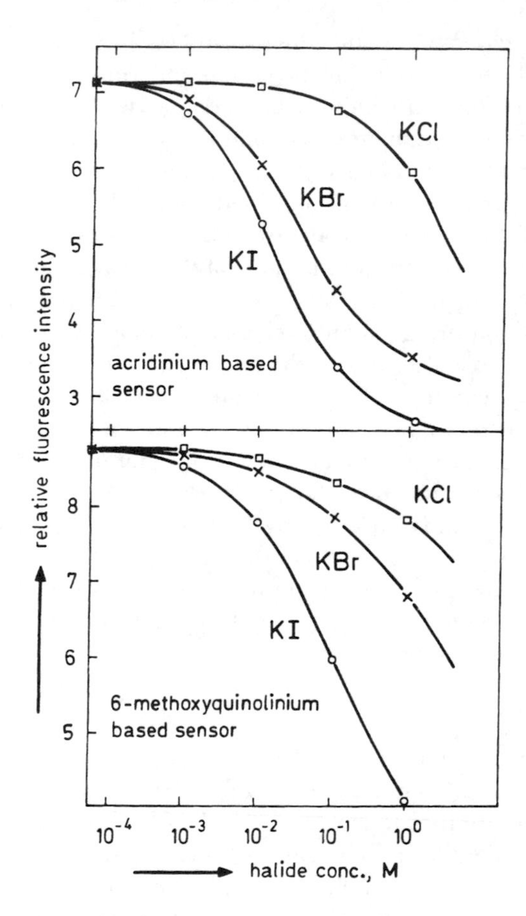

Figure 8.22 Dynamic quenching of the fluorescence of immobilized acridinium ion (top) and quinolinium ion (bottom) by chloride, bromide, and iodide, respectively.

recent work [22]. However, total signal changes do not exceed 50% in the best case, and a moderate pH-dependence still exists. The PSD-based scheme works best for such anions as nitrate [24], where the hydrophilicity-lipophilicity balance (HLB) seems to be better established (see Section 8.1.6). A schematic of the processes occurring at the sensor/sample interface is shown in Figure 8.7. A fairly specific carrier is known for nitrite anion and a sensor has been designed based on its use [191].

An apparently alternative sensing scheme for potassium was introduced by Kawabata and others [192,193] by making use of the same components except for another dye. It can be assumed, however, that the alkylacridinium dye used in this work in fact acts as a PSD and undergoes a voltage-induced partitioning. This is supported by the fact that all PSDs are both electrochromic and solvatochromic.

Sensors based on PSDs can be used to monitor, in a continuous and virtually pH-independent fashion, ions for which respective carriers are known. Both the plasticizer and the concentration of the borate counterion have a distinct effect on the work function [193]. The details of mechanisms that lead to a response of a PSD are still unknown and probably result from more than one single effect. Based on several findings, including turbidity effects, fast response, and relatively large signal changes, it was concluded that membranes have a microstructure that cannot be described by a model of a single homogeneous layer, but rather a complex microemulsion.

References

[1] Bishop, E., (Ed.), *Indicators*, Pergamon Press, Oxford, 1972.
[2] Cheng K. L., Ueno, K., and Imamura T., (Eds.), *Handbook of Organic Analytical Reagents*, CRC Press, Boca Raton, FL, 1982.
[3] Fernandez-Guttierez, A. and Munoz de la Pena, A., "Determination of Inorganic Substances by Luminescence Methods," in *Modern Luminescence Spectroscopy: Methods and Applications*, Vol. I, Schulman, S. G., (Ed.), John Wiley & Sons, New York, 1985.
[4] Haugland, R. D., (Ed.), *Handbook of Fluorescent Probes*, Mol. Probes, Inc., Eugene, OR, 1996.
[5] Leiner, M.J.P., and Wolfbeis, O. S., *pH Sensors*, in: *Fiber Optic Chemical Sensors and Biosensors*, Vol. 1, Wolfbeis, O. S., (Ed.), CRC Press, Boca Raton.
[6] Janata, J., *Anal. Chem.*, 59, 1987, p. 1351.
[7] Edmonds, T. E., et al., *Talanta*, 35, 1988, p. 103.
[8] Offenbacher, H., Wolfbeis, O. S., and Fürlinger, E., *Sens. Actuators*, 9, 1986, p. 73.
[9] Wolfbeis, O. S., and Offenbacher, H., *Sens. Actuators*, 9, 1986, p. 85.
[10] Oehme, I., Wolfbeis, O. S., *Mikrochim. Acta*, 1996, submitted.
[11] Dix, J. P., and Vögtle, F., *Angew. Chem. Intl. Ed. Engl.*, 17, 1978, p. 857.
[12] Sandanayake, K.A.R.S., and Sutherland, I. O., *Sensors & Actuators*, B11, 1993, p. 331.
[13] Wolfbeis, O. S., and Offenbacher, H., *Monatsh. Chem.*, 115, 1984, p. 647.
[14] Bourson, J., and Valeur, B., *J. Phys. Chem.*, 93, 1989, p. 3871.
[15] Bell, T. W., and Santora, V., *J. Am. Chem. Soc.*, 114, 1992, p. 8300.
[16] Czarnik, A. W., (Ed.) *Fluorescent Chemosensors for Ion and Molecule Recognition, ACS Symp. Ser.*, 538, Am. Chem. Soc., Washington (DC), 1992.
[17] Desvergne, J. P., et al., *Pure Appl. Chem.*, 64, 1992, p. 1231.
[18] Bryan, A. J., et al., Biosensors 4, 1989, p. 169.
[19] DeSilva, A. P., et al., *Angew. Chem.*, 107, 1995, p. 1889.
[20] DeSilva, A. P., Gunaratne, H.Q.N., and McCoy, C. P., *Nature*, 364, 1993, p. 42.
[21] Tsien, R. R., Chapter 8 in [16].
[22] Wolfbeis, O. S., *Sensors & Actuators* 29B, 1995, p. 140.
[23] Kawabata, Y.; Imasama, T., and Ishibashi, N., *Anal. Chim. Acta.*, 255, 1991, p. 97.
[24] Mohr, G. J., and Wolfbeis, O. S., *Anal. Chim. Acta*, 316, 1995, p. 239.
[25] Mohr, G. J., et al., *Fresenius J. Anal. Chem.*, in press.
[26] Zhujun, Z., and Seitz, W. R., *Proc. SPIE*, 906, 1988, p. 74.
[27] Lakowicz, J. R., *Principles of Fluorescence Spectroscopy*, Plenum Press, New York, 1984.
[28] Wolfbeis, O. S., *Oxygen Sensors*, in: *Fiber Optic Chemical Sensors and Biosensors*, Chapter 9, CRC Press, Boca Raton, FL, 1991.
[29] Draxler, S., et al., *J. Phys. Chem.*, 99, 1995, p. 3162.
[30] Klimant, I., and Wolfbeis, O. S., *Anal. Chem.*, 67, 1995, p. 3160.

[31] Leiner, M.J.P., et al., *Anal. Chim. Acta*, 205, 1988, p. 1.
[32] Bambot, S. B., et al., *Biosensors & Bioelectronics*, 10, 1995, p. 643.
[33] Keefe, G. O., et al., *Sensors & Actuators*, B29, 1995, p. 226.
[34] Holst, G. A ., Lübbers, D. W., and Voges, E., *Proc. SPIE*, 1885, 1993, p. 216.
[35] Papkovsky, D. M., et al., *Anal. Chem.*, 67, 1995, p. 4112.
[36] Urbano, E., Offenbacher, H., and Wolfbeis, O. S., *Anal. Chem.*, 56, 1984, p. 427.
[37] Wolfbeis, O. S., Posch, H. E., and Kroneis, H., *Anal. Chem.*, 57, 1985, p. 2556.
[38] Wolfbeis, O. S., and Trettnak, W., *Spectrochim. Acta*, 43B, 1987, p. 405.
[39] Bartl, H., and Falbe, J., (Eds.), *Houben Weyl's Methods of Organic Chemistry*, vol. E 20: *Macromolecular Materials*, Thieme Publ., Stuttgart, 1987.
[40] Davidson, R. L., *Handbook of Water-Soluble Gums and Resins*, McGraw-Hill, New York, 1980.
[41] Seymour, R. B., and Mark, H. F., (Eds.), *Applications of Polymers*, John Wiley & Sons, New York, 1988.
[42] Wall, L. A., *Fluoropolymers*, John Wiley & Sons, New York, 1972.
[43] Molyneux, R., *Water-Soluble Synthetic Polymers. Properties and Behaviors*, CRC Press, Boca Raton, 1984.
[44] Goethals, E. J., *Polymeric Amines and Ammonium Salts*, Pergamon Press, London, 1980.
[45] Batzer, H., (Ed.), *Polymeric Materials*, Thieme Publ., Stuttgart - New York, 1985.
[46] Yasuda, H., and Stannett, V., *Permeability Coefficients*, in *Polymer Handbook*, Brandrup, J., and Immergut, E. H., (eds.), John Wiley & Sons, New York, 1981, p. III–229.
[47] Robb, W. L., *Ann. New York Acad. Sci.*, 146, 1968, p. 119.
[48] Stern, S. A., Shah V. M., and Hardy, B. J., *J. Polym. Sci., Part B: Polym. Phys.*, 25, 1987, p. 1263.
[49] N. N., *Selection Guide to Dow Corning Organosilane Chemicals*, The Dow Corning Co., Midland, MI, 1986.
[50] Anderson, R., Arkles, B., and Larson, G. L., *Silicone Compounds. Register and Review*, Petrarch Systems, Bristol, PA, 1987.
[51] N.N., *The Pierce Handbook and General Catalog*, Pierce Chem. Co., Rockford, IL, 1987.
[52] Xu, W., et al., *Anal. Chem.*, 67, 1995, p. 3172.
[53] Papkovsky, D. M., Mohr, G. J., and Wolfbeis, O. S., unpublished results, 1995.
[54] Koncki, R., Mohr, G. J., and Wolfbeis, O. S., *Biosensors & Bioelectronics*, 10, 1995, p. 653.
[55] Leyden, D. E., and Collins, W. T., (Eds.), *Silylated Surfaces*, Gordon and Breach, New York, 1980.
[56] Plueddemann, E. P., *Silane Coupling Agents*, Plenum Press, New York, 1982.
[57] Wolfbeis, O. S., Reisfeld, R., and Oehme, I., "Sol-Gels and Sensors," in *Structure and Bonding*, in press.
[58] Rogowin, Z. A., and Galbraich, A., *Chemical Treatment and Modification of Cellulose*, Thieme Publ., Stuttgart, 1983.
[59] Saari, L., and Seitz, W. R., *Anal. Chem.*, 54, 1982, p. 821.
[60] Wegscheider, W., and Knapp, G., *CRC Crit. Rev. Anal. Chem.*, 11, 1981, p. 79.
[61] Seitz, W. R., *CRC Crit. Rev. Anal. Chem.*, 19, 1988, p. 135.
[62] Mohr, G., and Wolfbeis, O. S., *Anal. Chim. Acta*, 292, 1994, p. 41.
[63] Werner, T., and Wolfbeis, O. S., *Fresenius J. Anal. Chem.*, 346, 1993, p. 564.
[64] Leiner M.J.P., and Hartmann P., *Sensors & Actuators*, B11, 1993, p. 281.
[65] Peppas, N. A., (Ed.), *Preparation, Methods and Structures of Hydrogels;* CRC Press, Boca Raton, 1986.
[66] Ratner, B. D., "Biomedical Applications of Hydrogels: A Critical Appraisal," in *Biocompatibility of Clinical Implant Materials*, Vol. 2, CRC Ser. in Biocompatib., Wiliams D. F., Ed., CRC Press, Boca Raton, FL, 1981, p. 45.
[67] Guilbault, G. G., *Analytical Uses of Immobilized Enzymes*, Dekker, New York - Basel, 1994.
[68] Carr, P. W., and Bowers, L. D., *Immobilized Enzymes in Analytical and Clinical Chemistry*, John Wiley & Sons, New York, 1980.

[69] Peterson, J. I., et al., *Anal. Chem.*, 52, 1980, p. 864.
[70] Avnir, D., et al., 1994, in: *Sol-Gel Optics - Processing and Applications*, Klein, L. C. (Ed.), *Kluwer*, USA, 1994, Chapter 23, p. 539.
[71] Meier, B., et al., *Sensors & Actuators*, 29B, 1995, p. 240.
[72] Wolfbeis, O. S., Leiner, M.J.P., and Posch, H. E., *Mikrochim. Acta*, III, 1986, p. 359.
[73] Carraway, E. R., Demas, J. N., and DeGraff, B. A., *Anal. Chem.*, 63, 1991, p. 332.
[74] Seiler, K., and Simon, W., *Anal. Chim. Acta*, 266, 1992, p. 73.
[75] Charlton, S. C., Fleming, R. L., and Zill, A., *Clin. Chem.*, 28, 1982, p. 1857.
[76] Morf, W. E., et al., *Pure Appl. Chem.*, 61, 1989, p. 1613.
[77] Mills, A., and Chang, Q., *Sensors & Actuators*, B21, 1994, p. 83.
[78] Werner, T., Klimant, I., and Wolfbeis, O. S., *Analyst*, 120, 1995, p. 1627.
[79] Kirkbright, G. F., Narayanaswamy, R., and Welti, N. A., *Analyst*, 109, 1984, p. 15.
[80] Narayanaswamy, R., and Sevilla, F. S., *Anal. Chim. Acta*, 189, 1986, p. 365.
[81] Zhujun, Z. and Seitz, W. R., *Anal. Chim. Acta*, 160, 1984, p. 47.
[82] Harper, B. G., *Anal. Chem.*, 47, 1975, p. 348.
[83] Kadin, H., *Anal. Lett.*, 17, 1984, p. 1245.
[84] Munkholm, C., et al., 58, 1986, p. 1427.
[85] Wolfbeis, O. S., et al., *Mikrochim. Acta*, 108, 1992, p. 133.
[86] MacCraith, B. D., et al., *Sol-Gel Sci Technol.*, 2, 1994, p. 661.
[87] Grattan, K.T.V., et al., *Sensors & Actuators*, A26, 1991, p. 483.
[88] Neurauter, G., Diploma work; KFU Graz, 1996.
[89] Mohr, G. J., et al., unpubl. results, 1995.
[90] Schaffar, B.P.H., and Wolfbeis, O. S., *Proc. SPIE*, 990, 1989, p. 122.
[85] Wolfbeis, O. S., et al., *Fresenius' Z. Anal. Chem.*, 314, 1983, p. 119.
[91] De Marcos, S., and Wolfbeis, O. S., *Sensors & Mat.*, submitted 1996.
[92] Wolfbeis O. S., (Ed.), *Fiber Optic Chemical Sensors and Biosensors*, Vol. 1, Chapter 2, CRC Press, Boca Raton, FL.
[93] Suidan, J. S., et al., *Clin. Chem.*, 29, 1983, p. 1566.
[94] Monici, M., et al., *Proc. SPIE*, 798, 1987, p. 294.
[95] Boisde, G., and Prez, J. J., *Proc. SPIE*, 798, 1987, p. 238.
[96] Moreno, M. C., et al., *J. Mol. Struc.*, 143, 1986, p. 553.
[97] Goldfinch, M. J., and Lowe, C. R., *Anal. Biochem.*, 138, 1984, p. 430.
[98] Grattan, K.T.V., Mouaziz, Z., and Palmer, A. W., *Biosensors*, 3, 1987/88, p. 17.
[99] Attridge, J. W., Leaver, K. D. and Cozens, J. R., *J. Phys. E: Sci. Instrum.*, 20, 1987, p. 548.
[100] Boisde, G., Blanc, F., and Perez, J. J., *Talanta*, 35, 1988, p. 75.
[101] Kostov, Y., and Tzonkov, S., *Anal. Chim. Acta*, 280, 1993, p. 15.
[102] Bacci, M., Baldini, F., and Scheggi, A. M., *Anal. Chim. Acta*, 207, 1988, p. 343.
[103] Weigl, B. H., et al., *J. Biotechnol.*, 32, 1994, p. 127.
[104] Gehrich, J. L., et al., *IEEE Trans. Biomed. Eng.*, 33, 1986, p. 117.
[105] Kawabata, Y., et al., *Anal. Sci.*, 3, 1987, p. 7.
[106] Fuh, M.R.S., et al., *Analyst*, 112, 1987, p. 1159.
[107] Agayn, V. I., and Walt, D. R., *Biotechnology*, 11, 1993, p. 726.
[108] Tan, W., Shi, Z. Y., and Kopelman, R., *Sensors & Actuators*, B28, 1995, p. 157.
[109] Jordan, D. M., and Walt, D. R., *Anal. Chem.*, 59, 1987, p. 437.
[110] Draxler, S., Lippitsch, M. E., Leiner, M.J.P., *Sensors & Actuators*, B11, 1993, p. 421.
[111] Thompson, R. B., and Lakowicz, J. R., *Anal. Chem.*, 65, 1993, p. 853.
[112] Draxler, S., and Lippitsch, M. E., *Sensors & Actuators*, B29, 1995, p. 199.
[113] Kautsky, H., *Trans. Faraday Soc.*, 35, 1939, p. 216.
[114] Kautsky, H., and Müller, G. O., *Z. Naturforsch.*, 2A, 1947, p. 167.
[115] Zakharov, I. A., and Grishaeva, T. I., *Zh. Prikl. Spektrosk.*, 57, p. 1240, 1984.

[116] Shah, R., Margerum, S. C., and Gold, M., *Proc. SPIE*, 906, 1988, p. 65.
[117] Di Marco, G., Lanza, M., and Campagna S., *Adv. Mat.*, 7, 1995, p. 468.
[118] Lübbers, D. W., and Opitz, N., *Z. Naturforsch.*, 30C, 1975, p. 532.
[119] Bergman, I., *Nature*, 218, 1968, p. 396.
[120] Liu, Y. M., Pereiro-Garcia, R., Sanz-Medel, A., *Anal. Chem.*, 66, 1994, p. 836.
[121] Charlesworth, J. M., *Sensors & Actuators*, B22, 1994, p. 1.
[122] Lee, W.W.S., et al., *J. Mat. Chem.*, 3, 1993, p. 1031.
[123] Twarowski, A. J., and Good, L., *J. Phys. Chem.*, 91, 1987, p. 5252.
[124] Del Bianco, A., et al., *Sensors & Actuators*, B11, 1993, p. 347.
[125] Carraway, E. R., Demas, I. N., and DeGraff, B. S., *Anal. Chem.*, 63, 1991, p. 332.
[126] Carraway, E. R., et al., *Anal. Chem.*, 63, 1991, p. 337.
[127] Demas, I. N., DeGraff, B. A., and Xu, W., *Anal. Chem.*, 67, 1995, p. 1377.
[128] Lee, E. D., Werner, T. C., and Seitz, W. R., *Anal. Chem.*, 59, 1987, p. 279.
[129] Kroneis, H. W., and Marsoner, H. J., *Sensors & Actuators*, 4, 1983, p. 587.
[130] Wolfbeis, O. S., et al., *Mikrochim. Acta*, I, 1984, p. 153.
[131] Cox, M. E., and Dunn, B., *Appl. Optics*, 24, 1985, p. 2114.
[132] Sharma, A., and Wolfbeis, O. S., *Appl. Spectrosc.*, 42, 1988, p. 1009.
[133] Peterson, J. I., Fitzgerald, R. V., and Buckhold, D. K., *Anal. Chem.*, 56, 1984, p. 62.
[134] Okazaki, T., Imasaka, T., and Ishibashi, N., *Anal. Chim. Acta*, 209, 1988, p. 327.
[135] Moreno-Bondi, M. C., et al., *Anal. Chem.*, 62, 1990, p. 2377.
[136] Bacon, J. R., and Demas, J. N., *Anal. Chem.*, 59, 1987, p. 2780.
[137] Papkovsky, D. B., Ponomarev, G. V., and Wolfbeis, O. S., *Spectrochim. Acta*, submitted.
[138] Lübbers, D. W., Köster, T., and Holst, G. A., *Proc. SPIE*, 2388, 1995, p. 507.
[139] Gruber, W. R., O'Leary, P., and Wolfbeis, O. S., *Proc. SPIE*, 2388, 1995, p. 148.
[140] Holst, A. G., and Lübbers, D. W., *Sensors & Actuators*, B29, 1995, p. 231.
[141] Zhujun, Z., and Seitz, W. R., *Anal. Chim. Acta*, 160, 1984, p. 305.
[142] Heitzmann, H. A., and Kroneis, H. W., US Pat. 4,557,900, 1985.
[143] Marazuela, M. D., Moreno-Bondi M. C., and Orellana, G., *Sensors & Actuators*, B29, 1995, p. 126.
[144] Munkholm, Ch., Walt, D. M., and Milanovich, F. P., *Talanta*, 35, 1988, p. 109.
[145] Wolfbeis, O. S, et al., *Anal. Chem.*, 60, 1988, p. 2028.
[146] Vurek, G. G., et al., *Proc. Fed. Am. Soc. Exp. Biol.*, 41, 1982, p. 1484.
[147] Hirschfeld, T., et al., *J. Lightwave Technol.*, 5, 1987, p. 1027.
[148] Leiner, M.J.P., *Sensors & Actuators*, B29, 1995, p. 169.
[149] Weigl, B. H., et al., *Anal. Chim. Acta*, 282, 1993, p. 335.
[150] Moreno-Bondi, M. C., et al., *Proc. SPIE*, 1368, 1990, p. 157.
[151] Mills, A., Chang, Q., and McMurray, N., *Anal. Chem.*, 64, 1992, p. 1383.
[152] Mills, A., and McMurray, N., *Analyst*, 118, 1993, p. 839.
[153] Mills, A., and Chang, Q., *Sensors & Actuators*, B21, 1994, p. 83.
[154] Mills, A., and Chang, Q., *Anal. Chim. Acta*, 285, 1994, p. 113.
[155] Weigl, B. H., and Wolfbeis, O. S, *Sensors & Actuators*, B28, 1995, p. 151.
[156] Weigl, B. H., and Wolfbeis, O. S, *Anal. Chim. Acta*, 302, 1995, p. 249.
[157] Kawabata, Y., et al., *Anal. Chim. Acta*, 219, 1989, p. 223.
[158] Giuliani, J. F., Wohltjen, H., and Jarvis, N. L., *Opt. Lett.*, 8, 1983, p. 54.
[159] Blyler, L. L., et al., *Eur. Pat. Appl.*, 292, 1988, p. 207.
[160] Arnold, M. A., and Ostler, T. J., *Anal. Chem.*, 58, 1986, p. 1137.
[161] Wolfbeis, O. S., and Posch, H. E., *Anal. Chim. Acta*, 185, 1986, p. 321.
[162] Kov, S., and Arnold, M. A., *Anal. Chem.*, 64, 1992, p. 2438.
[163] Caglar, P., and Narayanaswamy, N., *Analyst*, 112, 1987, p. 1285.
[164] Potyrailo, R. A., and Talanchuk, P. M., *Sensors & Actuators*, B21, 1994, p. 65.
[165] Shahriari, M. R., Zhou, Q., and Sigel, G. H., *Optics Lett.*, 13, 1988, p. 407.

[166] Zhou, Q., et al., *Appl. Optics*, 28, 1989, p. 2022.
[167] Trinkel, M., et al., *Anal. Chim. Acta*, 320, 1996, p. 235.
[168] Werner, T., Klimant, I., and Wolfbeis., O. S., *J. Fluoresc.*, 4, 1994, p. 41.
[169] Brandenburg, A., et al., *Mikrochim. Acta*, 121, 1995, p. 95.
[170] Nakane, N., Sugata K., and Nagashima, K., *Anal. Chim. Acta*, 302, 1995, p. 201.
[171] Beer, P. D., "Molecular and Ionic Recognition by Chemical Methods," in: Edmonds, T. E., (Ed.), *Chemical Sensors*, Blackie, London, 1987.
[172] Lakowicz, J. R., (Ed.) *Topics in Fluorescence Spectroscopy, vol. IV: Probe Design and Chemical Sensing*, Plenum Press, New York, 1994.
[173] Charlton, S. C., Fleming, R. L., and Zipp, A., *Clin. Chem.*, 28, 1982, p. 1857.
[174] Kumar, A., et al., *Clin. Chem.*, 34, 1988, p. 1709.
[175] Charlton, S. C., et al., US Pat. 4,645,744, 1987.
[176] Ng, R. H., Sparks, K. M., and Statland, B. E., *Clin. Chem.*, 38, 1992, p. 1371.
[177] Gibb, I., *J. Clin. Pathol.*, 40, 1987, p. 298.
[178] Roe, J. N., Szoka F. C., and Verkman, A. S., *Analys,t* 115, 1990, p. 353.
[179] He, H., et al., *Anal. Chem.*, 65, 1993, p. 123.
[180] Rosatzin, Th., et al., *Anal. Chim. Acta*, 280, 1993, p. 197.
[181] Spichiger, U., et al., *Sensors & Actuat.*, B11, 1993, p. 1.
[182] Eugster, R., et al., *Anal. Chem.*, 63, 1991, p. 2285.
[183] Eugster, R., Spichiger, U. E., and Simon, W., *Anal. Chem.*, 65, 1993, p. 689.
[184] Suzuki, K., et al., *Anal. Chim. Acta*, 237, 1990, p. 155.
[185] Morf, W. E., et al., *Pure Appl. Chem.*, 61, 1989, p. 1613.
[186] Hisamato, H., Miyashita N., and Suzuki, K., *Sensors & Actuators*, B29, 1995, p. 378.
[187] Garcia, R. P., et al., *Clin. Chim. Acta*, 207, 1992, p. 31.
[188] Wolfbeis, O. S., and Schaffar, B.P.H., *Anal. Chim. Acta*, 198, 1987, p. 1.
[189] Schaffar, B.P.H., Wolfbeis, O. S., and Leitner, A., *Analyst*, 113, 1988, p. 693.
[190] Shimomura, M., et al., *Sensors & Actuat.*, B13, 1993, p. 629.
[191] Mohr, G. J., and Wolfbeis, O. S., *Analyst*, submitted May 1996.
[192] Kawabata, Y., et al., *Anal. Chem.*, 62, 1990, p. 1531 and 2054.
[193] Kawabata, Y., Yamamoto, T., and Imasaka, T., *Sensors & Actuat.*, B11, 1993, p. 341.

Chapter 9

Dynamic Light Scattering and Its Application in Concentrated Suspensions

C. J. Lloyd, E A Perkins and R. J. G Carr
Centre for Applied Microbiology and Research (CAMR), England

This chapter describes a variant of dynamic light scattering known as diffusing wave spectroscopy (DWS), which is capable of monitoring particle size in highly concentrated and optically opaque samples (> 40% (by volume) v/v) [1,2]. Our work [3,4] has shown the reproducibility and robustness associated with the homodyne method and indicates its potential in industrial process control.

9.1 DYNAMIC LIGHT SCATTERING IN DILUTE SUSPENSIONS

Dynamic light scattering (DLS) analysis of dilute (e.g., < 0.05% (v/v)) suspensions of submicron particles is a well-established technique for monitoring the mean sphere-equivalent hydrodynamic radius and obtaining an indication of particle size distribution and asymmetry.

DLS is based on the measurement of the Doppler frequency or dynamic phase shift imparted to light when it is scattered by a suspension of particles undergoing Brownian motion. The suspended particles are in constant random thermal motion due to collisions with energetic molecules of the liquid. The instantaneous particle speed, after collision with liquid molecules, will be linearly dependent on particle density, due to a momentum transfer. However, over timescales where a significant number of particle/liquid-molecule collisions have taken place (> nsecs), the effect of density is negligible and the diffusion rate is dependent on the frequency of collisions, which depends on particle area, and so is a function of the radius [2].

Smaller particles will travel larger distances between each liquid molecule collision and thus the average displacement in a given time is greater. The Doppler shift is typically only a few hertz, compared to the light's original frequency of 10^{14} Hz, making direct analysis via filtering impractical. Forrestor and others [5] described how small shifts may be analyzed using the principle of beating. Mixing light of two frequencies gives a resultant beat frequency that is the difference of the two; hence, mixing the light scattered by a suspension with scattered light obtained directly from the source gives a readily measurable signal. Forrestor [6] extended the use of optical beating as a spectroscopic tool, which was later utilized by Cummins and others [7,8] to measure the mean size of particles in suspensions.

In many earlier works, the signal was analyzed by the use of scanning filters; either Fabry-Perot types for optical predetection filtering or, more economically, signal analyzers for postdetection electronic filtering. The latter experiments have generally employed the classical heterodyne technique, where the scattered light is mixed with light from a different source, as shown in Figure 9.1(a). Using sources of slightly different frequency produces a low-frequency intensity spectrum at the detector, so suffers strong "1/f" noise. Also, a significant practical difficulty of classical heterodyne systems is in ensuring that the frequency between the two sources remains constant, fluctuations in either of the sources giving rise to a loss in coherence and a reduced signal to noise ratio (SNR).

Early workers in DLS alleviated the problem of frequency matching two separate sources by mixing the scattered light from a single source with a portion of the original beam, as shown in Figure 9.1(b), though the signal is now situated close to 0 Hz, and so will still suffer from high 1/f noise. Somewhat confusingly, DLS users continued to refer to this approach as a heterodyne technique, though in communications and other fields it is classically known as homodyne. Nevertheless, both methods suffer from noise due to environmental changes such as vibration, heating etc. Random variations in either of the optical paths leads to a false signal and a reduced signal to noise level.

These pathlength difference problems were shown to be capable of being significantly reduced by mixing scattered light with other scattered light that has traveled a similar path, as shown in Figure 9.1(c), giving a method that proves to be the most environmentally robust and least complex. Perversely, the DLS community now calls this a homodyne method despite the fact that the signals have the same mean frequency, so the alternative classical description of autodyne or self-beat spectroscopy is less confusing.

9.1.1 Postdetection Correlation

Significant developments in DLS have occurred through utilization of the Weiner Kirchoff principle [9]. The intensity of the light scattered from a constantly illuminated

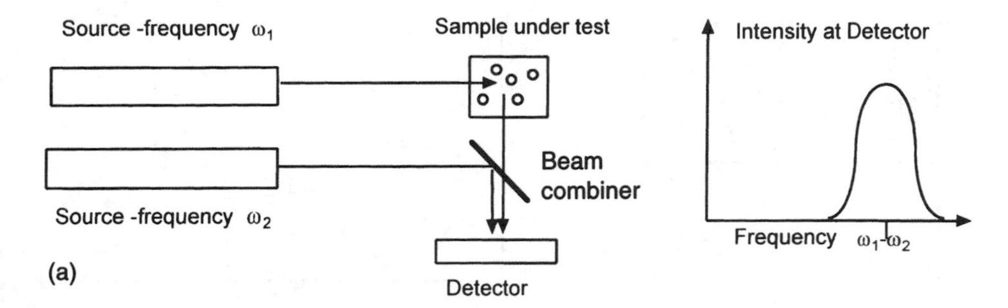

Figure 9.1 Mixing techniques: (a) classical heterodyne, (b) heterodyne, and (c) homodyne.

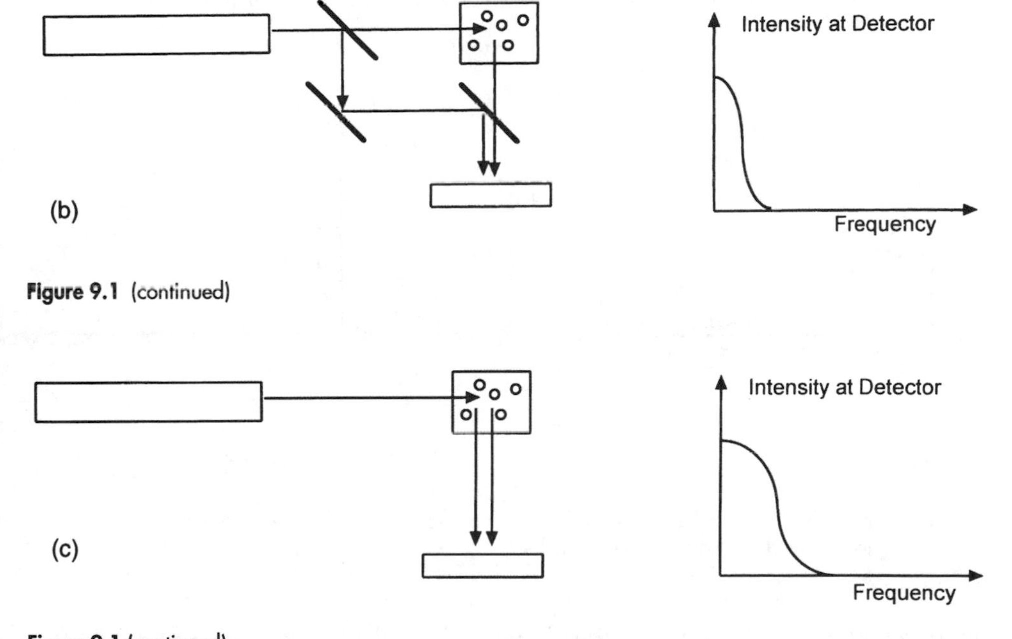

Figure 9.1 (continued)

Figure 9.1 (continued)

suspension fluctuates randomly as the movement of the particles gives rise to dynamic phase shifts between different optical paths through the system (constructive and destructive interference). This gives variations in the scattered intensity between 0 and $<I>^2$ and the rate of fluctuation is a measure of the diffusion coefficient of the particle. This value may be found accurately and rapidly by correlation, via the decay rate, Γ, (Figure 9.2) where $\Gamma = \delta\ [\log\ (I)]/\delta\tau$.

The Weiner Kirchoff principle is the formal statement that correlation (©) of the

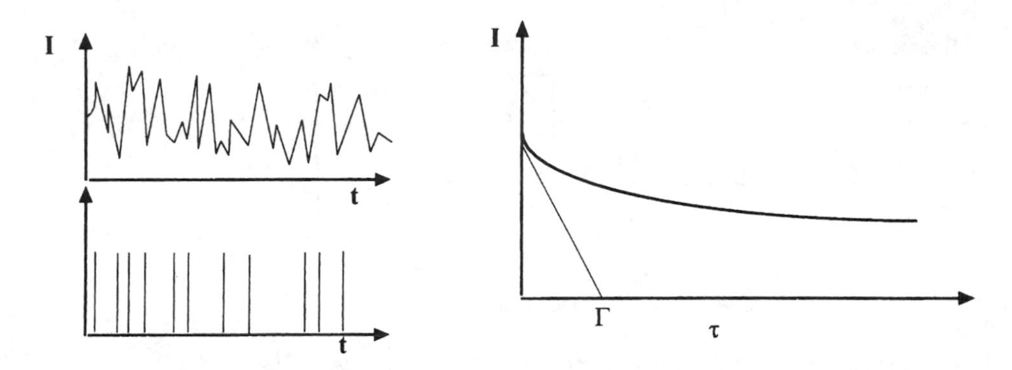

Figure 9.2 Electronic correlation: the intensity variation at the detector may either be measured as an analog signal (top LH) or a digital pulse train (lower LH). Correlation of the signal in the time domain gives an exponential function, the characteristic decay (Γ) of which is a measure of the particle's diffusion coefficient.

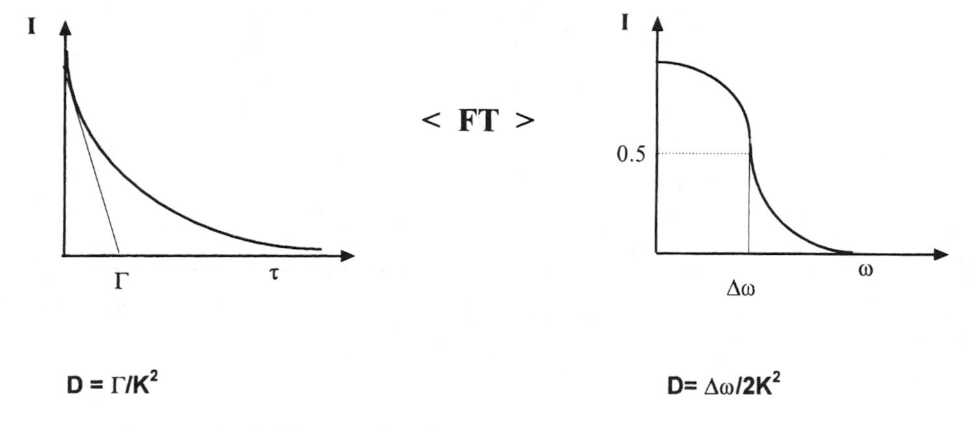

Figure 9.3 The Weiner Kirchoff theorem: the relationship between the amplitude correlation and intensity frequency spectrum via the Fourier transform.

temporal amplitude-fluctuations function, $f(t)$, is equivalent to a measure of the frequency power spectrum ($|F(v)|^2$), the two being a Fourier Transform pair (9.1), as shown in Figure 9.3.

$$T\{|F(v)|^2\} = f(t)\ C\copyright\ f^*(t) \tag{9.1}$$

Hence, analysis of the frequencies present within a correlation may be carried out by measurement of the correlation decay (Γ). However, it is not possible to measure amplitude fields, $f(t)$, directly. Photon detection provides only measurement of the square of the amplitude envelope or optical intensity. The resulting intensity correlation $g^{(2)}$ may be related to the amplitude correlation $g^{(1)}$ by (9.2), where I_s is the intensity of the scattered light, I_L is the intensity of the unscattered light, and I_T the total intensity, $I_s + I_L$,

$$g^{(2)}(t) = 1 + (<I_s>/<I_T>)^2[g_s^{(2)}(t) - 1] + (<I_s>I_L)/<I_T>^2 g_s^{(1)}(t) \quad (9.2)$$

The situation $I_s << I_L$, the heterodyne approximation (9.2a), allows a direct measure of the amplitude correlation. While an intensity signal is measured and correlated, the scattering signal is present in only a single path, as shown in Figure 9.1(b), resulting in information on the amplitude fields:

$$g^{(2)}(t) = + (<I_s>I_L)/<I_T>^2 g_s^{(1)}(t) \quad (9.2a)$$

In the homodyne case (9.2b), $I_L = 0$, the scattered light has mixed with other scattered light, as shown in Figure 9.1(c), during detection and no information on the amplitude fields is available directly.

$$g^{(2)}(t, t + t) = 1 + 1(g_s^{(2)}(t + t) - 1) \quad (9.2b)$$

The homodyne intensity correlation must be related to the amplitude correlation by means of the Siegert relationship (9.3), which is valid for Gaussian scattering fields and noncoupled particles:

$$g^{(2)}(t, t + t) = 1 + \left| g_s^{(1)}(t + t) \right|^2 \quad (9.3)$$

It is necessary to normalize the correlation to ensure the measurement is not a function of the incident intensity. The baseline can be either calculated or measured. Measurement of the baseline requires the correlation to be measured at long delay times and is subject to statistical error, while calculation assumes all noise is uncorrelated, which may give rise to significant bias. In the homodyne case, for an ideal laser and detector, the correlation intersect will be double the baseline. However, in the heterodyne case the correlation height will only be a small fraction of the baseline, as the majority of the detected light, I_L, contains no information on scattering (Figure 9.4). The reduced intersect makes heterodyne measurements more susceptible to errors in the calculated/measured baseline.

For a dilute and monodisperse sample, where the photons have undergone only single scattering, the normalized correlation function G may be described by an exponential (9.4). The free diffusion coefficient may be found directly from the correlation decay, taking into account the effect of wavelength (9.5). The mean hydrodynamic sphere-equivalent size is then calculated by equating the thermal energy kT of the particles with an effective frictional drag force on the scattering particles due to solvent interactions, via the Stokes-Einstein relationship (9.6).

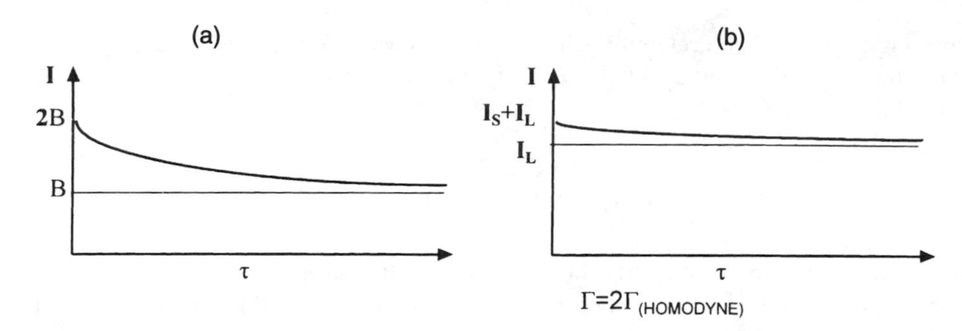

Figure 9.4 Heterodyne and homodyne correlations: (a) the homodyne correlation has an intersect = 2x baseline while in (b) the heterodyne case the correlation height is proportional to (IS/IL) of the baseline, typically 1/100.

$$\Gamma = [\log(G^1(t))]/t \tag{9.4}$$

$$D = \Gamma/K^2 \tag{9.5}$$

$$R = kT/6\Pi\eta D \tag{9.6}$$

where T is temperature, K is a scattering vector, k is Boltzman's constant, and η is viscosity.

If the particle suspension under analysis is polydisperse (i.e., exhibiting a range of sizes) or of variable optical properties, the resultant correlation will be a summation of several exponentials. Each particle type, within the suspension, generates a separate exponential decay and limited information on the particle size distribution (PSD) can be recovered by deconvolution of the complex function (e.g., by curve fitting or Fourier transform techniques). More simply, a single value describing the particle size distribution width can be estimated from the degree of multiexponentiality of the measured correlation function.

When the particle diameter is significantly smaller that the wavelength of light used (i.e., $\lambda/20$), the intensity of light scattered varies as the 6th power of the particle radius, making DLS particularly sensitive to larger which, when present as contaminating particles of dust, can totally mask the desired signal.

9.1.2 Photon Correlation Spectroscopy (PCS)

The full benefits of electronic correlation are best realized when digital circuits are employed [10]. Photon counting allows the advantages due to the quantum nature of light to be utilized to give an almost ideal SNR. At low light fluxes, typical of light scattering, the individual photon arrivals may be measured using low noise, high gain detectors such as photomultiplier tubes (PMTs) or the more recently developed solid-state devices, avalanche photodiodes (APDs). Via a pulse-amplifier discriminator, a pulse train of 1's and 0's can be correlated rather than a varying voltage (Figure 9.2), giving a dramatic increase in SNR.

The inclusion of this technology into DLS led to the technique of photon correlation spectroscopy (PCS), which has been used widely to study a range of materials [11–14]. This type of measurement is typically carried out with the detector aligned to receive light at 90 degrees to the plane polarized source, which reduces the effects of dust contamination and which can often lead to disproportionately large intensity fluctuations and depolarization (which can destroy the measurement) [15]. The degree of depolarization introduced by a particle in suspension is dependent on its size, shape, and scattering angle, and Cummins and others [16] have described the use of forward scattering in PCS to give an estimate of the aspect ratio of nonspherical particles.

The requirement to ensure that each photon is scattered by only a single particle restricts the use of the method to low concentrations, which can be a significant drawback (e.g., PCS can rarely be used for "online" industrial applications). Some work has been carried out on chromatography eluants, which are usually low concentration samples, comprising peaks of monodisperse macromolecular particles eluting separately and sequentially off the separation column (Carr and others [17,18]). However, in most cases, an industrial sample must be diluted significantly before measurement by PCS. Such extreme dilution raises questions about the validity of extrapolation of the measurements to industrial process concentrations. Furthermore, for detection of very small particles of weak refractive index, higher power lasers (e.g., > 100 mW) are normally required, significantly increasing safety requirements and costs.

9.1.3 The Effect of Increasing Concentration on Particle Diffusion

PCS theory assumes the infinite dilution limit, where only single scattering occurs and scattering particles do not interact. In practice, this typically limits the maximum sample concentration to below 10^{-4}% (v/v). If the concentration is increased beyond this, then multiple scattering may occur. Detection of photons that have been scattered from two or more particles leads to an increase in the frequency shift and hence the scattered linewidth (decrease in Γ). In this regime, analysis is difficult and requires significant prior knowledge of the system. The resulting measured diffusion coefficient, D_m, is a strong function of concentration, which can lead to increased measurement inaccuracy in poorly characterized samples. However, the rate of increase of D_m with concentration begins to reduce above 1 to 5% (v/v) as particle-particle interactions hinder their free diffusion. This leads to a turning point, with a maximum D_m of between 5 and 25% v/v, depending on measurement technique, after which D_m decreases with increasing concentration.

Using tracer particles added to samples containing high concentrations of index-matched material, Ottewell and others [19] have shown that at the glass transition point (≈50% (v/v)), D_m reduces to zero as the particles can no longer freely diffuse. When multiple scattering techniques are employed, D_m may be significantly reduced at the glass transition point, but a signal is still present. For PCS measurements at 90 degrees, the particles must move a distance comparable to half a wavelength to produce a π phase shift. However, because DWS measures the summation of many particle movements, to obtain the necessary π shift in phase, each particle needs individually to have moved much less, perhaps only a few nm, allowing dense gels and near-solid materials (in which diffusion is heavily constrained) to be monitored [20].

9.2 MULTIPLE SCATTERING TECHNIQUES

A number of light-scattering techniques have been utilized to measure concentrated fluids. The terms fiber-optic dynamic light scattering (FODLS) and fiber-optic quasi-elastic light scattering (FOQELS) have been used to describe numerous optical configurations based on temporal measurement of the signal, while the term fiber-optic Doppler anemometry (FODA) is used to describe systems employing signal analyzers.

Under ideal conditions, both techniques would give identical information (i.e., as in (9.1)), so only the benefits of different optical setups are considered here (a full review of the majority of systems has been given by MacFadyen and Jennings [20]). Most instruments have operated in retroreflection, often with a single multimode fiber-optic probe, and the outgoing and return optical paths are typically separated by means of a beamsplitter [21] or mirror with a hole in the center [22]. Bifurcated fiber bundles have also been utilized [23], as have monomode fibers [24] and closely spaced fiber pairs operating close to retroreflection to define a scattering cell [25].

9.2.1 Use of Fibers

The use of backscatter systems has allowed the possibility of a dip-in probe, employing economic, flexible, and miniature fiber optics. There have been significant discussions on the relative benefits and disadvantages of multi and monomode fiber for PCS. However, any arguments in favor of multimode fiber reduce significantly in multiple scattering techniques, especially for use in industrial environments.

While multimode fiber allows a significant amount of light to be guided, the vast majority of this must be excluded by means of a pinhole to ensure only a single coherence area is monitored; otherwise, any slight thermal or environmental fluctuation will tend to cause modal noise ,which can scramble the signal. Brown [25,26] has shown that a monomode fiber acts as an ideal single-coherence area detector. No extra alignment or optics are required and the final model is much simpler. Furthermore, the use of polarization-maintaining monomode fibers ensures a stable output even in noisy environments. Though single-mode fibers are more limited in the intensity they can transmit, multiple scattering techniques are rarely photon-limited and we have found a launch of only 2.5 mW of 633-nm HeNe laser light more than adequate for most applications.

9.2.2 Effect of Optical Mixing in Concentrated Systems

In single scattering techniques, the only real concern, when deciding on options for optical mixing, is whether the measurement of the amplitude correlation (hence absolute size) is of sufficient importance to warrant the added complexity. However, dense scattering systems give rise to very specific effects when heterodyne optical systems are employed, and these must be fully understood before meaningful data can be recovered.

9.2.2.1 Pathlength Aspects

Signals in typical heterodyne systems travel different optical paths, which can give rise to unacceptable sensitivity to vibration, temperature, or pressure fluctuations. Moreover, because it is not possible to define the photon path traveled in a multiple scattering system, the laser coherence requirement will increase significantly, particularly in the heterodyne case. While single-fiber emitter/detector probes overcome some of these problems, they can themselves cause further complications.

9.2.2.2 Weighting

All heterodyne techniques are, in practice, a heterodyne/homodyne mixture, as discussed by Bremer and others [27], and often require significant modeling [23,24]. Use of a purely heterodyne model requires that the standing oscillator be stable and accurately measured and that no homodyne signals are allowed to bias the results (homodyne signal decay rates are twice those of heterodyne). Furthermore, using a single fiber as both emitter and detector makes control and measurement of only the heterodyne signal more difficult to perform, leading to more ambiguous models [28].

9.2.2.3 Speed

The rapid decays typical of diffusing wave spectroscopy exaggerate limitations due to the correlator hardware. An insufficiently fast correlator will be unable to count at the rates necessary to obtain a sufficiently high number of photon-detection events within the available delay time, an effect known as "clipping." Clipping has little effect on the homodyne measurement, but excessive clipping limits the minimum delay time a correlator can employ when monitoring a heterodyne signal. The reduced correlation intersect of the heterodyne signal, and thus sensitivity to baseline errors, also make the heterodyne technique more (up to twentyfold) susceptible to noise when suffering clipping than the equivalent homodyne measurements [15].

9.3 DIFFUSING WAVE SPECTROSCOPY

Pine and Weitz [28] introduced the term diffusing wave spectroscopy (DWS), which was based on the concept of diffusion of photons. In highly concentrated opaque samples, where the photons have undergone a random scattering path or random "walk" in the medium, it was suggested that photon migration could be modeled as diffusion. Experimental work by Garcia and others [29] gave quantitative support for this assumption. The approximation representing purely diffuse scattering is given as [28]

$$G(t) = (1 + Y(6t/t_0)^{0.5})^{-1} \text{ and is valid for } L >> Y l_* \tag{9.7}$$

where l_* represents the mean free path of the photons and Y is a parameter that describes the crossover from ballistic to diffusive light.

Thus, in retroreflection, the resulting correlation function should be a single exponential when the log intensity versus the square root of the correlator delay time (as opposed to linear delay time for PCS) is plotted. The resulting straight line may be related directly to the diffusion coefficient of the particles, provided the laser has a coherence length greater than the maximum possible photon pathlength in the multiple scattering sample. The lack of a defined scattering volume and the random walk of the photon make this coherence length considerably longer than would be the case for conventional PCS [30].

In their consideration of samples of finite extent, where the assumptions in (9.7) are no longer valid, Pine and Weitz modeled their experiments as transmission measurements where the mean free path, l_*, must be measured independently. Conversely, Wolf and Maret's work [31] has suggested that when sizing with a pure homodyne instrument in infinite media, the basic model, based on approximation of diffusion theory, may serve to represent results more accurately than the formal solutions. Thus the effective size can be found in a similar way to PCS, except that the decay is now linearized by plotting against $\sqrt{\tau}$ instead of τ,

$$\Gamma = [\log(G^2(t)) - 1]/t^{0.5} \tag{9.8}$$

As the decays are more rapid than in PCS, the number of Γ measured in a given experiment time increases significantly [10], improving the SNR (9.9):

$$\text{Var } G(t) = G^2(t)\Gamma/n^2E \tag{9.9}$$

where n is the average number of correlator delay (sample) time, and E the experimental duration.

While a size estimate may be derived from a DWS measurement using the basic model (9.8), this can only be considered valid if the assumptions that all detected photons are diffuse and that no particle-particle interactions have occurred are true. While the experimental arrangement described here is designed to minimize detection of ballistic scatter and the effect of interactions, the absolute size cannot be stated with certainty. Accordingly, for most practical purposes, DWS should be considered predominantly as a tool for nonabsolute (i.e., qualitative) measurements, for instance, for determining relative particle size, studying the effects of high concentrations and changes in fluid rheology/structure, and for QA monitoring of process fluids to look for changes. Nevertheless, the simplicity and robustness of the technique will ensure its adoption in a wide range of applications where comparable analyses are lacking.

As described above, while PCS is capable of measurement of particle size distribution, in DWS analysis, such information is lost [29]. In DWS, each photon has seen many particles so each photon delay contains information about an average particle and, furthermore, any given photon has arrived at the detector from an unknown direction (i.e., with an undefined scattering vector, K). This represents a serious limitation to the future widespread application of DWS when compared to PCS.

We shall now discuss our own novel arrangement for measurement in the multiple scattering condition.

9.4 NOVEL EXPERIMENTAL ARRANGEMENT FOR INVESTIGATION OF MULTIPLE SCATTERING

In our arrangement, the probe head consisted of two low numerical aperture (3-deg half-cone angle in water), highly birefringent "bow tie" fibers (Figure 9.1). Plane-polarized light from a TEM_{00} HeNe laser, (633 nm), was launched into the fast (low NA) axis of one fiber using an achromatic lens, after passing through a laser line filter and polarizer. Cross-polarized light from the fast axis of the detection fiber was collimated with a GRIN lens prior to selection of light polarized orthogonal to the source. Interference from background light was reduced by using a bandpass filter prior to optical focusing onto a red-sensitive, electronically focused EMI photomultiplier tube (Figure 9.5). Correlation was carried out using a 256-channel, 25-ns minimum sample time correlator (on a plug-in PC-compatible board). Analysis of the correlation was carried out using Basic software written in the LabWindows development package. The fibers were aligned to give an extinction of 1,000:1, although this reduced to 500:1 when the probe was mechanically perturbed due to vibration when employed in an industrial situation.

The instrument was checked by pointing the clean probe into free space after each experiment. A background count rate of 100 counts per second was typical and included both the thermal noise and the heterodyne component due to Fresnel reflection at the fiber end. It should be noted that the lack of index-matching of the probe meant that an artificially large heterodyne component was measured. This allowed the majority of experiments to operate with a true experimental SNR of approximately 1,000:1, even though the incident light is 10^{10} times larger than the signal. The most suitable probe windows were found to be thin (0.1 mm) sheets of low refractive index glass, treated with a suitable nonstick silane coating [30].

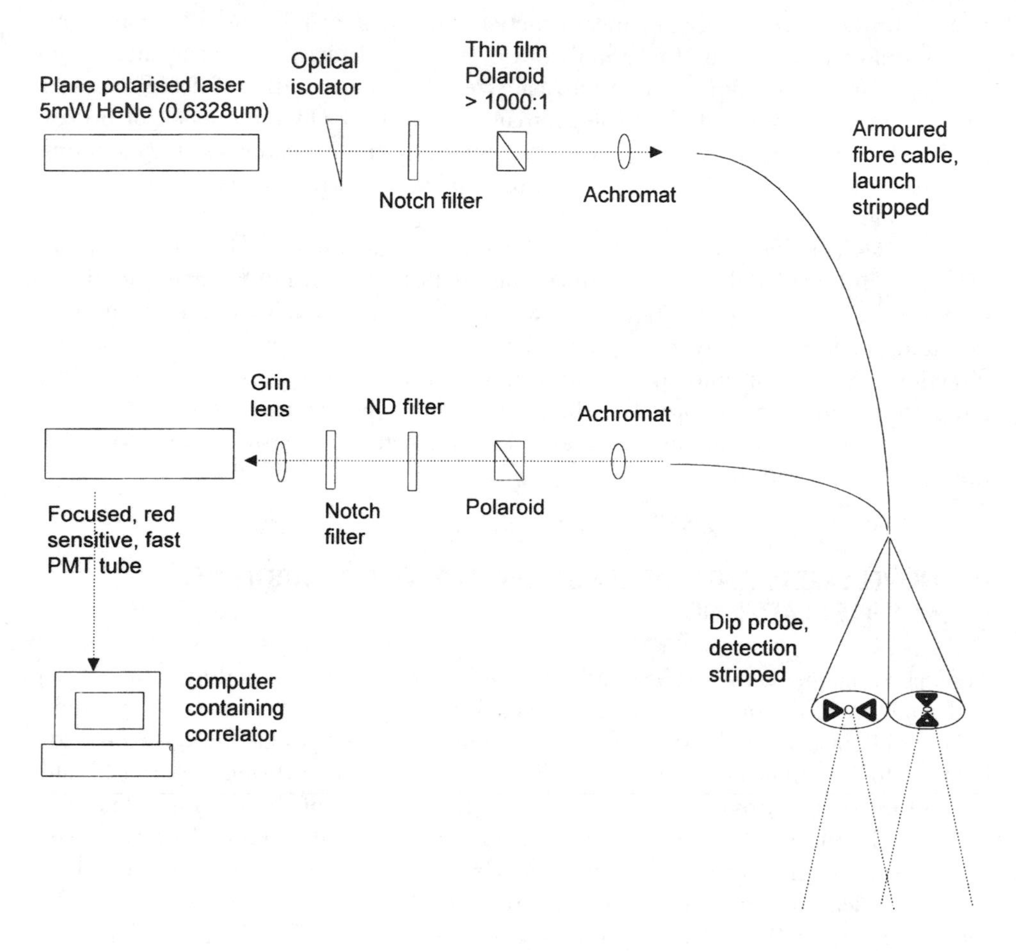

Figure 9.5 Simplified homodyne DWS instrument: practical limitations meant that the ND filter was placed, more conventionally, on the laser.

The probe described has a number of significant advantages:

1. Detection of only light polarized at 90 degrees with respect to the incident beam ensures that the detected light is maximally multiply scattered as described by Lilge and Horn [32] and Shmitt and others [33]. (Bruscaglioni and others [34] have shown that the scattering order must be significant for small spheres to totally depolarize the illuminating beam.)
2. Cross-polarization also ensures that Fresnel reflection from the probe face or window minimizes the heterodyne component, as the specularly reflected light does not couple into the orthogonal axis of the fiber.

3. The low NA of the fiber ensures the ballistic crossover point, the fiber acceptance cone crossover point (from which for single scattering would be detected), is many centimeters from the probe tip. This also ensures that the effect of large contaminating dust particles close to the probe will be minimized, even if they depolarize the scattered light.
4. Homodyne operation reduces excessive laser coherence requirements, allowing for example, a multilongitudinal-mode laser to be used.

The above system offers a near-pure homodyne method via a robust dip-in probe requiring no focusing or alignment, rapid measurements and low power, low-cost optoelectronic components (laser source and detector). No background measurements are required and analysis using the simple model is user friendly, allowing near real-time analysis.

9.4.1 Preliminary Experimental Results and Discussion

While the correlation should in principle be a straight line when plotted as log I versus $\sqrt{t}$, any nonideal instrument characteristics or sample properties tend to result in a departure from this. Our approach to solving this problem has been to force fit to the expected model and hence monitor the attributes we understand. In more complex procedures, the meaning of many of the fitting parameters may be uncertain or unknown. Analysis is similar to PCS, but differs in that the approximation $Ri_{\text{bulk}} = R_{\text{liquid}}$ will be in error because of the high-volume fraction of the sample. It is suggested the following refractive index be used:

$$Ri = (1 - \%(v/v)_{\text{solid}}) \cdot Ri_{\text{solid}} + Ri_{\text{liquid}} \tag{9.10}$$

However, a variation over the range 0 to 40% volume solids/liquids of polystyrene only gives a variation in Ri of around 5%, and in many cases only a very rough estimate of concentration is actually required to recover a reasonable estimate of D. To ensure our results are not limited by errors in the altered model, I is plotted directly.

9.4.1.1 Reproducibility

Figure 9.6(a) shows four sets of ten experimental measurements for an aqueous suspension of high refractive index ceramic powder suspended in water to 40% (v/v). The integration time was 30 sec and the probe was removed and cleaned between each set of measurements to ensure sample-to-sample reproducibility. The variation was found to be less than ±1% between each set of 10 experiments. Calculation of the error, using all the selected points, as shown in Figure 9.6(b), gave a reproducibility of 0.1%. It is suggested that this reproducibility is better than that

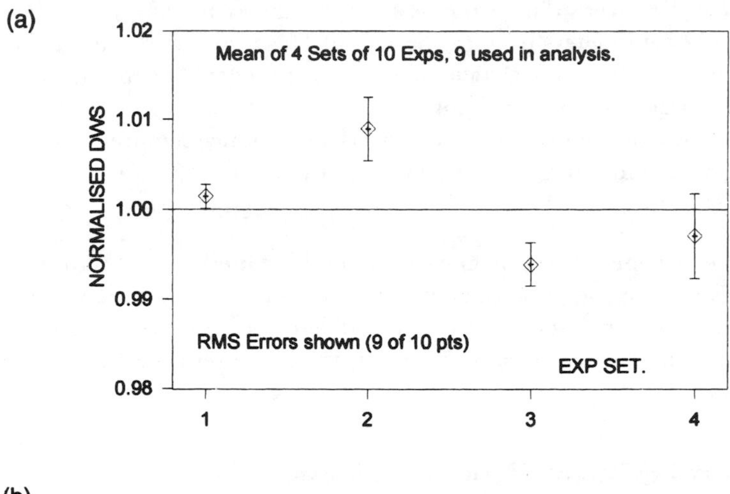

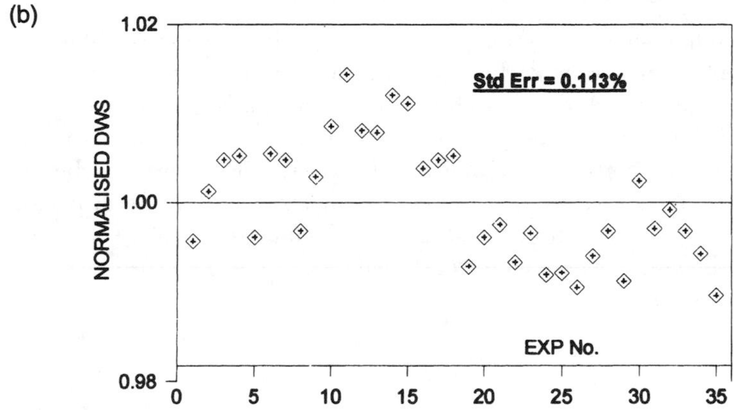

Figure 9.6 (a) Reproducibility test, with probe removed and cleaned between readings and (b) all selected sizes (35 from 40).

achievable in a heterodyne experiment, which would suffer from propagation of baseline errors.

9.4.1.2 Ranking

Figure 9.7 shows the relative sizing ability of DWS and indicates the accuracy of the experimental DWS instrument. Six different mixtures of 240-nm and 450-nm polystyrene were produced. The monodisperse samples were measured 100 times, with experimental durations of 3 and 5 sec for both sizes. The mixed samples were each

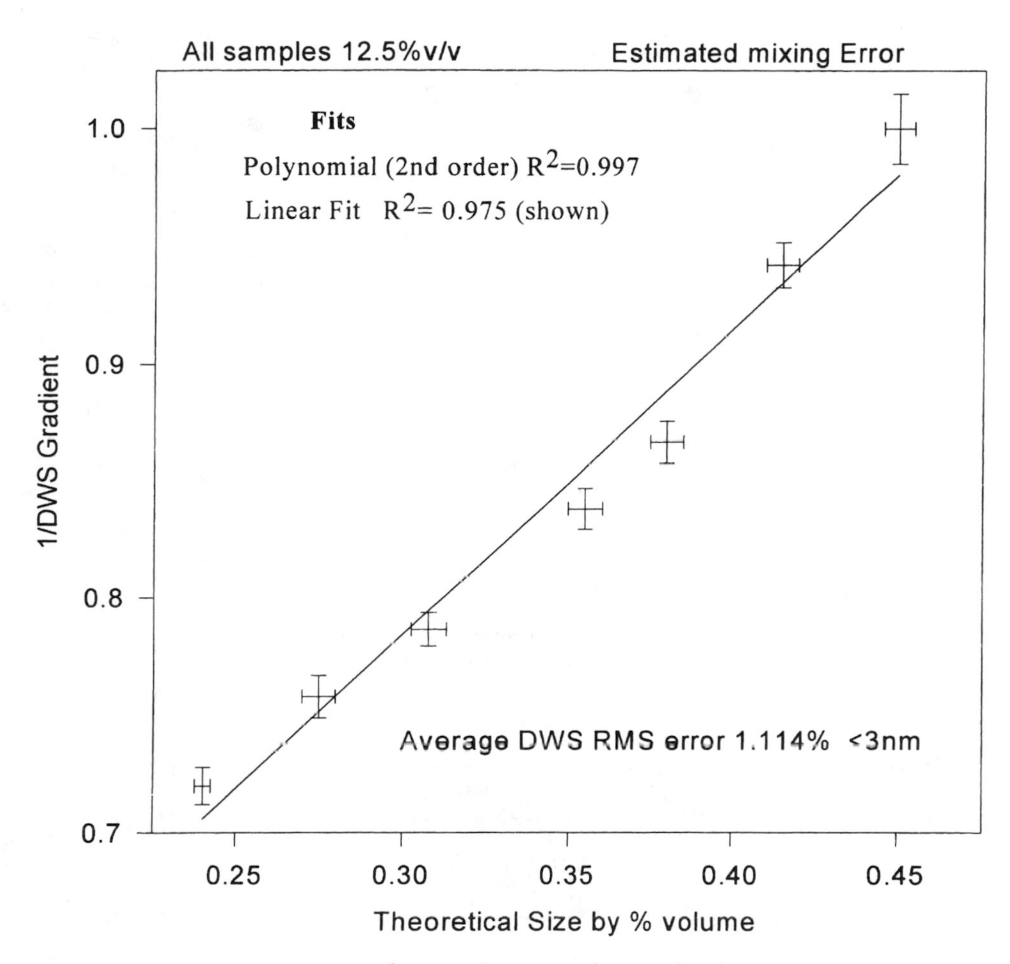

Figure 9.7 DWS response to mixtures of 240- and 450-nm polystyrene beads.

measured 50 times. The average error suggests an ability to differentiate between sizes at a resolution of better than 3 nm. Later work has confirmed this exceptionally high resolution (i.e., an average error of only ±1.3 nm when measuring a 0.3 μm diameter industrial ceramic particle suspension).

9.4.1.3 Concentration

From the basic theory, DWS is expected to be concentration-independent to at least 10% (v/v). However, this is not the case practically, as shown by Figure 9.8, where

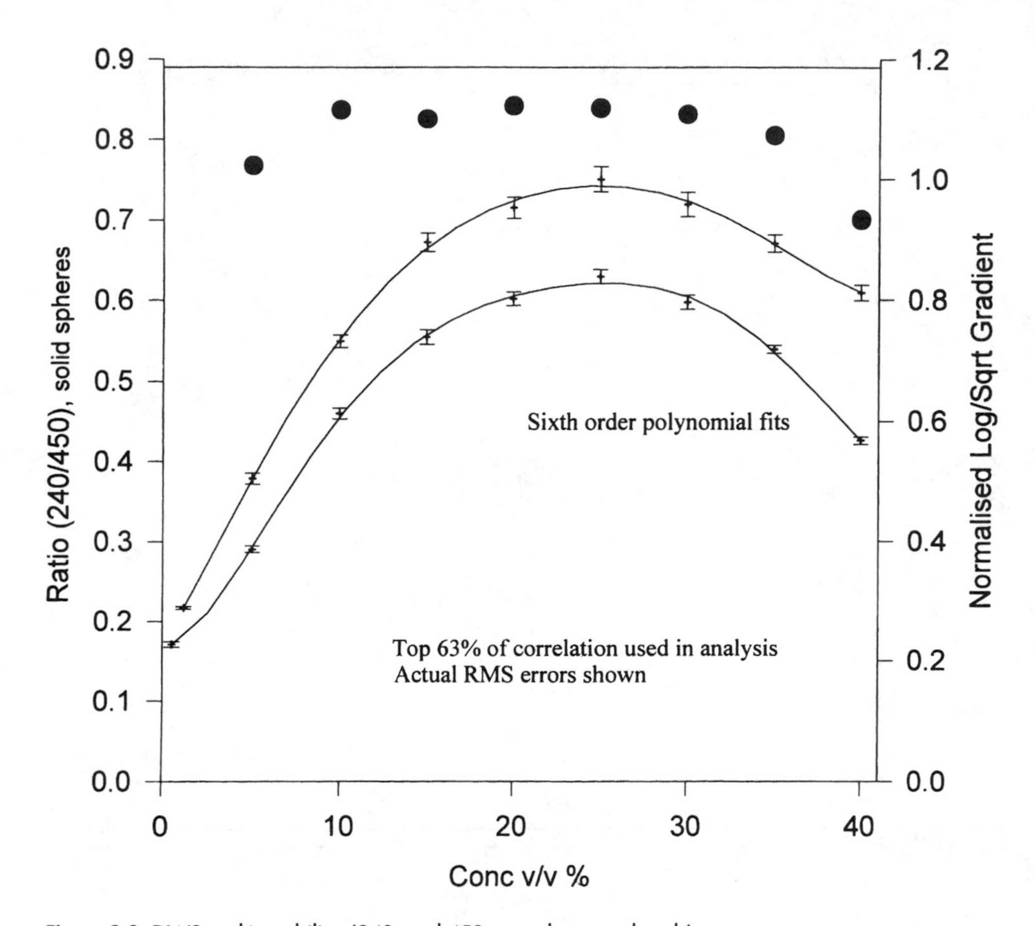

Figure 9.8 DWS ranking ability: (240- and 450-nm polystyrene beads).

240- and 450-nm polystyrene beads were measured at various concentrations up to 40% v/v, the upper 63% of the correlation being used in analysis. The curve appears to indicate an area of greater concentration independence above 12% (v/v), above which ordering [28] begins to reduce the intensity.

While the relative sizing and constant ranking shown by these results is promising, the high concentration dependence was less so, and did not agree with earlier results [3]. Later work has shown that the dependence on both particle/particle interactions and low-order scattering can be minimized by analyzing a shorter region of the measured correlation function (i.e., ignoring the longer "tail" of the correlation function when curve fitting). Dyott's [35] method of concentration calibration, using the count rate, may be useful in removing any residual dependence in well-characterized particulate systems.

The simple retroreflection model has been shown to be applicable to pure homodyne operation at any concentration above 1%, provided the material shows some fluidity. This may be above the glass transition, where diffusion itself is frozen (as discussed by Kaplan and others [36], Meller and Stavans [37], and Wang and Miller [19]). Work by Weitz and others [38] has shown that DWS may even be capable of measuring the initial rapid ballistic movement of particles after single water molecule collisions.

9.4.2 Applications

Our work has included studies on cosmetics, foodstuffs, industrial oil-in-water emulsions, ceramic materials, high-concentration microbial cell pastes, as well as whole blood undergoing clotting. Multiple scattering studies based on DWS or associated techniques have included blood flow *in vivo* [39], aqueous suspensions containing 40% (v/v) olive oil [40], as well as gelling and setting in dairy products [1,2].

The ability to work at high concentrations, as modeled by Maret and Frayden [41], to 45% (v/v), not only permits in-line monitoring for process control but also overcomes many of the problems normally associated with dynamic light-scattering analysis of dilute suspensions. The technique is unaffected by dust; the large volume interrogated by DWS minimizes sampling error; there is no requirement for sample dilution (therefore extrapolation of data to high-volume fraction concentrates is unnecessary); and the rapid decays allow theoretical SNRs better than 10^3:1, which may give actual resolutions of 1% or better, all from experiments of only a few seconds duration.

This suggests DWS may not only be useful as a process monitor, but can utilize feedback, as a rapid analytical method through which improved control of a process may be effected. The technique appears ideally suited to investigating the processes of drying, setting/gelling, polymerization, flocculation, and aggregation of optically opaque, dense systems.

The use of a specific instrument design and a particular data analysis method that, in combination, permit the selective detection of multiple scattering, reduces the dependence on l^*. The method is therefore insensitive to sample concentration when compared to techniques that define a fixed and frequently small (e.g., 100 μm^3) scattering volume. These advantages have allowed the system to be operated in particulate systems with significant optical absorption, greatly expanding the potential applications. Miniaturization of the present equipment to produce portable or even handheld devices for field use appears eminently feasible.

ACKNOWLEDGMENTS

We would like to thank Dr. A. Tinsley-Bown, Dr. H. Ajoula, and Mr. M. Hutchins for supporting the work.

References

[1] Horne, D. S., "Influence of polydispersity on dynamic light scattering measurements in concentrated suspensions," *J. Chem. Soc. Faraday Trans.*, 86, (7), 1990, pp. 1149–1150.
[2] Horne, D. S., "Particle size measurements in concentrated latex suspensions using fibre-optic photon correlation spectroscopy," *J. Phys. D. Appl. Phys.*, 22, 1989, pp. 1257–1265.
[3] Lloyd, C. J., "Diffusive wave spectroscopy," MSc Thesis, Newcastle Polytechnic, 1991.
[4] Lloyd, C. J., et al., "Diffusing wave spectroscopy," *Rank Prize Funds Mini-symposium on Optical Localisation and Slow Waves*, Grasmere, Sept. 1993, (Invited Presentation).
[5] Forrestor, A. T., Gudmundsen, R. A., and Johnson, P. O., "Photoelectric mixing of incoherent light," *Phys. Rev.*, 55, 1955, pp. 1691–1700.
[6] Forrestor, A. T., "Photoelectric mixing as a spectroscopic tool," *Opt. Soc. Am.*, 51, 1961, p. 253.
[7] Cummins, H. Z., and Yueh, K., "Observation of diffusion broadening of Rayleigh scattered light," *Phys. Rev. Lett.*, 16, (6), 1964, pp. 159–153.
[8] Cummins, H. Z., and Swinney, H. L., *Progress in Optics*, Wolf, E., (Ed.), Vol 8, North Holland Pub., 1970.
[9] Steward, E. G., *Fourier optics - an introduction*, Edition 2, Ellis Horwood Ltd., 1989.
[10] Pusey, P., "Detection of small polydispersities by photon correlation spectroscopy," *J. Chem. Phys.*, 80, (8), 1984, p. 3513.
[11] Rarity, J. R., "Measurement of radii of low axial ratio ellipsoids using cross correlation spectroscopy," *J. Chem. Phys.*, 85, (2), 1985, p. 733.
[12] Brown, R.G.W., and Pusey, P., "Photon correlation study of polydisperse samples of polystyrene in cyclohexane," *J. Chem. Phys.*, 62, (3), 1986, p. 1136.
[13] Beenakker, C. W .J., "Electromagnetic waves," *MazirPhysica*, 120A, 1983, pp. 388–410.
[14] Oliver, C. J., *Correlation techniques in Photon correlation and light beating spectroscopy*, Cummins, H. Z., and Pike, E. R., (Eds.), Plenum Press, 1973, pp. 151–223.
[15] Cummins, H. Z., et al., "Translational and rotational diffusion constants of tobacco mosaic virus from Rayleigh linewidths," *Biophysical J.*, 9, 1969, pp. 518–546.
[16] Carr, R. J. G., et al., "Determination of protein size in chromatography eluants by on-line photon correlation spectroscopy," *Analytical Biochemistry*, 175, 1988, pp. 492–499.
[17] Chow, K. M., et al., "On-line photon correlation spectroscopy using fibre-optic probes," *Journal of Physics E (Sci. Instrumen.)*, 21, 1988, pp. 1186–1190.
[18] Ottewill, R. H., and N. Williams, "Study of particle motion in concentrated dispersions by tracer diffusion," *Nature*, 325, (15), 1987, pp. 232–234.
[19] Wang, L, and Miller, W. G., "Short-time mobility of spherical particles in concentrated aqueous dispersions determined by diffusing wave spectroscopy," *Theor Chim Acta*, 82, 1992, pp. 419–423.
[20] Macfadyen A. J., and Jennings, B. R., "Fibre optic systems for dynamic light scattering - a review," *Optics & Laser Tech.*, 22, (3), 1990, pp. 175–187.
[21] Thomas, J. C., and Tjin, S. C., "Fibre optic dynamic light scattering (FODOLS) from moderately concentrated suspensions," *J. of Coll. and Interface Sci.*, 129, (1), 1989, pp. 15–31.
[22] Ross, D. A., Dhadwal, H. S., and Dyott, R. B., "The determination of the mean and standard deviation of the size distribution of a colloidal suspension of submicron particles using the fibre optic Doppler anemometer," *J. of Coll. and Interface. Sci.*, 64, (3), 1978, pp. 533–543.

[23] Van Der Meeren, P. V., et al., "Accurate determination of the short time self diffusion coefficient by fibre optic quasi-elastic light scattering: New methods for correcting the homodyne effect," *J. Coll. and Interface. Sci.*, 160, 1993, pp. 117–126.
[24] Ansari, R. R., et al., "Microemulsion characterisation by the use of a non-invasive backscatter fiber optic probe," *Appl. Optics*, 32, (21), 1993, pp. 3822–3827.
[25] Brown, R. G. W., "Dynamic light scattering using monomode fibres," *Appl. Optics*, 26, 1987, p. 4846.
[26] Brown, R. G. W., and Jackson, "Monomode fibre components for optical light scattering," *J. Phys. E. Sci. Instrum.*, 20, 1987, p. 1503.
[27] Bremer, L. G., et al., "Fibre optic dynamic light scattering, neither homodyne nor heterodyne." *Am. Chem. Soc.*, Langmuir, 9, 1993, pp. 2008–2014.
[28] Pine, D. J., et al., "Diffusing wave spectroscopy," *Phys. Rev. Lett.*, 60, (12), 1988, pp. 1134–1137.
[29] Garcia, N., Genack, A. Z., and Lisyansky, A. A., "Measurement of the transport mean free path of diffusing photons," *Phys. Rev. B*, 46, (22), 1992, pp. 14475–14479.
[30] Tinsley-Bown, A., PhD Thesis, Open University, 1994.
[31] Wolf, P. E., and Maret, G., "Dynamics of Brownian particles from strong multiple light scattering," *in Scattering in Volumes and Surfaces*, Nieto-Vesperinas, M., and Dainty, J. C., (Eds.), Elsevier Science Publishers, Amsterdam, 1990.
[32] Lilge, D., and Horn, D., "Diffusion in concentrated dispersions: a study with fiber–optic quasi-elastic light scattering (FOQELS)," *Colloid Polymer Sci.*, 269, 1991, pp. 704–712.
[33] Shmitt, J. M., Gandjbakhche A. H., and Boinner, R. F., "Use of polarised light to discriminate short-path photons in a multiple scattering medium," *Appl. Optics*, 31, (30), 1992, pp. 6535–6546.
[34] Bruscaglioni, P., Zaccanti G., and Wei, Q., "Transmission of a pulsed polarized light beam through thick turbid media: numerical results," *Appl. Optics*, 32, (30), 1993, pp. 6142–6150.
[35] Dhadwal, H. S., and Ross, D. A., "Size and concentration of particles in Syton using the fiber optic Doppler anemometer (FODA)," *J. of Coll. and Interface Sci.*, 76, (2), 1989, pp. 478–489.
[36] Kaplan, P. D., Yodh A. G., and Pine, D. J., "Diffusion and structure in dense binary suspensions," *Phys. Rev. Lett.*, 68, (3), 1991, pp. 393–396.
[37] Meller, A., and Stavens, J., "Glass transition and phase diagrams of strongly interacting binary colloidal mixtures," *Phys. Rev. Lett.*, 68, (24), 1992, pp. 3646–3649.
[38] Weitz, D. A., et al., "Nondiffusive Brownian motion studied by diffusing wave spectroscopy," *Phys. Rev. Lett.*, 63, (16), 1989, p. 1747.
[39] Tanaka, T., and Benedek, B., "Measurement of the velocity of blood flow (in vivo) using a fiber optic catheter and optical mixing spectroscopy," *Appl. Optics*, 14, (1), 1975, pp. 189–196.
[40] Floy, B. J., White, J. L., and Hem, S. L., "Fibre optic anemometry (FODA) as a tool in formulating emulsions," *J. of Coll. and Interface Sci.*, 125, (1), 1988, pp. 23–33.
[41] Zhu, J. X., et al., "Scaling of transient hydrodynamic interactions in concentrated suspensions," *Phys. Rev. Lett.*, 68, (16), 1992, pp. 2559–63.

Chapter 10

In Vivo Medical Sensors

Anna Grazia Mignani and Francesco Baldini
IROE-CNR, Italy

10.1 ADVANTAGES OF OPTICAL-FIBER SENSORS FOR MEDICAL USE

Few things affect our lives as closely and as consistently as healthcare. Consequently, as new healthcare concepts and technologies have been developed, they have been largely aimed at maximizing the efficiency and effectiveness of healthcare service, while minimizing the discomfort and suffering of patients. Among the newer technologies, fiber optics is a major contributor to the development of advanced diagnostic and therapeutic techniques.

Fiber-optic methods are proving ideal, not only for intracavity imaging and safe laser delivery, but also for monitoring critical physiological parameters. The many advantages to be gained from the use of fiber optics in medical sensing are related to the intrinsic nature of the sensor fiber:

- Uncabled fibers are so small and flexible they can be inserted inside very thin catheters and hypodermic needles, thereby ensuring highly localized and minimally invasive monitoring.
- Fibers are nontoxic, chemically inert, and intrinsically safe for the patient. In particular, their electromagnetically inert nature allows them to be safely used without any electrical interference in the presence of other electrical equipment, an aspect of vital importance for patient monitoring.
- The absence of crosstalk between neighboring fibers allows several sensors, controlled by a single electro-optic unit, to be grouped in a single catheter.

This chapter outlines the uses of fiber-optic sensors (FOSs) for *in vivo* monitoring. Special emphasis is placed on the positive results that can be achieved in a host of applications, including cardiovascular and intensive care, angiology, gastroenterology, ophthalmology, oncology, neurology, dermatology, and dentistry.

10.2 FUNDAMENTAL DESIGN CRITERIA AND CRITICAL ASPECTS

In vivo monitoring FOSs are sensors applied directly to a patient: They are ideal for measuring physical and chemical parameters as well as for performing spectral measurements directly on the patient. Obviously, noninvasive sensors are preferable, since they are safer and less traumatic for the patient and are capable of monitoring parameters by direct application on the patient's skin. However, the use of invasive sensors is unavoidable when parameters inside the body must be continuously monitored. The ideal FOS for medical applications should possess the following characteristics:

- Reliability;
- Automated or semiautomated operation that allows use by operators with little or no technical background;
- Simple implantation and low-cost maintenance.

These requirements limit the choice of operating principle of the sensor and restrict the complexity of the whole electro-optical system. Consequently, FOSs for medical applications are based on sensing mechanisms common to other types of fiber sensors, especially intensity changes (often characterized by wavelength) and time-domain modulation. They usually act as point sensors, either intrinsic or extrinsic, according to whether the modulation is produced in the fiber sensor itself, or whether there is an external transducer connected to the fiber. They can also be used as spectral sensors, where the fibers act simply as light guiders, and localized spectral analysis of tissues or blood may be performed at the far end without necessitating sampling. Owing to the requirements of minimal invasiveness and patient perturbation, the particular problems involved in the use of medical FOSs are mostly related to probe design and biocompatibility:

- *Probe design:* FOS probes must be designed so that no thrombosis or inflammatory effects of the blood vessels and tissues are produced, and so they are not toxic. Since these problems have only been solved for temporary use (typically a few minutes), long-term implantable sensors do not appear feasible in the near future.
- *Probe encapsulation:* Proper probe encapsulation is a crucial factor, so potential problems of selectivity, hysteresis, and stability must be effectively solved, and high backtransmitted signals and fast response times must be achieved. Unlike physical sensors, chemical and biochemical sensors cannot usually be

hermetically encapsulated, so there may be problems of interference with other substances or the environment. One solution, the use of membrane encapsulation, partially eliminates the interferences, but has the drawback of compromising response times. Another problem is the irreparable damage to the sensor that can result from sterilization procedures. To avoid this, disposable probes are preferable, and optical fibers afford the further benefit of being low cost as well.

- *Biocompatibility:* Special care must be taken to meet the requisites of biocompatibility (i.e., the sensor must not adversely affect the body, nor must it be adversely affected by it). The latter consideration should be taken with due regard: we normally consider only the harmful effects of the invasive sensor on the human body, without considering that the local environment can notably impair sensor performance, especially when we are dealing with a chemical sensors where an optrode is attached at the fiber end.

10.3 CIRCULATORY AND RESPIRATORY SYSTEMS

The earliest use of optical fibers to monitor the circulatory-respiratory systems dates back to 1964, when the invasive fiber-optic oximeter was introduced [1]. Since then, fiber-optic sensors (FOSs) have become increasingly popular in both invasive and noninvasive versions.

As already pointed out, invasive sensors must fulfill very exacting requirements in order to be applied safely to the patient, and this is specially true when the probe has to be located in the bloodstream. They must be suitably miniaturized so as not to slow down blood flow and cause clotting, and they must not be thrombogenic; moreover, they must be resistant to build-up of platelet and protein deposits. Since deposit resistance is primarily related to the sensor's shape and to its chemical and physical properties, probes with smooth surfaces and coated with materials such as anticoagulants, and antiplatelet agents (e.g., heparin) are commonly used.

10.3.1 Oximetry

Optical fibers were originally used to measure the oxygen saturation in blood (i.e., the fraction of oxygen carried by the hemoglobin (Hb) in the erythrocytes compared to its maximum capacity). This parameter, which provides essential information on how the cardiovascular and cardiopulmonary systems are functioning, is optically measured by exploiting changes in the absorption spectra of the Hb and the oxyhemoglobin (OxyHb) in the near-infrared region.

Following the introduction of the first invasive oximeter [1], numerous artery- and vein-insertion models have become commercially available (Oximetric Inc., Mountain View, CA; BTI, Boulder, CO; Abbott Critical Care, Mountain View, CA). In the simpler version, reflected or absorbed light is collected at two different wavelengths and the oxygen saturation is calculated (via the ratio technique) on the basis of the isosbestic regions of Hb and OxyHb absorption. On the other hand, the presence of other hemoglobin derivates, such as carboxyhemoglobin, (carbon monoxide hemoglobin), methemoglobin, and sulfhemoglobin, makes preferable the use of multiple wavelengths, or of the whole spectrum, which allows for their discrimination [2,3].

Noninvasive optical oximeters, which calculate oxygen saturation via the light transmitted through the earlobes, toes, or fingertips, have also been developed, primarily for neonatal care [4]. In this case, particular attention has to be paid to differentiating between the light absorption due to arterial blood and that due to all other tissues and blood in the light path. This implies the use of multiple wavelengths, such as the Hewlett Packard ear oximeter, where no less than eight wavelengths are monitored [5].

Such a drawback can be avoided by using a pulse oximeter. This original approach is based on the assumption that a change in the light absorbed by tissue during systole will be caused primarily by the passage of arterial blood. By an appropriate choice of two wavelengths, it is possible to measure noninvasively the oxygen saturation by analyzing the pulsatile rather than the absolute level of transmitted or reflected light intensity [6–8]. The detection of blood absorbance fluctuations that are synchronous with systolic heart contractions is called photoplethysmography.

Spectrophotometric measurements performed directly on the skin tissue and on the organ surface at wavelengths between 700 and 1,100 nm in order to penetrate the tissue and provide essential information on the microcirculation in tissue and skin and on the metabolism of an organ, respectively [9]. For example, online monitoring of the oxygen supply in peripheral organs has considerable importance. It is apparent that a correct perfusion of the whole organism is basic to the safety of the patient. In the presence of pathological changes in the oxygen transport chain, the organism, by itself, decreases the perfusion in peripheral organs (e.g., skin, skeletal muscles, gut) in favor of central organs such as the brain and the heart: this automatic mechanism, according to which the most important internal organs are supplied with oxygen to the detriment of the most peripheral ones, is called centralization. Such a mechanism is one of the most effective and important pieces of equipment during different forms of clinical shock in the patient. Spectra from biological tissues are able to detect the onset of centralization before any external, physical and more dangerous symptoms become visible. On the other hand, during the early stages of shock, such an alteration of oxygen transport does not occur homogeneously on the tissue surface; therefore, only optical fibers offer a sufficient spatial resolution for immediate detection.

A special algorithm is generally used, since the spectra of the Hb and its derivatives are unevenly distributed in a highly scattering medium, and thus are notably altered [3].

With the spectrophotometer analyzer developed by BGT (Überlingen,

Germany) [10], important parameters such as intracapillary hemoglobin oxygenation and concentration, local oxygen uptake rate, local capillary blood flow, changes in subcellular particle sizes, and capillary wall permeability (via the injection of exogenous dyes) can be measured in real time. Light from a xenon arc lamp illuminates the tissue via a bifurcated fiber bundle. The backscattered light, filtered by a monochromator, impinges on a photomultiplier and a computer records the spectra returning from the biological tissue. Thanks to the use of fibers, only small volumes of tissue are investigated, thus making possible the resolution of spatial variations. The instrument is able to record spectra of high quality, even in moving organs.

A new type of noninvasive sensor has been proposed to measure the local oxygen uptake through the skin. Direct measurement of the oxygen flow on the skin surface provides information regarding the oxygen flow inside the tissue, which can help physicians diagnose circulatory disturbances and their consequences. Two optrodes measure the difference in oxygen pressure across a membrane placed in contact with the skin. *In vivo* tests performed on the left lower forearm of a patient gave good results [11].

10.3.2 Blood Gases and Blood pH

Real-time monitoring of the blood pH and the blood oxygen (pO_2), and carbon dioxide (pCO_2) partial pressures to determine the quantity of oxygen delivered to the tissues and the quality of the perfusion, is of paramount importance in operating rooms and intensive care units. These parameters are conventionally measured by benchtop blood gas analyzers on handdrawn blood samples. However, significant changes can occur in blood samples after the removal from the body, before the measurements are carried out by the blood gas analyzer. Due to their ability to provide continuous monitoring, FOSs represent a welcome and significant improvement in patient management. The pH is detected by a chromophore, which changes its optical spectrum as a function of the pH; absorption-based indicators or fluorophores are usually used. The carbon dioxide is detected indirectly, since its diffusion in a carbonate solution fixed at the optical fiber tips alters the pH, so that the carbon dioxide content can be determined by measuring the pH. Measurement of oxygen cannot, unfortunately, always be performed using hemoglobin as the *in situ* indicator due to its full saturation at $\approx$100 torr; this fact prevents, for example, the use of this method in the case of the respiration of gas mixtures with an O_2 content larger than 20%, as is routinely used in anesthesia. Therefore, the use of a separate chemical transducer becomes necessary; oxygen is usually detected via fluorescence techniques that exploit the quenching produced by oxygen on fluorophores.

The first optrode-based oxygen sensor was described by Peterson [12]. The fluorophore, perylene-dibutyrate, adsorbed on amberlite resin beads, was retained at the distal end of two plastic optical fibers with a hydrophobic membrane permeable to oxygen. The probe described was tested *in vivo* for the measurement of the arterial pO_2 level in dog eyes [13].

The first intravascular sensor for simultaneous and continuous monitoring of the pH, pO_2, and pCO_2 was developed by CDI-3M Health Care (Tustin, CA) [14], based on a system designed and tested by Gehrich and others [15]. Three optical fibers (∅ = 125 mm) are encapsulated in a polymer enclosure, along with a thermocouple embedded for temperature monitoring (Figure 10.1). Three fluorescent indicators are used as chromophores. The pH and pCO_2 are measured by the same fluorophore (hydroxypyrenetrisulfonate):

- *pH measurement*: The fluorophore is covalently bonded to a matrix of cellulose attached to the fiber tip. Both the acidic (λ_{exc} = 410 nm) and alkaline (λ_{exc} = 460 nm) excitation bands of the fluorophore are used, since their emission bands are centered on the same wavelength (λ_{em} = 520 nm) The ratio of the fluorescence intensity for the two excitations is measured, to render the sensor relatively insensitive to fluctuations of optical intensity.
- *Carbon dioxide pressure measurement*: The fluorophore is dissolved in a bicarbonate buffer solution, encapsulated in a hydrophobic silicon membrane permeable to CO_2 and attached to the fiber tip.
- *Oxygen pressure measurement*: A specially synthesized fluorophore, a modified decacyclene (λ_{exc} = 385 nm, λ_{em} = 515 nm), is mixed with a second reference fluorophore that is insensitive to oxygen, and both fluorophores are incorporated into a hydrophobic silicon membrane that is permeable to oxygen.

The optoelectronic system is composed of three modules, one for each sensor. A

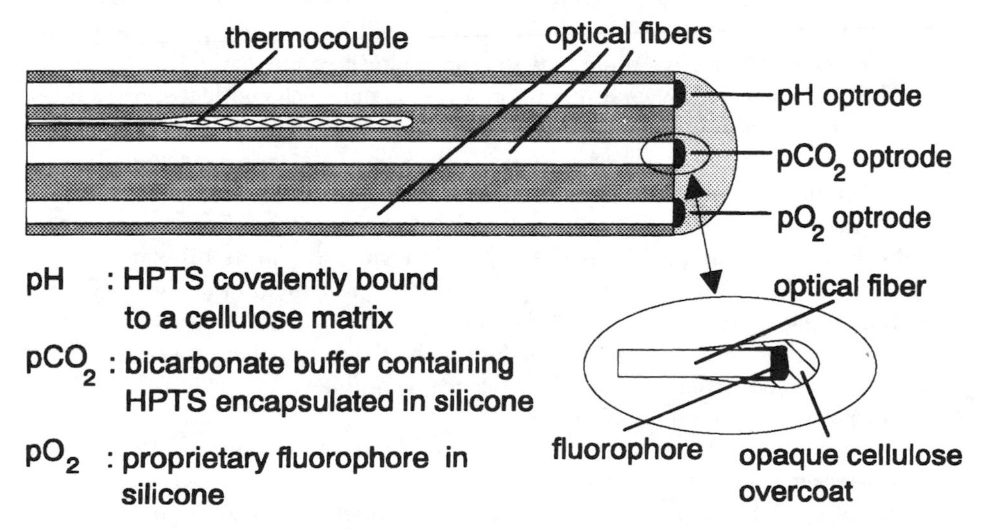

Figure 10.1. Sketch of the fiber-optic probe for *in vivo* simultaneous detection of pH, pO_2, and pCO_2 in human blood.

suitably filtered xenon lamp, modulated at 20 Hz, provides illumination for the modules. The source light is focused through a GRIN lens onto a prism beamsplitter and coupled through a fiber to the sensor tip, while the deflected light is collected by a reference detector for source control. The returning fluorescence is deflected by the prism beamsplitter onto the signal detector through a filter to select the fluorescence light. A microprocessor processes all the detected signals, giving the readout of the three parameters on a monitor. The probe (OD = 0.6 mm) has been tested *in vivo* on animals [16,17], and has shown satisfactory correlation with data obtained *ex vivo* from electrochemical blood gas analyzers.

On the other hand, some problems regarding the use of intravascular FOSs have emerged during clinical trials on volunteers in critical care and on surgical patients [18] and remain at the moment unsolved:

- A blood flow decrease due to peripheral vasoconstriction lasting for several hours after surgical operations; this can give rise to a contamination by flush solutions, which can seriously affect the measurements.
- The so-called wall effect, which primarily affects the oxygen count (if the fiber tip is very close to or touches the arterial wall, it measures the tissue oxygen, which is lower than the arterial blood oxygen).
- The formation of a thrombus (clot) around the sensor tip, which alters the value of all the analyte values.

Other intravascular-probe systems have been proposed by Abbott (Mountain View, CA) [19] and by Optex Biomedical (Woodlands, TX) [20], where the structure of the probe is essentially similar to the CDI one previously described, that is, three different multimode fibers, each of which is associated with the specific chemistry and charged with the detection of a single measurand. In the Optex Biomedical approach, the configuration is modified slightly, as each single fiber is bent and a side window sample chamber is built up to contain the appropriate chemistry (see Figure 10.2). The use of plastic fibers assures that the bundle does not break during insertion, routine patient manipulations, and removal. There are basically three advantages in this lateral configuration: 1) the real optrode does not suffer any mechanical stress during insertion through the arterial catheter, 2) the *wall effect* can be avoided by rotating the probe into areas where the blood flow is good, and 3) the washability of the sensing element in the presence of a thrombus or other fouling phenomena is improved.

Notwithstanding these technical improvements, invasive sensors for blood gas measurements remain at the research level and the fundamental drawbacks appear difficult to overcome to the exacting standards necessary.

These problems are clearly avoided in a system working in an extracorporeal blood circuit, developed by CDI-3M, which has been commercially available since 1984. Figure 10.3 shows the connection between the fiber link and the blood circuit. A disposable probe that uses the same chemistry as the previously described intravascular optrode is inserted online in the blood circuit on one side and connected to the

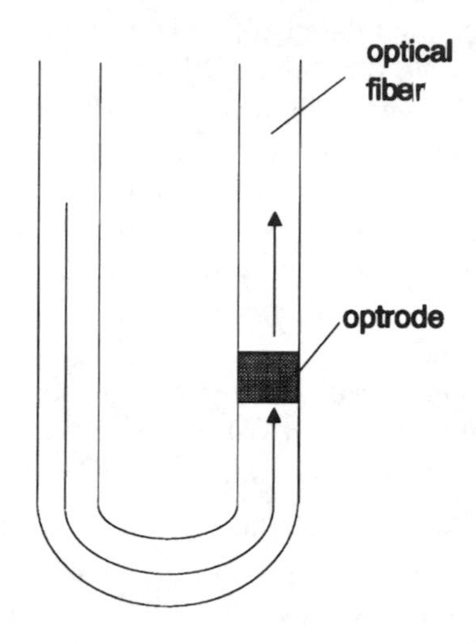

Figure 10.2 Intravascular probe with bent fiber and side window sample chamber.

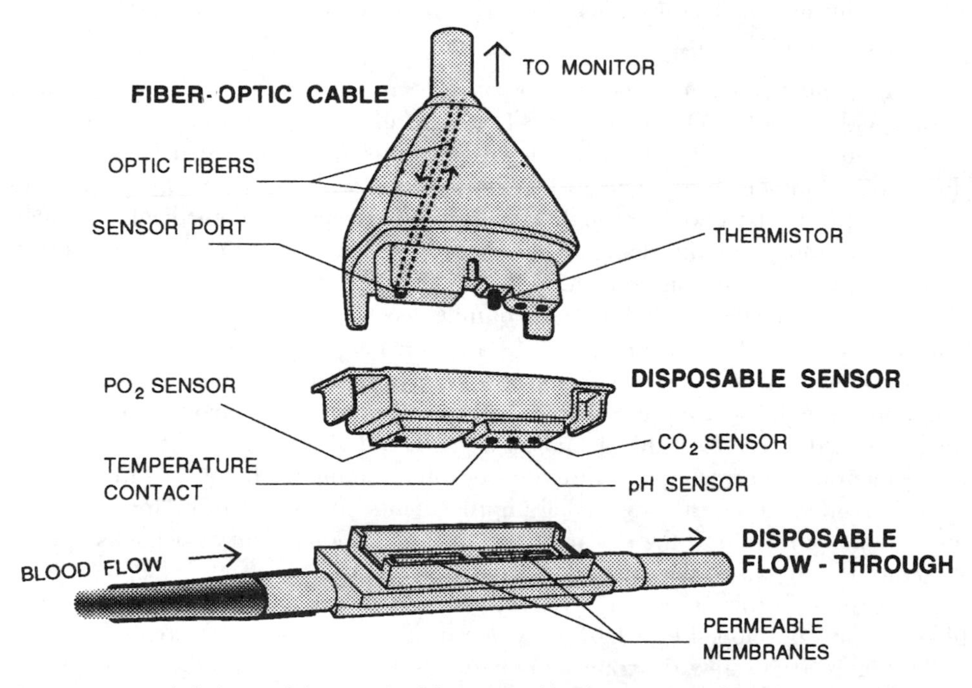

Figure 10.3 Exploded view of the optrode of the CDI-3M blood gas analyzer for extracorporeal analysis.

fiber bundle on the other. The system is currently employed in open heart measurements, with more than 10,000 disposable probes produced monthly by CDI-3M.

A different approach has recently been described [21]: a fiber-optic blood gas and pH monitoring system, capable of performing *paracorporeal* measurements for use at the patient's bedside, has been developed. The developed probe, consisting of three different fluorescent-based sensors for pH, pCO_2, and pO_2, respectively, is placed in series with a standard arterial line. Measurements have been performed on withdrawn samples of arterial blood, which return to the patient after the detection.

Though this procedure does not allow continuous monitoring, samples can be withdrawn continuously without any risk of inducing anemia (which may otherwise occur with the use of benchtop analyzers), since no blood loss results from such a paracorporeal system. Such an approach seems very promising, since it is capable of quasi-continuous monitoring without the problems affecting intravascular sensors.

10.3.3 Respiration Monitoring

A major requirement for intensive care is the continuous monitoring of breathing condition (i.e., the cough, sneeze, and breathing count). It should be possible to monitor from the nurse's station, so patients can be kept under observation without the nurse being physically present at their bedsides. An optical fiber with a moisture-sensitive cladding has been developed for this purpose. It is simple to use and has given good performance results. The cladding of the sensitive fiber section is a plastic film doped with the umbelliferon dye, which is a moisture-sensitive fluorescent material when pumped with UV light. The sensitive fiber section is placed over the patient's mouth and excited with a He-Cd laser or a halogen lamp. Since the water vapor content in human exhalation exceeds that in the room, the patient's exhalation produces a fluorescent signal that is detected by an electro-optical unit at the nurses' station. This kind of monitoring is particularly useful in detecting abnormal breathing in bedridden patients [22,23].

Another very simple, yet exceedingly interesting, FOS has been developed for the continuous monitoring of respiratory rate. The sensor is based on the change of light reflection at the fiber end, arising from a change in the condensed humidity from the airways during respiration. An optical fiber is simply clasped inside the nostril. During inspiration, the air is dry and cool, and there is maximum of backreflected light. During expiration, a water film deposits on the fiber and the backreflected light is reduced by about 50%, so that the respiration rate can be monitored. Sensor output thus consists of a rhythmic signal, the frequency of which represents the respiratory rate [24]. Figure 10.4 shows a view of the optical-fiber end passing from the wet to the dry phase.

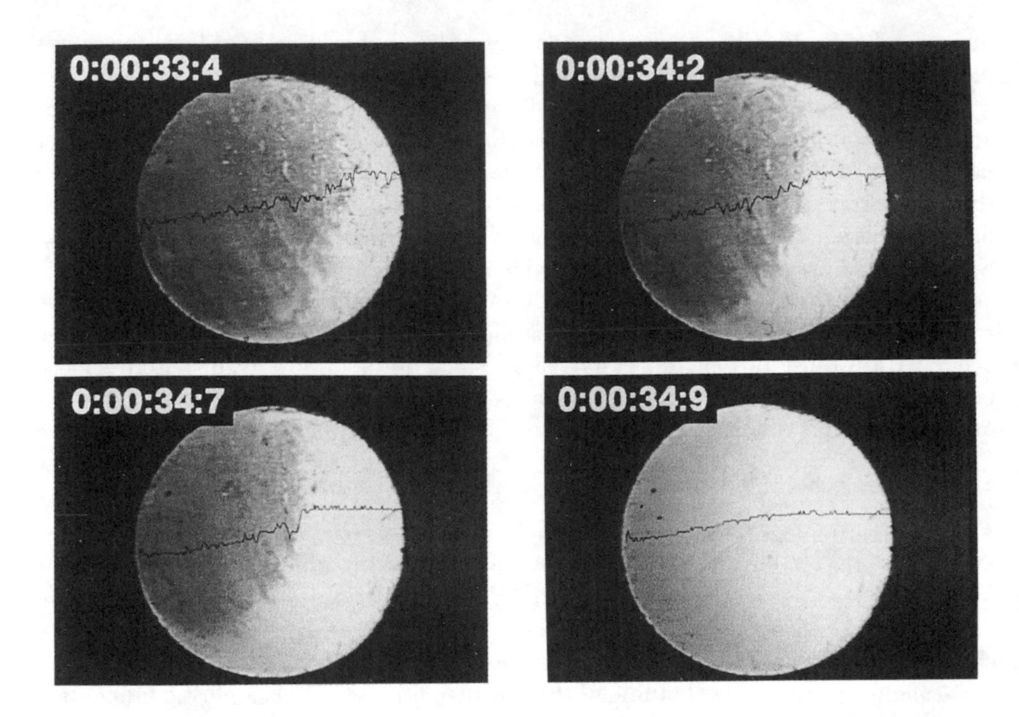

Figure 10.4 Optical-fiber end passing from the wet to the dry phase for respiratory rate monitoring.

10.4 ANGIOLOGY

Blood vessels obstructed by atherosclerotic plaques can be recanalized by means of pulsed excimer laser radiation, guided by an optical fiber (laser angioplasty) [25]. Despite its widespread use, this technique implies a high risk of accidental perforation of the vessel wall, estimated to occur in 26 to 42% of patients. A convenient way to minimize the risk, by identifying the target under irradiation, is laser-induced fluorescence diagnosis of the vessel wall, a measurement that can be performed via the same fiber used for therapy [26]. Its major limitation lies in the difficulty in controlling the distance between the fiber end and the tissue, which influences fluorescence intensity and hence tissue composition determination. To overcome this problem, hybrid photoacoustic systems have been proposed that are capable of recognizing the fiber-tissue distance by means of an ultrasound emitter/receiver placed adjacent to the fiber end [27].

An all-optical approach to target identification is suggested by the fact that short subablation threshold optical pulses, when absorbed by the tissue, generate ultrasonic thermoelastic waves. The amplitude and temporal characteristics of the acoustic signal are dependent upon the target composition and can be detected by a pressure FOS. For this purpose, an optical fiber tipped with a Fabry-Perot cavity is inserted in the lumen of the artery, and the sensor is placed in contact with the tissue [28]. As shown in Figure 10.5, two signals are guided by the optical fiber: the 0.3 mJ pulsed light (10 ns) of a Q-switched frequency-doubled Nd:YAG laser used for generating the thermoelastic wave in the tissue, and the low-power continuous-wave light of a tunable laser diode used for sensor interrogation. The Fabry-Perot cavity at the fiber tip is formed by a polyethylene-terepthalate (PET) polymer film, 50-μm thick, which is transparent at the Nd:YAG wavelength and withstands the optical pulses without being damaged. A water layer, placed between the PET film and the fiber end, provides a fiber-polymer acoustic impedance match. As the polymer film is in contact with the tissue, the stress due to the thermoelastic wave modulates its thickness and hence the optical phase difference between the interfering Fresnel reflections from both sides of the film.

As the sensor head is coaxially positioned with the delivery fiber, measurements are made from the center of the acoustic source, thus giving an improved targeting accuracy. This kind of photoacoustic spectroscopy has been experimentally tested on

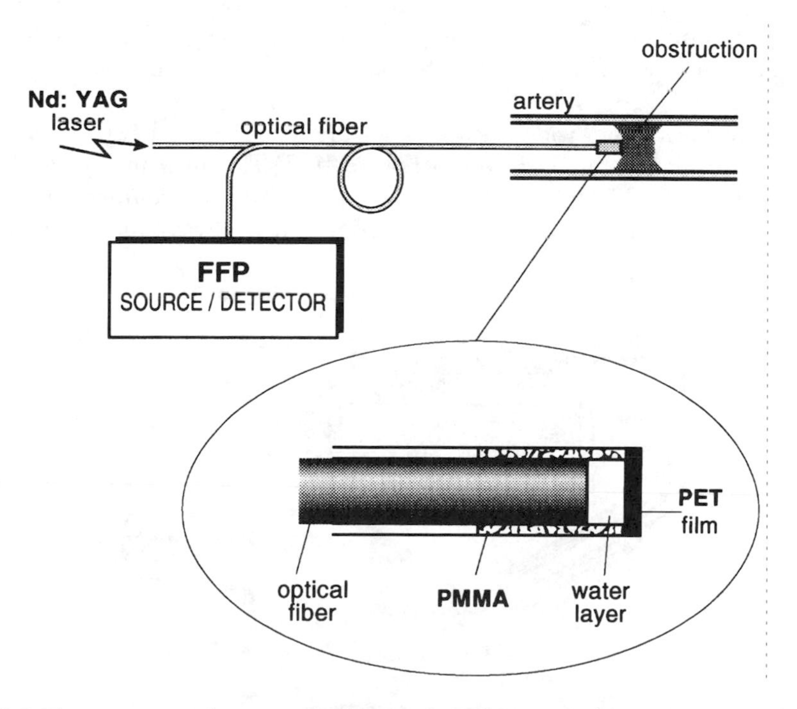

Figure 10.5 Fiber-optic system for target identification during laser angioplasty.

postmortem human aortas: the shape and amplitude of the detected thermoelastic wave being notably different in the case of normal arterial tissue and atherosclerotic plaques. This system, which is at present only a diagnostic tool, could be implemented as a combined diagnostic/therapeutic system by coupling the therapeutic excimer laser radiation via the same fiber.

Apart from disturbing the central circulation, cardiovascular diseases may also influence the peripheral circulation by affecting the microvascular perfusion in tissue. Insufficient peripheral circulation may produce ischemia and ulcers, causing immobilization or hospitalization of the patient. The best method for assessment of microvascular perfusion is laser Doppler flow monitoring [29], where the use of optical fibers can improve the possibilities of both invasive and contact measurements [30,31].

The basic concept of fiber-optic laser Doppler flow measurement is illustrated in Figure 10.6. Light from a He-Ne laser is guided by an optical fiber to the tissue or vascular network being studied. The light is diffusely scattered and partially absorbed within the illuminated volume. Light hitting a moving blood cell undergoes a small Doppler shift, due to the motion of the scattering particles. The blood flow rate is derived by spectral analysis of the frequency shifts in the backscattered signal, which presents a flow-dependent Doppler-shifted frequency. The frequency shift is derived by homodyne or heterodyne detection of scattered light, using the laser light as a local oscillator. Various types of probes have been developed, using either a single fiber for illumination and detection, or one fiber for illumination and two or three for detection [32].

An instrument that is widely used in clinical practice is produced by the Swedish company Perimed AB, and is shown in Figure 10.7. This instrument has a unique detection scheme that is able to extract the Doppler-shifted components from the background noise and to give an output signal that is always proportional to the

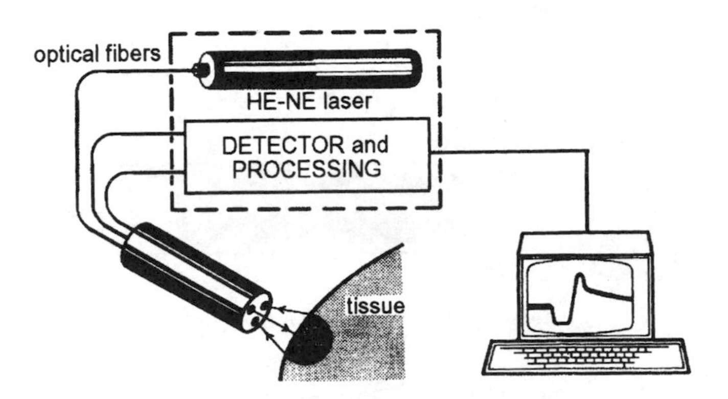

Figure 10.6 Basic scheme of fiber-optic laser Doppler flowmetry.

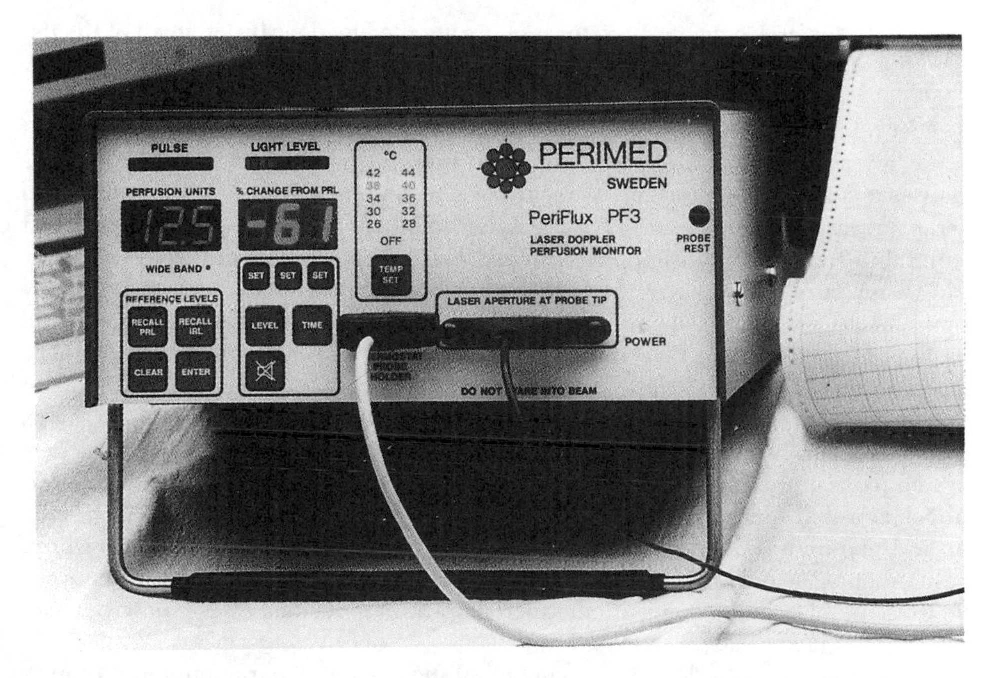

Figure 10.7 Periflux PF3: the Perimed instrument for fiber-optic laser Doppler flowmetry.

perfusion, even at very high blood cell concentrations [33]. Perimed AB also offers a wide selection of contact and endoscopic probes, with straight or angular tips, as well as with channels for liquid flushing. Since many diseases can influence tissue perfusion, optical-fiber laser Doppler flow measurement has been used to advantage in a wide variety of clinical applications, as listed below [34]:

- *Dermatology:* Contact probes for studies of skin physiology and for the testing of skin irritations (scleroderma, psoriasis, drug- or cosmetic-induced irritation/inflammation);
- *Gastroenterology:* Endoscopic probes for blood perfusion measurements in the stomach, duodenum, colon, and rectum;
- *Dentistry:* Contact probes for measuring blood flow in teeth and gingiva, facilitating the assessment of pulp vitality and the state of the gums;
- *Internal medicine:* Contact probes for measuring the vasoconstrictive or vasodilatational responses in the microcirculation of the skin, to assess functional and structural microangiopathies;
- *Vascular surgery:* Contact probes for monitoring the microvascular blood flow during vascular reconstruction and the degree and location of atherosclerosis in the arteries of the lower limbs;

- *Orthopedics:* Contact probes for monitoring the blood perfusion in all of the tissue of the musculoskeletal system during surgery and the postoperative period.

In addition to microvascular perfusion monitoring, an overall assessment of blood flow in the lower limbs is also useful, especially in bedridden patients. This analysis is frequently performed by means of a technique known as venous occlusion plethysmography, which consists of inducing limb expansion using an air-filled cuff with a limb volume-expansion measurement performed downstream, conventionally using a mercury strain gauge. Now, the environmental protection agencies of many countries are prohibiting the marketing of mercury sensors, and this has motivated the implementation of a FOS replacement dedicated to this application [35]. The microbending sensor used consists of two mechanically coupled circular brass disks, each engraved with a surface ridge pattern with a period of 6.6 mm and a step-index plastic fiber sandwiched between the ridged surfaces of the plates. Detection is performed by a second fixed-displacement microbending device equipped with an optical detector that sandwiches the fiber just after the microbending structure. The sensor limb fixture is a flexible, thin steel spring that acts as a mechanical integrator of the nonhomogeneous limb volume expansion. The results of a comparison between the pressure FOS and a conventional mercury strain gauge were extremely satisfactory during *in vivo* tests.

A knowledge of the lipoprotein content of the blood is of paramount importance, as this protein is the carrier of cholesterol. Epidemiological studies have indicated that a reduction in blood cholesterol levels significantly reduces the risk of atherosclerosis, ischemia, myocardial infarction, and premature death. On the other hand, it has been well demonstrated that measuring the total cholesterol in the blood is not sufficient for predicting the risk of cardiovascular diseases; however, a knowledge of the type of lipoproteins (which can differ in size, density, and their lipid and apolipoprotein composition) is essential. For example, high-density lipoproteins, the second largest carriers of plasma cholesterol, have been recognized as being able to remove cholesterol from tissue and, consequently, of being capable of protective action against arterial disease.

The use of optical fibers allows an *in vivo* investigation of arterial lesions. Fluorescence spectroscopy, via optical fibers, permits the imaging of the internal surfaces of arteries. On the other hand, interferences from the whole blood suggest the use of near-IR wavelengths, so such interferences are noticeably reduced. Collection of many broadband near-IR spectra, from one location at a time, has recently allowed the chemical analysis of arterial lesions in living tissues, with the aid of a new experimental clustering technique, for processing of the data using a parallel-vector algorithm [36]. Such an *in vivo* qualitative (recognition of the different types of lipoproteins) and quantitative (concentration evaluation) analysis appears fundamental, both for monitoring changes in arterial wall composition and for testing important new hypotheses describing the formation, growth, and regression of a lesion.

10.5 GASTROENTEROLOGY

The demand for FOSs for *in vivo* monitoring of functional diseases of the foregut is notably increasing. The first, and to date only, commercially available FOS is the Bilitec 2000, commercialized by Synectics Medical AB [37] for the detection of enterogastric and nonacid gastroesophageal refluxes [38] (Figure 10.8). These are considered contributing factors to the development of several pathological conditions, such as gastric ulcers, "chemical" gastritis, upper dyspeptic syndromes, and severe esophagitis. Under certain conditions, the enterogastric reflux may also increase the risk of gastric cancer. Optical detection is based on the optical properties of bile, which is always present in such refluxes [39].

Basically, the Bilitec 2000 instrument utilizes two light-emitting diodes sources at a = 465 nm and 570 nm (reference) and an optical-fiber bundle to transport the light from the sources to the probe (which is actually a miniaturized spectrophotometric cell, of 3-mm external diameter) and the returning light from the probe to the detector. The instrument evaluates the logarithm of the ratio between the light intensities collected by the detection system. According to the Lambert-Beer law, the

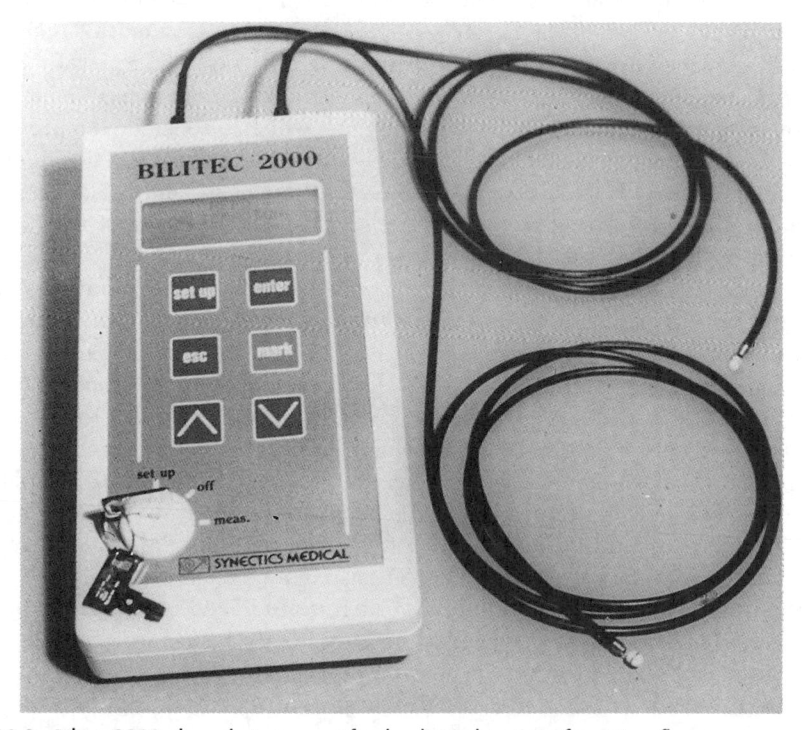

Figure 10.8 Bilitec 2000: the only instrument for the direct detection of gastric refluxes.

difference in the logarithms measured in the sample and those measured in a pure water reference is proportional to the bilirubin concentration. This is related to bile-containing reflux in the stomach and/or esophagus. The method was validated on numerous patients by inserting the optical-fiber bundle into the stomach or esophagus via the nasal cavity [40]. The sensitivity of the sensor was 2.5 μmol/L (bilirubin concentration) and the working range was 0 ÷ 100 μmol/L. This range fits well with that encountered in the stomach or esophagus (although bilirubin concentration in pure bile can be up to 10 mmol/L, it is progressively diluted to its final concentration in the refluxate by pancreatic enzymes, duodenal secretion and, finally, by gastric content). These characteristics clearly refer to *in vitro* tests, whereas, in the case of *in vivo* measurements the nonhomogeneity between the gastric content and the mucus and the solid particles in suspension represents a serious impediment. In such a case, although absorbance values could numerically express bilirubin concentration, they can only allow an approximate quantitative assessment of the overall bile-reflux concentration. However the sensor is able to accurately measure the contact time between the refluxate and the gastric and/or esophageal mucosa.

Another important parameter when studying the human foregut is the gastric and esophageal pH. Monitoring gastric pH for long periods (for example, 24 hours) serves to analyze the physiological pattern of acidity. It provides information regarding changes in the course of the peptic ulcer and enables assessment of the effect of gastric antisecretory drugs. In the esophagus, the gastroesophageal reflux, which causes a pH decrease in the esophagus contents from 7 to 2, can determine esophagitis with possible strictures and Barrett's esophagus, which is considered a preneoplastic lesion. In addition, in measuring the bile-containing reflux, the bile (generally slightly alkaline) and pH should be measured simultaneously, since, due to a shift in the bilirubin absorption peak to lower wavelengths, the accuracy of the bile measurement decreases by about 30% for values of pH < 3.5 [41].

The current practice is to insert a miniaturized glass electrode mounted in a flexible catheter into the stomach or esophagus through the nostrils. This electrical system is, however, impractical, owing to the size and rigidity of the glass electrode, and has the added drawback of being subject to electromagnetic interference. FOSs overcome these problems, although the broad range of interest (from 1 to 8 pH units) requires the use of more than one chromophore, thereby complicating the design and construction of the optrode. This is probably the reason why almost all the pH FOSs developed for biomedical applications have been proposed for blood pH detection, with only a few intended for the detection of gastric or esophageal pH [42–44].

The first sensor proposed for detecting gastric and esophageal pH [42], made use of two fluorophores (fluorescein and eosin) immobilized onto fibrous particles of amino-ethyl cellulose, fixed on polyester foils. Only tested *in vitro*, the sensor reveals a satisfactory response time of around 20 sec. *In vivo* tests have been reported very recently [43–46], but none of the proposed pH FOSs appears completely satisfactory.

A sensor proposed by Peterson and others [46] is based on two absorbance dyes (meta-cresol purple and bromophenol blue) bound to polyacrylamide microspheres. The dyed particles are enclosed in cellulosic dialysis tubing of 300-μm inside diameter, attached to a 250-μm diameter acrylic optical fiber (Figure 10.9). The laboratory optical system arrangement was composed of a lamp plus filters, a fiber-coupled probe, a CCD spectrometer, and a personal computer. The sensor was tested on samples of human gastric fluid and was also tested *in vivo* after inserting the optical probe into the stomach of a dog. The accuracy (better than 0.1 pH units) satisfies clinical requirements, but the response time to each pH step was longer than desired, ranging between 1 and 6 minutes. Such a long response time would prevent the detection of fast changes in pH, and makes the sensor useless for the detection of gastroesophageal reflux, where pH changes are usually extremely rapid (less than 1 minute).

Another sensor makes use of two dyes (bromophenol blue (BPB) and thymol blue (TB)) to cover the range of interest [45]. The chromophores, immobilized on controlled pore glasses, are fixed at the end of plastic optical fibers. The distal end of the fibers is then heated and the CPGs form a very thin pH-sensitive layer on the fiber tips (Figure 10.10). The probe has four fibers (two for each chromophore). A Teflon diffuse reflector was held in front of the fibers using a small fine steel wire in order to improve the return coupling of the modulated light. An optoelectronic unit, similar to that used for bilimetric monitoring, was developed. It consists of two identical channels, separately connected to each of the two fibers carrying TB and BPB for the detection of pH in the ranges 1 to 3.5 and 3.5 to 7.5, respectively. The use of LEDs as sources, solid-state detection, and as an internal microprocessor make it a truly portable, battery-powered sensor (Figure 10.11) [47]. Response time was less than

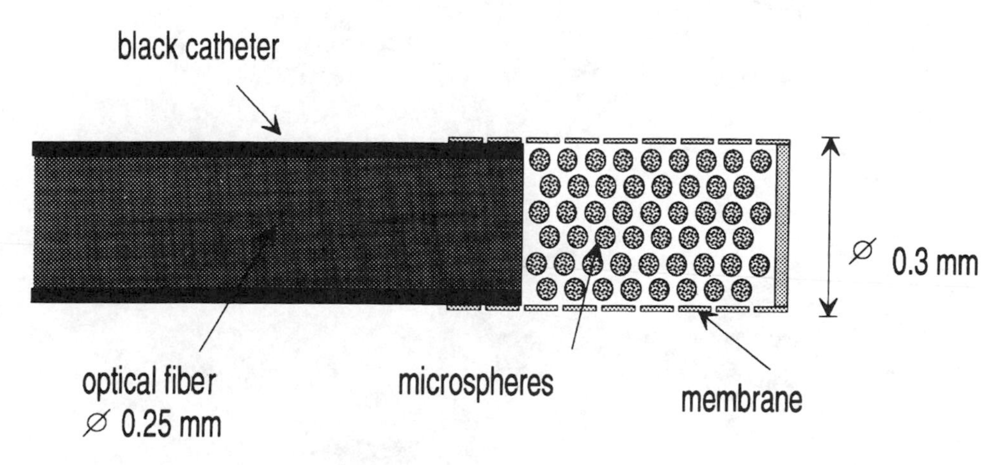

Figure 10.9 Fiber-optic probe for pH monitoring.

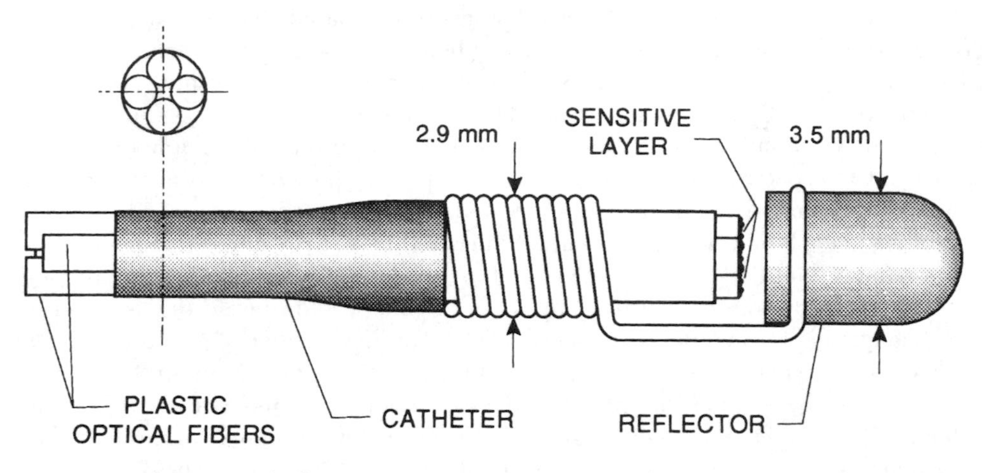

Figure 10.10 Fiber-optic probe for *in vivo* gastric pH monitoring.

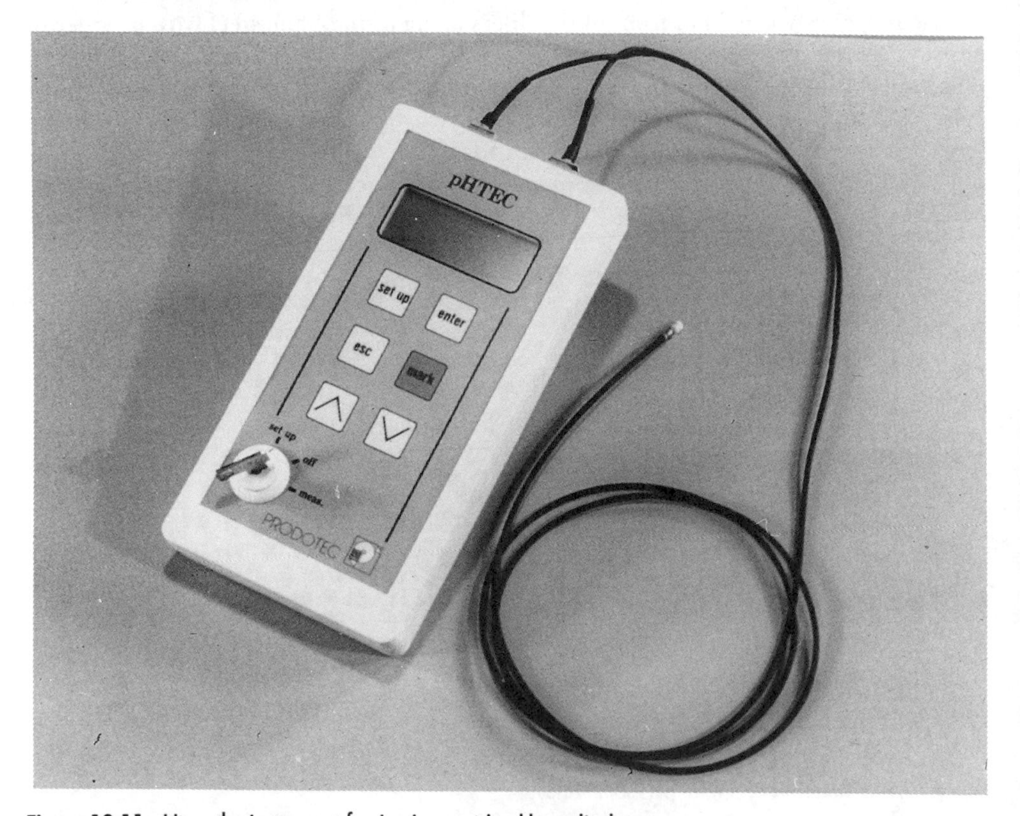

Figure 10.11 pHtec: the instrument for *in vivo* gastric pH monitoring.

1 minute for every pH step. Although the *in vitro* accuracy was 0.05 pH units, satisfactory *in vivo* accuracy has not been attained since, in some cases, a step of some tenths of pH is present between the response of the optical sensor and of the pH electrode.

10.6 OPHTHALMOLOGY

FOSs are attractive noninvasive diagnostic tools for ophthalmology. They have been experimented *in vivo* for monitoring the onset of cataract formation and for the measurements of intraocular distances.

10.6.1 Cataract Onset

A major ophthalmological application of FOSs is the recognition of the onset of eye-lens opacification, commonly known as cataract formation. In addition to being caused by aging, cataracts can be caused by diseases such as hyperglycemia or injury such as exposure to radiations, so the early detection of a warning condition is essential in preventing cataracts and for testing the new therapies that are now becoming available. As opposed to current clinical diagnostic methods, which can only detect cataracts when they have become so severe as to be nearly irreversible, fiber-optic monitoring makes their detection possible at a much earlier stage, in time for reversal.

The lens of the eye is a water-protein system composed of 65 weight percentage water and 35 weight percentage proteins. About 10% of the proteins, called the albuminoid fraction, are insoluble, whereas the remaining 90% of proteins (soluble) are divided into α, β, and γ crystallins. Cataract onset is attributed to crystallin aggregation and can be recognized by periodic measurement of the crystallin dimensions.

A powerful and versatile tool for measuring particle size distribution in fluid systems is the dynamic light-scattering (DLS) technique, also known as photon correlation spectroscopy (see Chapter 9 for more details). In this case, the Brownian motion of crystallins in the cytoplasm illuminated by a coherent light source produces temporal fluctuations in the scattered light. The measured autocorrelation function of the detected light intensity shows an exponential decay, with a decay constant related to the hydrodynamic radius of the scattering particles. Cataract onset is recognized from DLS measurements of the α-crystallin size, which will change during aggregation since the α-crystallins are of greater molecular weight and size, and thus scatter more light than the β- and γ-crystallins.

The noninvasive fiber-optic probe is a stainless steel capillary (OD $\approx$ 5 mm) ending in a faceplate housing two monomode fibers, each fiber sloped at a fixed angle with respect to the capillary axis (Figure 10.12). One of the fibers is coupled to an He-Ne laser to illuminate the scattering region inside the lens of the eye. The other collects the backward scattered light, which is measured by a photomultiplier and processed by a digital correlator. Autocorrelation measurements of the detected light intensity can be processed to show a nearly bimodal distribution of particle size, one maximum corresponding to α-crystallins and the other to their aggregation. The probe can be easily incorporated into conventional ophthalmological instruments, such as an applanation tonometer mount, that is normally used for eye-pressure measurements [48–50].

In addition to allowing prompt, noninvasive monitoring of cataract onset, optical fibers can also enhance the efficiency of the DLS technique. By providing a beam of small numerical aperture and dimensions, monomode fibers are able to produce an extremely small angle of coherence and hence an optimal spatial coherence factor ($\approx$0.9), which would be difficult to obtain with bulk optics systems [51].

The successful results obtained for analyzing the lens of the eye suggested the use of optical-fiber DLS measurements for analyzing the vitreous body or for monitoring cholesterol levels in the aqueous humor. Preliminary tests performed on animals have demonstrated the possibility of measuring the intensity autocorrelation

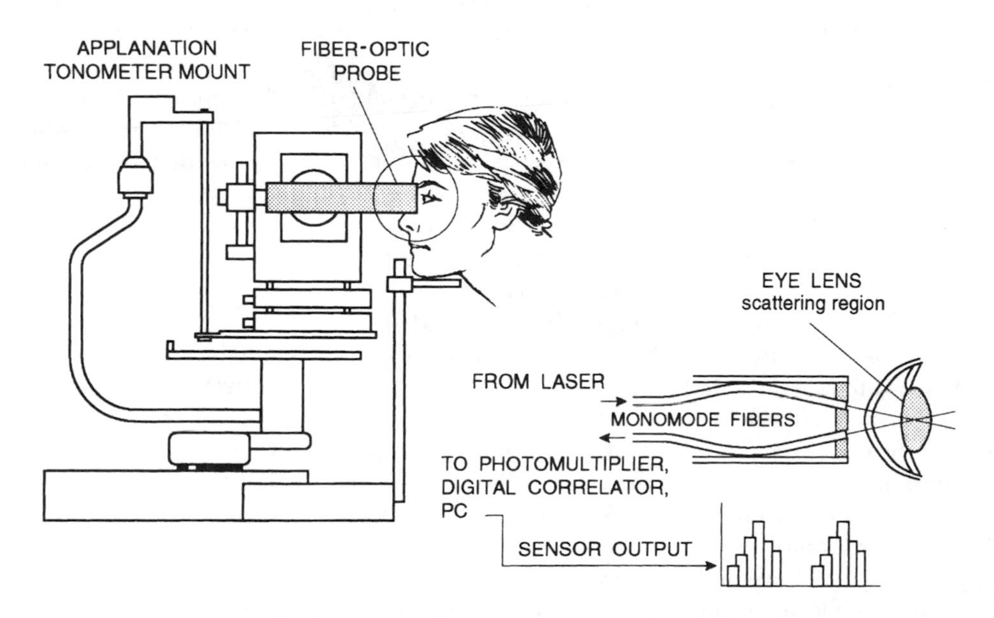

Figure 10.12 Assembly for cataract onset monitoring by optical fibers and dynamic light scattering measurements.

function of scattering particles within the vitreous body of the eye, although, at the present stage, a medical interpretation of the results has still to be defined [52].

10.6.2 Intraocular Distances

The noninvasive measurement of the internal dimensions of the eye is another major problem in ophthalmology. Measurements of corneal thickness are useful in keratorefractive surgery, while measurements of the anterior chamber depth or of the eye axial length are important in determining the refractive power of intraocular lens implants. Standard methods are based on ultrasonic probes, but these need a mechanical contact between the eye and the probe and are not fully satisfactory in terms of the available transversal resolution. Interferometric optical ranging, however, permits high-resolution noncontact measurements.

An optical-fiber Michelson interferometer has been implemented and tested for measuring corneal thickness and anterior chamber depth [53]. As shown in Figure 10.13, a superluminescent diode is coupled to a fiber-optic coupler, one arm of which is used for illuminating a reference mirror and the other of which is used for focusing the light beam onto the cornea and for guiding the reflected beams from the eye layer interfaces to the photodetector. By varying the reference path length by translation of a mirror, coherent interference peaks are measured at the detector when the eyelayer optical pathlengths and the reference pathlengths are equal (within the source coherence length). Therefore, recording the interference signal magnitude as a function of the reference mirror position provides an indication of the axial separation of reflective interfaces within the eye. Unfortunately, since the light reflected by the retina is too weak to be detected, this FOS gives only the measurement of the corneal thickness, of the anterior chamber depth, and of the eyelens thickness.

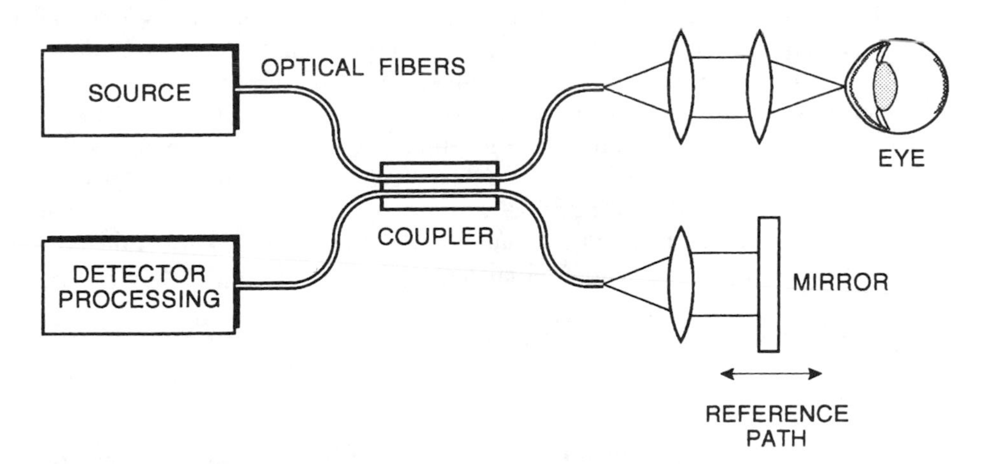

Figure 10.13 Optical-fiber interferometry for corneal thickness measurements.

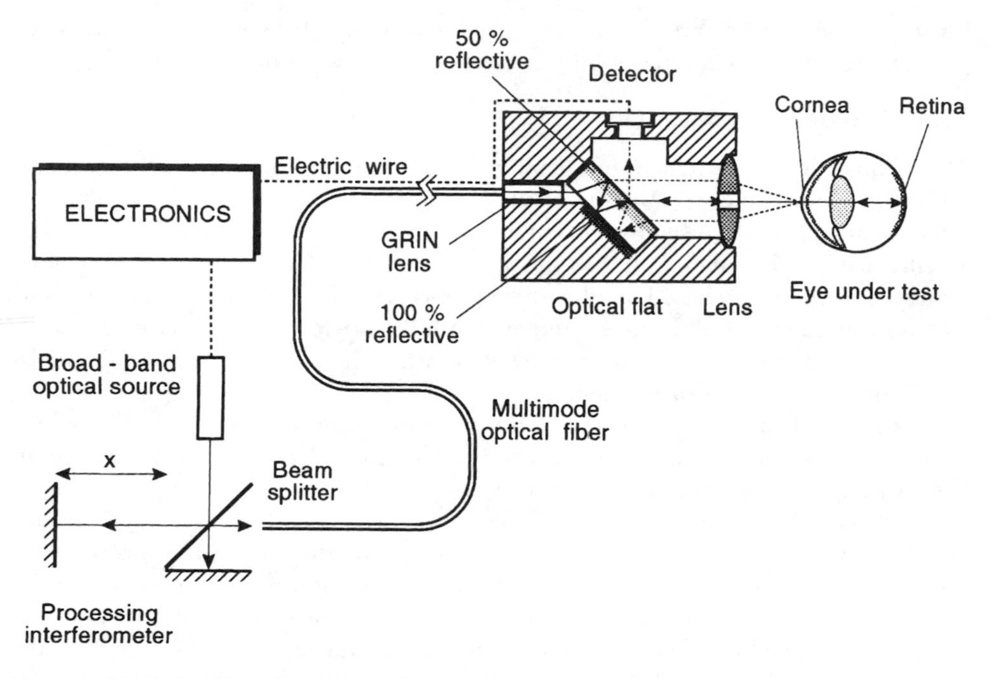

Figure 10.14 Optical fiber white-light interferometry for eye axial-length measurements.

Another FOS has been proposed, capable of extending the detection range to the full axial length of the eyeball [54,55]. This sensor is based on white-light interferometry and makes use of the measuring system illustrated in Figure 10.14. Light from a visible low-coherence source was launched into a Michelson interferometer that acted as a processing interferometer, and the output beam was guided to the ocular probe by means of a multimode optical fiber. The ocular probe housed a detector and a combination of micro-optical components that were capable of both focusing the light on the cornea and illuminating the retina by means of a collimated beam. In addition, they are also capable of collecting the reflections of such beams on the detector. The measurement of the eye length is achieved by scanning the reference Michelson interferometer until a high-contrast interference signal appears, when the optical path difference of the reference interferometer is equal to the length of the eye.

10.7 ONCOLOGY

10.7.1 Therapy

The success of radiotherapy without adverse side effects relies on the online monitoring of the dose to which the tumor and the adjacent tissues are exposed. Conventional thermoluminescence dosimeters only provide offline monitoring, since they determine the radiation exposure after completion of exposure. A short length of heavy-metal-doped optical fiber coupled to a radiation-resistant fiber is an optimal system for the continuous monitoring of radiation dosage in both invasive and noninvasive applications. The light propagating in the suitably doped fiber section undergoes an intensity attenuation in the presence of radiation. By suitable choice of dopant, the attenuation coefficient is nearly linearly related to the radiation dose. Two-wavelength differential attenuation measurement allows compensation cable and connector losses [56].

Another frequent tumor therapy, hypothermia, requires tissue temperature monitoring during radiofrequency or microwave irradiation. The fiber-optic thermometer commercialized by the American company Luxtron [57], one of the first developed for this purpose to be insensitive to electromagnetic radiation, stands out for technical performance and patient safety (Figure 10.15). New sensors based on a thermally expanded fiber Fabry-Perot (FFP) cavity coupled at the fiber tip [58] and midinfrared (MIR) fibers used as light-gathering extensions to pyrometers [59] are currently under development.

Figure 10.15 Luxtron 3000: an eight-channel fiber-optic temperature sensor.

10.7.2 Diagnostics

Minimally invasive methods for oncological diagnostics make use of optical fibers to measure tissue fluorescence. Two types, simple twin-fiber systems and more sophisticated optimized fiber ring catheters for insertion in fiber endoscopes, have been developed:

- *Twin-fiber probes:* These probes are generally of a pH sensing type and are used for *in vivo* mapping of normal and tumoral areas, since malignant tumors induce a decrease in the pH of the interstitial fluid and depression of pH is caused by the administration of glucose. Dual-wavelength fluorimetry via optical fibers provides a new diagnostic tool for highly localized measurements. A nontoxic pH-dependent indicator based on fluorescein derivatives is injected in the tissues to be analyzed, and the twin-fiber probe illuminates the tissue and measures the fluorescence intensities at 465 and 490 nm. The ratio of these fluorescence intensities is related to the pH of the tissue, thus providing normal-neoplastic tissue mapping [60,61].
- *Fiber-ring catheters:* These probes are used for spectral autofluorescence analysis of tissues during endoscopy. Tissue autofluorescence excited by a He-Ne laser is observed and spectrally analyzed in the 660- to 850-nm range. The differences in spectral distribution between normal, cancerous, and necrotic tissues allow for real-time diagnostics without necessitating biopsy. Stomach, bronchial, and lung tissues have been diagnosed by fiber catheters, combining imaging fiber bundles with custom bundles optimized for simple light collection for fluorescence measurements [62].

MID-IR fiber-based systems have been tested for *in vitro* tissue spectra measurements with evanescent fiber probes coupled to Fourier Transform spectrometers. They have demonstrated the possibility of recognizing differences in spectra between normal and malignant tissue (kidney, stomach, lungs) [63] and their suitability for diagnosing the condition of atherosclerosis [64].

10.8 NEUROLOGY

Head trauma patients require continuous monitoring of intracranial pressure. During the postoperative and drainage-monitoring phases, it is essential to know both the subdural and ventricular pressures and their temporal variations. Nonoptical instruments make use of catheters tipped with miniaturized piezoresistive or capacitive transducers, but their main drawbacks are long-term drifts, electrical shock hazard, fragility, and expense. FOSs, being relatively easy to manufacture and therefore inexpensive enough to be disposable, can overcome these drawbacks. In addition, they can perform as well as or better than electrical instrumentation. Main measurement requirements are 1) a working range from −50 to 300 mmHg; 2) a sensitivity of at least 0.1 mmHg; 3) an accuracy of at least 1%; 4) a flat frequency response up to 1 kHz; and 5) no shock hazard.

Among the many proposed pressure FOSs, two types fulfill the low-cost, high-performance requirements. One has a Fabry-Perot cavity at the fiber tip [65], the other has a small diaphragm in front of the fiber-optic link [66–68].

The Fabry-Perot cavity for pressure sensing is formed in a glass cube, having a partially etched face to produce the cavity, which is then covered by a pressure-sensitive silicon diaphragm (Figure 10.16). The pressure, deflecting the diaphragm, alters the cavity depth and thus the optical cavity reflectance at a given wavelength. If an LED source is used, the spectral variation in the reflected light can be observed by splitting it into two wavebands by a dichroic mirror. The ratio of the two signals provides a pressure measurement immune to the light level changes that can occur in FOS systems. A commercial sensor of this type, together with the complete intracranial pressure monitoring system, is currently available from the American company Innerspace [69].

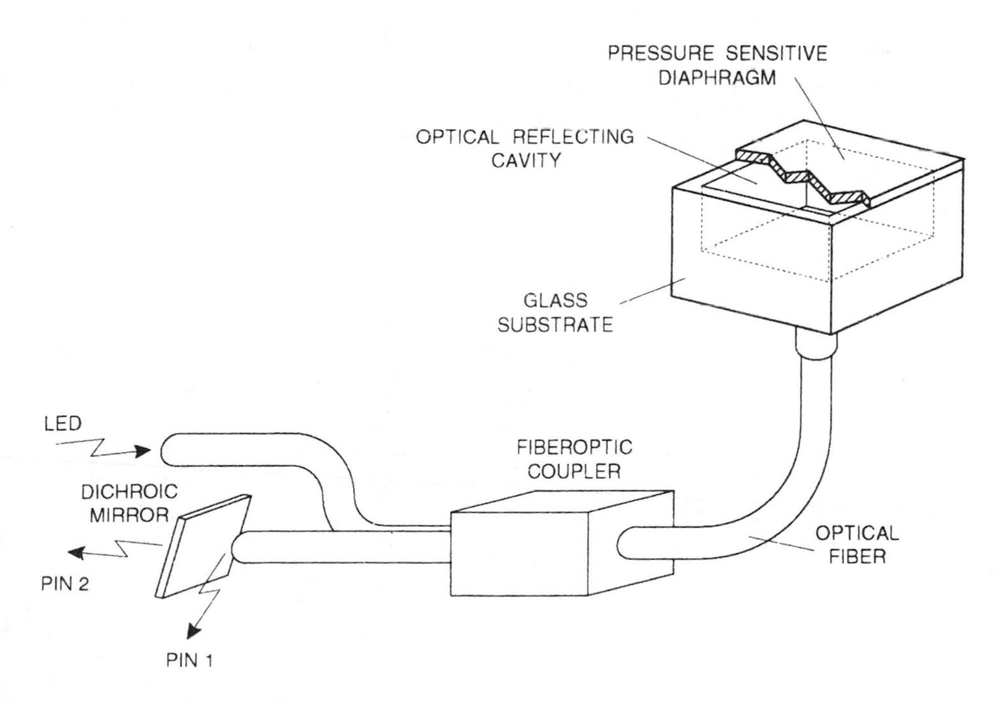

Figure 10.16 Fabry-Perot fiber-optic pressure sensor.

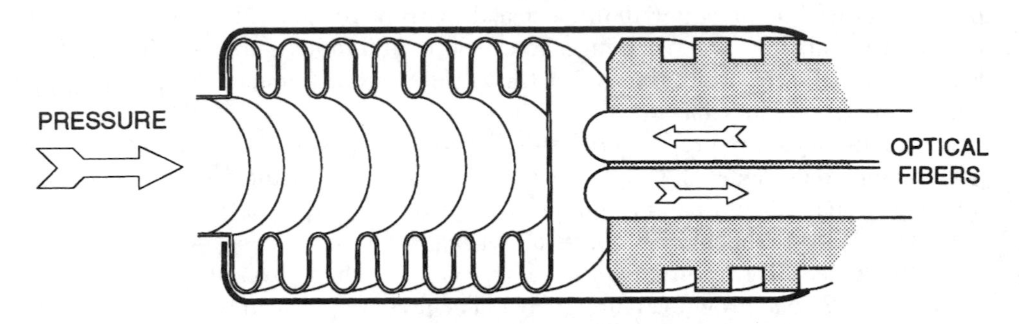

Figure 10.17 Fiber-optic pressure sensor based on a mechanical transducer.

The other pressure sensing approach, which uses a diaphragm close to the fiber-optic link itself, is again based on the light intensity modulation of the reflected light caused by the pressure-induced position of the diaphragm. Two schemes can give immunity to false fluctuations: 1) a dichroic coating on the diaphragm and detection with a dual-wavelength referencing technique, and 2) a dual-beam referencing using a second fiber-optic path, which is joined to the pressure measuring fiber link to render it insensitive to pressure fluctuations. A low-cost disposable FOS, manufactured by the American company Camino Labs [70], uses a bellows to create a pressure-sensitive displacement that is interrogated optically by the intensity modulation technique with dual-beam referencing (Figure 10.17) [71].

10.9 DERMATOLOGY

An important medical requirement is a knowledge of skin condition. The health of skin can be roughly determined by monitoring the skin's hydration state and, more accurately, by determining the lipid content. The latter parameter is of particular importance in acne studies. An inexpensive fiber-optic refractometer has been developed for both measurement tasks.

The sensor comprises a plastic fiber, with a short section stripped of its cladding and bent into an U shape. With the U-shaped tip placed against the skin, the skin surface acts as the fiber cladding. Since the skin surface is not perfectly smooth, some gaps are created around the fiber core. At the time of initial probe contact, the gaps are filled with air, but while the probe is held against the skin, they fill up with skin moisture. Consequently, the sensor transmission decreases with time, at a rate dependent on the hydration state. The skin moisture content is obtained after calibration of the probe [72].

It has been noted that, the higher the lipid contents, the higher the light reflection, thus reducing sensor output. This disturbance has been used for estimating lipid quantity from a comparison of the different sensor outputs from solvent-defatted and unprepared skin. Again, with calibration, the lipid content can be successfully monitored with the experimental results in good agreement with those obtained by the traditional gravimetric technique [73]

Another FOS developed for dermatological applications aids in quantifying erythema, an acute inflammatory skin disease that is characterized by localized redness of the skin in areas of variable size. The optical-fiber erythema meter (Figure 10.18) has been developed for allergy and irritancy testing, as well as for objectively studying UV-induced erythema. Skin color is mostly the result of blood quantity and melanin content. Skin color measurements are performed by a fiber-optic bundle coupled to a dual-wavelength reflectometer, which measures the ratio of skin reflectance at 555 nm (which is modulated by blood quantity), to that at 660 nm (which is melanin-sensitive but not influenced by blood quantity and hence is used as a reference) [74,75].

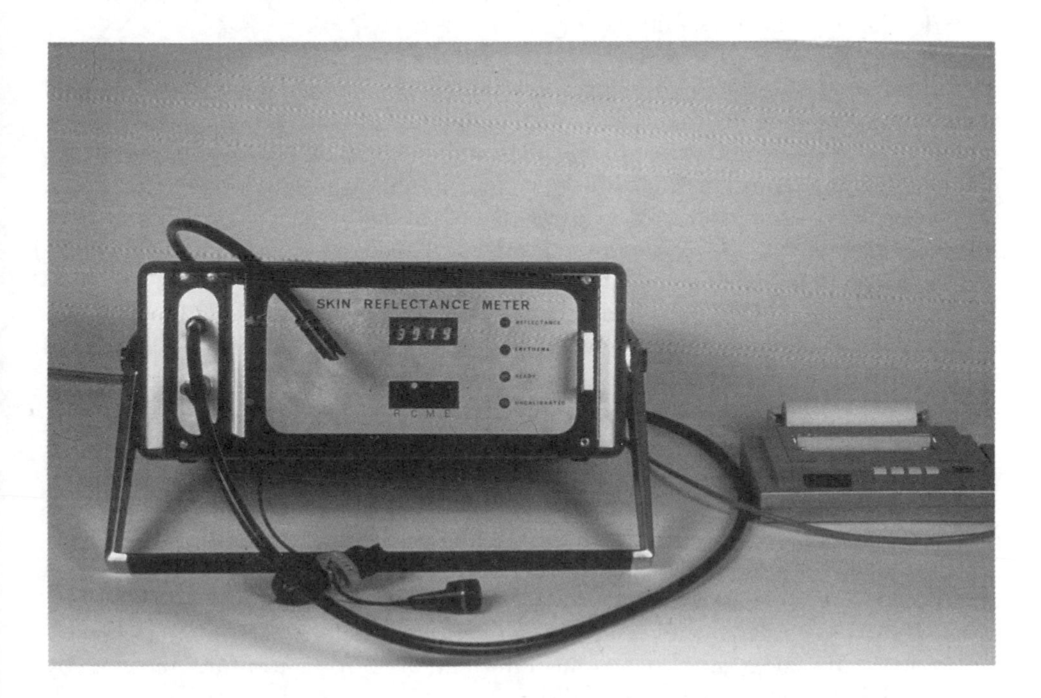

Figure 10.18 Fiber-optic erythema meter.

10.10 DENTISTRY

Matching of the color of teeth is a constant problem in restorative dentistry. The disadvantages of subjective visual comparison of natural tooth color with standard colors can be overcome by use of an objective instrument using a fiber-optic reflectance spectrometer. This sensor, originally developed for tissue diagnosis, can be a great help in selecting the material to best replicate the appearance of natural teeth. A slender, flexible polymethylmethacrylate-plastic (PMMA) fiber-optic bundle is used to illuminate a portion of the tooth with white light and to guide reflected light from the tooth to a portable grating spectrometer. The reflectance spectrometer can also be used for diagnosing gingiva using bent probe heads inserted in the patient's oral cavity [76].

A major problem in dentistry is the assessment of the condition of dental pulp. Pulp inflammation is a very common pathology, as pulp necrosis is the most common complication following dental trauma. While tooth vitality can be evaluated by conventional diagnostic techniques, such as radiographic examination or painful subjective tests of neural response, more attractive optical methods that determine pulp condition by noninvasive measurement of blood perfusion or of blood oxygenation have been developed and successfully tested.

A simple diagnosis that provides information on whether the pulp is vital or necrotic is based on photoplethysmography (i.e., the detection of fluctuations of blood absorbance occurring in synchronism with systolic heart contractions). The tooth is squeezed between two reflective prisms placed at the end of a fiber-optic bundle to illuminate the tooth and to collect the light transmitted through it. The ratio between the recorded ac (@ 0.5 Hz) and dc components of the detected signal at 570 nm, constitutes sensor output, i.e. the tooth plethysmogram. Vitality is indicated by a rhythmic plethysmogram, necrosis by a random plethysmogram [77].

Using the laser Doppler technique via optical fibers to assess pulp blood perfusion provides an extremely accurate diagnosis. However, the instrument is considered too expensive for routine diagnosis, and the use of laser equipment requires specific training for personnel as well as special precautions in usage.

A simpler dental pulp vitalometer for everyday dentistry is based on measuring oxygen saturation of the pulp, since inflammations cause deoxygenation of hemoglobin. The oxygen saturation content of blood in the teeth is obtained by a fiber-optic probe, which is placed in contact with the tooth to perform reflectance measurements at 660 nm and 850 nm. The signal ratio is processed together with a nonoptical plethysmogram signal (pulse pressure versus time) from one of the patient's fingers, which is used as a reference in determining the timing of the blood pulse. The resultant tooth plethysmogram also contains information on oxygen saturation, thereby allowing pulp vitality to be assessed [78].

Another FOS for dentistry measures biting force, which is an essential parameter in studying disturbances of the masticatory system and to examine the capability of patients with osseointegrated implants or dentures. Unlike conventional monitors, based on piezoelectric crystals, quartz crystals, and resistive strain gauges-all of which are unsuitable due to the high conductivity of the oral cavity, optical fibers are intrinsically safe. The FOS consists of a mouthpiece made of two stainless steel plates with a microbending FOS placed in between. The system is managed by a personal computer with software providing display, zeroing, and calibration. The dynamic range for measurement of biting force was 1 to 1,000N with a resolution of 10N, which is satisfactory for the application [79].

10.11 FUTURE TRENDS

The FOSs described above have given good results in *in vivo* testing. Some of these are already commercially available, while others are being developed in response to demands from the medical profession and from encouraging market studies. There appears to be a particularly strong need for optical biochemical sensors and multi-function sensors.

10.11.1 Biochemical Sensors

As far as chemical sensors are concerned, a recent survey has evidenced the main areas in which *in vivo* sensors would be helpful [80]. Table 10.1 reports the results of this survey. The number in parentheses for each analyte shows the number of physicians who answered the survey.

Some of the required analytes are already being monitored online with FOSs and have been described here. Other FOSs (glucose, potassium, urea, lactate)

Table 10.1
Main Clinical Problems and Related Analytes for Which a Continuous *In Vivo* Monitoring Should Be Performed

Clinical Problem	*Analyte*
Diabetes mellitus	glucose (24), K^+ (3), ketones (2), insulin (2), lactate (1), pH (1)
Vital function monitoring in intensive care/anaesthetics/prolonged surgery	O_2 (15), CO_2 (10), pH (8), haemoglobin (3), K^+ (2), glucose (2), electrolytes (unclassified) (1), gases (unclassified) (1), Na^+ (1), osmolality (1), lactate (1)
Renal failure/monitoring dialysis	urea (4) creatinine (2), K^+ (2), atrial natriuretic peptide (1), pH (1)

have either been proposed or seem promising, but none has as yet undergone *in vivo* testing.

Glucose is the analyte most in demand, because its control is of vital importance in diabetic patients, and a glucose sensor is an integral part of an artificial pancreas, which requires controlled insulin release. In patients affected by diabetes mellitus, a monitoring of blood glucose levels is needed during the whole day; however, so far no continuous sensor is available, obliging patients to monitor their own blood glucose by periodic sampling and use of chemistry reagent strips (color indicator strips).

Alternative approaches have been presented using fiber optics, but at the moment some problems for their *in vivo* utilization still exist. Reversible competitive binding of glucose with fluorescence labeled dextran has been proposed for monitoring the sugar binding sites of a lectin, concanavalin A [81,82]. The probe consists of an appropriate hollow fiber coupled to fiber optics, capable of preventing the exit of reagents but permeable to glucose (selection made on the basis of the molecular weight). Glucose, if present, takes the place of fluorescence labeled dextran in the bond with concanavalin A and an increase of fluorescence is detected. There are still problems with the poor lifetime of the probe and excessive response time. Another approach makes use of glucose oxidase (an enzyme that catalyses the oxidation of glucose).

By exploiting the enzymatic reaction, it is possible to indirectly monitor glucose levels from a measurement of oxygen [83] or pH [84]. Although sensitive, and in some aspects satisfactory (for example, having a fast response time of less than 1 minute), this approach seems more appropriate for laboratory diagnostics than for *in vivo* sensing. This is mainly due to the deterioration of the chemicals utilized, which strongly limits the lifetime of the probes. An interesting alternative is to exploit the optically active properties of the glucose, which is capable of inducing a rotation of the polarization state of light. The amount of rotation can be related to the glucose concentration. Although not so far used with optical fibers, this optical method seems very promising and has also been tested for the noninvasive monitoring of the glucose in the aqueous humor of the eye [85,86].

The advantage of this direct spectrophotometric technique is apparent and enormous: detection of the analyte is made without chemical reactions; the probe manufacturing is simplified, and the problem of the lifetime of the probe is eliminated.

Identical advantages are presented by infrared spectroscopy. The advent of IR transmitting fibers provides the possibility of observing the rotational/vibrational bands of molecules via fiber-optic probe.

The use of infrared spectroscopy via IR fibers for *in vivo* monitoring in biomedicine is now taking its first steps. Preliminary results are very encouraging. *In vitro* spectral analysis of human blood serum (performed with nontoxic IR-transmitting silver-halide fibers connected to an FTIR spectrometer) has recently been proposed [87], making possible the simultaneous monitoring of cholesterol, urea, total protein, uric acid, and calcium after suitable spectral analysis. Continuous monitoring of respiratory gases could be performed *in situ* by tunable diode laser spectroscopy (TDLS) via IR fibers. with. TDLS has recently been successfully applied to the determination of trace gases, such as CO, CO_2, NH_3, CH_4 and NO, in the exhalation of both humans and animals [88].

10.11.2 Multifunction Sensors

Another important area where improvement in technology is needed is to extend the ability for simultaneous monitoring of many parameters with a single optoelectronic unit. Multifunction or multiplexed FOSs for medical application are in the development stage and many others are being encouraged by doctors, who greatly appreciate the advantage of a multitest portable unit with low-cost disposable probes.

10.11.2.1 Multianalyte Sensors

In this direction, there is pioneering work being carried out at Tufts laboratory under the direction of Professor D. Walt. His approach, still not tested *in vivo* [89], makes possible simultaneous multianalyte detection with a single bundle of imaging fibers (1,500 individual 10-μm fibers, forming a bundle with an overall diameter of 400 μm). Spatial discrimination is obtained by creating, at the distal end of the bundle, separate zones, each having different indicating chemistry. These zones are formed by means of a program of several photopolimerization stages. The same optoelectronic system is used (but in opposite directions) for both photodeposition and detection (Figure 10.19). Fluorescein for pH and pCO_2 detection and a ruthenium complex for pO_2 were immobilized on different zones of the same bundle with this method. Since imaging fibers are used, combined *in vivo* imaging and chemical sensing (which would represent a noticeable progress in clinical diagnostics) are possible with the same fiber bundle, the only requisite being the thickness of the sensitive layer on the distal surface of the fiber so as not to alter the imaging capabilities. First laboratory results on pH sensing appear encouraging [90,91].

10.11.2.2 Temperature, Flow Rate, and Blood Pressure

The simultaneous and independent measurement of these three parameters has been achieved by means of a single optical fiber and a combination of three distinct technologies in the sensing tip [92]. Pressure is measured by a compressible dome-shaped optical element attached to the fiber end, which is covered by a reflective coating to provide pressure-dependent reflection due to a pressure-induced deformation of the dome shape. The pressure dome is covered by a fluorescent material that provides temperature measurements using photoluminescence decay-time monitoring [93]. Finally, a near-infrared absorbing layer covers the probe tip, converting the temperature sensor into a flow sensor. Illumination of the tip by means of fiber-coupled

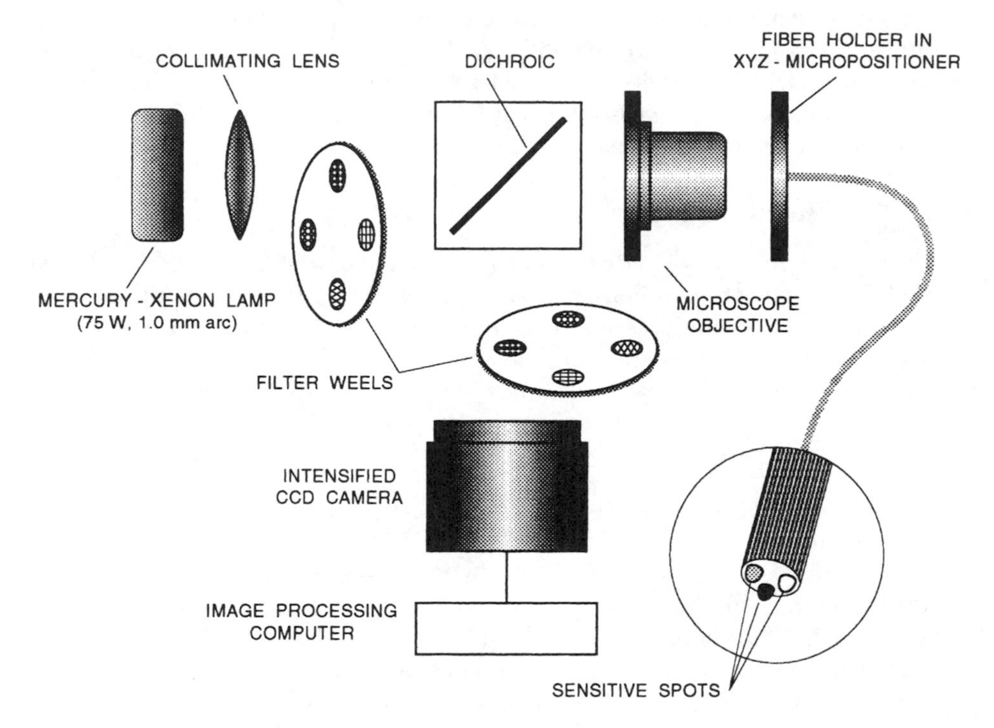

Figure 10.19 Multianalytic detection by means of an imaging fiber bundle: Scheme of the optoelectronic system for both optrode photodeposition and detection.

near-infrared radiation produces a temperature rise. When the tip is exposed to a flowing fluid, heat is carried away by the fluid and a temperature reduction occurs. The flow rate is measured by modulating the intensity of the near-infrared laser diode illuminator in order to maintain a constant thermal modulation on the fiber tip. The tiny dimensions of the sensing tip make it compatible with internal catheters such as those for angioplasty or for use in cardiac output monitoring using the thermodilution technique.

10.11.2.3 Temperature and Pressure

An example of a dual temperature and pressure sensor has been recently reported [94]. The Fabry-Perot cavity pressure sensor is as the one described in the Section 10.8. The temperature sensor is a long-wavelength pass-absorptive filter glass, which exhibits a temperature-dependent absorption-edge position at about 670 nm. The detection scheme for the temperature sensor and the Fabry-Perot pressure sensor are similar, based on the ratioing of longpass and shortpass wavebands.

One side of a cube of the temperature-sensitive glass is coupled to the optical fiber, while the other side has a reflecting dichroic layer to reflect the red light (for thermal sensing) and transmit the near-infrared light used for pressure-sensor interrogation. The near-infrared radiation lies in a spectral region in which the transmission of the temperature transducer is flat and hence it does not affect temperature measurements.

10.11.2.4 Oxygen Saturation and Blood Pressure

This dual sensor has been realized using two distinct sensors integrated in a single catheter. The pressure sensor is based on a reflective diaphragm at the end of a fiber probe that is mechanically displaced by pressure. The oxygen saturation is determined by a spectral attenuation measurement of blood using a short piece of unclad optical fiber to form an evanescent transmission probe [95].

10.12 CONCLUDING REMARKS

The motivation for developing new types of FOSs for *in vivo* monitoring, as well as for the further experimentation of existing sensors, mainly depends on the demand and decisions on the part of physicians. In fact, FOSs have a place and a well-defined importance when they are presented as the only instruments capable of performing a certain type of diagnostics, possibly with additional applications spinning off as a result. Two prominent examples are the following.

1. The FOS for bile-reflux monitoring, addressed to an area in which no instrument for similar diagnostics exists, which made possible new interpretations of the monitoring being carried out. For example, it has become possible to detect a reflux during the early morning hours, even in healthy patients: this previously unknown aspect is currently in the discussion and interpretation phase.
2. The FOS for monitoring the onset of cataract formation, which makes possible a detection that no other techniques can perform and, in addition, a verification check of the efficacy of drugs that would otherwise not be possible.

Conversely, where FOSs are not the only available method, acceptance is not automatic and, owing to the unpredictable difficulties that can arise in testing and/or results differing from those forecast in the laboratory, even FOSs that have already demonstrated their technical-scientific validity have not always received the acceptance they deserve. For example, the utility of laser angioplasty is constantly being discussed since operators do not attribute to it any marked advantage over the traditional technique of balloon angioplasty. Consequently, FOSs for laser angioplasty are lagging behind in application.

As far as a long-range forecast is concerned, multifunctional sensors indubitably represent the most sought after frontier for the field of fiber-optic sensors. Diagnostic areas in which no satisfactory instrument exists, for instance, are those of haematic analyses in general, and of the measurement of glucose in diabetic patients, both of which can be carried out with sensors providing nonstop and minimally invasive monitoring.

ACKNOWLEDGMENTS

The authors are grateful to Professor Annamaria Scheggi for her many helpful discussions and constructive suggestions. Thanks are also due to Dr. Klaus Frank, BGT, Überlingen, Germany; Professor Harri Kopola of the University of Oulu, Oulu, Finland; Dr. Gordon Mitchell of Future Focus, Woodinville, WA; Dr. John Peterson of the National Institutes of Health, Bethesda, MD; Dr. Mei Sun of Luxtron, Santa Clara, CA; and, lastly, to Professor Takashi Takeo of the Nagoya Municipal Industrial Research Institute, Nagoya, Japan, for providing illustrative material.

References

[1] Kapany, N. S., and Silbertrust, N., "Fiber optics spectrophotometer for *in-vivo* oximetry," *Nature*, 208, 1964, pp. 138–145.

[2] Milano, M. J., and Kim, K. Y., "Diode array spectrometer for the simultaneous determination of hemoglobin in whole blood," *Analytical Chemistry*, 49, 1977, pp. 555–561.

[3] Lubbers, D. W., "Chemical *in-vivo* monitoring by optical sensors in medicine," *Sensors and Actuators B*, 11, 1993, pp. 253–262.

[4] Scoggin, C., Nett, L., and Petty, T. L., "Clinical evaluation of a new ear oximeter," *Heart and Lung*, 6, 1977, pp. 121–126.

[5] Merrick, E. B., and Hayes, T. J., "Continuous, non-invasive measurements of arterial blood oxygen levels," *Hewlett Packard J.*, 28, 1976, pp. 2–9.

[6] Mendelson, Y., and Ochs, B. D., "Noninvasive pulse oximetry utilizing skin reflectance photoplethysmography," *IEEE Trans. Biomedical Engineering*, BME-35, 1988, pp. 798–805.

[7] Cui, W., Ostrander, L. E., and Lee, B. Y., "*In-vivo* reflectance of blood and tissue as a function of wavelength," *IEEE Trans. Biomedical Engineering*, BME-37, 1990, pp. 632–639.

[8] Ugnell, H., and Oberg, P. A., "The time variable photoplethysmographic signal. Its dependance on light wavelength and sample volume," *Proc. SPIE Medical Sensors II and Fiber Optic Sensors*, 2331, Los Angeles, CA, Jan. 24–27, 1994, pp. 89–97.

[9] Frank, K. H., et al., "*In-situ* monitoring of organs," in *Handbook of Critical Care*, J. L. Ber and J. E. Sampliner (Eds.), Boston, MA: Little, Brown and Company, 1990, pp. 145–159.

[10] Frank, K. H., et al., "The Erlangen micro-lightguide spectrophotometer Empho I," *Phys. Med. Biol.*, 34, 1989, pp. 1883–1900.

[11] Holst, G. A., et al., "FLOX-an oxygen-flux-measuring system using a phase modulation method to evaluate the oxygen dependent fluorescence lifetime," *Sensors and Actuators B*, 29, 1995, pp. 231–239.

[12] Peterson, J. I., Fitzgerald, R. V., and Buckhold, D. K., "Fiber-optic probe for *in-vivo* measurement of oxygen partial pressure," *Analytical Chemistry*, 56, 1984, pp. 62–67.

[13] Stefansson, E., Peterson, J. I., and Wang, Y. H., "Intraocular oxygen tension measured with a fiber optic sensor in normal and diabetic dogs," *American Journal of Physiology*, 256, 1989, pp. H1127–33.

[14] CDI-3M Health Care, 1311 Valencia Avenue, Tustin, CA 92680.

[15] Gehrich, J. L., et al., "Optical fluorescence and its application to an intravascular blood gas monitoring system," *IEEE Trans. Biomedical Engineering*, BME-33, 1986, pp. 117–132.

[16] Miller, W. W., et al., "Performance of an *in-vivo*, continuous blood-gas monitor with disposable probe," *Clinical Chemistry*, 33, 1987, pp. 1358–1365.

[17] Hansmann, D. R., and Gehrich, G. L., "Practical perspectives on the *in-vitro* and *in-vivo* evaluation of a fiber optic blood gas sensor," *Proc. SPIE Optical Fibers in Medicine III*, 906, Los Angeles, CA, Jan. 13–16, 1988, pp. 4–10.

[18] Khalil, G., and Malin, S. F., "*In-vivo* blood gases: promises and realities," *Proc. SPIE Advances in Fluorescence Sensing Technology II*, 2388, San Jose, CA, Feb. 6–8, 1995, pp. 502–506.

[19] Khalil, G., Yim, J., and Vurek, G. G., "*In-vivo* blood gases: problems and solutions," *Proc. SPIE Biomedical Fiber Optic Instrumentation*, 2131, Los Angeles, CA, Jan. 24–27, 1994, pp. 437–451.

[20] Schlain, L., and Spar, S., "Continuous arterial blood gas monitoring with transmitted light sensors and LED light sources," *Proc. SPIE Biomedical Fiber Optic Instrumentation*, 2131, Los Angeles, CA, Jan. 24–27, 1994, pp. 452–458.

[21] Martin, R. C., et al., "Performance and use of paracorporeal fiber optic blood gas sensors," *Proc. SPIE Biomedical Fiber Optic Instrumentation*, 2131, Los Angeles, CA, Jan. 24–27, 1994, pp. 426–436.

[22] Muto, S., et al., "Breathing monitor using dye-doped optical fiber," *Japanese J. of Applied Physics*, 29, 1990, pp. 1618–1619.

[23] Muto, S., Sato, H., and Hosaka, T., "Optical humidity sensor using fluorescent plastic fiber and its application to breathing-condition monitoring," *Japanese J. Applied Physics*, 33, 1994, pp. 6060–6064.

[24] Öberg, P. Å., et al., "Evalution of a new fibre-optic sensor for respiratory rate measurements," *Proc. SPIE Medical Sensors II and Fiber Optic Sensors*, 2331, Lille, France, Sept. 9, 1994, pp. 98–109.

[25] Margolis, J. R., et al., "Excimer laser coronary angioplasty: American multicenter experience," *Hertz*, 15, 1990, pp. 223–232.

[26] Andersson-Engels, S., Johansson, J., Stenram, U., Svamberg, K., and Svamberg, S., "Malignant tumor and atherosclerotic plaque diagnosis using laser-induced fluorescence," *J. Quantum Electronics*, QE-26, 1990, pp. 2207–2217.

[27] Warren, S., et al., "Combined ultrasound and fluorescence spectroscopy for physico-chemical imaging of atherosclerosis," *IEEE Trans. Biomedical Engineering*, BME-42, 1995, pp. 121–132.

[28] Beard, P. C., and Mills, T. N., "An optical fibre sensor for the detection of laser generated ultrasound in arterial tissues," *Proc. SPIE Medical Sensors II and Fiber Optic Sensors*, 2331, Lille, France, Sept. 9, 1994, pp. 112–122.

[29] Stern, M. D., "*In-vivo* evaluation of microcirculation by coherent light scattering," *Nature*, 254, 1975, pp. 56–58.

[30] Nilsson, G. E., Tenland, T., and Öberg, P. Å., "A new instrument for continuous measurement of tissue blood flow by light beating spectroscopy," *IEEE Trans. Biomedical Engineering*, BME-27, 1980, pp. 12–19.

[31] Nilsson, G. E., Tenland, T., and Öberg, P. Å., "Evaluation of a laser Doppler flowmeter for measurement of tissue blood flow," *IEEE Trans. Biomedical Engineering*, BME-27, 1980, pp. 597–604.

[32] Shepherd, A. P., and Öberg, P. Å., *Laser Doppler Blood Flowmetry*, Dordrecht, The Netherlands: Kluwers Academic Press, 1990.

[33] Nilsson, G. E., "Signal processor for laser Doppler tissue flowmeters," *Medical & Biological Engineering & Computing*, 22, 1984, pp. 343–348.
[34] Perimed Literature Reference List n. 10, 1990. Perimed AB, box 5607, S-114 86 Stockholm, Sweden.
[35] Stenow, E. N. D., and Öberg, P. Å., "Venous occlusion plethysmography using a fiber-optic sensor," *IEEE Trans. Biomedical Engineering*, BME-40, 1993, pp. 284–289.
[36] Cassis, L. A., and Lodder, R. A., "Near-IR imaging of atheromas in living arterial tissue," *Analytical Chemistry*, 65, 1993, pp. 1247–1256.
[37] Synectics Medical AB, Renstiernas Gata 12, S-116 28 Stockholm, Sweden.
[38] Falciai, R., et al., U.S.A. Patent 4,976,265, 1990.
[39] Falciai, R., et al., "Bile enterogastric reflux sensor using plastic optical fibers," *Fiber and Integrated Optics*, 12, 1993, pp. 215–222.
[40] Bechi, P., et al., "Long-term ambulatory eterogastric reflux monitoring. Validation of a new fiberoptic technique," *Digestive Diseases and Sciences*, 38, 1993, pp. 1297–1306.
[41] Vaezi, M. F., Lacamera, R. G., and Richter, J. E., "Validation study of Bilitec 2000: an ambulatory duodenogastric reflux monitoring system," *American Journal of Pysiology*, 1994, pp. G1050–G1057.
[42] Posch, H. E., Leiner, M. J. P., and Wolfbeis, O. S., "Towards a gastric pH-sensor: an optrode for the pH 0–7 range," *Fresenius Z. Analytical Chemistry*, 334, 1989, pp. 162–165.
[43] Baldini, F., et al., "CPG embedded in plastic optical fibres for gastric pH sensing purposes," *Applied Spectroscopy*, 480, 1994, pp. 549–552.
[44] Netto, E. J., Peterson, J. I., and Wang, B., "Fiber-optic pH sensor for gastric measurements—preliminary results," *SPIE Proc. Fiber Optic Sensors in Medical Diagnostics*, 1886, Los Angeles, CA, Jan. 21, 1993, pp. 109–117.
[45] Baldini, F., et al., "*In-vivo* optical fibre pH sensor for gastro-oesophageal measurements," *Sensors and Actuators B*, 29, 1995, pp. 164–168.
[46] Netto, E. J., et al., "A fiber-optic broad-range pH sensor system for gastric measurements," *Sensors and Actuators B*, 29, 1995, pp. 157–163.
[47] Prodotec, Viadotto Indiano 50145, Firenze, Italy.
[48] Dhadwal, H. S., Ansari, R. R., and Della Vecchia, M. A., "Coherent fiber optic sensor for early detection of cataractogenesis in a human eye lens," *Optical Engineering*, 32, 1993, pp. 233–237.
[49] Ansari, R. R., et al., "Microemulsion characterization by the use of a noninvasive backscatter fiber optic probe," *Applied Optics*, 32, 1993, pp. 3822–3827.
[50] Hamano, K., et al., "Dynamic light scattering measurement for a salt-induced cataract in the eye lens of a chicken," *Physical Review A*, 43, 1991, pp. 1054–1060.
[51] Macfayden, A. J., and Jennings, B. R., "Fibre-optic systems for dynamic light scattering—a review," *Optics & Laser Technology*, 22, 1990, pp. 175–187.
[52] Könz, F., et al., "Dynamic light scattering in the vitreous: performance of the single-mode fiber technique," *Optical Engineering*, 34, 1995, pp. 2390–2395.
[53] Swanson, E. A., et al., "High-speed optical coherence domain reflectometry," *Optics Letters*, 17, 1992, pp. 151–153.
[54] Chen, S., Palmer, A. W., Grattan, K. T. V., and Meggit, B. T., "Extrinsic optical-fiber interferometric sensor that uses multimode optical fibers: system and sensing-head design for low-noise operation," *Optics Letters*, 17, 1992, pp. 701–703.
[55] Chen, S., et al., "An optical fibre sensor for eye-length measurement," *Proc. 9th International Conf. on Optical Fiber Sensors*, Firenze, Italy, May 4–6, 1993, pp. 321–324.
[56] Bueker, H., Haesing, F. W., and Gerhard, E., "Physical properties and concepts for applications of attenuation-based fiber optic dosimeters for medical instrumentation," *Proc. SPIE Fiber Optic Medical and Fluorescent Sensors and Applications*, 1648, Los Angeles, CA, Jan. 23–24, 1992, pp. 63–70.

[57] Luxtron, 2775 Northwestern Parkway, Santa Clara, CA 95051-0941.

[58] Wolthuis, R. A., et al., "Development of medical pressure and temperature sensors employing optical spectrum modulation," *IEEE Trans. Biomedical Engineering*, BME-38, 1991, pp. 974–981.

[59] Shenfeld, O., et al., "Silver halide fiber optic radiometry for temperature monitoring and control of tissues heated by microwave," *Optical Engineering*, 32, 1993, pp. 216–221.

[60] Mordon, S., et al., "Study of normal/tumorous tissue fluorescence using a pH-dependent fluorescent probe *in-vivo*," *Proc. SPIE Fiber Optic Medical and Fluorescent Sensors and Applications*, 1648, Los Angeles, CA, Jan 23–24, 1992, pp. 181–191.

[61] Devoisselle, J. M., et al., "Measurement of *in-vivo* tumorous/normal tissue pH by localized spectroscopy using a fluorescent marker," *Optical Engineering*, 32, 1993, 239–243.

[62] Baryshev, M. V., and Loshohenov, V. B., "Optimization of optical fibre catheter for spectral investigations in clinics," *Proc. SPIE Biomedical Optoelectronic Devices and Systems*, 2084, Budapest, Hungary, Sept. 1–3, 1993, pp. 106–118.

[63] Artjushenko, V. G., et al., "Medical applications of MIR-fiber spectroscopic probes," *Proc. SPIE Biochemical and Medical Sensors*, 2085, Budapest, Hungary, Sept. 1–3, 1993, pp. 137–142.

[64] Baraga, J. J., Feld, M. S., and Rava, R. P., "Infrared attenuated total reflectance spectroscopy of human artery: a new modality for diagnosing atherosclerosis," *Lasers in Life Sciences*, 5, 1992, pp. 13–29.

[65] Wolthuis, R. A., et al., "Development of medical pressure and temperature sensors employing optical spectrum modulation," *IEEE Trans. Biomedical Engineering*, 38, 1991, pp. 974–981.

[66] Lindström, L. H., "Miniaturized pressure transducer intended for intravascular use," *IEEE Trans. Biomedical Engineering*, 17, 1970, pp. 207–219.

[67] Hansen, T. E., "A fiberoptic micro-tip pressure transducer for medical applications," *Sensors & Actuators*, 4, 1983, pp. 545–554.

[68] He, G., and Wlodarczyk, M. T., "Catheter-type disposable fiber optic pressure transducer," *Proc. 9th International Conf. on Optical Fiber Sensors*, Firenze, Italy, May 4–6, 1993, pp. 463–466.

[69] Innerspace Inc., 1923 Southeast Main Street, Irvine, CA 72714.

[70] Camino Laboratories, 5955 Pacific Center Blvd., San Diego, CA 92121.

[71] Trimble, B., "Fifty thousand pressure sensors per year: a successful fiber sensor for medical applications," *9th International Conf. on Optical Fiber Sensors*, Firenze, Italy, May 4–6, 1993, pp. 457–462.

[72] Takeo, T., and Hattori, H., "Application of a fiber optic refractometer for monitoring skin condition," *Proc. SPIE Chemical, Biochemical, and Environmental Fiber Sensors III*, 1587, Boston, MA, Sept. 4–5, 1991, pp. 284–287.

[73] Takeo, T., and Hattori, H., "Quantitative evaluation of skin surface lipids by a fiber-optic refractometer," *Sensors & Actuators*, B-29, 1995, pp. 318–323.

[74] Kopola, H., et al., "Two-channel fiber optic skin erythema meter," *Optical Engineering*, 32, 1993, pp. 222–226.

[75] Myllylä, R., Marszalec, E., and Kopola, H., "Advances in color measurement for biomedical applications," *Sensors & Actuators*, B-11, 1993, pp. 121–128.

[76] Ono, K., et al., "Fiber optic reflectance spectrophotometry system for *in vivo* tissue diagnosis," *Applied Optics*, 30, 1991, pp. 98–105.

[77] Schmitt, J. M., Webber, R. L., and Walker, E. C., "Optical determination of dental pulp vitality," *IEEE Trans. Biomedical Engineering*, 38, 1991, pp. 346–352.

[78] Makinlemi, M., et al., "A novel fibre optic dental pulp vitalometer," *Proc. SPIE Medical Sensors II and Fiber Optic Sensors*, 2331, Lille, France, Sept. 9, 1994, pp. 140–148.

[79] Kopola, H., et al., "An instrument for measuring human biting force," *Proc. SPIE Medical Sensors II and Fiber Optic Sensors*, 2331, Lille, France, Sept. 9, 1994, pp. 149–155.

[80] Pickup, J. C., and Alcock, S., "Clinicians' requirements for chemical sensors for *in vivo* monitoring: a multinational survey," *Biosensors and Bioelectronics*, 6, 1991, pp. 639–646.

[81] Schultz, J. S., Mansouri, S., and Goldstein, I. J., "Affinity sensor: a new technique for developing implantable sensors for glucose and other metabolites," *Diabetes Care*, 5, 1982, pp. 245–252.
[82] Meadows, D. L., and Schultz, J. S., "Design, manufacture and characterization of an optical fiber glucose affinity sensor based on an homogeneous fluorescence energy transfer assay system," *Analytica Chimica Acta*, 280, 1993, pp. 21–30.
[83] Schaffar, B. H., and Wolfbeis, O. S., "A fast responding fiber optic glucose biosensor based on an oxygen optrode," *Biosensors and Biolelectronics*, 5, 1990, pp. 137–148.
[84] Trettnak, W., Leiner, M. J. P., and Wolfbeis, O. S., "Fibre-optic glucose sensor with a pH optrode as the transducer," *Biosensors*, 4, 1988, pp. 15–26.
[85] Rabinovitch, B., March, W. F., and Adams, R. L., "Noninvasive glucose monitoring of the aqueous humor of the eye: Part II. Animal studies and the scleral lens," *Diabetes Care*, 5, 1982, pp. 259–265.
[86] Coté, G. L., Fox, M. D., and Northrop, R. B., "Noninvasive optical polarimetric glucose sensing using a true phase measurement technique," *IEEE Trans. Biomedical Engineering*, 39, 1992, pp. 752–756.
[87] Simhi, R., et al., "Fiberoptic evanescent spectroscopy (FEWS) and its applications for multicomponent analysis of blood and biological fluids," *Proc. SPIE Medical Sensors II and Fiber Optic Sensors*, 2388, San Jose, CA, Feb. 6–8, 1995, pp. 493–500.
[88] Stepanov, E. V., et al., "Detection of small trace molecules in human and animal's exhalation by tunable diode lasers," *Proc. SPIE Medical Sensors II and Fiber Optic Sensors*, 2331, Lille, France, Sept. 9, 1994, pp. 173–183.
[89] Walt, D. R., and Bronk, K. S., "Spatially resolved photo-polymerized image-ready single fiber sensor for blood gas analysis," *Proc. SPIE Fiber Optic Medical and Fluorescent Sensors and Applications*, 1648, Los Angeles, CA, Jan. 23–24, 1992, pp. 12–14.
[90] Pantano, P., and Walt, D. R., "Analytical applications of optical imaging fibers," *Analytical Chemistry*, 1995, pp. 481A–487A.
[91] Walt, D. R., et al., "Concurrent imaging and sensing optical imaging fiber," *Proc. SPIE Advances in Fluorescence Sensing Technology*, 2388, San Jose, CA, Feb. 6–8, 1995, pp. 554–557.
[92] Sun, M. H., and Kamal, A., "A small single sensor for temperature, flow, and pressure measurement," *Proc. SPIE Optical Fibers in Medicine VI*, 1420, Los Angeles, CA, Jan 23–25, 1991, pp. 44–52.
[93] Wickersheim, K. A., "A new fiberoptic thermometry system for use in medical hyperthermia," *Proc. SPIE Optical Fibers in Medicine II*, 713, Cambridge, MA, Sept. 17–19, 1986, pp. 150–157.
[94] Wolthuis, R., Mitchell, G., Hartl, J., and Saaski, E., "Development of a dual function sensor system for measuring pressure and temperature at the tip of a single optical fiber," *IEEE Trans. Biomedical Engineering*, BME-40, 1993, pp. 298–302.
[95] Anderson, C. D., Vokovich, D., and Wlodarczyk, M. T., "Fiber optic sensor for simultaneous oxygen saturation and blood pressure measurement," *Proc. SPIE Fiber Optic Medical and Fluorescent Sensors and Applications*, 1648, Los Angeles, CA, Jan. 23–24, 1992, pp. 116–129.

Chapter 11

Fiber-Optic Gyros

Kazuo Hotate
University of Tokyo, Japan

11.1 INTRODUCTION

In the family of optical-fiber sensors, the fiber-optic gyro, or FOG for short, has become the most sophisticated type [1–5]. It provides a rotation detection function with respect to an inertial frame. This sensor is based on a physical principle called *the Sagnac effect* [1–5]. Two lightwaves, propagating in opposite directions in the same closed optical path, show a traveling time difference, which, as expected from relativistic theory, is proportional to the rotation rate of the optical path with respect to the inertial frame. Unfortunately, the sensitivity of the Sagnac effect, which is in proportion to the area surrounded by the path, is quite low. To overcome this difficulty, a long length of low-loss fiber is introduced into the optical system in the form of a multiturn sensing fiber coil with a diameter of about 10 cm. Even when using a fiber of 1-km length, the induced optical phase difference is still as small as 1 μrad, for a rotation rate of 0.01 deg/h, corresponding to the resolution required for the aircraft navigation. The fiber-optic gyro therefore needs extreme reduction of noise factors arising from the optical system in order to detect these levels.

FOGs have been studied and developed for about 20 years. Since the initial proposal in 1976 [6], almost all the noise and error factors have been clarified, and for the first generation FOG, which we shall call the interferometer FOG (I-FOG), satisfactory precautions and countermeasures have already been invented. This version has now passed the turning point towards meeting the performance for practical applications [7–10]. I-FOGs have already been used in the traditional application fields of the gyro, such as aircraft navigation, rocket control, and ship navigation. One of the most significant recent advances is the selection of an intermediate grade I-FOG for the navigation system of a modern aircraft, the Boeing 777 [11–13]. The fiber-optic gyros have many significant advantages over the traditional spinning mass gyros, in particular short warm-up time, lightweight, maintenance-free and reliable operation, wide dynamic range, large bandwidth, low power consumption, and low cost. These features have already created new application fields [7–10], including automobile navigation, antenna/camera stabilizer, crane control, unmanned vehicle control, and so forth with a range of sensitivities (Figure 11.1). However, to produce superior grade models having a sensitivity better than 0.01 deg/h, additional research is still continuing, for example, to reduce light-source noise, thermally-induced drift, and cost. This chapter will firstly review the fundamentals of the I-FOG, then cover its applications and describe recent work to improve the performance.

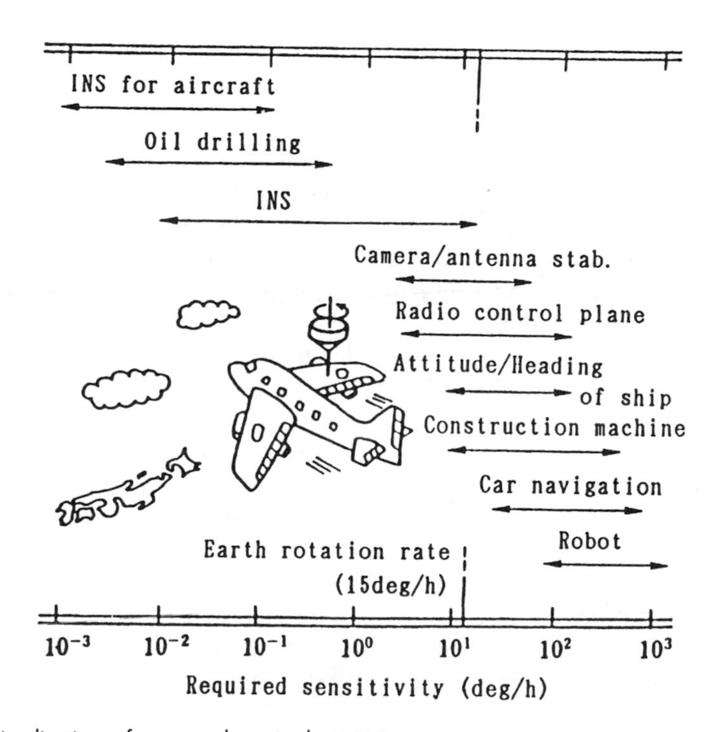

Figure 11.1 Applications of gyros and required sensitivity.

Other configurations, namely the resonator FOGs (R-FOGs) and Brillouin FOGs (B-FOGs), may have the capability to solve the inherent problems in I-FOGs mentioned above. The R-FOG uses a fiber ring resonator having a high finesse, the resonant frequency of which is changed via the Sagnac effect [1,3–5,14–16]. Because of the steeper slope in the resonant characteristics, the sensitivity is improved, which allows a shorter fiber length. Theoretical prediction tells us that only a 5 to 10m length of the fiber will be enough for aircraft navigation. For R-FOGs, extensive research on noise and error sources and the necessary countermeasures has already been conducted [7,9,17–19], and ways to integrate these countermeasures within a small device size must be developed. The B-FOG is an active configuration in which stimulated Brillouin scattering in the fiber resonator results in lasing action. The change in lasing frequency due to the Sagnac effect now provides the gyro function in an analogous manner to the earlier He-Ne ring laser gyros (RLG) [20,21]. For B-FOGs, the characteristics of the Brillouin fiber laser itself have been studied, as well as the gyro configuration. These candidates for possible future generation FOGs are also discussed in this chapter.

11.2 CLASSIFICATION OF FIBER-OPTIC GYROS

Figure 11.2 shows the basic configurations of three types of fiber-optic gyros: I-FOG, R-FOG and B-FOG. In this section, the principles of rotation sensing in each configuration are described and the current states of the technologies are reviewed. The components shown in Figure 11.2 are all of the all-fiber type, but waveguide/integrated circuits types have also been used in experimental versions.

The basic principle used for *all* these methods of optical rotation sensing is the Sagnac effect [1–5]. For a fiber length, L, and a fiber coil diameter, a, the phase difference, $\Delta\Phi$, induced between the two lightwaves traveling in the opposite directions in the fiber coil, by a rotation rate, Ω, between the two lightwaves traveling in the opposite directions in the fiber coil, is given by

$$\Delta\Phi = \frac{4\pi La}{c\lambda}\,\Omega, \tag{11.1}$$

where c denotes the light velocity in vacuum and λ is the wavelength of the light, respectively. Equation (11.1) can be rewritten as

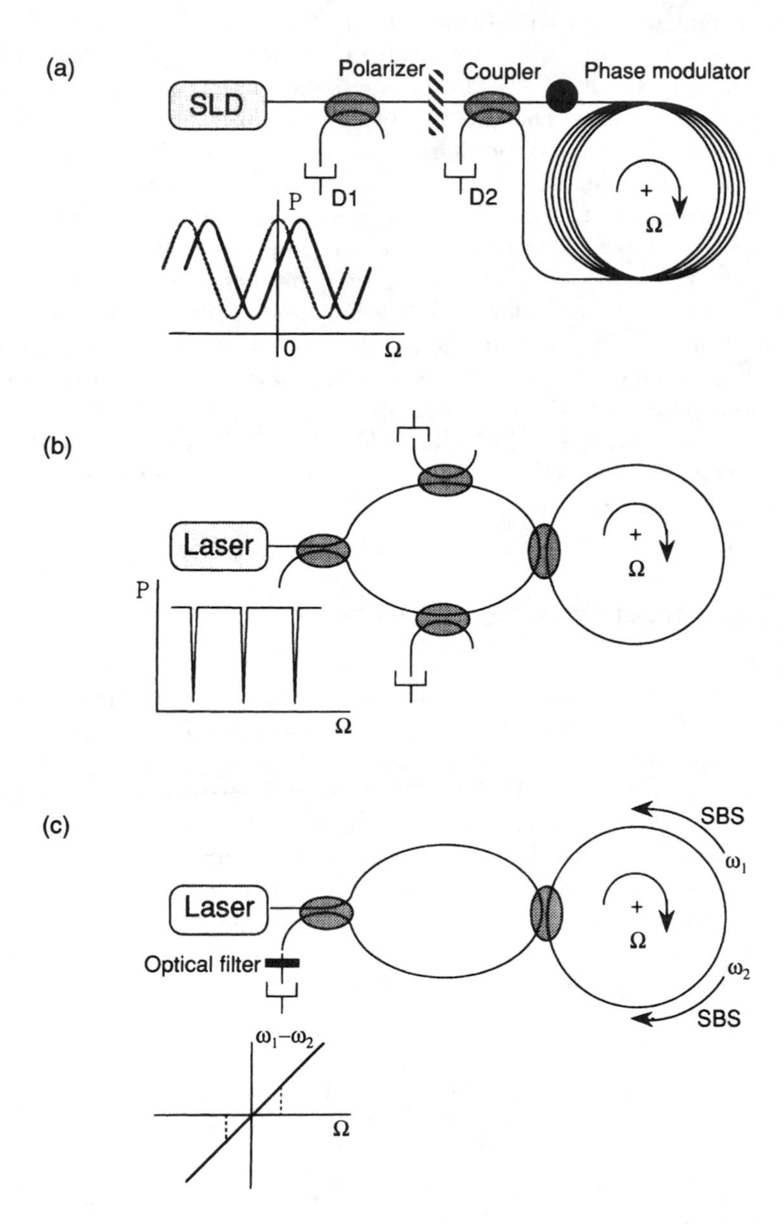

Figure 11.2 Classifications of fiber-optic gyros: (a) interferometer fiber-optic gyro (I-FOG), (b) resonator fiber-optic gyro (R-FOG), and (c) Brillouin fiber-optic gyro (B-FOG).

$$\Delta L = \frac{2La}{c}\,\Omega, \tag{11.2}$$

where ΔL is the effective fiber length difference between the two lightwaves. By measuring the phase difference, $\Delta\Phi$, in (11.1) or the effective fiber length difference, ΔL, in (11.2), the input rotation rate, Ω, can be measured. However, the magnitude of the quantity $4\pi La/c\lambda$ or $2La/c$, which give the scale factor of the gyros, is rather small.

In the I-FOG, a long length of the fiber is used to enhance the scale factor, and hence the sensitivity, as shown in Figure 11.2(a). By measuring the change in the detected intensity after interference between the two counterpropagating lightwaves in the fiber coil, a phase difference, $\Delta\Phi$, is obtained. Even when using a l-km length of the fiber, with a 10-cm coil diameter, the induced phase difference corresponding to a rotation rate of 0.01 deg/h is only about 1 μrad. During the research and development of the I-FOG, many ways to detect such a small phase shift have been invented [l–5]. To reduce parasitic noises (see next section), a low-coherence light source, such as a semiconductor superluminescent diode (SLD) is used. Moderate-grade gyros (with about 100-deg/h resolution and bias stability) and intermediate-grade gyros (requiring about l deg/h) have already been used in practical applications. For high-grade gyros, requiring about 0.01 deg/h, developments are currently in their final stage, and polarization maintaining (PM) fiber has been used to reduce several kinds of noise. However, use of long lengths of PM fiber results in a significant increase in cost. Drift due to time-variant temperature distribution in the coil is another problem. Changes in the wavelength of the light source, due to temperature, and relative intensity noise (RIN) represent additional performance deterioration factors for high- or superior-grade I-FOGs.

In the R-FOG, a shorter length of the fiber is used as a sensing coil, as shown in Figure 11.2(b). This is no longer a simple two-pass interferometer, but a multipass ring resonator with high finesse [l,3-5,14–16]. The resonant frequency, corresponding to the "dip" position in the response curve, is changed by rotation, again due to the Sagnac effect. Because the slope of the resonant dip for this resonator is steeper than that of the sinusoidal curve (characteristic of the normal two-way Sagnac loop), the sensitivity of the R-FOG is much higher than that of the I-FOG. A coil of only 5 to 10m of fiber is enough to meet aircraft navigation requirements. This configuration also has the potential to reduce the drift that can arise from a time-variant temperature distribution in the coil, and to reduce the cost, even though more expensive PM fiber is used. The light source must be a highly coherent laser, so the RIN and spectral variations of the source are reduced (compared with the SLD, whose output is the amplified spontaneous emission). However, the nature of the systematic errors are now very different from those in the I-FOG, primarily because of the different nature of the light sources. Studies on precautions and countermeasures for the error factors have already been carried out, and technologies to integrate new configurations into a miniaturized unit are now being researched.

Figure 11.2(c) shows the basic layout of the B-FOG [20,21]. By launching a relatively high power into the fiber ring resonator, compared with the R-FOG, stimulated Brillouin scattering can be deliberately induced in the resonator. As a result of this gain phenomenon, a fiber ring laser can be realized. The threshold power is dependent on both the finesse and the length of the resonator, but about 1-mW pump power is usually enough. The Brillouin scattering is induced only in a direction opposite to the direction of propagation of the input lightwave because of the phase-matching condition. Therefore, using two guided-wave pump signals traveling in opposite directions in the coil, two essentially independent counterpropagating Brillouin lasers are obtained. The result is similar to the He-Ne RLG [22]. The laser eigenfrequencies of oscillation must correspond to the length of the cavity. According to the Sagnac effect (i.e., (11.2)), the lasing frequency difference, Δf, is given by

$$\Delta f = \frac{2a}{n\lambda}\,\Omega \tag{11.3}$$

is given, where n is the effective refractive index of the fiber. Counting the beat note, the gyro function can be obtained. Because the beat note output is a frequency output, the method intrinsically provides a wide dynamic range. This output format is suitable for a strapped-down navigation system, for example, for aircraft navigation that requires seven decades dynamic range. In the I-FOG and the R-FOG, additional optical and electronic devices are required to provide such a convenient frequency output. The B-FOG is therefore expected to give an ideal output format with a very simple configuration. However, several error factors must be overcome, such as the "lock-in" effect (well-known with the traditional He-Ne RLG), where the two frequencies can remain locked (i.e., identical) at low rotation rates due to coupling between the two lasers. Studies to overcome the performance deterioration factors have been recently carried out, with particular emphasis on methods to avoid the lock-in.

Table 11.1 summarizes the features and the state of the technology of the three different types of FOG.

Table 11.1
Fiber-Optic Gyros—Features and State of the Technology

Interferometer Fiber Optic Gyro (I-FOG): Practical Application Stage
Interferometer with long length of fiber (100 ~ 1,000m). # Low coherence light source, for reducing several kinds of noises. + Practical application stage: Boeing 777, TR1-A rocket, Automobile-navi. etc. + Open-loop for Rate Gyro, Closed-loop for Rate Integration Gyro. – For high grade: • Cost increase due to long-length polarization maintaining fiber. • Scale factor fluctuation due to SLD wavelength drift. • Limitation of the resolution due to RIN in the light source. • Drift due to time-variant temperature distribution in the fiber coil.
Resonator Fiber Optic Gyro (R-FOG): Accumulation of the Studies on Noise Factors
Resonator with short length of fiber (5 ~ 10m): + Possibility to reduce the drift due to time-varient temperature distribution. + Possibility to reduce the sensing coil cost. # High coherent light source: + Reduction of the scale factor fluctuation due to wavelength drift. – Several schemes to reduce noise factors must be integrated.
Brillouin Fiber Optic Gyro (B-FOG): Basic Research Stage
Frequence change of Brillouin fiber ring laser. + Wide dynamic range with simple setup because of the frequency output. – Noise factors in the laser itself must be studied as well as gyro system configuration.

11.3 FUNDAMENTALS OF I-FOG

11.3.1 Minimum Configuration

The configuration shown in Figure 11.2(a) has been discussed in detail in [1–5]. To reliably determine the rotation rate from the tiny phase difference (on the order of mrad) between the two lightwaves, these must travel in the fiber in exactly the same path and in the same propagation mode. Therefore, the fiber must effectively be a true single-mode one. In general, the output phases of the clockwise (CW) and the counterclockwise (CCW) traveling waves in the fiber coil differ even in the rest state, corresponding to different propagation delays for each state of polarization. This induces a bias fluctuation [23,24]. To avoid the problem, a polarizer is required in the system, as shown in Figure 11.2(a) [24]. As the polarizer transmits only one linear polarization component, the modes of propagation of the two waves, although not linearly polarized through the coil, become exactly the same provided the polarizer is ideal.

The transmission and coupling characteristics of the fiber coupler are different, so the two lightwaves must experience *both* characteristics. To satisfy this requirement, two couplers must be present in the system, as shown in the figure, and the detector Dl should be used. If the detector D2 were to be used instead, the difference between the transmission and the coupling characteristics of the coupler causes gyro drift [25].

When the system is in the rest state, the output detected by Dl shows a maximum, as shown by the dashed curve in Figure 11.2(a). Unfortunately, the zero gradient of the curve at a peak means no sensitivity for a small rotation rate. Therefore, a phase modulator is placed in the end of the sensing fiber coil and is driven by a sinusoidal or square wave signal to provide a phase bias [24]. Due to the timing difference of the modulation imparted on the CW and the CCW waves, the two lightwaves now have a phase difference at the detector, giving a signal of the same frequency as the modulation signal. By synchronous detection of the output at the modulation frequency, the signal shown by the solid curve in Figure 11.2(a) can be obtained, which now changes in a sensitive manner when the coil is rotated.

The system shown in Figure 11.2(a) provides the ideal "reciprocity" between the CW and the CCW waves [25], and is called the "minimum configuration for the I-FOG."

11.3.2 Noise Factors

Table 11.2 summarizes the dominant noise factors in I-FOGs. Details of these factors have been discussed in [l–5]. Here, the outline is shown. All the factors have been studied in detail, and extensive precautions and countermeasures have already been invented [l–5].

Table 11.2
Noise Factors and Countermeasures in I-FOG

Noise Factors in Fiber Coil	*Countermeasures*
Backscattering	Low coherence source
Optical Kerr Effect	Low coherence source
Uneven thermal fluctuation	Special winding
Faraday effect	PM fiber / Depolarizers
Polarization fluctuation	Use of polarizer

Rayleigh backscattering and/or back reflections induced by one sensing lightwave in the fiber coil can act as a noise source when it interacts with the other sensing lightwave propagating in the opposite direction. This is the dominant noise in the I-FOG. To reduce it, a low-coherence light source, such as SLD is used [26]. Because the coherence length of the source is only several tens of microns, only the backscattering from a tiny region near the middle point of the sensing coil can interfere effectively. By using a short-coherence length SLD, this noise is therefore dramatically reduced.

The refractive index of the core material of the fiber is changed in proportion to the intensity of the lightwave due to an effect called the optical Kerr effect. The phase of the CW or the CCW lightwave is therefore changed by its own intensity (self-phase modulation)! Also, the phase is changed by the intensity of the counterpropagating lightwave (cross-phase modulation). The coefficient for the latter effect is twice that for the former when the light source is highly coherent. In this case therefore, gyro bias is induced in proportion to imbalance of the intensity between the CW and the CCW lightwaves [27]. Change in the imbalance results in gyro output drift. A 50% imbalance typically corresponds to a rotation rate error of 10 deg/h. The self-phase and the cross-phase modulation have the same coefficient when the light source is lowly coherent. Therefore, the most effective way to reduce this phenomenon is again to use a low-coherence light source [28].

Figure 11.3 shows a mechanism by which further errors can be induced by time-variant temperature changes in the fiber coil [29]. When the temperature distribution changes asymmetrically with respect to the center of the coil, as shown in Figure 11.3, the CW and the CCW lightwave experience relative changes in propagation delay; that is, they experience different phase changes for their complete propagation around the coil. This results in an output error or drift. When we consider the worst case shown in the figure, a temperature change of 0.01 deg C/s can cause a drift of the order of 10 deg/h [30]. Special coil-winding technologies have been devised such that all portions of the coil, located at symmetrical positions with respect to the fiber center, lie side by side [31]. This makes the temperature distribution symmetrical with respect to the fiber center. Such technologies have been applied to make a high-grade model having 0.01-deg/h sensitivity, but gyro output compensation from a monitored value of the temperature is still needed for best performance. Even after applying such methods, this phenomenon still represents the major barrier to making practical high- or superior-grade I-FOGs.

The Faraday effect in the sensing fiber coil is another error factor in the I-FOG, making it sensitive to the Earth's magnetism [32]. The Earth's magnetic field is parallel, so the effect caused in one half of the coil is ideally canceled by that in the other half. However, birefringence exists in the fiber coil due to bending and/or lateral

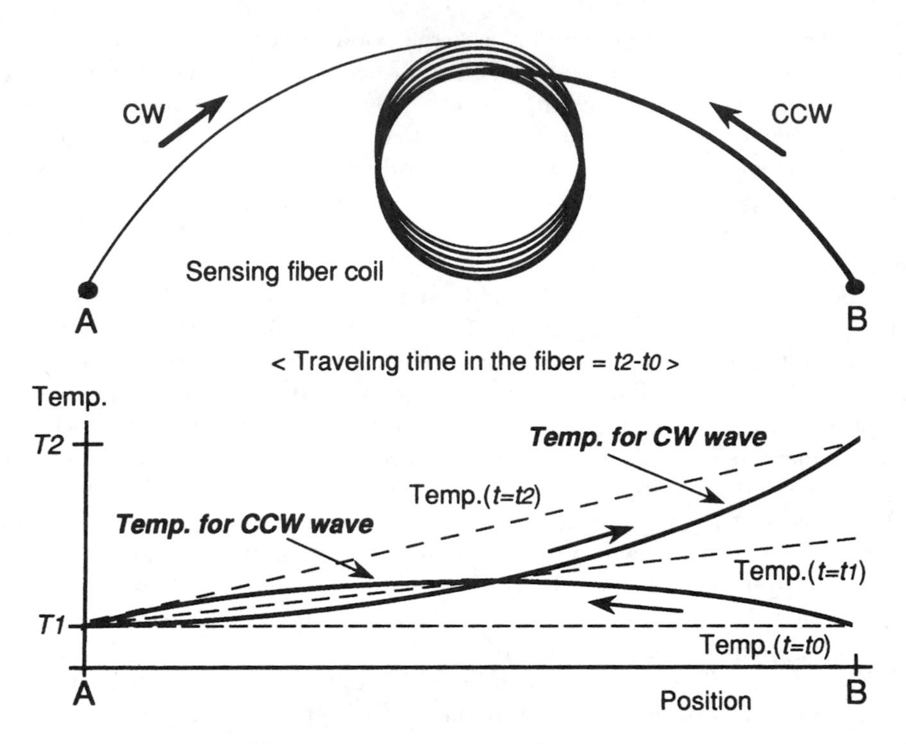

Figure 11.3 Mechanism to induce the drift due to the time-variant temperature distribution in the sensing fiber coil.

stress. When the two factors, the magnetic field and the birefringence, coexist, the I-FOG will show the drift. Magnetic shielding is one way to reduce it. Another way is to use PM fiber for the sensing coil [33–35]. This way has been proved to be effective, as shown in Figure 11.4 [34]. The PM fiber, having a high birefringence, $\Delta\beta$, can effectively reduce the drift by preventing the lightwave suffering from polarization rotation due to the Faraday effect. Additionally, it is shown in the figure that even a fiber with quite a low birefringence can also be effective in reducing the drift. However, such an ideal fiber cannot be realized in practice. Theoretical formulation tells us that reduction of the drift by the PM fiber is related not only to the birefringence but also to the twist in the sensing fiber coil [34]. The drift is in proportion to the specific ac twist component whose period is just equal to the one-turn length of the coil [34]. Other twist components, including the dc one, do not make the drift. This has been confirmed by experiment [35]. When the fiber is coiled without such a twist component, the Faraday effect-induced drift can be reduced more effectively. Recently, another way of reducing the drift has been shown in which two depolarizers are

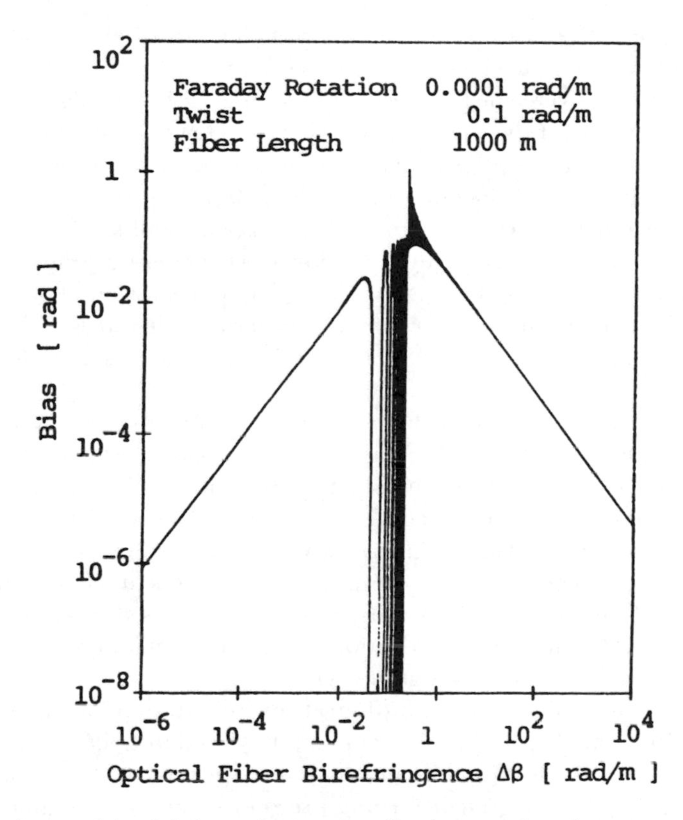

Figure 11.4 Reduction of the drift due to the Faraday effect by using the polarization maintaining fiber coil [34].

used in the system, enabling the gyro to be constructed using an ordinary single-mode fiber coil [36]. This method will be discussed again in Section 11.5.

Next, the polarization problem in the I-FOG system will be discussed. Ideally, the drift due to the polarization fluctuation in the sensing fiber coil can be removed by using one polarizer as shown in Figure 11.2(a). However, in practice, there are still problems. The first one is polarization fading. When the state of the polarization fluctuates, the detected light power changes, which can result in either output drift or a reduction in signal to noise ratio (SNR). In the worst case, the signal can totally disappear! To avoid this problem, a depolarizer is inserted in the end of the sensing fiber coil, as shown in Figure 11.5(a), and a single-mode fiber coil is used [26]. A Lyot-type fiber depolarizer is often used, which simply consists of two PM fiber portions, connected with a relative 45-degree rotation between their polarization axes. The length of one portion is chosen to be twice that of the other. Each portion induces a propagation delay difference between the two principle polarization modes in the PM fiber, which is larger than the coherence time of the SLD. Using this depolarizer, the polarization fading is overcome by effectively "scrambling" the polarization, averaging all possible states over the waveband of the SLD. However, an undesirable polarization component, which is perpendicular to the polarization axis of the polarizer, is induced. Due to the finite extinction ratio of the polarizer, this configuration cannot therefore realize a high resolution. Figure 11.5(a) is called all the "single-mode fiber configuration," and is suitable only for a low-cost moderate-grade gyro. In this configuration, the phase modulator is usually a fiber portion wound on a PZT cylinder. The specially designed PM fiber acts as a polarizer when it is coiled in a small diameter. Other small polarizers with fiber pigtails have also been developed for this configuration.

To increase the sensitivity for high-grade types, PM fiber coil is used [33,37]. Figure 11.5(b) shows a system called the *all PM fiber configuration*. In this system, the unwanted polarization component is not significantly excited. Moreover, the two polarization modes in the fiber have different propagation constants, so the unwanted component has a large delay compared with the desired component. Therefore, when using a low-coherence source, the unwanted component cannot interfere with the desired one. Consequently, the sensitivity can be increased using this configuration [38–40], and of course, the polarization fading problem disappears. This configuration is used for the intermediate-grade gyro.

Figure 11.5(c) shows the configuration for the high-grade I-FOG. A LiNbO3(LN) integrated optical (IO) circuit is introduced into the Sagnac loop [41]. This circuit is called the I-FOG chip, in which one optical Y branch and one or two phase modulator(s) are integrated. When a proton-exchanged waveguide is used, it also acts as a polarizer having quite a high extinction ratio, typically higher than 60 dB [42]. When a titanium-diffused waveguide is used, an additional polarizer, such as a metal-coated waveguide, must usually be integrated into the circuit. The phase-modulation frequency applied to the IO chip should correspond to the fundamental eigenfrequency of the loop, which is inversely proportional to the fiber coil length. Otherwise, the intensity modulation associated with the phase modulation will induce gyro drift [43]. When the fiber length is 500m, the eigenfrequency is 200 kHz. The PZT phase modulator must be very small to operate efficiently (i.e., resonate) at this frequency. On the contrary, it is easy to construct and drive a typical LN modulator. More sophisticated ways of using an LN modulator in a gyro circuit are shown in the next section.

Recently, the configuration shown in Figure 11.5(d) has been researched and occasionally used in commercially available models [44,45]. The aim of this configuration is to reduce the cost as compared to designs using long lengths of PM fiber. The point of the configuration is to use two short lengths of PM fiber between the chip and the single-mode sensing fiber coil. These two PM fibers act as depolarizers, again

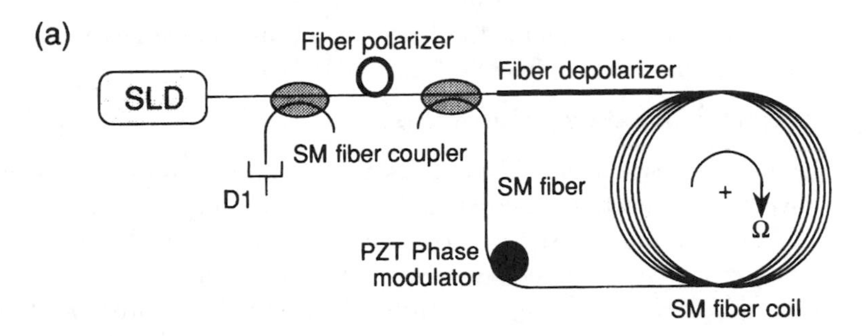

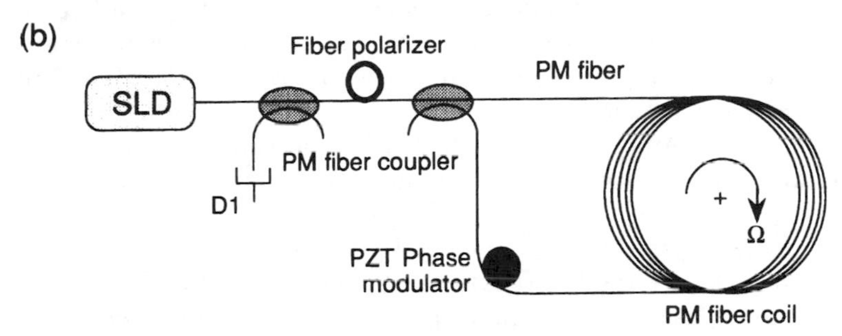

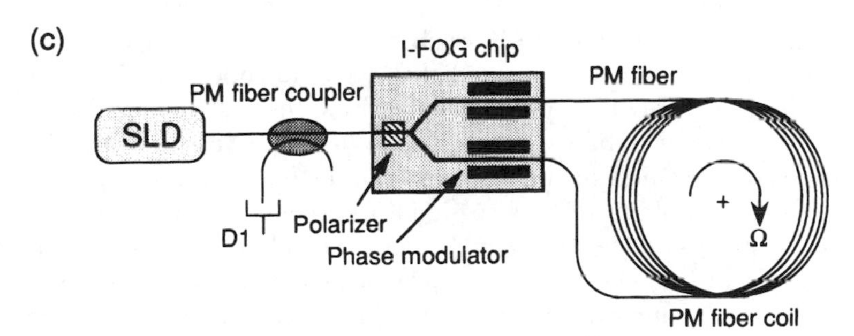

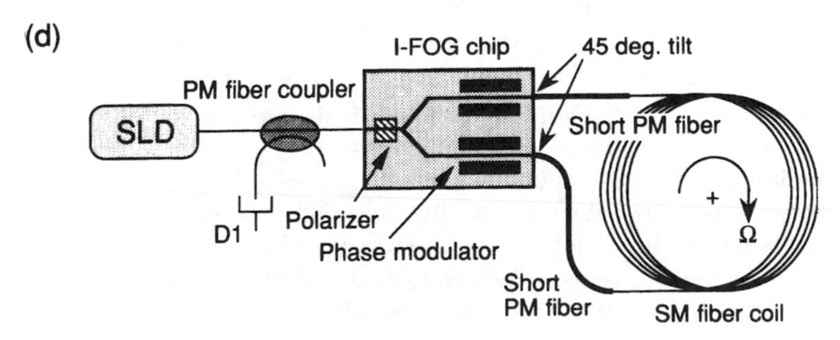

Figure 11.5 Configurations of I-FOGs: (a) all single-mode fiber configuration, (b) all PM fiber configuration, (c) configuration with the I-FOG chip and the PM fiber coil, and (d) configuration with the I-FOG chip and the single-mode fiber coil.

based on a similar principle to that of the Lyot depolarizer. Recent studies on this configuration will be again discussed in Section 11.5.

11.3.3 Open-Loop and Closed-Loop Operations

By applying the precautions shown in Table 11.2, the resolution and the bias stability can be improved to approach the theoretical limit given by the detector shot noise [46,47]. Besides a resolution/bias stability requirement of about 0.01 deg/h, about 10 ppm scale-factor stability and seven decades dynamic range are additionally required for aircraft navigation. An analog output is unsuitable to satisfy such high requirements. To overcome these difficulties, a closed-loop system is adopted where the output of the detector is fed back to some optical element inside the gyro system to allow a readout of the input rotation rate by monitoring an electronic drive frequency. Only frequency readout can provide the precision necessary to achieve seven decades dynamic range. Among various trial methods of realizing a frequency output [48–51], an attractive method using phase modulation has been most commonly adopted [2,3,52,53].

The schematic configuration is shown in Figure 11.6. To induce a phase difference between the CW and the CCW lightwaves, the phase-modulation waveform shown in the figure is used. This is known as the *digital serrodyne* waveform. This waveform has an approximate "sawtooth" shape, but with a fine-detail staircase slope that is generated by a D/A converter [52]. Each step in the stair is adjusted to be equal to one round-trip traveling time, τ, for the lightwave in the fiber coil. The amplitude is kept at 2π. Because of the fiber propagation delays, the CW and the CCW waves are modulated with a relative time delay of τ, and these two waves have a phase difference at the detector equal to the phase step ϕ in the waveform, as shown in Figure 11.6. Using this phase-difference signal as a control loop error signal, the Sagnac phase is compensated by feedback. Under the condition that the amplitude of the waveform is 2π, the phase difference f is proportional to the frequency of the serrodyne waveform. Consequently, the input rotation rate is converted to a drive frequency that can be monitored. If the amplitude is exactly 2π, the transient change in the phase difference between the two waves, occurring at the flyback in the serrodyne waveform, is also 2π. Of course, a phase change of 2π does not result in any change in the detected output. On the contrary, however, if the amplitude is not 2π, the detected signal will show a change. This can conveniently be used to adjust automatically the amplitude to the ideal value.

The digital serrodyne waveform can be simply generated using a digital logic circuit [52,53] in which the parameters in the waveform, such as the flyback amplitude, can be controlled to suppress the error. A/D and D/A converters with 12-bit accuracy are precise enough to obtain seven decades dynamic range, because, with noisy detected signals, statistical averaging is performed in the processor logic and the noise is monitored at a higher sampling rate than the required gyro bandwidth

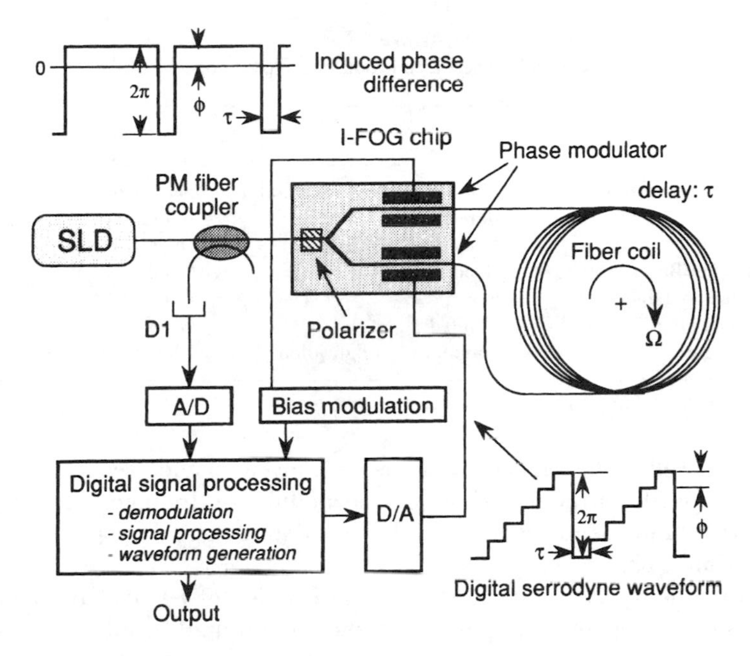

Figure 11.6 Basic configuration of a typical closed-loop I-FOG-digital serrodyne configuration.

[2,3,52,53]. A digital signal processor (DSP) can be used to perform all this processing. The digital serrodyne processing scheme is the most sophisticated type used for the I-FOG. Another method has, however, also been developed where an analog serrodyne waveform, having a straight slope (i.e., no staircase) is used with an analog circuit for signal processing.

The serrodyne method is a phase-nulling technique in which neither fluctuations of light intensity nor loss in the optical circuit affect the output. Therefore, the scale factor, that is the ratio between the output and the input rotation rate, remains constant. The requirement specification for aircraft navigation has already been achieved. In this method, the typical frequency needed to compensate for the Earth's rotation rate is about 5 Hz, which means that a serrodyne waveform of about 50 kHz, having a fine shape, must be generated for such high-grade applications. Therefore, an LN phase modulator with a good frequency response should be used. As shown in Figure 11.6, the I-FOG chip has been used for this method. The serrodyne method is called the "closed-loop operation" mode.

On the other hand, the configuration without the feedback is called the *open-loop operation mode*. Comparison between the two methods is shown in Table 11.3. In the open-loop operation, the scale factor is directly affected by the intensity fluctuation and by the phase-modulation efficiency. A means of monitoring these two

Table 11.3
Comparison Between Open-Loop and Closed-Loop I-FOG

Items	*Open-Loop*	*Closed-Loop*
System configuration	Phase modulation method	Serrodyne method
Function	Rate gyro	Rate integration gyro
Output	Analogue	Digital (Frequency)
Resolution/Bias stability	Good	Good
Scalefactor stability	Fair	Good
Dynamic range	Limited	Good
Architecture	Minimum Configuration	Ultimate Method

parameters as well as the Sagnac effect has been invented and used in which the second and the fourth harmonics of the phase-modulation frequency are detected in addition to the fundamental component. By calculating these three parameters, a stable output can be obtained [54].

Both the moderate- and the intermediate-grade of I-FOGs are usually constructed using the open-loop operation, whereas the high-grade I-FOGs use the closed-loop operation.

11.4 APPLICATIONS OF I-FOGS

Table 11.4 summarizes the developments and applications of the I-FOG. In addition to the traditional applications, the advantages of the I-FOG, such as maintenance-free, short warm-up time, and low cost, have created new applications [9,10]. Figures 11.7 and 11.8 show examples of the I-FOG products. The former is the closed-loop I-FOG with about l-deg/h resolution/stability, and the latter is the open-loop I-FOG for automobile navigation.

Japanese gyro manufacturers have developed the gyros for automobile navigation, attitude control and navigation of autonomous vehicles, antenna/camera stabilization, radio-control helicopter, crane control, agricultural machine control, borehole routing, motion measurements of automobiles, trains (and even the human body), as a north finder, and so forth, as shown in Table 11.4.

Hitachi Cable Ltd. has developed the I-FOGs for industrial applications, including automobile navigation systems, in a liaison with Hitachi Laboratories [55–58]. They commenced research on I-FOGs for automobile navigation in the 1980s, and developed the first practical systems for taxis and police cars. The first mass-produced I-FOGs for commercially available consumer automobiles were produced by Sumitomo Electric Industries in June 1991 for the sedans of the Nissan Motor Company [59].

Table 11.4
Developments and Applications of I-FOGs

Moderate Grade: Open-Loop, Resolution/Bias: ~ 200 deg/h, Scale Factor: ~ 0.1%
+ Car navigation + Antenna/camera stabilizer + Navigation of autonomous vehicles + Cleaning robots + Crane control + Control and navigation of agricultural machines + Bore hole routing + Motion monitoring of automobile, train and human body etc.
Intermediate Grade: Open/Closed-Loop, Resolution/Bias: ~ 1 deg/h, Scale Factor: ~ 0.1%
+ Aircraft/rocket navigation: Boeing 777, TR-1A rocket + Control of radio controlled helicopter + Antenna/camera stabilizer + Navigation of autonomous vehicles + Cleaning robots + Crane control + Control and navigation of agricultural machines + Bore hole routing + Motion monitoring of automobile, train and human body etc. + Gyro compass (North finder)
High Grade: Closed-Loop, Resolution/Bias: ~ 0.01 deg/h, Scale Factor: ~ 10ppm
+ Aircraft navigation + Ship navigation (higher bias stability) + Navigation and/or attitude control of space crafts (higher grade: 0.001 deg/h)

Toyota started to sell the sedans with an I-FOG navigation system in November 1992, using an I-FOG from Hitachi Cable Ltd. In early 1996, Hitachi Cable Ltd. produced 2,500- to 3,000 I-FOGs per month for such automobile navigation in sedans. Automobile navigation systems are classified into several grades. The lowest grade has only global positioning system (GPS) and map-matching software. The intermediate grade has an additional gyro function, but, because of cost, this is not the I-FOG, but a vibration gyro. The highest grade has an I-FOG, and now the gyro acts as the main sensor of the system. The GPS only corrects the data occasionally in case of unlikely error. The system with the I-FOG operates where the GPS does not work and/or map information is lacking. The resolution of the gyro used in this system is around 0.05 deg/sec. By 2000, the number of automobile navigation products is estimated to

Figure 11.7 Example of the intermediate-grade I-FOG products. This I-FOG employs the configuration with the I-FOG chip and the single-mode fiber coil, and the closed-loop operation. Bias drift: 0.5 deg/h max, scale factor stability: 0.05%, max, range: ±200 deg/h. (Product of Japan Aviation Electronics Industries Ltd.: by courtesy of Dr. K. Sakuma of JAE.)

be around 2 million per year in Japan alone. The author has bought an automobile with an I-FOG, which allows navigation even on zigzag paths and complicated roads in the Tokyo suburban area.

As mentioned in the introduction, I-FOGs for autonomous vehicles have been developed by several gyro makers. Cleaning robots, forklifts, agricultural machines, and unmanned dump trucks operating in hazardous environments are examples of the applications. Japan Aviation Electronics Industry Ltd. (JAE) [44,45,60] has a camera-stabilizer product that provides stable TV pictures from a helicopter and has produced I-FOGs for crane controllers. The I-FOG is also used in radio-control helicopters for agricultural spraying. Hitachi Cable provides about 1,000 I-FOGs per year for this application. Tamagawa Seiki and Tokyo Aircraft Instrument [61] have also been developing I-FOGs for industrial applications in Japan.

As for the traditional applications, several activities have also been presented. JAE has provided the inertial sensor packages for the TR1-A rockets of the National Aerospace Development Agency, Japan, whose mission was to perform microgravity

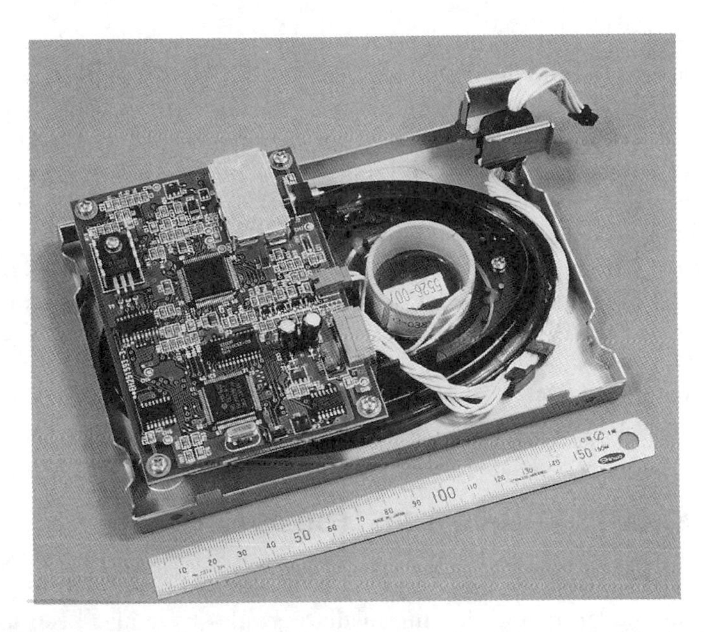

Figure 11.8 Open-loop I-FOG for automobile navigation. This I-FOG is used in the highest grade automobile-navigation system. (Product of Hitachi Cable Ltd.: by courtesy of Dr. H. Kajioka and T. Yuhara of Hitachi Cable Ltd.)

experiments [59]. The I-FOG was particularly suitable for this mission because it is a silent vibrationless gyro. The first launching was done in 1991, which was the world's first application of the I-FOG in space. Mitsubishi Precision Company Ltd. has been developing I-FOGs for inertial navigation systems (INS) in space and for other applications [62,63]. These gyros were tested using the M-3S2-7 rocket of the Institute of Space and Astronautical Science, Ministry of Education and Culture, Japan, in 1993. A launching of the rocket with the I-FOG for INS is now planned. Tokimec Inc. has been developing gyros for ship applications, using I-FOGs with a resolution of 0.003 deg/h [9,64].

In the United States, Honeywell [11,13,65], Litton [66], Smith [67], Draper Lab. [68], Northrop [69], Allied Signal [70], and Andrew [71] have developed several grades of I-FOGs for aircraft, space, and other applications. The Boeing 777 aircraft uses a new INS composed of RLGs and all-PM fiber, open-loop I-FOGs of 0.5 deg/h grade [12,13], which are products of Honeywell. In this navigation system, six RLGs of 0.01 deg/h grade and four intermediate-grade I-FOGs are used. Traditionally, aircraft navigation systems have used three sets of navigation units to avoid total failure. Each unit has three gyros of high grade. However, by developing new navigation software, the number of high-accuracy and high-cost RLGs can now be reduced while keeping the same total accuracy. Honeywell has also provided this I-FOG for the attitude and heading reference system (AHRS) in the new regional airliner, the Dornier 328. As of June 1995, over 1,000 units of this open-loop I-FOG had been delivered by Honeywell for these applications [13]. Activities in the United States seem to be mainly in aircraft and space applications. However, industrial applications have also been investigated. Andrews Corporation has been developing I-FOGs for land vehicles and north finders [71].

In Europe, Photonetics [41,52,53] has shown significant progress, including the all-digital signal processor for the I-FOG. This is the signal-processing method that is most suitable for closed-loop digital serrodyne operation. Alcatel SEL [72] has developed several types of moderate- and intermediate-grade gyros and provided an inertial measurement unit (IMU), an AHRS, and a GPS/INS. Deutsche Aerospace has presented a low-cost I-FOG using a 3 by 3 coupler to detect the Sagnac phase shift [73].

11.5 TECHNOLOGIES TO IMPROVE THE I-FOG SENSITIVITY

As shown in the previous sections, I-FOGs have already been used in practical applications. The moderate and the intermediate grades have also been used in some applications. However, for the high grade with 0.01-deg/h sensitivity, no practical applications have so far been realized, though several companies have already presented prototypes designed to meet the requirements for high-grade applications [59,62,65,66,74]. To achieve the required accuracy under the expected environmental conditions (such as a temperature range of −55°C to −90°C, and a vibration level up to 10 g rms) is the remaining hurdle to practical application. However, the research is now in the final stages and appears promising. Additionally, application fields requiring more sensitivity (~0.001 deg/h) exist [75,76], such as space applications and ship navigation. Potential applications include navigation of deep-space and precision spacecraft, space pointing, and stabilization (e.g., Hubble Space Telescope) [13]. The major items to be studied for the development of the high- or the superior-grade I-FOGs are discussed below.

The first item is cost reduction, with the most expensive aspect being the long length of the sensing coil. To allow use of cheaper single-mode fiber, a single-mode fiber coil having two short PM fibers at both ends of the coil has been investigated, in conjunction with a proton-exchanged LN I-FOG chip [44,45], as illustrated in Figure 11.5(d). The cost of PM fiber is about 100 times higher than that of ordinary single-mode communications fiber. The 45-degree tilted connection between the polarization axes of the chip and the PM fibers provides the depolarizer function necessary to reduce polarization-induced fluctuation in the gyro output. One PM fiber is twice as long as the other to achieve a depolarization function when averaged over the source spectrum. This configuration has already been used in commercially-available intermediate-grade I-FOGs with about l-deg/h stability. Figure 11.7 shows a product using this technique. Because of the high polarization extinction ratio of the proton-exchanged LN chip, the sensitivity is increased even when the ordinary single-mode fiber is used. Recently, this configuration has been demonstrated to achieve a sensitivity better than 0.01 deg/h under stable temperature conditions [13,77]. As was shown in Section 11.3.2, an effective way of reducing the Faraday effect is to use a PM fiber coil [34,35]. However, it has been shown experimentally that an optical system with two depolarizers can reduce the effect, even when using a single-mode fiber coil [36]. The configuration shown in Figure 11.5(d) can also reduce the drift without using the PM fiber [13,36].

The next important topic in I-FOG research is the use of wideband doped-fiber light sources, such as laser-diode-pumped Er-doped fibers [78,79]. When both ends of the doped fiber are coated by mirrors and excited by a pump source, a laser source with a narrow linewidth is obtained. However, if one of the mirrors is replaced by a tilted endface to suppress reflection, the output consists of amplified spontaneous emission with a wide bandwidth of about 15 nm. The wavelength drift in the SLD due to temperature change is about 300 ppm/°C. Compared with this device, extremely high wavelength stability of, typically, about several ppm/°C, can be obtained with an Er-doped fiber source. This means that the device can meet the requirements for space and for aircraft applications [78,79]. Besides its greater wavelength stability, the higher source intensity and lower RIN are other reasons for obtaining a higher grade performance. Methods of detecting the intensity noise of the source and to compensate the gyro output using the detected data have been studied. A resolution better than 0.001 deg/h has already been demonstrated using such methods under temperature-stabilized conditions [13,68,70].

Another important precaution used in the superior-grade model is reduction of drift, arising from the time-variant change of the temperature distribution in the fiber coil by suitable coil winding [29], as was described in Section 11.3.2. Methods of winding the coil [31] and compensating for residual drift effects by monitoring the temperature in the coil must be studied in detail to achieve the higher performance. The effects of the change in the stress distribution along the fiber due to thermal expansion or contraction should also be considered.

11.6 RESONATOR FIBER-OPTIC GYRO

11.6.1 Noise Factors

Resonator FOGs have the potential to solve the difficulties that still exist in high-grade I-FOGs [1,3–5,14–16]. As mentioned above, the length of the fiber is particularly short in the R-FOG. A coil of only 10m length is, in terms of the detector shot noise, enough to meet the requirements for aircraft application [80,81], assuming the use of a highly coherent laser of 100-kHz spectral width. To use this shorter length of the fiber results in a reduction of drift due to the time-variant temperature distribution [82] and also a reduction in the cost of the sensing coil. Moreover, the light source is not a low-coherence one but a highly coherent laser, having a much better wavelength stability.

In our group at RCAST, significant recent research advances on the R-FOG have been made [1,17–19]. In the R-FOG system, similar noise factors to the I-FOG are present, but their behavior is significantly different because of differences in the light source characteristics. So far, almost all the factors have been investigated, and the most useful precautions and countermeasures have already been clarified, as shown in Table 11.5 [17–19].

Figure 11.9 shows the basic experimental setup of an R-FOG made by our group [18,83]. The 1.32-mm light source is a diode-pumped mini-YAG laser having a spectral width less than 100 kHz. The resonator is composed of 10m of PM fiber connected via a fiber coupler with about a 99:1 coupling ratio. The finesse approaches a value of 100. To track the resonant frequency changes due to the Sagnac effect, two frequency shifters (acousto-optic modulators (AOM)) are used with feedback loops. One AOM is mainly for compensating the temperature change (which also changes the resonant frequency). The error signal for feedback control is obtained using synchronous detection by lock-in amplifiers while wobbling the resonator length or the relative phase of the input lightwaves to create the modulation. The difference between the driving frequencies for the two AOMs gives the input rotation rate. The relation between the rotation rate, Ω, and the resulting frequency difference, Δf, is also given by (11.3). Referring to this setup, the noise and error factors and their countermeasures will now be discussed.

Table 11.5
Noise Factors and Countermeasures in R-FOG

Noise Factors		*Proposed Countermeasures*
Polarization	---	90-degree polarization axis rotation at splice
Back scattering	---	Reduction of carrier in CW or CCW wave (b-PSK)
	---	Different wobbling frequencies for CW and CCW waves
Temperature drift	---	Small influence of time variant temperature distribution
	---	Frequency tracking by Partially Digital Feedback
Kerr effect	---	Scheme to distinguish Kerr effect form Sagnac effect with control of CW or CCW intensity
Faraday effect	---	Reduction of specific twist component in PM fiber coil

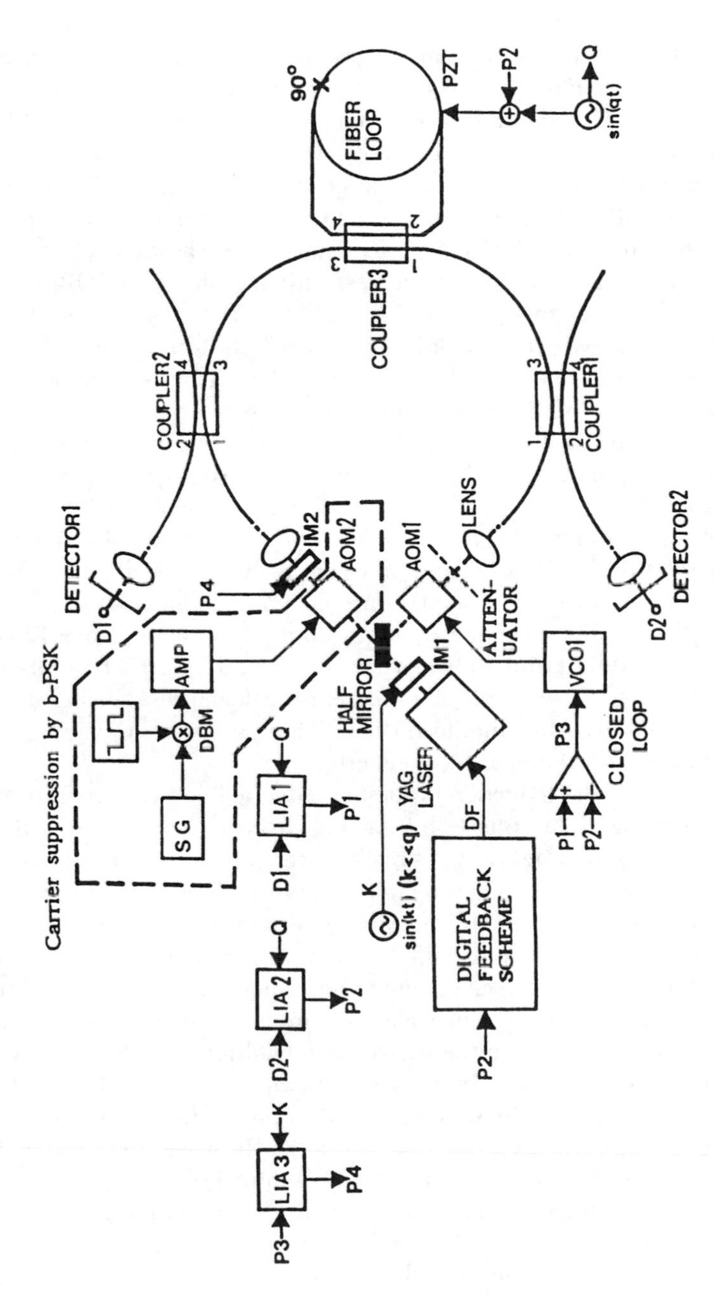

Figure 11.9 Experimental arrangement of the R-FOG with several schemes to reduce noise/error [18,83].

Figure 11.10(a) shows the resonant characteristics, taking the polarization properties in the fiber resonator into account. In the resonator, there are two special states of polarizations (SOP) that result in a high-quality resonance. These SOPs do not change after one round trip in the resonator, which is the condition required for efficient resonance [84,80]. These are called the eigenstate of polarization (ESOP) [80]. Two ESOPs are the eigenvectors of the transfer matrix of the polarization in the resonator, so the two ESOPs are orthogonal to each other. Then, we have the two independent resonance conditions corresponding to the two ESOPs, as shown in Figure 11.10(a). The resonance point of the two ESOPs can cross as a result of temperature change, which can cause a large gyro drift [85,86]. Even when using PM fiber, the problem could not be solved [85,86]. The reason is mentioned below. Even if the maladjustment of the polarization axis of the fiber at the splice point in the resonator is less than 0.5 degrees, the shape of the ESOP can differ significantly from the desired linear polarization [87]. Therefore, it is extremely difficult to excite only one ESOP selectively. Moreover, the two ESOPs have large phase differences due to the large birefringence of the fiber. As a result, two distinct sets of resonant dips, corresponding to the two ESOPs, exist and they can easily cross each other. The most effective way to solve this problem is to orientate the polarization axes by 90 degrees at the splice [88–90]. In this way, the resonant points belonging to one ESOP are located at just the middle point between the periodical resonant points of the other ESOP. Figure 10(b) is the result of a theoretical calculation showing the stable gyro output obtained by this method [90]. Though other ways have been considered [91,92], this technique is most effective.

Figure 11.11 shows the reduction of the noise due to backscattering in the resonator achieved by using binary phase shift keying (B-PSK) modulation of the light source [93]. When one signal lightwave is backscattered or reflected, it disturbs the phase of the other signal wave propagating in the opposite direction. Figure 11.11(a) shows the effects of such a noise [93]. One way to overcome this problem is to modulate one of the two signal lightwaves with a binary modulation signal to reduce the carrier component, transferring the energy into phase-modulation sidebands [16,94]. Under this condition, the interference component between the signal and the backscattering does not exist within the narrow bandwidth of the detected gyro signal. Then, the noise can be removed. In the B-PSK modulation method, the lightwave phase is modulated alternatively by 0, p, 0, p, and so forth when the carrier component is exactly canceled. In the experimental setup, the RF waveform driving AOM2 is modulated by the B-PSK signal, which is easily obtained by multiplication of the RF frequency by a low-frequency square waveform, as shown in Figure 11.9. The diffracted lightwave at the AOM2 is modulated by the B-PSK. Figure 11.11(b) shows the reduction of noise achieved using B-PSK modulation [93].

The backscattering causes two noise components [81]. One is the interference term between the scattering and the signal, which was discussed above. The other is the noise intensity itself, which has been analyzed and shown to result in a reduction

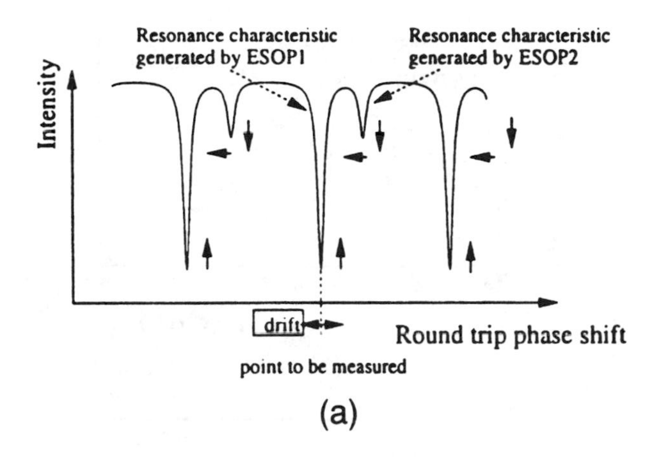

(a)

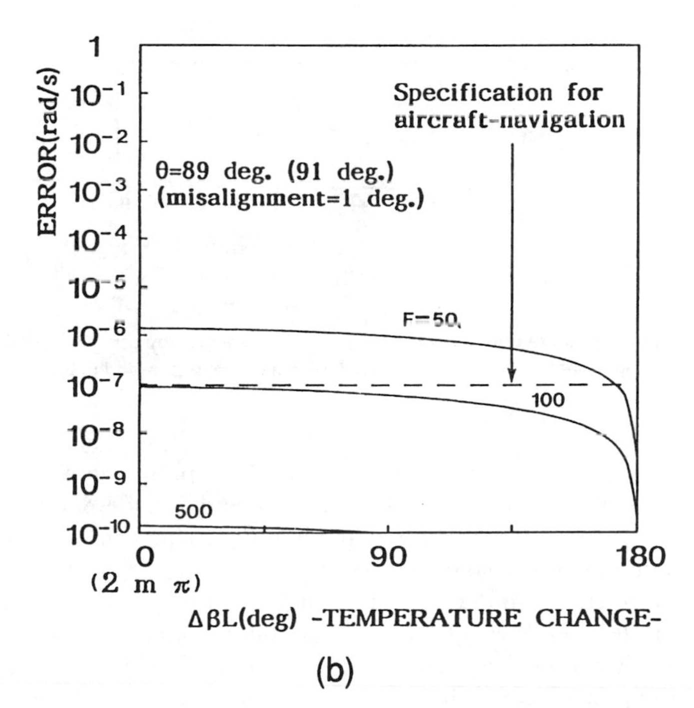

(b)

Figure 11.10 (a) Drift mechanism due to polarization fluctuation in the resonator of the R-FOG and (b) its reduction by using the resonator configuration with 90-degree polarization axis rotation at the slice point [90].

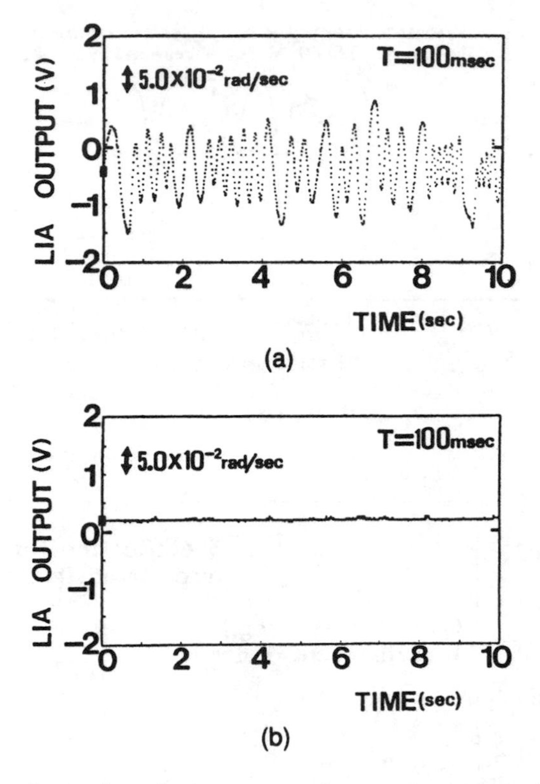

Figure 11.11 Reduction of noise due to backscattering in the resonator, by using the B-PSK modulation of one input signal lightwave. Gyro output (a) without, and (b) with the B-PSK modulation [93].

of the linearity of the gyro output [81,95]. To avoid this problem, the "wobbling" frequencies for the CW and the CCW signal waves should be different. By synchronous detection of the output with two different wobbling frequencies, this noise term is masked to obtain a clean signal component [81].

Temperature change induces too large a drift in the resonant frequency to be able to compensate using analog feedback techniques. A large feedback error appears because of insufficient feedback gain. To overcome this problem, a digital feedback scheme has been proposed [96]. In this scheme, a small value is defined as an allowed error and the error signal in the feedback loop is then compared with this value. When the absolute value of the error signal crosses this value, the frequency of the laser is changed upward or downward by an amount just corresponding to the value. By adjusting the value to be identical to the practical noise given by the other noise factors, this scheme can precisely compensate the huge resonant frequency drift due to temperature changes.

With these schemes to reduce noise, the experimental setup of Figure 11.9 has been tested. A drift of about 10 deg/h has been achieved, a value just below the Earth's rotation rate [18]. The dominant source of the drift is the optical Kerr effect in the resonator. The drift due to the optical Kerr effect is proportional to the imbalance between the CW and the CCW intensity in the resonator [96,97]. The first step to overcome this noise factor is to distinguish the optical Kerr effect from the Sagnac effect. We have proposed a method [83] in which the source intensity is modulated at a low frequency, but at one that is still high compared to the gyro signal bandwidth, and this frequency component is measured by synchronous detection of the gyro output at this frequency. When there is no imbalance between the two intensities, the output no longer has a component at the source intensity modulation frequency (otherwise, the output will have such a component). Therefore, feedback of this component to control the intensity, for example, of the CCW lightwave as shown in the figure, allows the effect to be compensated. Figure 11.12 demonstrates the results of drift reduction using this method [83].

The Faraday effect can be reduced in the same way as for the I-FOG (i.e., by using a PM fiber coil and reducing twist in the fiber) [98].

Again, it should be emphasized that almost all the factors causing deterioration of the R-FOG performance have already been clarified, and useful countermeasures have been proposed. The next step in developing the R-FOG is integration into a small size optical circuit, reducing the number of the optical components.

11.6.2 Integration of Components

Figure 11.13 shows the R-FOG system that our group originally proposed and recently constructed [99]. In this system, the I-FOG chip is used as a multifunctional device to control the input frequencies into the resonator and can also be used to implement countermeasures against the noise factors. Digital serrodyne modulation has, for both the R-FOG and the I-FOG, been proved to be useful in tracking the resonant frequency change of the resonator [99]. If each stair step in the digital serrodyne waveform (shown in Figure 11.13) is equal to one round-trip traveling time of the light in the resonator, and the amplitude of the waveform is just 2π, it has been proved that a digital serrodyne modulation of frequency f has exactly the same function for the resonator as a frequency shifter having a frequency shift also of f [99]. The system is quite tolerant to adjustment of the width of the stair. To track the resonant point, two-frequency digital serrodyne modulation is introduced, as shown in the figure. For a CCW lightwave, the serrodyne frequencies are f_1 and $2f_1$ with a repetition frequency of p. The error signal component obtained by the first lock-in amplifier, LIA1, at the frequency p, is fed back to control the laser frequency and hence track the resonant point. When the amplitudes differ from 2π, the detector output shows an error in the desired serrodyne frequency [100]. By feeding back this error signal to control the step of the stair in the serrodyne waveform, its amplitude condition is automatically

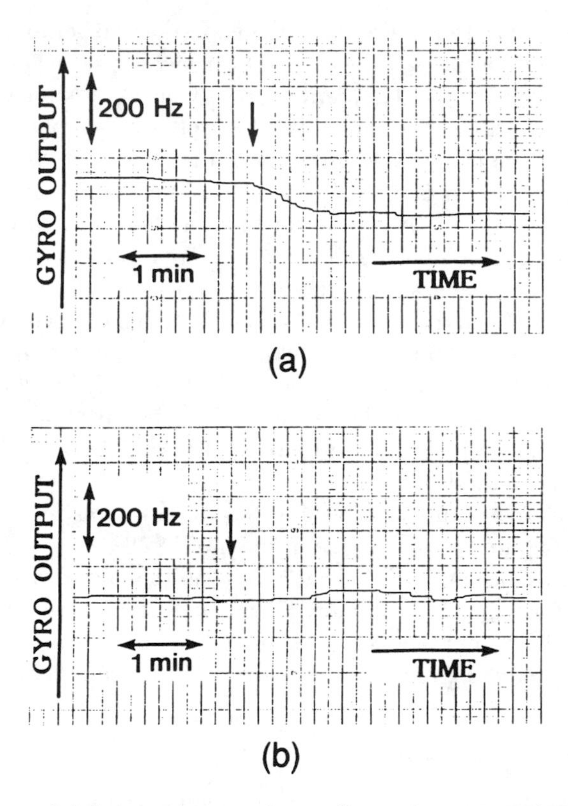

Figure 11.12 Reduction of drift due to the optical Kerr effect in the resonator by first distinguishing the Sagnac effect from the optical Kerr effect, followed by control of the input intensity [83]. Gyro output (a) without and (b) with the scheme, applying the intentional input intensity change at the point indicated by the arrow.

held at the ideal level. For the CW wave, the two serrodyne frequencies are f_2 and $f_2 + f_1$. The output of LIA2 is fed back to control the step of the stair in the waveform at a frequency f_2, which means the serrodyne frequency f_2 is changed to track the change in the resonant point due to the Sagnac effect. The waveform is reset by comparing the reference voltage corresponding to the ideal amplitude 2π. To adjust this reference automatically, additional synchronous detection by LIA2 is adopted. Finally, the frequency difference, $f_2 - f_1$ gives the gyro output.

Figure 11.14 shows preliminary experimental results for rotation detection using the arrangement shown in Figure 13 [99]. The input rotation is fed back to control the step of the stair in the serrodyne waveform, namely the serrodyne frequency, f_2, which is counted to give the gyro output. The fluctuation in the result was mainly due to errors in the homemade rotation table used for the measurement.

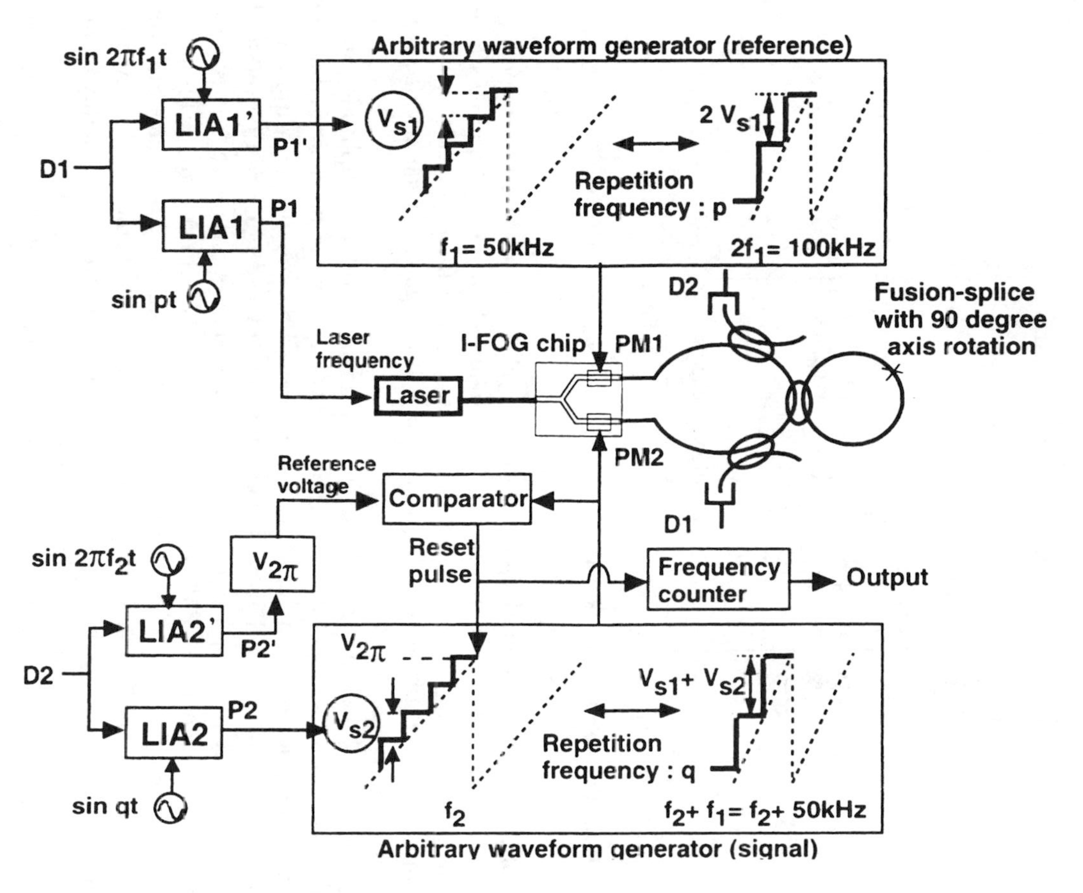

Figure 11.13 Basic configuration of the digital serrodyne R-FOG [99].

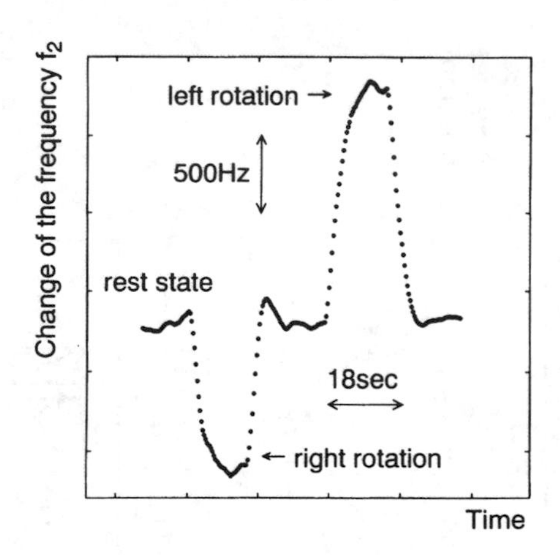

Figure 11.14 Rotation detection by an R-FOG using the digital serrodyne configuration [99].

In this scheme, all-digital signal processing scheme can again be adopted, as with the I-FOG [2,3,53]. The number of "bits" required for the A/D and D/A converter in the system has been analyzed. It is shown that in the presence of detected noise, about 12 bits is precise enough for the aircraft navigation grade having seven decades dynamic range.

In the setup shown in Figure 11.13, the serrodyne waveform amplitude can be controlled to its ideal value of 2π. Therefore, the value of π can be generated in the signal-processing logic circuit, which can generate B-PSK modulation with a high accuracy [93]. This means that the backscattering-induced noise can be reduced without adding any other optical elements. By changing the difference between the two serrodyne frequencies for PM2, the intensity of the CW lightwave in the resonator can easily be changed, which can be used to adjust the intensity to reduce the drift induced by the optical Kerr effect [83]. Consequently, this scheme has the potential to include ways of reducing the errors shown in Table 11.5, but only by modifying the modulation waveform for the two phase modulators in the I-FOG chip.

Recently, methods of reducing the polarization-induced drift in the resonator, even when using a single-mode fiber, have been proposed and studied from a cost-reduction viewpoint [101–103].

11.7 BRILLOUIN FIBER-OPTIC GYRO

Brillouin FOGs (B-FOGs)have also been investigated as a candidate for a next-generation FOG [20,21] that can directly provide a frequency output suitable for the strap-down INS system. However, research is still at a basic stage, with the key areas indicated in Table 11.6.

We have shown that the use of a splice with a 90-deg polarization axis misalignment, as used in the PM fiber resonator, is also useful to achieve stable operation of the Brillouin laser [104,105]. Also, use of a single polarization fiber can, like the fiber polarizer, be effective in overcoming problems due to polarization fluctuation [106]. The lock-in phenomenon is the most serious problem in active laser gyros. In the conventional He-Ne RLG, a well-known mechanical dithering technique has been applied to solve this problem [22]. For the B-FOG, push-pull phase modulation of laser light in the resonator has been proposed, and has been demonstrated successfully [107]. However, modification of the resonator itself can induce another cause of instability in the laser oscillation. An alternative way has been proposed [108] in which the CW and the CCW Brillouin lasers have a frequency difference due to the use of different pump frequencies. This, however, causes a gyro bias error. We have proposed a way to use the optical Kerr effect in the fiber resonator to avoid lock-in [109]. By modulating the intensity of one of the two pump waves, equivalent optically induced dither can be introduced through the Kerr effect. Therefore, by both adjusting the frequency and the amplitude of the modulation, the lock-in may be removed. Figure 11.15(a) shows the lock-in that is obtained in the R-FOG system of Figure 11.9, with all closed-loop operation but without use of B-PSK modulation [109]. In such a case, lock-in takes place, even in the R-FOG [93] because of backscattering in the resonator. The lock-in can be clearly observed in Figure 11.15(a). When modulating the CCW intensity, the lock-in could be

Table 11.6
Key Research Items for Brillouin FOG

Items	*Descriptions*
Polarization	90-degree polarization axis rotation at slice
	Use of a single polarization fiber
Rock-in (scattering)	Use of the Kerr effect in resonator
	Push-pull phase modulation in resonator
	Frequency bias
Rotation direction	Phase diversity detection
	Push-pull phase modulation in resonator
Kerr effect	Observation in experiment
	Possibility to use similar countermeasure to R-FOG
Faraday effect	Use of polarization maintaining fiber
Noise factors in Brillouin laser	Temperature / Strain effect, Spectral width, Frequency stability, Intensity noise, Amount of backscattering, etc.

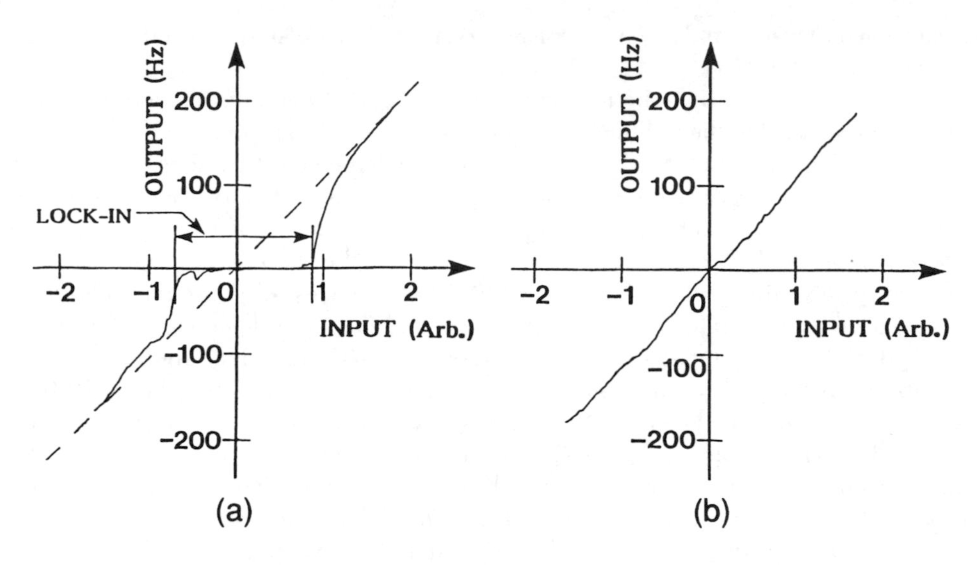

Figure 11.15 Removal of the lock-in by using the optical Kerr effect in the fiber resonator, as demonstrated in an R-FOG system with closed-loop operation [109]. Gyro output (a) without and (b) with the intensity modulation.

suppressed by the optical Kerr effect, as shown in Figure 11.15(b) [109]. This experiment shows the attractive possibility of removing the lock-in in B-FOG systems by the optical Kerr effect.

Figure 11.16 shows our recent setup of the B-FOG [110]. In this setup, a splice with 90-degree polarization axis rotation has been used to overcome the problem due to polarization fluctuation in the resonator [104,105]. In addition, a scheme using optical phase diversity to detect the direction of the rotation is proposed and implemented [110]. When the two outputs of this resonator pass through the polarizers (whose polarization coincides with that of the PM fiber), the intensities can be stable in this resonator configuration. Then, the SOP can be stably adjusted using wave plates to produce the 45-degree tilted linear and the circular state, respectively, as shown in the figure. By combining the two waves with a polarization beamsplitter, a pair of beat notes between the CW and the CCW waves are obtained. These two beat notes have a phase difference of ±90 degrees and the sign of the phase difference indicates the rotation direction. This operation has no influence on the fiber coil nor on the frequency bias, so it results in stable operation. This first experimental system was demonstrated using bulk elements, but this scheme could also be implemented using waveguide devices or integrated optics. Figure 11.17 shows the detection of rotation direction by this scheme [110].

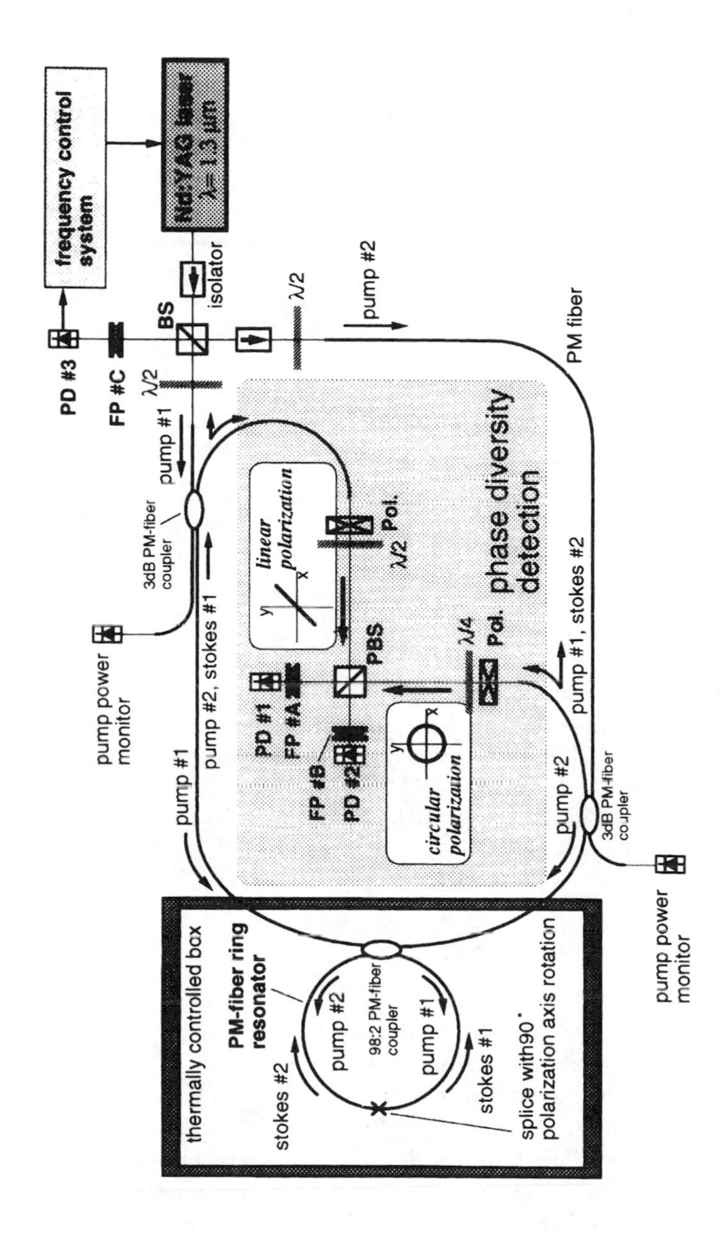

Figure 11.16 An experimental setup of the B-FOG with directional sensitivity employing the phase diversity detection [110].

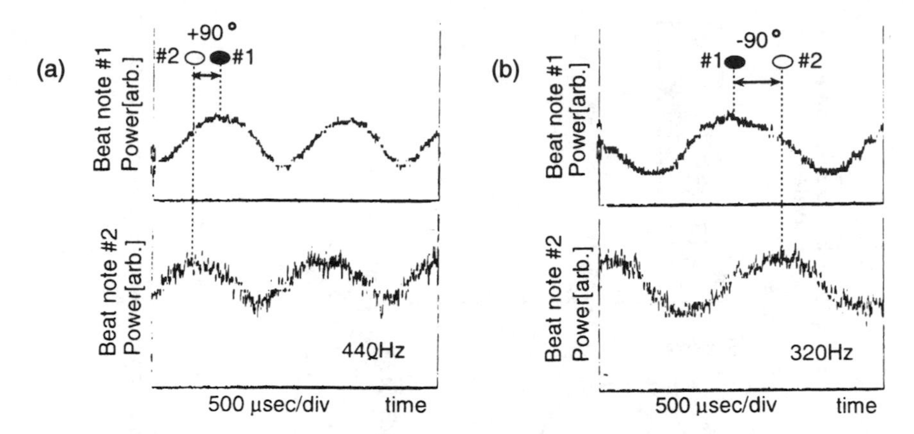

Figure 11.17 A pair of beat notes showing the rotation direction obtained with a B-FOG system using phase diversity detection [110]: (a) +90-degree phase difference under the CW rotation, and (b) −90-degree phase difference under the CCW rotation.

The drift induced by the optical Kerr effect has been investigated experimentally [111,112]. The change in lasing characteristics due to temperature change has been studied in detail [113], and the influence of the Brillouin gain distribution along the fiber has been analyzed and measured [114]. In the future, research on the B-FOG system configuration and studies on the behavior of the Brillouin fiber ring laser itself will be carried out in greater detail.

11.8 CONCLUSIONS

Three types of the fiber-optic gyros, the interferometer FOG, the resonator FOG, and the Brillouin FOG have been described.

I-FOGs have already shown advantages in real applications. The moderate (~ 100 deg/h) and the intermediate (~ 1deg/h) grades have been produced on a commercial basis. The Boeing 777 has used I-FOGs in its navigation system. Industrial applications of I-FOGs have also been developed, including autonomous vehicle control, antenna/camera stabilizer, radio-controlled helicopter, agricultural machine control/navigation, and automobile navigation. High-grade gyros having 0.01-deg/h resolution/stability are in the final stage of development for practical applications, such as inertial navigation of aircraft. To realize superior grades (~0.001 deg/h), for example, for space applications, is still difficult. However, much useful research has already been carried out. Particular problems of drift due to environmental perturbations (such as time-variant temperature distribution in the fiber coil), and the cost increase when using expensive polarization maintaining fiber must be solved. Temperature drift in the light source wavelength and its relative intensity noise contribution must also be reduced for superior grades. The doped fiber source is an effective means of reducing wavelength drift. Also, many ways to monitor and compensate for source intensity noise have also been studied.

Basic researches on the R-FOG have provided alternative ways to solve the difficulties in I-FOGs. Almost all the noise and error factors have been studied, both in theory and by experiment, and research to reduce the size of the system has been carried out. B-FOGs have the potential to achieve wide dynamic range measurement with a simple configuration, and basic studies, including those on the Brillouin laser itself, have already been carried out.

When the author started research on fiber-optic gyros [115,116] late in 1970s, the FOG seemed to be one of the most intractable research subjects among the various optical-fiber sensors (OFS). However, as described in this chapter, it has since become one of the most mature types of OFS. No other optical-sensing technique has achieved a sensitivity of 1-μrad phase measurement! This sensor is based on relativistic theory, uses advanced photonic devices, and includes a number of most interesting physical phenomena. To gyro researchers, the FOG is the king of photonic sensors. In [117], a number of the milestones in this research field are gathered to enable the interested reader to learn more of the history of the fiber-optic gyros.

References

[1] Hotate, K., "Fiber optic gyros," in *Optical Fiber Sensors*, Okoshi, T., (Ed.), OHM-sha, (in Japanese), 1986, pp. 207–252.

[2] Lefevre, H., "Fiber optic gyroscope," in *Optical Fiber Sensors Vol. II*, Culshaw, B., and Dakin, J., (Ed.), Artech House, 1989, pp. 381–429.

[3] Lefevre, H., The Fiber-Optic Gyroscope, Artech House, 1993).

[4] Ezekiel, S., and Arditty, H. J., (Eds.), *Fiber-optic rotations sensors and related technologies*, Springer-Verlag, New York, 1982.

[5] Bergh, R. A., Lefevre, H. C., and Shaw, H. J., "An overview of fiber-optic gyroscopes," *J. Lightwave Technol.*, 2, (2), April 1984, pp. 91–107.

[6] Vali, V., and Shorthill, R. W., "Fiber Ring Interferometer," *Appl. Opt.*, 15, (5), May 1976, pp. 1099–1100.

[7] Hotate, K., "Future evolution of fiber optic gyros," *Int. Conf. on Optical Fiber Sensors*, Sapporo, May 21–24, 1996, Tu3-S3.

[8] Hotate, K., "Fiber sensor technology today," *Conference on Lasers and Electro-Optics, CLEO/QELS'96*, Anaheim, June 2–7, 1996.

[9] Hotate, K., "Recent research and applications of fiber optic gyros," *Australian Conf. on Optical Fiber Technol.*, Sunshine Coast, Dec. 1995, pp. 367–370.

[10] Hotate, K., "Fiber optic gyro, Technologies and applications in Japan," *Int. Conf. on optical Fiber Sensors*, Firenze, May 1993, Tu4.1, pp. 89–95.

[11] Sanders, G. A., Liu, R-Y., and Strandjord, L. K., "Progress in interferometer and resonator fiber optic gyros," *Int. Conf. on optical Fiber Sensors*, Monterey, CA Jan. 1992, W2.1, pp. 26–29.

[12] MaClary, C. R., and Walborn, J. R., "Fault-tolerant air data inertial reference system development results," *IEEE PLANS*, 1994, pp. 31–36.

[13] Sanders, G. A., et al., "Fiber-optic gyro development for a broad range of applications," *Fiber Optic and Laser Sensors XIII, SPIE 2510*, Munich, June 1995, pp. 2–11.
[14] Ezekiel, S., and Balsamo, S. R., "Passive ring resonator laser gyroscope," *Appl. Phys. Lett.*, 30, (9), May 1977, pp. 478–480.
[15] Sanders, G. A., Prentiss, M. G., and Ezekiel, S., "Passive ring resonator method for sensitive inertial rotation measurements in geophysics and relatively," *Optics Lett.*, 6, (11), Nov. 1981, pp. 569–571.
[16] Meyer, R. E., et al., "Passive fiber-optic ring resonator for rotation sensing," *Optics Lett.*, 8, (12), Dec. 1983, pp. 644–646.
[17] Hotate, K., "Noise sources and countermeasures in optical passive ring-resonator gyro," *Int. Conf. on Optical Fiber Sensors*, Sydney, Dec. 1990, TUO1.1, pp. 11–17.
[18] Hotate, K., and Takiguchi, K., "Drift reduction in an optical passive ring-resonator gyro," *Fiber optic gyros, 15th Anniversary Conference, SPIE 1585*, Boston, Sept. 1991, pp. 116–127.
[19] Hotate, K., "Polarization problem and countermeasures in passive/active resonator fiber optic gyros," *Fiber Optic and Laser Sensor XII, SPIE 2292*, San Diego, July 1994, pp. 227–239.
[20] Kadiwar, R. K., and Giles, I. P., "Optical fiber Brillouin ring laser gyroscope," *Electron. Lett.*, 25, (25), 1989, pp. 1729–1731.
[21] Zarinetchi, F., Smith, S. P., and Ezekiel, S., "Stimulated Brillouin fiber-optic laser gyro-scope," *Optics Lett.*, 16, (4), 1991, pp. 229–231.
[22] Aronowitz, F., "The laser gyros," *Laser Applications*, 1, 1971, pp. 133–200.
[23] Ulrich, R., and Johnson, M., "Fiber-ring interferometer, polarization analysis," *Optics Lett.*, 4 (5), May 1979, pp. 152–154.
[24] Kintner, E. C., "Polarization control in optical-fiber gyroscope," *Optics Lett.*, 6, (3), March 1981, pp. 154–156.
[25] Ulrich, R., "Fiber-optic rotation sensing with low drift," *Optics Lett.*, 5, (5), May 1980, pp. 173–175.
[26] Boehm, K., et al., "Low-drift fiber gyro using a superluminescent diode," *Electron. Lett.*, 17, (10), May 1981, pp. 352–353.
[27] Bergh, R. A., Lefevre, H. C., and Shaw, H. J., "Compensation of the optical Kerr effect in fiber optic gyroscope," *Optic Lett.*, 7, (6), June 1982, pp. 282–284.
[28] Bergh, et al., R. A., "Source statistics and the Kerr effect in fiber optic gyroscope," *Optics Lett.*, 7, (11), Nov. 1982, pp. 563–565.
[29] Shupe, M. D., "Thermally induced nonreciprocity in the fiber-optic interferometer," *Appl. Opt.*, 19, (5), 1980, pp. 654–655.
[30] Mohr, F., and Kiesel, P., "Thermal sensitivity of sensing coils for fiber gyroscopes," *Int. Conf. on Optical Fiber Sensors*, Stuttgart, Sept. 5–7, 1984, pp. 305–308.
[31] Frigo, N. J., "Compensation of linear sources of non-reciprocity in Sagnac Interferometers," *Fiber Optic and Laser Sensor I, SPIE 412*, Arlington, April 5–7, 1983, pp. 302–305.
[32] Boehm, K., Petermann, K., and Weidel, E., "Sensitivity of a fiber-optic gyroscope to environmental magnetic fields," *Optics Lett.*, 7, (4), April 1982, pp. 180–182.
[33] Auch, W., and Schlemper, E., "Drift behaviour of a fiber optic rotation sensor using polarization-preserving fiber," *1st Int. Conf. on Optical Fiber Sensors*, London, UK, April. 26–28, 1983,
[34] Hotate K., and Tabe, K., "Drift of an fiber gyroscope caused by the Faraday effect, Influence of the earth's magnetic field," Appl. Opt., 25, (7), 1986, pp. 1086–1092.
[35] Hotate, K., and Tabe, K., "Drift of an optical fiber gyroscope caused by the Faraday effect," *J. of Lightwave Technol.*, 5, (7), July 1987, pp. 987–1001.
[36] Blake, J. N.,"Magnetic field sensitivity of depolarized fiber optic gyros," *Fiber Optic and Laser Sensors VIII, SPIE 1367*, San Jose, Sept. 1990, pp. 81–86.
[37] Okamoto, K., et al., "All PANDA-fiber gyroscope with long term stability, " *Electron. Lett.*, 20, (10), May 1984, pp. 429–430.
[38] Burns, W. K., and Moeller, R. P., "Polarizer Requirements for fiber gyroscope with high-birefringence fiber and broad-band sources," *J. Lightwave Technol.*, 2, (4), Aug. 1984, pp. 430–435.

[39] Fredricks, R. J., and Ulrich, R., "Phase error bounds of fiber gyro with imperfect polarizer/depolarizer," *Electron. Lett.*, 20, (8), April 1984, pp. 330–332.
[40] Kozel, S. M., et al., "Effect of random inhomogeneities in a fiber lightguide on the nullshift in a ring interferometer," *Optics and Spectroscopy*, 61, (6), 1986, pp. 814–816.
[41] Bourbin, Y., et al., "Fiber-optics rotation sensing using integrated optics," *Symp. Gyro Technol.*, Stuttgart, F.R. Germany, Sept. 1982.
[42] UTP Catalogue, "APE fiber optic gyro circuit," March 1993.
[43] Bergh, R. A., Lefevre, H. C., and Shaw, H. J., "All single-mode fiber-optic gyroscope with long-term stability," *Optics Lett.*, 6, (10), Oct. 1981, pp. 502–504.
[44] Motohara, S., and Ohno, A., "Fiber optic gyroscope with single mode fiber coil," *Optical Fiber Sensors*, Monterey, Jan. 1992, W2.2, pp. 30–33.
[45] Ohno, A., et al., "Intermediate and moderate grade fiber optic gyroscope for industrial applications," *Fiber Optic and Laser Sensor XII, SPIE 2292*, San Diego, July 1994, pp. 166–176.
[46] Lefevre, H. C., Bergh, R. A., and Shaw, H. J., "All fiber gyroscope with inertial-navigation short-term sensitivity," *Optics Lett.*, 7, (9), Sept. 1982, pp. 454–456.
[47] Moeller, R. P., Burns, W. K., and Frigo, N. J., "Open-loop output and scale factor stability in a fiber-optic gyroscope," *IEEE J. Lightwave Technol.*, 7, (2), Feb. 1989, pp. 262–269.
[48] Cahill, R. F., and Udd, E., "Phase-nulling fiber-optic laser gyro," *Optics Lett.*, 4, (3), Mar. 1979, pp. 93–95.
[49] Davis, J. L., and Ezekiel, S., "Techniques for shot-noise-limited inertial rotation measurement using a multiturn fiber Sagnac interferometer," *SPIE 157*, Aug. 30–31, 1978, pp. 131-136.
[50] Hotate, K., et al., "Rotation detection by optical heterodyne fiber gyro with frequency output," *Optics Lett.*, 7, (7), July 1982, pp. 331–333.
[51] Hotate K., and Sambuca, S., "Drift reduction in an optical heterodyne fiber gyro," *Appl. Opt.*, 29, (9), March 1990, pp. 1345–1349.
[52] Lefevre, H. C., et al., "Double closed loop hybrid fiber gyroscope using digital phase ramp," *Int. Conf. on optical Fiber Sensors*, San Diego, Feb. 13–14, 1985, PDS7-1~PDS7-5.
[53] Lefevre, H. C., et al., "Latest advances in fiber-optic gyroscope technology at Photonetics," *Fiber Optic and Laser Sensor XII, SPIE 2292*, San Diego, July 1994, pp. 156–165.
[54] Boehm, K., et al., "Direct rotation-rate detection with a fiber optic gyro by using digital data processing," *Electron. Lett.*, 19, (13), Nov. 1983, pp. 997–999.
[55] Oho, S., et al., "Optical fiber gyroscopes for automobiles," SAE SP-805, *Sensors and Actuators*, Detroit, Feb. 26-March 2, 1990, pp. 113–120.
[56] Kajioka, H., et al., "Fiber-optic gyro production at Hitachi," *Fiber Optic Gyros, 15th Anniversary Conference, SPIE 1585*, Boston, Sept., 1991, pp. 17–29.
[57] Yuhara, T., et al., "Fabrication of integrated optical gyro chip, pigtailed with elliptical-core polarization-maintaining optical fiber for industrial applications," *Fiber Optic and Laser Sensor XII, SPIE 2292*, San Diego, July 1994, pp. 218–226.
[58] Oho, S., Kajioka, H., and Sasayama, T., "Optical fiber gyroscope for automotive navigation," *IEEE Trans. on Vehicular Technol.*, 44, (3), 1995, pp. 698–705.
[59] Nishi, Y., et al., "Single mode fiber based fiber optic gyroscope for automobile navigation system," *Int. Conf. on Optical Fiber Sensors*, Firenze, May, 1993, pp. 109–112.
[60] Sakuma, K., "Fiber optic gyro productization at JAE," *Fiber Optic Gyros, 15th Anniversary Conference, SPIE 1585*, Boston, Sept. 1991, pp. 8–16. <Invited>
[61] Imai, T., et al., "Development of resonator fiber optic gyros," *Int. Conf. on Optical Fiber Sensors*, Sapporo, May 1996, Ex2-1.
[62] Hayakawa, Y., and Kurokawa, A., "Fiber-optic gyro production at Mitsubishi Precision Co.," *Fiber Optic Gyros, 15th Anniversary Conference, SPIE 1585*, Boston, Sept. 1991, pp. 30–39
[63] Kurokawa, A., "Space applications of I-FOG," *Int. Conf. on Optical Fiber Sensors*, Sapporo, May 1996, Ex2-3.
[64] Yamamoto, K., and Okada, Y., "High resolution I-FOGs for ship applications," *Int. Conf. on Optical Fiber Sensors*, Sapporo, May 1996, Ex2-4.

[65] Liu, R. Y., El-Wailly, F. T., and Dankwort, R. C., "Test results of Honeywell's first generation high-performance interferometric fiber optic gyroscope," *Fiber Optic Gyros, 15th Anniversary Conference, SPIE 1585*, Boston, Sept. 1991, pp. 262–275.

[66] Pavalath, G. A., "Production of fiber gyros at Litton guidance and control systems," *Fiber Optic Gyros, 15th Anniversary Conference, SPIE 1585*, Boston, Sept. 1991, pp. 2–6.

[67] Scholten, K. C., "Fiber optic gyro inertial measurement unit for fly-by light advanced systems hardware," *Fiber Optic and Laser Sensor XII, SPIE 2292*, San Diego, July 1994, pp. 264–271.

[68] Laznicka, O., et al., "I-FOG technology achievements at draper laboratory," *Fiber Optic and Laser Sensor XII, SPIE 2292*, San Diego, July 1994, pp. 177–191.

[69] Perlmutter, M. S., "A tactical fiber optic gyro with all-digital signal processing," *Fiber Optic and Laser Sensor XI, SPIE 2070*, Boston, Sept. 1993, pp. 192–205.

[70] Killian, K., Burmenko, M., and Hollinger, W., "High performance fiber optic gyroscope with noise reduction," *Fiber Optic and Laser Sensor XII, SPIE 2292*, San Diego, July 1994, pp. 255–263.

[71] Dyott, R. B., and Allen, D. E., "A fiber optic gyroscope North finder," *Int. Conf. on Optical Fiber Sensors*, Glasgow, Oct. 1994, pp. 442–448.

[72] Auch, W., "Progress in fiber-optic gyro development and application," *Fiber Optic and Laser Sensor XI, 2070*, Boston, Sept. 1993, pp. 104–112.

[73] Trommer, G., Hartl, E., and Muller, R., "Progress in passive fiber optic gyroscope development," *Int. Conf. on optical Fiber Sensors*, Glasgow, Oct. 1994, pp. 438–441.

[74] Asami, E., et al., "Inertial grade IFOG Development," *Int. Conf. on Optical Fiber Sensors*, Sapporo, May 1996, Tu3-6.

[75] Bielas, M. S., and Taylor, W. L., "Progress in interferometric fiber optic gyroscopes for space inertial reference units," *Fiber Optic and Laser Sensor XI, SPIE 2070*, Boston, Sept. 1993, pp. 132–141.

[76] Cordova, A., et al., "Interferometric fiber optic gyroscope with inertial navigation performance over extended dynamic environments," *Fiber Optic and Laser Sensor XI, SPIE 2070*, Boston, Sept. 1993, pp. 164–180.

[77] Szafraniec, B., et al., "Performance improvements in depolarised fiber gyros," *Fiber Optic and Laser Sensors XIII, SPIE 2510*, Munich, June 1995, pp. 37–48.

[78] Digonnet, M.J.F., "Status of broadband rare-earth doped fiber sources for FDG applications," *Fiber Optic and Laser Sensors XI, SPIE 2070*, Boston, Sept. 1993, pp. 113–131.

[79] Hall, D. C., Burns, W. K., and Moeller, R. P., "High-stability Er^{3+}-doped superfluorescent fiber sources," *IEEE J. of Lightwave Technol.*, 13, 1995, pp. 1452–1460.

[80] Iwatsuki, K., Hotate, K., and Higashiguchi, M., "Eigenstate of polarization in a fiber ring resonator and its effect in an optical passive ring-resonator gyro," *Appl. Opt.*, 25, (15), Aug. 1986, pp. 2606–2612.

[81] Iwatsuki, K., Hotate, K., and Higashiguchi, M., "Effect of Rayleigh backscattering in an optical passive ring-resonator gyro," *Appl. Opt.*, 23, (21), Nov. 1984, pp. 3916–3924.

[82] Shupe, D. M., "Fiber resonator gyroscope, Sensitivity and thermal nonreciprocity," *Appl. Opt.*, 20, (2), 1981, pp. 286–289.

[83] Takiguchi, K. ,and Hotate, K., "Method to reduce the optical Kerr-effect-induced bias in an optical passive ring-resonator gyro," *IEEE Photon. Technol. Lett.*, 4, (2), 1992, pp. 202–206.

[84] Lamouroux, B., Prade, B., and Orszag, A., "Polarization effect in optical-fiber ring resonators," *Optics Lett.*, 7, (8), Aug. 1982, pp. 391–393.

[85] Takahashi, M. , Tai, S., and Kyuma, K., "Effect of polarization coupling on the drift characteristics of a fiber-optic passive ring-resonator gyro," *Int. Conf. on Integrated Optics and Optical Fiber Communication*, Tokyo, 1989, 20B2-26, pp. 138–139.

[86] Sanders, G. A., et al., "Evaluation of polarization maintaining fiber resonator for rotation sensing applications," *Int. Conf. on Optical Fiber Sensors*, New Orleans, Jan. 27-29, 1988, FBB7, pp. 409–412.

[87] Takiguchi, K., and Hotate, K., "Bias of an optical passive ring-resonator gyro caused by the misalignment of the polarization axis in the polarization-maintaining fiber resonator," *IEEE J. of Lightwave Technol.*, 10, (4), April 1992, pp. 514–522.

[88] Sanders, G. A., Smith, R. B., and Rouse, G. F., "Novel polarization-rotating fiber resonator for rotation sensing applications," *Fiber Optic and Laser Sensor VII, SPIE 1169*, Boston, Sept. 1989, 1169-74, pp. 373–381.
[89] Mouroulis, P., "Polarization fading effects in polarization-preserving fiber ring resonators," *Fiber Optic and Laser Sensor VII, SPIE 1169*, Boston, Sept. 1989, 1169-46, pp. 400–412.
[90] Takiguchi, K. ,and Hotate, K., "Evolution of the output error in an optical passive ring-resonator gyro, with a 90° polarization-axis rotation in the polarization-maintaining fiber resonator," *IEEE Photon. Technol. Lett.*, 3, (1, Jan. 1991, pp. 88–90.
[91] Takiguchi K., and Hotate, K., "Reduction of a polarization-fluctuation-induced error in an optical passive ring-resonator gyro by using a single-polarization optical fiber," *J. of Lightwave Technol.*, 11, (10), Oct. 1993, pp. 1687–1693.
[92] Shi, C.-X., and Hotate, K., "Bias of a resonator fiber optic gyro composed of a polarization-maintaining fiber ring resonator with the photoproduced birefringent grating," *J. of Lightwave Technol.*, 13, (9), Sept. 1995, pp. 1853–1857.
[93] Hotate, K., Takiguchi, K., and Hirose, A., "Adjustment-free method to eliminate the noise induced by the backscattering in an optical passive ring resonator gyro," *IEEE Photon. Technol. Lett.*, 2, (1), Jan. 1990, pp. 75–77.
[94] Zarinetchi, F., and Ezekiel, S., "Observation of lock-in behavior in a passive resonator gyroscope," *Optics Lett.*, 11, (6), June 1986, pp. 401–403.
[95] Iwatsuki, K., Hotate, K., and Higashiguchi, M., "Backscattering in an optical passive ring resonator gyro, Experiment," *Appl. Opt.*, 25, (23), Dec. 1986, pp. 4448–4451.
[96] Takiguchi, K., and Hotate, K., "Partially digital-feedback scheme and evolution of optical Kerr-effect induced bias in optical passive ring-resonator gyro," *IEEE Photon. Technol. Lett.*, 3, (7), July 1991, pp. 679–681.
[97] K., Iwatsuki, K., Hotate, and M., Higashiguchi, "Kerr effect in an optical passive ring-resonator gyro," *J. Lightwave Technol.*, 4, (6), June 1986, pp. 645–651.
[98] Hotate, K., and Murakami, M., "Drift of an optical passive ring-resonator gyro caused by the Faraday effect," *Int. Conf. on Optical Fiber Sensors*, New Orleans, Jan. 1988, FBB6, pp. 405–408.
[99] Hotate, K., and Harumoto, M., "Resonator fiber optic gyro using digital serrodyne modulation," *Int. Conf. on Optical Fiber Sensors*, Sapporo, May 1996, Tu3-5.
[100] Standjord, L. K., and Sanders, G. A., "Effects of imperfect serrodyne phase modulation in resonator fiber optic gyroscope," *Fiber Optic and Laser Sensor XII, SPIE 2292*, San Diego, July 1994, pp. 272–282.
[101] Hotate, K., and Kurakake, T., "Manner to reduce the drift due to polariza-tion fluctuation in a resonator fiber optic gyro composed of a single mode fiber," *Fiber Optic and Laser Sensor XI, SPIE 2070*, Boston, Sept. 1993, pp. 234–245.
[102] Hotate, K., and Kurakake, T., "Optical fiber ring-resonator composed of an ordinary single mode fiber for resonator fiber optic gyros, Experiment," *Int. Conf. on optical Fiber Sensors*, Glasgow, Oct. 1994, pp. 434–437.
[103] Hotate, K., and Ito, T., "A fiber-ring-resonator with stable eigenstate of polarization using twisted single-mode optical fiber," *CLEO/Pacific Rim'95*, Makuhari, April 1995, TuA3, 2.
[104] Hotate, K., and Tanaka, Y., "Analysis on state of polarization of stimulated Brillouin scattering in an optical fiber ring-resonator," *J. of Lightwave Technol.*, 13, (3), 1995, pp. 384–390.
[105] Tanaka, Y., and Hotate, K., "Fiber Brillouin ring laser without instability due to interaction between the polarization lateral modes," *IEEE Photon. Technol. Lett.*, 7, (5), 1995, pp. 482-–84.
[106] Hotate, K., and Tanaka, Y., "Fiber Brillouin laser gyro composed of single-polarization single-mode fiber," *Fiber Optic and Laser Sensor XI, SPIE 2070*, Boston, Sept. 1993, pp. 206–215.
[107] Huang, S., et al., "Lock-in reduction technique for fiber-optic ring laser gyros," *Optics Lett.*, 18, (7), 1993, pp. 555–558.
[108] Raab, M., and Quast, T., "Two-color Brillouin ring laser gyro with gyro-compassing capability," *Optics Lett.*, 19, (18), 1994, pp. 1492–1494.
[109] Takiguchi, K., and Hotate, K., "Removal of lock-in phenomenon in optical passive ring-resonator

gyros by using optical Kerr effect in fiber resonator," *IEEE Photon. Technol. Lett.*, 4, (7), 1992, pp. 810–812.

[110] Tanaka, Y., Yamasaki, S., and Hotate, K., "Brillouin fiber optic gyro with directional sensitivity," *Int. Conf. on Optical Fiber Sensors*, Sapporo, May 1996, Tu3-7.

[111] Nicati, P.-A., et al., "Frequency pulling in a Brillouin fiber ring laser," *IEEE Photon. Technol. Lett.*, 6, (7), 1994, pp. 801–804.

[112] Nicati, P. A., Toyama, K., and Shaw, H. J., "Frequency stability of a Brillouin fiber ring laser," *J. of Lightwave Technol.*, 13, (7), 1995, pp. 1445–1451.

[113] Nicati, P.-A., et al., "Temperature effects in a Brillouin fiber ring laser," *Optics Lett.*, 18, (24), 1993, pp. 2123–2126.

[114] Tanaka, Y., and Hotate, K., "Lasing characteristics of optical fiber Brillouin ring laser with gain-spectrum distribution along fiber axis," *Int. Conf. Integrated Optics and Optical Fiber Communications*, Hong Kong, June 1995, ThA3-6, pp. 30–31.

[115] Hotate, K., et al., "Rotation detection by optical fiber laser gyro with easily introduced phase-difference bias," *Electron. Lett.*, 16, (25/26), Dec. 1980, pp. 941–942.

[116] Hotate, K., et al., "Fiber optic laser gyro with easily introduced phase-difference bias," *Appl. Opt.*, 20, (24), Dec. 1981, pp. 4313–4318.

[117] R. B. Smith, Editor, *Selected Papers on Fiber Optic Gyroscopes*, SPIE Milestone Series, MS8, SPIE Optical Engineering Press, 1989.

Chapter 12

Fiber-Optic Sensors for Condition Monitoring and Engineering Diagnostics

Julian D. C. Jones and James S. Barton
Heriot-Watt University, Scotland

12. INTRODUCTION

12.1.1 Overview

For condition monitoring and engineering diagnostics, optical measurement techniques have long occupied a special role. Their ability to operate without contacting or mechanically loading the measurement volume is their most powerful advantage. Optical measurement techniques such as photogrammetry and photoelastic stress analysis have been well known for more than 50 years. Perhaps the greatest stimulus to the use of optics in engineering metrology was the advent of the laser, making interferometric techniques very much more practical. These were initially for distance and dimensional measurement, but they were rapidly adopted for applications such as laser velocimetry, first reported within a few years of the appearance of the continuous wave laser. This chapter emphasizes the impact of the second optical technology that has had such a positive influence on the practicality of laser-based engineering measurements-namely fiber optics.

12.1.2 Scope: Techniques

This chapter restricts its scope to physical measurements, and especially those of optical path length and its derivatives, thus encompassing velocity and acceleration. In the majority of cases, the measurement volume is interrogated by a free-space laser beam. However, in others, the beam is guided at the point of measurement, and its path length is modulated by the effect of the measurand on the properties of the optical-fiber guide. Such devices, defined here as *intrinsic* optical-fiber sensors, are useful not only for dimensional measurements but also for related properties such as temperature and strain that also affect the path length in the fiber. The chapter concentrates on interferometry, well-suited to the majority of the applications described, which require sub-nm to μm path-length resolution.

12.1.3 Scope: Measurands

12.1.3.1 Optical Path Length

Interferometry naturally measures optical path length. Its application in dimensional metrology is well-known and long-established. Even in this classical application, optical fibers and related technology (notably diode lasers) are being used more extensively [1]. However, fundamental metrology lies beyond the scope of this chapter. Nevertheless, there are relevant applications in which it is important to measure slowly varying air path lengths. An example is in the use of interferometric techniques for surface profiling by scanning a measurement point across a test surface. More conveniently, the shape of a complete surface can be measured using a full-field method. True full-field interferometry is suitable only for surfaces smooth on the scale of the wavelength of light (e.g., in the testing of optical surfaces). For rougher specimens, holographic or speckle methods are appropriate. Shape measurements can also be extended to measure the deformation of a surface under load.

12.1.3.2 Velocity

Most of the applications discussed are suitable for dynamic measurements. The interferometer is as useful for the measurement of the velocity of a point on a test surface as for measuring its distance. The technique is also used to measure the velocity of a fluid by comparing the optical phase of light scattered from particles entrained in the flow with that of a plane reference wave derived from the same laser: the velocity is proportional to the observed Doppler shift beat frequency, and the technique is called *reference beam laser Doppler anemometry* (LDA) [2].

In a refinement of the technique, no separate reference beam is used. Instead, the measurement volume is illuminated by two laser beams conditioned to intersect at an angle at their waists: viewed from the receiver, the Doppler shift from the two input beams differs (because of the different angles of incidence) to give a beat frequency again proportional to velocity. This is the *Doppler difference* technique.

Time-gated anemometry need not employ interferometry, and various *transit anemometry* techniques have evolved, the simplest of which uses two adjacent laser beam waists to define a measurement volume. From transit anemometry has developed *particle image velocimetry* [3]—effectively double-exposure photography using a pulsed-laser light sheet to derive a velocity map over an extended field by measuring the separations within image pairs. *Full-field anemometry* is also feasible by coherent techniques, for example *Doppler global velocimetry*, in which the variation of spectral absorption with optical frequency in an iodine-vapor cell is exploited to measure the Doppler shift over an extended field [4,5].

12.1.3.3 Vibration

Two-beam interferometers are often employed for noncontact measurements of the out-of-plane vibration of a point on a surface. The experimental arrangement is the same as for a reference beam laser anemometer. Laser anemometers and *vibrometers* are often known collectively as *velocimeters*. To measure the in-plane component of vibration, the Doppler difference arrangement can be used.

A special case of vibration measurement is for the detection of *acoustic emission*, revealed as high-frequency out-of-plane surface vibrations arising from stress waves. Reference beam velocimeters are suitable measuring instruments. However, the vibration-induced motion can also be detected directly from the Doppler shift generated in light scattered from a laser beam by the surface, resolved by using an interferometer as a frequency discriminator [6].

In many situations, it is desirable to be able to image the vibration of a complete surface-for example, to reveal the behavior of vibrational modes. Here, similar techniques to those used for the full-field measurement of surface shape can be used, especially holography and speckle pattern interferometry [7].

12.1.3.4 Other Measurands

Interferometric optical-fiber sensors were alluded to above, defined as devices in which the optical path length of a fiber-sensing element was modulated in response to an external measurand. Fiber interferometers in general were reviewed in Chapter 10 of Volume 2, with applications such as electrical and acoustic measurements (using hydrophones) in Chapters 18 and 19. However, some types of fiber-interferometric sensors are of special importance in the kind of engineering test facilities where the other types of optical diagnostics described in this chapter are often used, and particularly in aerodynamics test facilities where fiber interferometers are used for high-bandwidth measurements of temperature, heat transfer, and pressure.

12.2 SINGLE-POINT MEASUREMENTS

12.2.1 Reference Beam Interferometers

12.2.1.1 The Basic Interferometer

Interferometers formed the subject of Chapter 10 of Volume 2; however, it is helpful to summarize the most important features here. Figure 12.1 shows a Mach-Zehnder

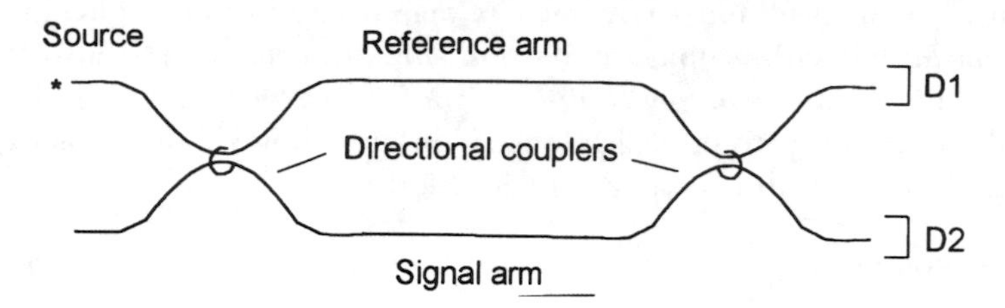

Figure 12.1 Optical fiber Mach-Zender interferometer. D1 and D2 are photodetectors.

interferometer constructed from single-mode optical fiber. Light from the source is coupled into the input fiber and amplitude is divided at the first directional coupler into the two fiber arms of the interferometer, here arbitrarily designated as the signal arm and reference arm. The arms of the interferometer terminate in a second directional coupler, thus combining the guided signal and reference beams that are then carried to the two photodetectors such that the power at the ith detector is given by

$$I_i = \frac{1}{2} I_o[1 + (-1)^i V \cos\phi] \tag{12.1}$$

where I_o is a constant (equal to the power of the source in lossless optical systems), V is the *visibility* of the interference, and ϕ is the phase difference between the interferometer arms. The phase can be written

$$\phi = \frac{2\pi}{\lambda_o} L = \frac{2\pi}{\lambda_o} [\int_{sig} ndl - \int_{ref} ndl] \tag{12.2}$$

where the integrals represent the optical path lengths of the signal (sig) and reference (ref) arms, L is the difference between the optical path lengths of the arms, n is the effective refractive index of the fiber, dl is a length element of a fiber and λ_o is the *in vacuo* wavelength of the source.

The visibility is a product of three terms, such that

$$= V_{pow} V_{pol} V_{coh} \tag{12.3}$$

representing the effect of the relative optical powers in the two arms, their relative states of polarization at the detectors, and the coherence properties of the source. It can be shown that

$$V\text{pow} = \frac{\sqrt{2 I_{sig} I_{ref}}}{I_{sig} + I_{ref}} \tag{12.4}$$

where I_{sig} and I_{ref} are the optical powers in the signal and reference arms, and that

$$V_{pol} = \hat{\mathbf{E}}_{\mathbf{sig}} \cdot \hat{\mathbf{E}}_{\mathbf{ref}} \tag{12.5}$$

where $\hat{\mathbf{E}}_{\mathbf{sig}}$ and $\hat{\mathbf{E}}_{\mathbf{ref}}$ represent the states of polarization of the recombining signal and reference beams expressed in terms of their Jones vectors. The form of V_{coh} is discussed more fully by [8], but for our purposes, it is sufficient to write

$$V_{coh} \approx 1, L << l_c; \text{ and } V_{coh} \approx 0, L >> l_c \tag{12.6}$$

where l_c is the coherence length of the source.

12.2.1.2 Signal Recovery Techniques

In all interferometric measurement systems, the measurand is recovered in the same way. From a measurement of the optical power or intensity, the relative phase of the interfering signals is found (for example, from (12.1)) and then related to the optical path length via (12.2). A difficulty arises from the periodic form of (12.1), leading to variable sensitivity $(\partial I/\partial \phi)$in the interferometer and $2\pi N$-rad ambiguities in the determination of phase, where N is an integer. Hence many techniques have evolved for signal processing in interferometers. All make use of phase control in the reference arm, such that (12.1) can be re-written as

$$I = \frac{1}{2} I_o[1 + \cos(\phi_o \pm \phi_{sig} + \phi_{ref})] \tag{12.7}$$

where ϕ_o is the nominal phase difference between the interferometer arms, ϕ_{sig} is the phase change induced by the measurand, and ϕ_{ref} is the extra phase change induced by control of the reference arm.

Processing techniques may be classified as homodyne or heterodyne. Heterodyne techniques demand an optical frequency difference between signal and reference arms, while in homodyne techniques the optical frequencies are equal.

Homodyne techniques can be subdivided into closed loop and open loop. In a closed-loop system, a feedback system and phase modulator are used to keep the interferometer phase at a constant value, usually at quadrature, where $\phi = (N + 1/2)\pi$, so that the sensitivity $(\partial I/\partial \phi)$ is a maximum. If the feedback loop bandwidth is high in comparison with the measurand bandwidth, then the reference phase modulation is exactly complementary to the signal phase modulation, and the signal phase is hence found; this is the high-gain bandwidth product (HGBWP) mode of operation,. However, the closed-loop homodyne system is often used for small-amplitude, high-frequency measurands, beyond the bandwidth of the reference phase modulation. In this low-gain bandwidth (LGBWP) product mode, we can write

$$I_i = \frac{1}{2} I_o \,(1 \pm V \sin \phi_{sig}) \tag{12.8}$$

and in the small-signal limit ($\phi_{sig} << 1$) then

$$I = I_1 - I_2 = I_o V \phi_{sig} \tag{12.9}$$

and ϕ_{sig} is readily found.

Open-loop homodyne schemes involve deriving a set of outputs from the interferometer, which are separated in phase by a fixed magnitude, often called *phase-stepping*. One example uses four phase steps, separated by steps of $\pi/2$ rad, giving four interferometer outputs of the form

$$I_i = \frac{1}{2} I_o \left[1 + V \cos\left(\phi + \frac{i\pi}{2} \right) \right] \tag{12.10}$$

where $i = 1, 2, 3, 4$, such that

$$\tan\phi = \frac{I_4 - I_2}{I_1 - I_3} \tag{12.11}$$

so that ϕ is found independently from I_o and V.

Heterodyne techniques involve shifting the frequency of the reference beam so that from (12.1)

$$I_1 - I_2 = I_o V \cos(\phi + \omega t) \tag{12.12}$$

which has the form of a phase-modulated heterodyne carrier; hence ϕ may be found by electronic techniques such as frequency discriminators or phase-locked loops.

The choice of a processing scheme is application-specific, often depending on the type of modulator that is acceptable. Simple active phase modulators can involve mechanical modulations of path length, perhaps by moving a mirror in a bulk-optic system or by straining a fiber in a fiber-optic system. More sophisticated modulators are based on electro-, magneto-, and acousto-optic techniques. Alternatively, phase modulation can be effected in unbalanced interferometers by tuning the source wavelength [9]. Passive phase modulation, useful in simultaneously generating a number of outputs from an interferometer with different phases, is possible by using birefringent linear retarders to create a phase difference between orthogonal states of polarization [10]. Fiber directional couplers also produce a phase difference between outputs, and 3 by 3 couplers are sometimes used in *phase-stepped* processing schemes, with three outputs with mutual phase differences of $2\pi/3$ rad [11].

Both homodyne and heterodyne processing techniques are capable of tracking phase changes greater than 2π rad, but neither are able to yield the initial value of the order of interference (i.e., there is an ambiguity that is a multiple of 2π-rad phase change from the path-balanced condition). In situations demanding the initial phase value, several techniques exist for extending the unambiguous range. In the first of these, the interferometer is illuminated with two slightly different wavelengths λ_1 and λ_2, and the fractional phase values at each wavelength, ϕ_1 and ϕ_2 are found [12]. It is easily shown that

$$\Delta\phi = \phi_1 - \phi_2 = \frac{2\pi L}{\lambda_{eff}} = 2\pi L\,\frac{\lambda_1 - \lambda_2}{\lambda_1\lambda_2} \tag{12.13}$$

so extending the unambiguous range by a factor $\lambda_2/(\lambda_1 - \lambda_2)$ in comparison with the single-wavelength scheme. Analogous extending techniques are available in interferometers using dual-mode optical-fiber arms, exploiting differential phase sensitivities between the modes. Either two polarization modes [13] or two spatial modes may be used [14].

The unambiguous range may be extended further by using more than two wavelengths. Multiple-wavelength interferometry is, for example, the final step in the chain of transferring the fundamental length standard to physical artifacts: a Twyman-Green interferometer is used to determine the height of metal gauge blocks, in turn used to calibrate other physical length gauges [15].

Multiple-wavelength interferometry reaches its ultimate development in so-called "white-light" or broadband interferometry, where the coherence length the source is short in comparison with the nominal path imbalance of the interferometer [16]. Hence the visibility of interference is small until the path difference is adjusted to near-zero, where fringes are seen modulated by the visibility envelope. The position of the central fringe thus corresponds to the $L = 0$ condition, and all other path imbalances can be measured relative to that reference point.

A difficulty in the practical implementation of broadband interferometry is that it inevitably requires a path-length modulator in the reference arm capable of modulation over the full range of possible changes in the signal path. It is thus often convenient to separate a sensing interferometer, compact through its lack of path-length modulator, from a receiving interferometer containing the demodulator, to produce a *tandem* arrangement [17]. In the example shown in Figure 12.2, the two interferometers have nominal path imbalances L_{sen} and L_{rec}, respectively, arranged so that L_{sen}, $L_{rec} >> l_c$. In operation, L_{rec} is modulated until $L_{sen} - L_{rec} < l_c$ so that interference is observed, and continues to be adjusted until the central fringe is located where $L_{sen} = L_{rec}$; thus L_{sen} is found.

We shall now consider some practical examples of the use of interferometers.

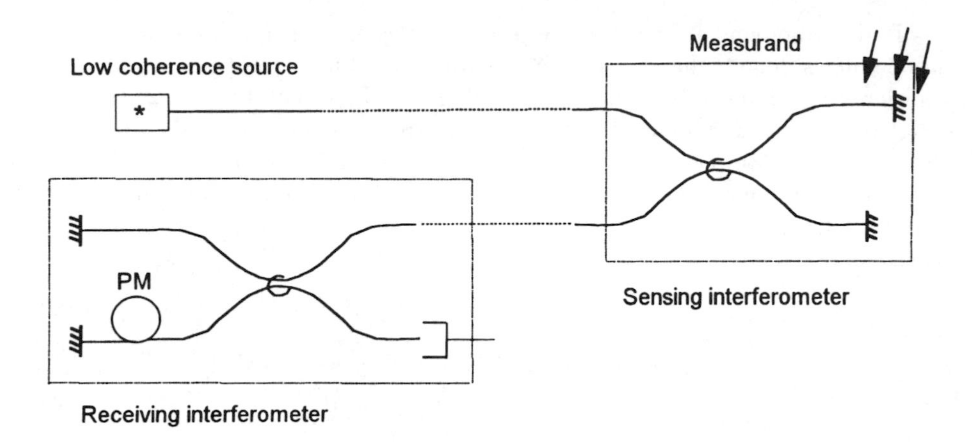

Figure 12.2 Tandem arrangement of Michelson interferometers. PM is the phase demodulator by means of which the path imbalance in the sensing interferometer can be recovered.

12.2.1.3 Surface Profiling

Consider the application of a two-beam interferometer in the measurement of slowly varying optical path differences. Such a situation occurs in the measurement of the surface profile of test objects by traversing a focused beam of light across the surface. Single-point traversing optical profilers are well-established in classical optical metrology. It must be ensured that changes in optical path length measured by the interferometer are due to changes in surface height rather than unwanted environmental effects in the interferometer. Bulk-optic profilers are necessarily substantial and delicate pieces of equipment, which normally requires that the test sample is brought to the instrument. However, there are many requirements for *in situ* surface metrology demanding some type of flexible profiling probe that can be brought to the sample. An example is shown in Figures 12.3 and 12.4 [18]. Its topology is common with many fiber-optical instruments of the "probe" type: it comprises two modules connected by a single-mode optical fiber that serves as a downlead. The first module contains the optical source and detector, with electrical connections to the signal processing system; the second is the probe itself.

The design of the profiler requires a compact and robust probe and an environmentally insensitive downlead. The optical path length in the downlead is sensitive to the environment, and if it were an uncompensated interferometer arm, then changes in temperature or vibration would generate signals indistinguishable from those produced by height changes in the surface. Hence, the measurement interferometer must be contained within the probe. However, to be compact and robust, the probe should ideally be passive, thus limiting the choices for signal processing schemes.

In [18], the profiler probe was a Fizeau interferometer. The reference reflection

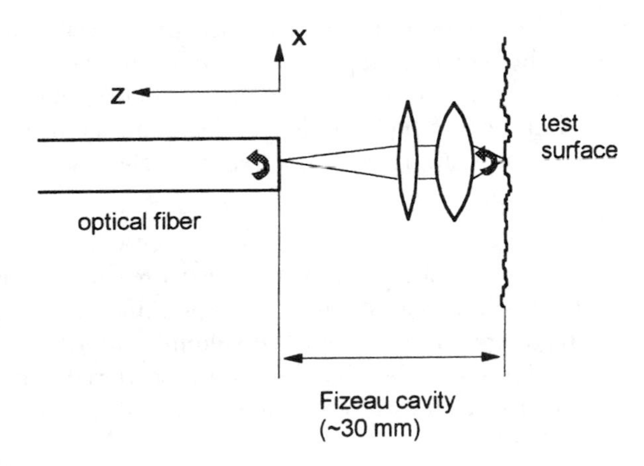

Figure 12.3 Fizeau interferometer arrangement for profiling a rough surface.

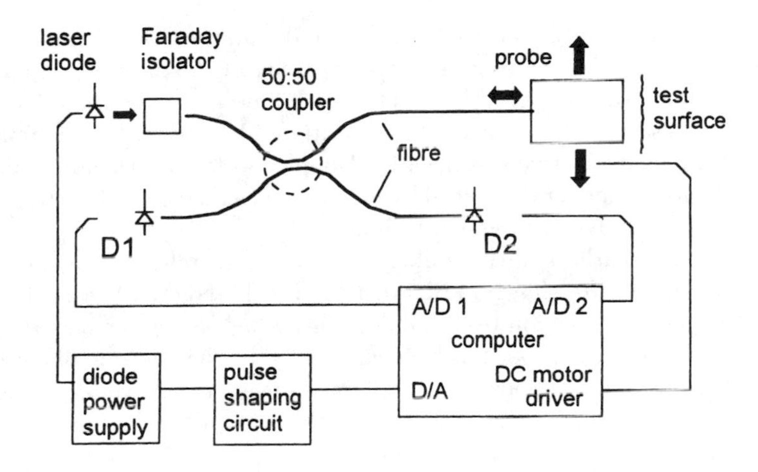

Figure 12.4 Complete surface profiler system with the probe shown in Figure 12.3. (*Source:* [24].)

was derived from the Fresnel reflection at the distal face of the downlead fiber. The signal reflection is generated by scattering at the test-surface. The intensity signals were processed by generating phase steps by modulating the laser wavelength via control of its injection current. Four steps were required, each of $\pi/2$ in phase, to accommodate changes in visibility and surface reflectances encountered with rough test surfaces. Hand and others [18] report variations in reflectivity of typically 10^5. The probe comprises little more than the end of the fiber and a lens, either of conventional or graded-index type. Out-of-plane resolution of 10 nm was achieved with in-plane resolution limited by the size of the laser spot on the test surface to about 7 μm.

The size of the traverse steps in the plane of the surface is chosen to avoid large changes in surface height between steps, since height changes leading to phase changes greater than a substantial fraction of 2π rad would lead to errors in the measured profile. The technique can be extended by using a second wavelength (see Section 12.2.1.2). For very rough surfaces, broadband interferometry is suitable [19].

12.2.1.4 Velocimetry

The basic principles of reference beam laser velocimetry are shown in Figure 12.5. A laser beam illuminates a small measuring volume within a fluid flow. Particles entrained by the flow traversing the measurement volume scatter laser light, and as a result of their movement impress a Doppler shift on the scattered light. The Doppler shift Δv_D is revealed by mixing the scattered light with a reference beam at a photodetector where

$$\Delta \nu_D = \frac{2\pi}{\lambda}(\hat{\mathbf{k}}_\mathbf{i} - \hat{\mathbf{k}}_\mathbf{r}) \cdot \mathbf{v} \tag{12.14}$$

for nonrelativistic velocities, where $\hat{\mathbf{k}}_\mathbf{i}$ and $\hat{\mathbf{k}}_\mathbf{r}$ are the unit vectors in the direction of the incident and scattered beams and $\mathbf{v}$ is the velocity of the particle, and assuming plane waves in the measurement volume. Hence, the component of the velocity in the direction of the bisector of the incident and scattered light directions is found. The inability to distinguish positive and negative Doppler shifts implies ambiguity in the sign of the velocity component, resolved by applying a frequency shift to the reference beam and using heterodyne processing techniques.

One of the very earliest extrinsic fiber sensors was the reference beam velocimeter reported by Dyott [20], shown in Figure 12.6. It is based on a Fizeau interferometer using the internal reflection from the fiber distal face as a reference. In developments of the technique, the probe has been elaborated by the addition of lenses at the end of the fiber to produce a more clearly defined measurement volume, and by the

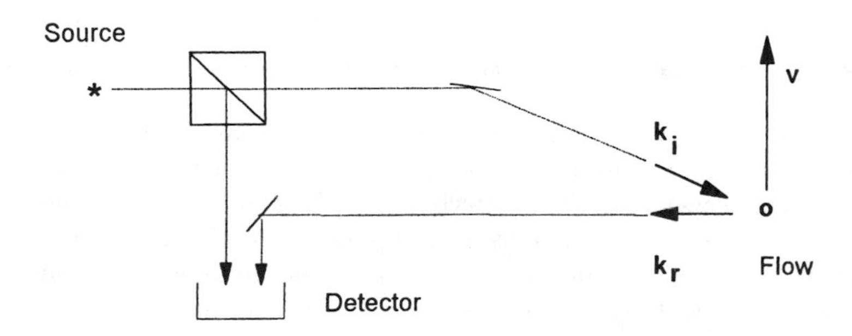

Figure 12.5 Schematic arrangement of a reference beam laser velocimeter. Light scattered by a particle in the flow is combined with a reference beam derived directly from the source.

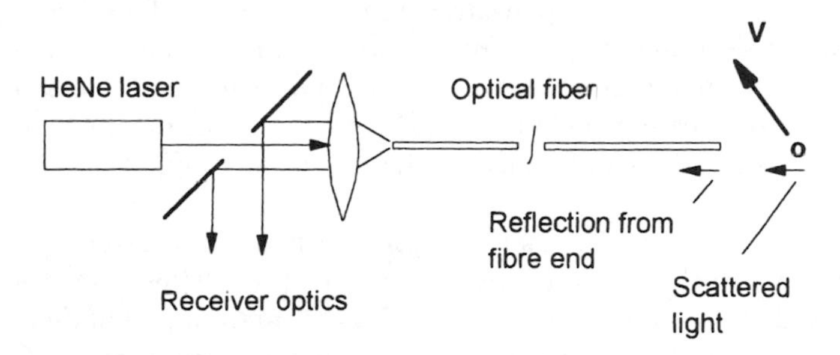

Figure 12.6 Reference beam Doppler anemomoter incorporating optical fiber. (*Source:* [20].)

use of frequency shifters to give directionality. The ultimate form of the reference beam velocimeter operated in a reflection configuration is the "self-mixing" interferometer, which uses a diode laser source where Doppler-shifted scattered light returning to the laser cavity beats with the laser output to produce a difference frequency, detectable using the laser's internal photodiode [21]. A fiber-optic reference beam velocimeter, suitable for incorporation into a three-dimensional laser Doppler velocimeter, has been described by James and others [22]. Both continuous and pulsed operation were demonstrated to illustrate the potential for either wavelength or time-division multiplexing schemes.

12.2.1.5 Vibration

Laser velocimeters are as suitable for measurements of solid surfaces as they are for use in fluids. Fiber-based velocimeters (of the Doppler difference type, see Section 12.2.3) are used, for example, in measuring the linear velocities in industrial processes such as the rolling of steel strip or in paper making, where only noncontact measurements are feasible. However, an important application is in vibration measurement. Some authors define laser velocimeters used for the measurement of vibration as vibrometers.

Laser vibrometers are used as alternatives to contacting accelerometers, and fiber-based commercial products are available, working on the same principle as the reference beam velocimeter. In early research papers, Fizeau [23], Michelson [24], and Mach-Zehnder [25] designs have all been used. The Fizeau is simple, and because the downlead forms no part of the interferometer, it is insensitive to environmental effects. Conversely, it is impractical to include an active modulator within the interferometer, limiting the choice of demodulation schemes. The Michelson and Mach-Zehnder can be used with modulators, and the Mach-Zehnder has the particular advantage of easy access to two complementary optical outputs, which can be subtracted or divided to compensate for intensity noise. True Mach-Zehnders are unsuitable for deriving a signal from a target in reflection, so that the modified arrangement shown in Figure 12.7 must be used instead. A 50% loss of power occurs, unless a polarizing arrangement is used. However, polarization techniques in vibrometry are never fully efficient because of changes in state of polarization as a result of scattering from the target.

Commercial laser vibrometers often use heterodyne signal processing with Bragg cells to give the necessary frequency shift [26]. Elsewhere, heterodyning has been achieved by injection current-induced wavelength tuning of diode laser sources, which gives a heterodyne signal in an unbalanced interferometer [23].

Phase-stepping is not well-suited to vibrometry, as coping with fast transient movements would require impractically rapid phase steps. However, in one implementation, a passive phase-diversity technique has been used [27]. The arrangement is shown in Figure 12.8. A ($\lambda/4$) waveplate is used in the probe beam to generate a $\pi/2$-rad phase difference between the orthogonal linear polarization eigenstates. A polarizing beamsplitter is used in the detector to give two pairs of complementary (i.e., π-rad out of phase) beams, with a $\pi/2$ phase difference.

12.2.1.6 Acoustic Emission

A special application of vibration measurement is in the detection of acoustic emission, which is an important condition monitoring technique, for example, for the detection of crack growth in engineering structures such as pressure vessels [28]. Acoustic emission is observed by monitoring the propagation of stress waves in the structure at ultrasonic frequencies typically in the 0.1- to 1.0-MHz range [29]. The surface-propagating stress waves are detectable by a laser vibrometer.

Acoustic emission detection is challenging because the amplitude of the surface waves is so small (in the order of 0.1 to 1.0 nm in the out-of-plane direction), and the

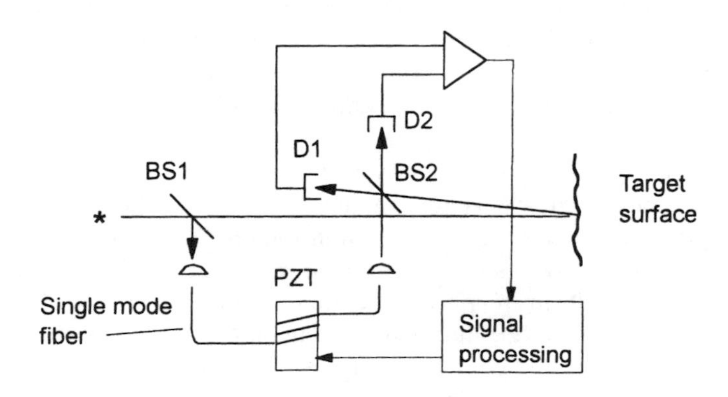

Figure 12.7 Mach-Zender vibrometer with fiber optic reference arm. (*After:* [25].)

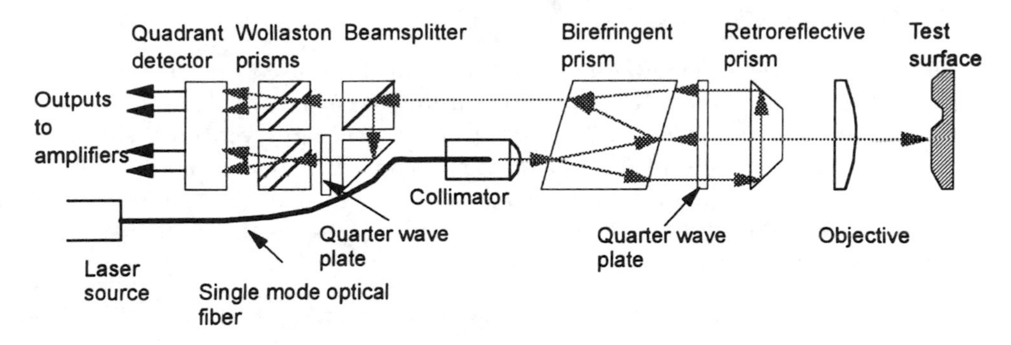

Figure 12.8 Laser Doppler anemometer with orthogonally polarized probe beams. (*After:* [27].)

vibration is inherently broadband. Of the various laser vibrometer architectures, the Michelson or Mach-Zehnder using active homodyne processing in the LGBWP mode are probably the most suitable. Unlike the Fizeau design, their path lengths can be accurately balanced to minimize the effect of laser phase noise, and hence give the best possible resolution necessary for the detection of small signals. The small signals mean that the small-angle approximation demanded by the signal-processing scheme is well-justified, and the active homodyne scheme by its simplicity adds the minimum of noise to the demodulated signal. Such interferometers have been used for detection of acoustic emission in composite materials [30] and for the diagnosis of progressive wear in machine tools [31,24]. Despite these successes, a velocity interferometer is often more appropriate than these displacement interferometer designs for the detection of acoustic emission.

12.2.2 Differential Interferometers

12.2.2.1 Differential Profilers

In the differential interferometer, both beams are derived from the target. The advantage when working with solid target surfaces is passive compensation for whole-body motion of the target. An example is a profiler with a passive, fiber-linked sensing head in which two beams, incident at different points on the target surface, have orthogonal polarization states [32]. At the source, orthogonally polarized beams traverse an unbalanced Mach-Zehnder interferometer and are launched into eigenstates of a highly birefringent fiber downlead to the sensing head. The difference in phase between two photodetector outputs is a measure of the surface height difference between the probe spots, independent of phase changes in the downlead or Mach-Zehnder module. The resulting profile is not absolute, and shows only the difference in height of adjacent points. However, the spacing and size of the spots can be optimized to reveal spatial frequencies that best indicate the surface structures to be monitored.

12.2.2.2 Torsional Vibrometer

An adaptation of the differential interferometer has been used to create a special type of vibrometer for the measurement of variations in the angular velocity of a shaft, usually called a *torsional vibrometer* [33]. A fiber-optic version is shown in Figure 12.9. It is essentially a Michelson interferometer, where both beams are reflected from the surface of the shaft in the plane normal to the axis. By considering the relative Doppler shifts of the light scattered from the two spots, the beat frequency, $\Delta\nu$, at the detector is

$$\Delta\nu = \frac{2\Omega d}{\lambda} \tag{12.15}$$

where Ω is the angular velocity of the shaft and d the perpendicular separation of the two beams. Hence, the angular velocity is determined independent of the shape of the rotating shaft or of the displacement of the shaft relative to the velocimeter. Liu and others [34] describe two modes of operation of the Michelson-based fiber-optic torsional vibrometer. In differential mode, the dominant signals are those backscattered by the rotating target. In reference mode, internal reflection in each arm combines with the backscattered signal to give the individual frequency shifts $\Delta\nu_1$, $\Delta\nu_2$ of each beam. To avoid ambiguity in this case, both beams should scatter from the same side of the center line of the shaft. Reference-mode operation tolerates lower power illumination and poorer surface reflectivity than in the differential case. Both systems were demonstrated on shafts rotating up to 19,000 rpm.

12.2.3 Doppler Difference Velocimeters

The reference beam laser velocimeter is almost as old as the laser itself, but has largely been superseded by the Doppler difference technique. The geometry of the Doppler difference arrangement can be adjusted to give a convenient optical beat frequency

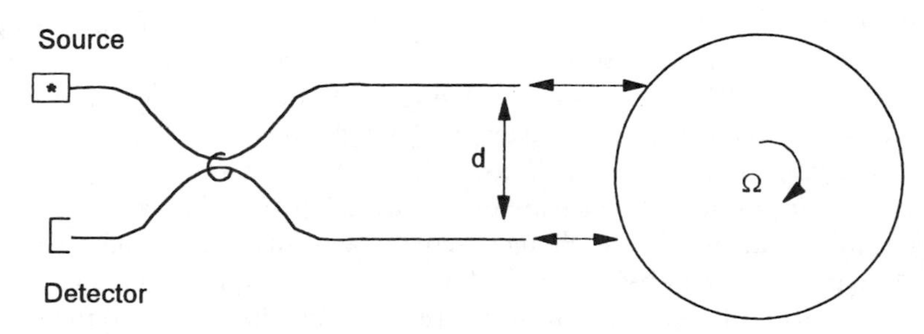

Figure 12.9 Principle of the fiber-optic torsional vibration sensor.

irrespective of the target velocity, and larger-aperture detectors can be used. The Doppler difference laser velocimeter is a very well-established commercial product, developed principally for use in aerodynamics test facilities. The majority of systems sold at present are equipped with some type of optical-fiber beam delivery system.

The basic instrument is shown in Figure 12.10. Both beams are incident on the measurement volume, focused to intersect at their waists at some angle θ. A particle entrained by the flow crossing the measurement volume scatters light independently from the two beams. Because the beams propagate at different angles, the light scattered from each beam has a different Doppler shift, and the two components beat together at the detector to yield a difference frequency.

$$\Delta \nu = \left| \frac{2V}{\lambda} n \sin \frac{\theta}{2} \right| \tag{12.16}$$

where V is the particle's transverse velocity component and n is the refractive index in the measurement volume. The same result may be derived from the so-called fringe model. If a screen were placed in the measurement volume, then plane-parallel interference fringes would be visible. A particle passing through that fringe set would scatter light intensity modulated at the frequency at which the particle crosses the fringes.

The Doppler difference design offers excellent spatial resolution because of the small intersection region defined by the sharply focused beams (see the shaded region in Figure 12.10). A consequent disadvantage is that if either of the incident beams is slightly defocused, then the beams fail to cross at their waists and the fringe spacing becomes nonuniform across the measurement volume. Thus the Doppler difference frequency varies (chirps) as a particle crosses the measurement volume, leading to a

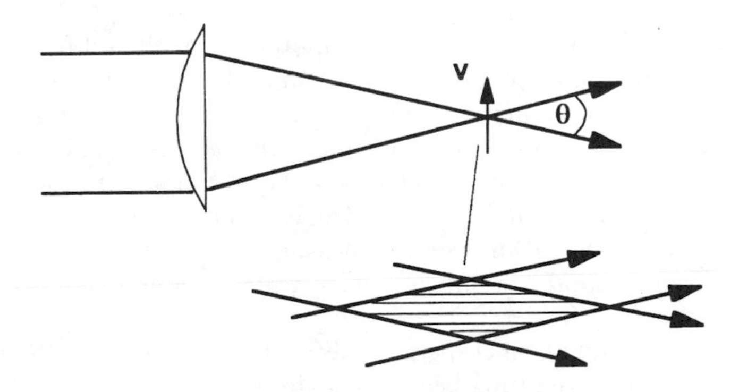

Figure 12.10 Principle of the Doppler difference technique: a seeding particle within the flow traversing the measurement volume scatters at a frequency dependent of velocity, *V*.

broadening of the apparent velocity spectrum, which can be erroneously interpreted as an increase in the level of turbulence in the flow [2].

Equation (12.16) shows that with incident beams of the same frequency, the sign of the velocity component is ambiguous (as was the case for the reference beam instrument, see (12.14)). Thus, most Doppler difference anemometers use heterodyne signal processing, with a frequency shifter placed in one of the arms of the transmission optics.

Various architectures for the use of optical fibers in Doppler difference anemometers are feasible, which we have classified into the groups shown in Figure 12.11. These will now be discussed individually.

12.2.3.1 Fiber Links

In fiber-linked designs, two optical fibers are used, one for transmitting and one for receiving. The purpose of the transmitting fiber is solely to allow remote operation of the laser. The fiber delivers the light to a bulk-optic assembly where the beam is divided, frequency shifted, and conditioned to form the measurement volume. In order that the beam can be focused to form a small measurement volume, it must have good spatial coherence. This is achieved by using a single-mode delivery fiber. The purpose of the receiving fiber is to guide light collected from the measurement volume by a lens to a remote photodetector. Good spatial coherence is not required in the return signal, and multimode fiber can be used to ease alignment and maximize light collection. It is only necessary that the dispersion in this multimode fiber be small enough to accommodate the bandwidth of the Doppler signal. Step-index fiber is often preferred because its acceptance angle is uniform over the core cross-section. Graded-index fiber has less intermodal dispersion, but the acceptance angle varies over the core area, being maximum on axis and zero at the core-cladding interface.

12.2.3.2 Fiber Probes

In a fiber-probe system, the laser output is amplitude divided and frequency shifted *before* being launched into fiber. The two transmitted beams are guided separately to the probe, normally by using two separate single-mode optical fibers. For high-visibility interference in the measurement volume, the recombining beams must have similar states of polarization. This is achieved either through polarization control in conventional fibers or by using highly birefringent fibers. The probe contains only passive optics to focus the beams into the measurement volume and to collect light scattered from it. A separate multimode receiving fiber is used to guide light back to the photodetector.

The two transmitting fibers experience different environments, so that the relative phase of the two transmitting beams is environmentally modulated [35]. Temperature effects are generally too slow to be of significance, but vibration and acoustic perturbations may cause motion of the fringes within the measurement volume with

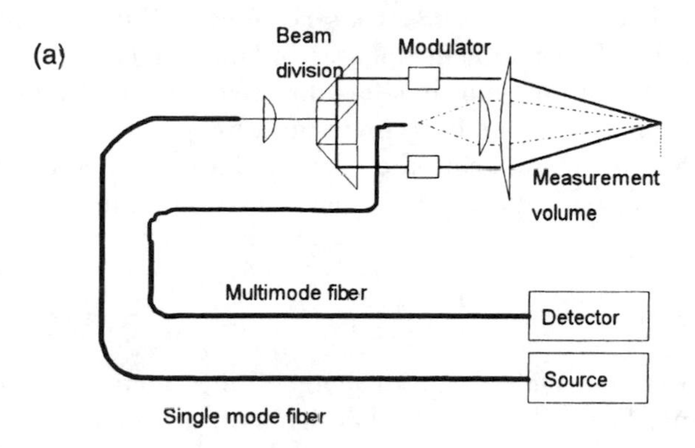

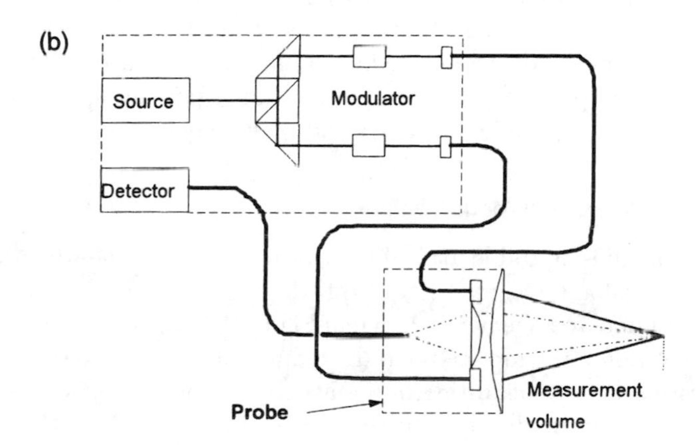

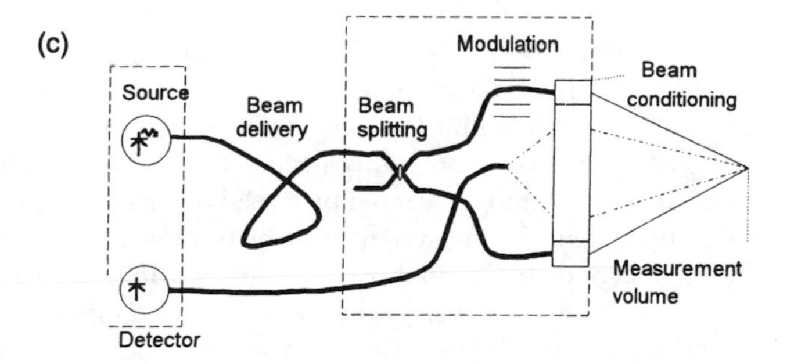

Figure 12.11 Fiber-optic Doppler difference anemometers: (a) fiber linked, (b) fiber probe, and (c) all-fiber designs.

frequencies comparable to the Doppler shift. The most serious effect is that of vibration, which induces strain in the fibers and hence phase modulation. The perturbation is generally of an oscillatory nature, and thus has the effect of broadening the observed Doppler spectrum, equivalent to an increase in the apparent measured turbulence of the flow. One technique for reducing the relative phase modulation is to transmit the two beams as the orthogonal polarization eigenmodes in a *single* highly birefringent fiber [36].

12.2.3.3 All-Fiber

In all-fiber systems, the probe contains fiber-optic elements for the functions of beam division and frequency shifting. The earliest reported design made use of a directional coupler for beam division, and a piezoelectric phase modulator for frequency shifting [37]. More recently, integrated optic modulators have been used for frequency shifting. Other alternatives include deriving the two input beams from two different laser sources of controlled frequency difference, using stabilized diode lasers [38], or by using a single frequency-swept diode laser source with transmitting optics containing a path imbalance so that a pseudo frequency shift can be generated [39].

12.2.3.4 Multiple Velocity Component Measurement

In many aerodynamic studies, it is desirable to be able to measure simultaneously more than one component of the velocity vector by multiplexing two or even three laser velocimeters. The most common arrangement, shown in Figure 12.12, is the measurement of the two orthogonal velocity components in the plane transverse to the optical axis of the instrument. Two sets of transmitting optics are used to produce two sets of orthogonal interference fringes; normally, a single set of receiving optics is used.

Various techniques are used to distinguish the two fringe sets [40]. Probably the commonest is wavelength division multiplexing, where the two sets of transmitted beams have different wavelengths, generated from-for example-different lines of an argon ion laser or different diode laser sources. Dichroic optics are then used in the receiver to separate the two wavelengths. Polarization division multiplexing is also used, where the two fringe sets have orthogonal states of polarization [41]. Some crosstalk between the velocity components is inevitably induced by changes in the state of polarization occurring when the light is scattered by a seed particle. Frequency division multiplexing is feasible in heterodyne systems in which frequency shifters are used; the different components are distinguished by being given different carrier frequencies. The velocity components are then distinguished in the signal processing. Time division multiplexing has been used with pulsed lasers, where the fringe patterns are illuminated sequentially [42], but cannot give a truly simultaneous measurement of the various velocity components, and may fail to reveal structures in the flow.

In the study of highly complex flows, it is necessary to be able to measure all

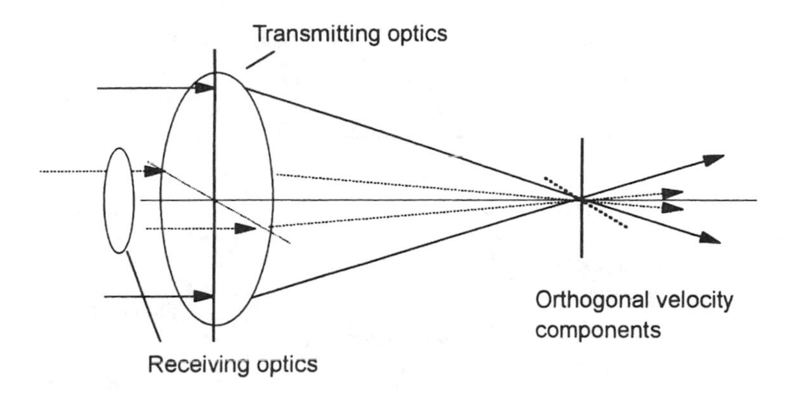

Figure 12.12 LDA arrangement of two orthogonal sets of intersecting beams.

three of the orthogonal velocity components, sometimes achieved using three Doppler difference velocimeters (Figure 12.13), and it is possible to arrange the transmitting optics to produce three mutually orthogonal fringe sets. An alternative is to measure the two transverse velocity components using Doppler difference velocimeters, but to measure the longitudinal component with a reference beam instrument, accepting the inevitable sacrifice in spatial resolution for the longitudinal component.

12.2.4 Velocity Interferometers

In velocimeters of either the reference beam or Doppler difference type, the target forms part of the arms of the interferometer to produce an optical phase change proportional to displacement, so that the optical output is in the form of an intensity-modulated signal, where the velocity is proportional to the modulation frequency and is found electronically. Alternatively, unbalanced interferometers can be used simply as optical-frequency discriminators for the direct detection of Doppler shift. Then, the phase shift is directly proportional to the velocity and the arrangement is called a *velocity interferometer*. Confocal Fabry-Pérot interferometers have also been used in a similar manner. However, the optical frequency shifts produced in practical situations are relatively small, so that the technique has only been applied to supersonic flows [43].

A disadvantage of the velocity interferometer is that it cannot distinguish Doppler shifts from frequency fluctuations of the light source. The effect of *frequency noise* scales with path imbalance. Hence the sensitivity cannot be increased arbitrarily by increasing path imbalance. Consequently, the resolution is limited by the frequency noise, whereas the reference beam velocimeter (in which the optical path

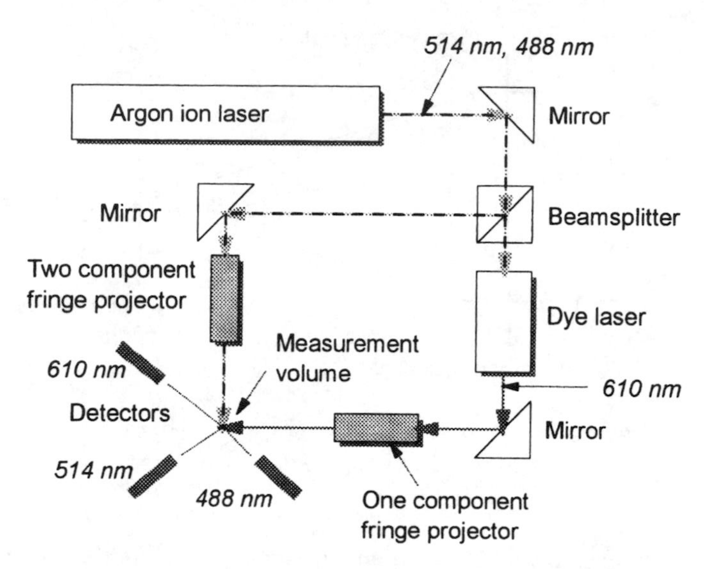

Figure 12.13 Three-color laser velocimeter for three-component measurement. (*After:* [40].)

lengths can be balanced so that source frequency noise has no effect) is typically shot noise-limited.

Velocity interferometers have greatest value when applied to acoustic emission measurements (Section 12.2.1.6) where, in conventional interferometry, the unwanted low-frequency motion produces such large optical phase changes that they swamp those due to the measurand. However, velocity interferometers effectively discriminate against low-frequency motion. Consider the unbalanced Michelson interferometer of Figure 12.14 used to measure the motion of a vibrating target surface whose displacement is described by

$$z = z_o \cos\Omega t \tag{12.17}$$

where z_o is the amplitude of the vibration with angular frequency Ω. Hence light of wavelength λ scattered from the target experiences a phase modulation

$$\phi = \frac{4\pi z_o}{\lambda} \cos\Omega t \tag{12.18}$$

or, equivalently, a frequency modulation

$$\Delta\nu = \frac{\partial\phi}{\partial t} = -\frac{4\pi z_o}{\lambda} \Omega\sin\Omega t \tag{12.19}$$

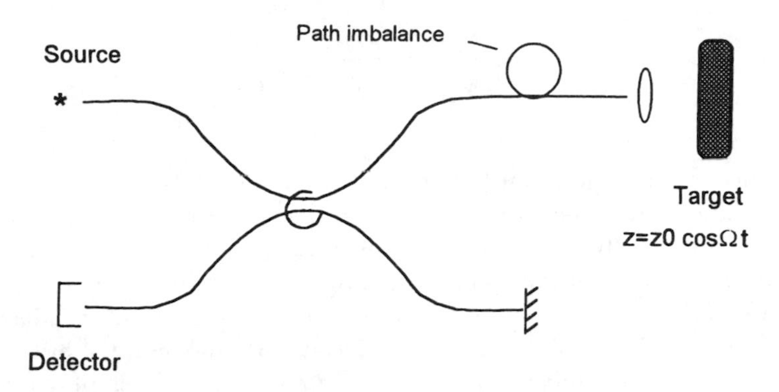

Figure 12.14 Unbalanced Michelson interferometer for measurement of the velocity of a vibrating target.

which generates a phase change in the interferometer given by

$$\psi = \frac{2\pi L \Delta\nu}{c} = \frac{8\pi^2 z_o L\Omega}{\lambda c}\sin\Omega t \tag{12.20}$$

so that the intensity at the interferometer output is

$$I = \frac{1}{2} I_o\left[1 + V\cos\left(\frac{8\pi^2 z_o L\Omega}{\lambda c}\sin\Omega t + \psi_o\right)\right] \tag{12.21}$$

where L is the optical-path difference in the interferometer and ψ_o is a constant phase offset, controlled to satisfy the quadrature condition $\psi_o = (N + 1/2)\pi$ where N is an integer. Hence,

$$I = \frac{1}{2} I_o\left[1 + \sin\left(\frac{8\pi^2 z_o L\Omega}{\lambda c}\sin\Omega t\right)\right] \tag{12.22}$$

assuming unit visibility. It is instructive to compare (12.22) with the response of a standard reference beam interferometer where, at quadrature and with unit visibility, the output intensity is

$$I' = \frac{1}{2} I_o\left[1 + \sin\left(\frac{4\pi z_o}{\lambda}\cos\Omega t\right)\right] \tag{12.23}$$

so that the ratio of the phase sensitivities of the velocity interferometer and the reference beam velocimeter is

$$\frac{\psi_{\text{vel}}}{\psi_{\text{ref}}} = \frac{2\pi L\Omega}{c} = \frac{fL}{c} \tag{12.24}$$

where $f = 2\pi\Omega$.

Hence, at high vibration frequencies, the velocity interferometer can be more sensitive than the reference beam instrument. Because the sensitivity scales with frequency, it intrinsically discriminates against unwanted low-frequency motion. The sensitivity also scales with the path-length imbalance, which can be adjusted to suit a particular application. At a practical ultrasound frequency, say 10 MHz, the velocity interferometer begins to be more sensitive at path imbalance of 30m. It is unusual to use such a large imbalance in practice, and instead to employ a high-finesse multiple-beam interferometer (normally a confocal Fabry-Pérot) of more modest dimensions.

In a reference beam instrument with homodyne processing, the interferometer must be actively stabilized to within << 1 rad (to satisfy the small angle approximation). Table 12.1 shows that in a velocity interferometer, such active stabilization will often not be required. That is, the interferometer path imbalance must be adjusted to quadrature and stabilized against internal path-length changes, but will then be relatively insensitive to unwanted target motion.

Optical fibers have been used extensively in acoustic emission measurements, including their use for beam delivery in laser-induced ultrasound experiments. However, single-mode optical fiber probes of the confocal type (i.e., light scattered from the target is reimaged into the fiber) collect relatively little light, effectively imaging only one laser speckle from the target. Conversely, bulk-optic interferometers can capture an entire speckle pattern.

However, optical-fiber interferometers offer some unique advantages in the design of velocity interferometers. One important special case is the application of the Sagnac interferometer to velocimetry, where a phase shift proportional to velocity is produced without sensitivity to source frequency variations by exploiting time-of-flight effects [44]. Time-of-flight effects in the Sagnac interferometer are the basis of

Table 12.1
Comparison of the Phase Sensitivity of a Reference Beam and Velocity Interferometer of Path Imbalance 0.3m Operated at a Wavelength of 850 nm

$z_o (m)$	$f (Hz)$	ψ_{ref} (rad)	ψ_{vel} (rad)
10^{-5}	10^2	1.48×10^2	$.48 \times 10^{-5}$
10^{-8}	10^7	$.48 \times 10^{-1}$	1.48×10^{-3}

its applications in hydrophones [45] and as a distributed sensor [46]. An arrangement for its use as a velocimeter is shown in Figure 12.15. The single-mode fiber loop is interrupted near to one end, and a probe is interposed. The probe takes light from the loop, reflects it from the target (a moving particle in the measurement volume or a solid surface), and returns it to the loop to continue in the same direction. Both the clockwise and counterclockwise propagating beams are reflected from the target, and they interfere at the detector.

Because the two beams are incident on the target at different times, then if the target is moving and assuming normal incidence and reflection from the target, a phase shift is produced, given by

$$\phi_D = 2(2\pi/\lambda)\bar{v}\Delta t \tag{12.25}$$

where $\bar{v}$ is the mean target velocity during the loop delay Δt. Because the interferometer is common path, it is (to first-order) insensitive to environmental phase perturbations and source frequency fluctuations.

For maximum sensitivity, it is necessary to bias the interferometer at one of its quadrature points, which can be achieved in the fiber Sagnac arrangement through control of the fiber birefringence. Hence, a polarization controller is included in

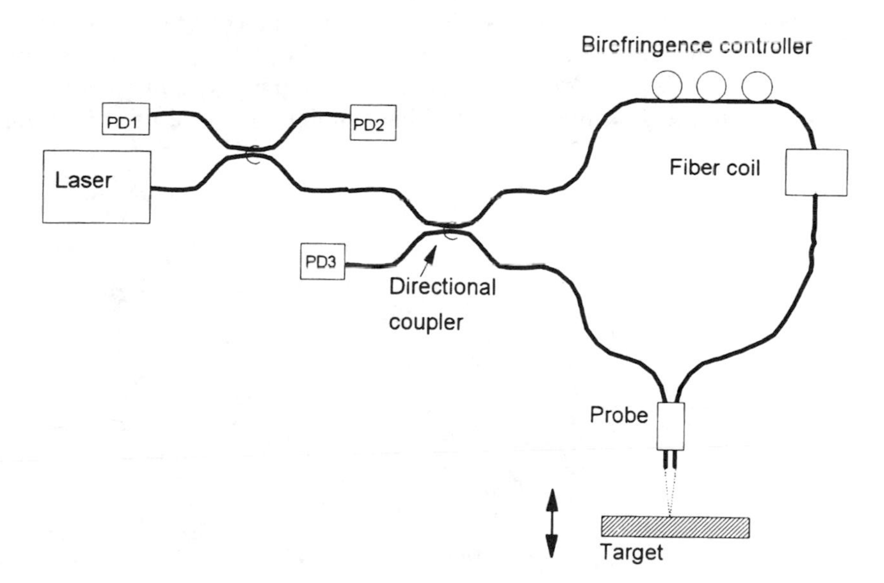

Figure 12.15 Doppler velocimeter based on the fiber-optic Sagnac interferometer.

the loop. The fiber Sagnac interferometer has been exploited for high-frequency vibration measurement. In one application, the noise-floor-limited velocity resolution was 50 nm/s^{-1}/Hz$^{-1/2}$ [47].

12.2.5 Transit Velocimeters

Of those instruments discussed so far, the Doppler difference arrangement has the highest spatial resolution, defined by the dimensions of the intersecting laser beam waists that form the measurement volume. The minimum size of the measurement volume is set by the need to have sufficient fringes in the measurement volume. Reducing the number of fringes increases the width of the Doppler difference spectrum-for example, making the measurement of higher order moments of the velocity distribution difficult.

To improve further the spatial resolution, "transit velocimetry" techniques have been devised that record the time interval taken for a particle to travel from a first focused light spot to a second one. In its simplest form, the transit velocimeter involves transmission optics that produce two laser beam waists symmetrically disposed about the optical axis of the receiver optics (Figure 12.16). Only a small percentage of particles pass through both waists, so that the data rate is low. Nevertheless, the direction of the particle velocity is well-defined by the relative geometry of the waists. It is possible to adjust the azimuth of the waists so that by rotating the spots the local velocity vector can be found.

Similar fiber architectures to those discussed for Doppler difference velocimetry have been devised for transit instruments (for example, [48]). More complex types of "optical gates" have also been devised to extract simultaneously a number of velocity components, using multiplexing techniques similar to those used in Doppler

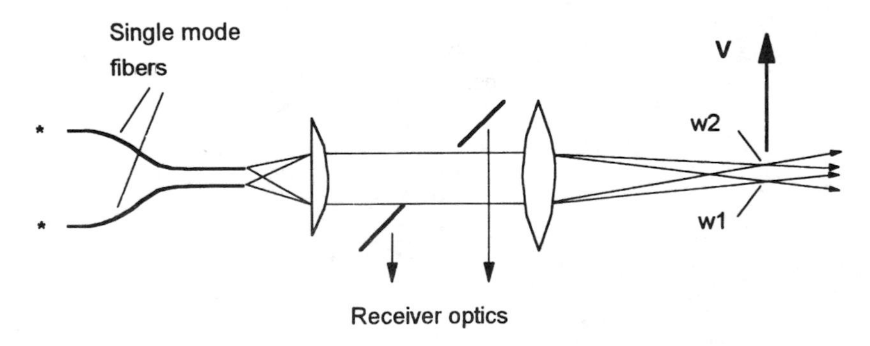

Figure 12.16 Schematic laser transit velocimeter; seeding particle passes through beam waists w1 and w2.

difference velocimetry. For example, laser diode arrays have been used to produce "stripes" of light to act as gates, rather like a Doppler difference arrangement, but without using interferometry. Some transit anemometers with large numbers of beams have been devised, sometimes known as "mosaic velocimeters" [49].

12.2.6 Intrinsic Sensors

The concept of intrinsic sensors was outlined in Section 12.1.2. In this section, examples of intrinsic sensors specifically devised for applications in experimental aerodynamics are described.

12.2.6.1 Thermal Measurements

In aerodynamics test facilities, such as wind tunnels, a primary role of experimental measurements is in the validation of the predictions of computational fluid dynamics (CFD). Velocity measurements and derived quantities such as turbulence index are of the greatest importance, although the accuracy of CFD predictions are improving to the extent that experimental measurements are required increasingly only for regions of complex flow. Attention is now being turned to the prediction of the scalar fields, notably temperature and pressure, where prediction is less reliable.

In turbomachinery research and design, two types of thermal measurements are required: heat transfer rates and total gas temperature. Gas turbines are designed to operate at the highest possible temperatures to optimize efficiency, so that adequate heat transfer is essential if the materials used in the construction are not to melt. In the gas flow itself, whether in the turbine or compressor, fluctuations in total temperature are indicative of energy losses, and are hence implicated in the overall efficiency of the machine.

Heat flux is often measured using electrical sensors in the form of thin-film resistance thermometers applied to the surface of model components [50]. The heat flux is defined as the heat energy supplied to the surface per unit area and per unit time, $\dot{Q}$. From the surface temperature-time history, the heat transfer rate is inferred. In some situations, an optical-fiber interferometer offers improved performance. An example is shown in Figure 12.17 in the form of a single-mode fiber Fizeau interferometer embedded with its axis normal to the surface of a test component, operated in a transient flow wind tunnel [51]. When the tunnel is fired, hot (or cold) gas flows over the test component for the duration of the runtime, heat transfer takes place and is recorded as a changing temperature by the fiber sensor.

Interference occurs between light reflected at the interface between sensor and downlead (forming a reference beam) and light reflected at the distal end of the fiber at the surface of the test component (forming the signal beam). The reflectivities are low, so that the transfer function is similar to that of any two-beam interferometer

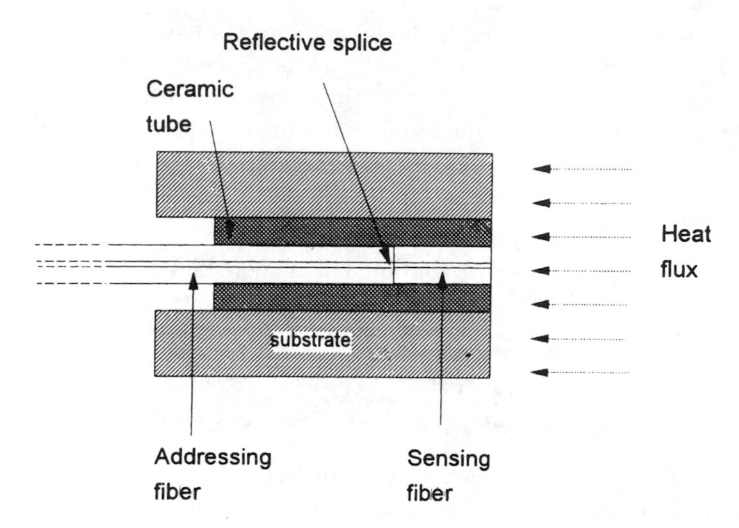

Figure 12.17 Intrinsic fiber Fizeau interferometer embedded in a test model for heat transfer measurements in a transient flow wind tunnel.

(see, for example, (12.1)). The optical path length, and hence the phase of the interference, is a function of the temperature. The predominant cause is the thermo-optic effect-the dependence of refractive index on temperature ($\partial n/\partial T$). Hence the time-dependent phase change induced by the heat transfer is given by

$$\Delta\phi(t) = \frac{2\pi}{\lambda}\frac{\partial n}{\partial T}\int_{z=0}^{l(T)} \Delta T(t, z)dz \tag{12.26}$$

where z is the axial distance along the fiber of length l.

The operation of the sensor can be illustrated by making the simplifying assumption that heat flows parallel to the z-axis, and that the thermal diffusion length during the tunnel run-time is short in comparison with the length of the sensor. For example, for a fused-silica sensor in a tunnel with a run-time in the order of 1 sec, a length of a few millimeters is appropriate. Under these conditions, the heat transferred to the sensor is balanced by a rise in its mean temperature (that is, the sensor behaves as a calorimeter) such that

$$\int_{t'=0}^{t} \dot{Q}(t')dt' = \rho c_o \int_{z=0}^{l} \Delta T(z, t)dz \tag{12.27}$$

where ρ and c_o are the density and specific heat capacity of the sensor. Hence

$$Q(t) = \frac{\rho c_o \lambda}{2\pi(\partial n/\partial T)} \frac{\partial}{\partial t} (\Delta\phi) \tag{12.28}$$

Thus, from a measurement of the optical phase, the heat flux rate is found. The optical-fiber sensor, through its calorimetric operation, has intrinsic calibration; that is, unlike the surface temperature sensor, its sensitivity is independent of its dimensions. However, its most striking advantage is one of spatial resolution, set by the dimensions of the core of the fiber, which is typically 5 μm.

Practical designs and applications of fiber heat transfer sensors have been reported [52], in which the phase-stepping techniques based on laser diode wavelength tuning were used to demodulate the interferometer phase.

In heat transfer measurement, a relatively long sensing element was required to give calorimetric operation. For unsteady gas temperature measurements, high sensor bandwidth is required, implying the need for short sensing elements. The principal aerodynamic application is for measuring the temperature field behind blade rows in compressor test rigs, where it is required to resolve features within the blade-passing frequency, implying bandwidths in the order of 50 kHz. Such test rigs, unlike those used for heat transfer studies, run continuously.

For temperature measurement, short sensors are desirable to give sufficient sensitivity at high frequencies. By shortening the length of the sensor, the effect of source frequency noise is reduced, improving resolution. However, at the lowest thermal frequencies the thermal diffusion length is greater than the sensor length and some energy is lost. However, sensitivity at these low frequencies remains adequate. Typically, a length of a few meters is suitable. Rather than fabricating such short sensors in fiber, it is more convenient to apply them as coatings onto a downlead fiber, with the interferometric sensor formed by the internal and external reflections. By choosing materials with a higher thermo-optic coefficient than fused silica (e.g., zinc selenide or titanium dioxide), improved sensitivity is achieved.. For such short sensors, phase-stepping by wavelength tuning of a single laser diode is no longer feasible, as the required wavelength interval is too great to be practical. Instead, two lasers of different wavelength may be used.

Practical versions of these sensors have been constructed and have been demonstrated in full-scale compressor test rigs, showing sensitivity at frequencies as high as 70 kHz [53].

12.2.6.2 Pressure Measurements

There is considerable demand for high spatial resolution pressure measurements in aerodynamics. Piezoresistive pressure sensors are often used, with millimeter spatial resolution. The resolution of a few μm, potentially available with single-mode fiber sensors, is hence attractive. Many designs have been reported, but have not met simultaneously the criteria of high measurement bandwidth and high spatial resolution. Sensitive high-frequency pressure transducers (essentially acoustic and ultrasonic sensors) have been developed extensively, generally driven by hydrophone applications, but they are formed from considerable lengths of fiber, and hence have relatively poor spatial resolution. Conversely, various types of "fiber microphones" have been developed, often using a pressure-responsive element, such as a diaphragm, to form one mirror of an optical cavity whose length is modulated by acoustic pressure. These designs have great spatial resolution, but their frequency response is usually limited by the relatively low resonant frequency of the pressure-responsive element.

Intrinsic fiber Fabry-Pérot interferometers have been embedded in aluminum for use as pressure sensors in internal combustion engines [54]. Four sensors were spatially multiplexed using a star coupler and illuminated by a distributed feedback laser. A 2-kHz linear ramp modulation was applied to the laser to chirp the optical frequency. The output phase was recovered by digitizing the time elapsed between the start of the linear ramp and a threshold crossing for each sensor signal. The output signal bandwidth was 1 kHz with full scale corresponding to 172-bar pressure. Eight-bit digitization limited the resolution to 0.34 bar, but the shot noise limit for four multiplexed sensors was 6 mbar.

12.3 FULL-FIELD MEASUREMENTS

Full-field measurements are used in the study of both surfaces and fluid flows. They are complementary to single-point measurements and have the advantage of providing simultaneous information over an extended field of view—essential, for example, in vibration modal analysis where there is a need to preserve the phase relationships between the motions of individual points, and in measurements on fluid flows where there is need to identify large-scale coherent structures. However, full-field techniques inevitably have less resolution than their single-point counterparts.

Full-field interferometry is a classical inspection technique for optically smooth surfaces, but is inapplicable to the overwhelming majority of practical surfaces that are rough on the scale of the wavelength of light. Holography offers a means for interferometry on rough surfaces. Holography is essentially a technique for recording wavefronts, and so offers a means for measuring the deformation of an object by comparing wavefronts derived from it in its undeformed (reference) and deformed states. Holographic interferometry is used extensively for deformation and vibration analysis, and for measurements in fluids (where density variations are revealed by the interference fringes produced by local variations in refractive index) [55]. However, there are inhibitions to the widespread use of holographic interferometry, set by the expense and complexity of the optical arrangements, and the indirect derivation of the experimental data that generally involves first the photographic processing of the holographic plates, whose fringes are then digitized and analyzed. Holographic interferometry may become more widespread in the future with the development of new recording materials and lower cost pulsed lasers. Already, thermoplastic and photorefractive materials, not requiring conventional photographic processing, are in common use for holography.

Despite innovations in holography, the closely related technique of electronic speckle pattern interferometry (ESPI) is much more commonly used in engineering applications. ESPI is a technique in which speckle fields, rather than holographic images, are recorded and analyzed [56]. Speckle fields have relatively coarse spatial resolution and can be recorded using television cameras. Hence, ESPI provides a means for full-field interferometry on rough surfaces with electronic recording, and almost instantaneous signal processing and display.

12.3.1 Electronic Speckle Pattern Interferometry

Laser speckle, the granular appearance of light scattered from a rough surface illuminated by an expanded laser beam, is a familiar phenomenon. In ESPI, the rough surface is imaged onto a television camera, typically a CCD. The aperture of the imaging lens is adjusted such that the resolution (i.e., the size of a speckle) is approximately the same as the spatial resolution of the camera (one pixel). As shown in Figure 12.18, a beamsplitter is used to overlay the speckle image with a smooth reference beam. At each pixel (or speckle), the light from the surface and the reference beam interfere. Thus each pixel behaves similarly to the detector in a standard two-beam interferometer, and the surface appears as an array of speckles ranging randomly in intensity from dark to bright. If the surface were moved steadily towards the camera, then each speckle would cycle in brightness, following the two-beam transfer function. Hence, by observing the change in brightness of a particular speckle as the surface is moved from a reference state to a deformed state, the displacement of the surface at the position of the speckle can be determined.

12.3.1.1 Formation of a Speckle Correlogram

The intensity distribution in the detector plane is stored with the object in its reference state; the object is then deformed and a second frame is stored. The two frames are subtracted electronically to give a resultant having the appearance of an interferogram showing fringes corresponding to the local displacement but modulated by a superimposed speckle pattern. The intensity distribution resulting from subtracting the frames is given by

$$\Delta I = \frac{1}{2} I_o V[\cos(\phi + \phi_d) - \cos\phi] \tag{12.29}$$

All terms in the equation vary with position; ϕ describes the surface profile, and

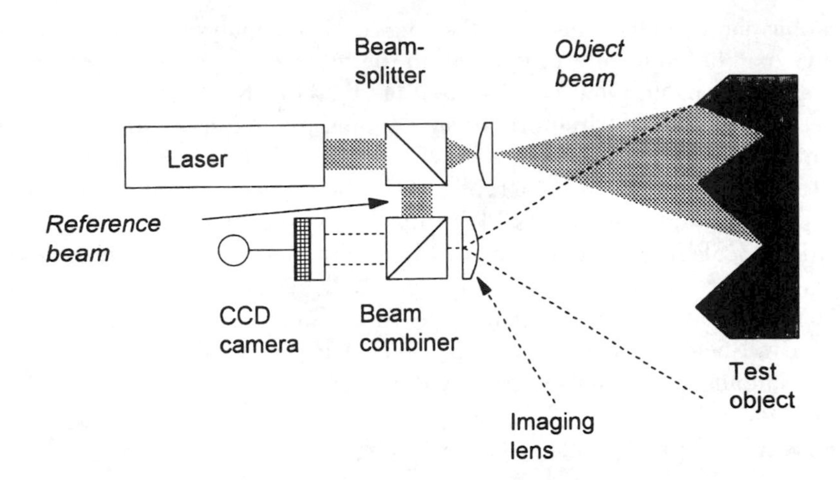

Figure 12.18 Optical arrangement for electronic speckle pattern interferometry (ESPI).

because the surface is rough ϕ varies by >> 2π rad on length scales in the order of a wavelength; ϕ_d describes the out-of-plane optical deformation.

Thus the object deformation or motion is revealed, yielding similar information to that obtained by holography but with the operational advantages of rapid viewing on television. ESPI is sometimes called *television holography*, although the principles on which it is based are those of correlation interferometry rather than true holography.

Rearranging (12.29),

$$\Delta I = -I_o V \sin\frac{1}{2}(2\phi + \phi_d)\sin\frac{1}{2}\phi_d \tag{12.30}$$

so that the term in ϕ_d yields the deformation information, while the term in $(2\phi + \phi_d)$ represents the speckle pattern that varies rapidly with position.

There are many ways in which (12.30) can be processed to yield ϕ_d and hence the out-of-plane deformation. In one simple example, the signal corresponding to ΔI is first squared to avoid negative values and then lowpass filtered. The term $\sin^2\frac{1}{2}(\phi + \phi_d)$ varies much more rapidly with position than $\sin^2\frac{1}{2}\phi_d$; the effect of filtering is thus to replace the first term with its averaged value, leaving the second term unchanged. Thus (12.30) becomes

$$\Delta I' = \frac{1}{2}I_o^2 V^2(1 - \cos\phi_d) \tag{12.31}$$

The analogy with (12.1) is evident, and similar techniques may be used to determine the phase.

The object deformation and phase ϕ_d can be related as follows. The phase change is given by

$$\phi_d = (\mathbf{k_i} - \mathbf{k_r}) \cdot \mathbf{n} d_n \tag{12.32}$$

where $\mathbf{k_i}$ and $\mathbf{k_r}$ are the propagation vectors of the light for the incident and scattered beams, $\mathbf{n}$ is the unit vector normal to the test surface, and d_n is the displacement of the surface in the direction of $\mathbf{n}$. Thus,

$$\phi_d = \frac{4\pi d_n}{\lambda} \cos \alpha \tag{12.33}$$

where λ is the wavelength of the light and α is the angle between the normal to the surface and the bisector between the incident and scattered beams. Most often in practice the normal incidence and viewing are used to a reasonable approximation, so that

$$\phi_d = \frac{4\pi d_n}{\lambda} \tag{12.34}$$

12.3.1.2 Measurement of In-Plane Motion

When used in the *reference beam* arrangement, described above, ESPI is sensitive to out-of-plane motion. For in-plane measurement, two beams are used to illuminate the test object in the form of plane waves intersecting at an angle. Hence, the *relative* phase change between light scattered from the two beams occurring when the object is deformed is proportional to the motion normal to the bisector of the illuminating beams, and in the plane of the beams. In general, the bisector is arranged to be normal to the plane of the test surface, so that the differential phase is sensitive to motion in the plane of the test surface. The similarity with Doppler difference velocimetry (Section 12.2.3) is evident.

12.3.1.3 Time-Averaged Vibration Measurement

Speckle pattern interferometry can also be applied to the measurement of vibration. Consider that the test object is illuminated by a continuous wave laser and imaged using a television camera with a frame rate of, typically, 25 Hz. The object is then set into vibration at frequencies much higher than the camera frame rate. Once again, a speckle interferogram is produced; the fringes created are averaged over many cycles of vibration, and their form is given by a Bessel function (of argument proportional to the amplitude of the vibration) rather than a sine. An effect of the averaging is that the brightness of the fringes decreases for regions of high vibration amplitude, where the surface moves through distances of many wavelengths in each vibration cycle.

Although the amplitude of the vibration is found, its phase is not. *Heterodyne* techniques have been developed in which the reference phase is modulated by a sine wave. By adjusting the amplitude, frequency, and phase of the modulation, points vibrating either in phase or out of phase with the modulation can be identified. An alternative to time-averaging is to illuminate the object with (stroboscopic) pulsed laser illumination, synchronized with the vibration frequency. It is then possible to make accurate measurements of the phase of the vibration.

Consider now that the test object is vibrating, so that d_n becomes time-dependent. If the vibration is harmonic, then d_n is given by

$$d_n = a \sin(\omega t + \alpha) \tag{12.35}$$

where the local amplitude, phase, and circular frequency of the vibration are a, α and ω. Hence, remembering that $\phi_d = 4\pi d_n/\lambda$ (for normal illumination and observation), then by substituting in (12.31) and integrating over the framing period to produce *time-averaged fringes*, we see that

$$\Delta I \propto J_o\,(4\pi a/\lambda) \tag{12.36}$$

where J_o is the zeroth order Bessel function.

12.3.1.4 Extraction of Phase Maps

The basic signal-processing problem in ESPI is to recover the phase ϕ_d from the intensity map described by (12.37) in the form

$$\Delta I' = \frac{1}{2} I_o^2 V^2 (1 - \cos\,\phi_d) \tag{12.37}$$

The problem is exacerbated because both the mean intensity I_o and visibility V vary strongly with position over the test surface. The most common technique for recovering ϕ_d is based on *phase-stepping*. The technique was described with respect to single-point interferometry in Section 12.2.1.2, and may be applied to ESPI as follows. A reference frame is acquired for the undeformed object, as before. A number of frames are acquired for the object in its deformed state, with an optical phase modulation between each one. The phase modulation is usually achieved by changing the path length of the reference arm of the interferometer. Hence, the phase is *stepped* between each acquired frame of the object in its deformed condition. Different versions of the technique use varying numbers of frames and phase modulation amplitudes, but a common and successful form uses four frames with a phase modulation of $\pi/2$ radians between each frame. This is the version that was described in Section 12.2.1.2.

The frames are processed as follows. The reference frame is subtracted from each of the frames for the object in its deformed condition, to form the intensity differences $\Delta I_i'$ for the ith frame of the deformed object. Hence, following the analysis of Section 12.2.1.2,

$$\tan \phi_d = \frac{I_4 - I_2}{I_1 - I_3} \tag{12.38}$$

A variation is to acquire phase-stepped reference frames, so that only a single frame in the deformed condition is required (useful in studying transient deformations).

Heterodyne phase-extraction techniques involve imposing the heterodyne carrier in the form of a *spatial* (rather than temporal) modulation on the speckle correlogram, for example by introducing a tilt into the reference beam to produce a constant rate of change of reference phase across the image of the test surface, thus generating the desired spatial frequency offset. The phase ϕ_d is thus revealed as a phase modulation imposed on the spatial carrier frequency and can be demodulated by Fourier transform techniques [57].

Both phase-stepped and heterodyne fringe analysis techniques are only capable of recovering the phase modulo 2π. In both cases, it is common to display their outputs in the form of a gray-scale image, going typically from fully dark for a phase of $-\pi$ to fully bright at $+\pi$ (modulo 2π in each case). These gray-scale plots are often called *wrapped fringe maps*. In ideal cases, the boundaries between individual interference fringes are clearly delineated by the abrupt dark-to-light transition, and relatively simple processing algorithms are capable of detecting the interfaces and interpreting the direction corresponding to an increment in fringe count. Such algorithms are thus capable of *unwrapping* the fringes, and resolving the 2 radians ambiguity. However, with real test objects the noise inherent in speckle fields obscures the fringe boundaries and has required the development of sophisticated fringe unwrapping techniques [58]. However, their discussion is beyond the scope of this chapter.

12.3.1.5 Fiber-Optic ESPI Systems

Many authors have discussed the advantages of using optical fibers in the construction of practical ESPI systems [59–62]. A typical arrangement is shown in Figure 12.19. The laser source is coupled into a single-mode fiber and divided into the signal and reference beams at a directional coupler. Typically, much more power is required in the signal beam than the reference beam. Commonly, the signal and reference beams are allowed to expand freely from the distal faces of their respective fibers, although lenses may sometimes be used to increase the divergence, particularly of the object beam in order to reduce working distances for a particular field of view. A significant advantage of using single-mode fibers over conventional optics is the high beam quality produced by the spatial-filtering effect of the fibers. The usual means of combining the signal and reference beam is to interpose a beamsplitter between the objective lens and the camera, albeit at the cost of wasted light from the unused beamsplitter port. Alternatively, the beamsplitter is omitted and the reference fiber itself is introduced between objective and camera to illuminate the camera directly. There is no need for the reference wavefronts to be plane: the shape of the reference wavefront is canceled in the subtraction process inherent in producing the correlogram.

It is necessary for the states of polarization of the scattered light from the object and the reference beam to be similar. While this could ideally be achieved by using highly birefringent fibers, it is more commonly achieved in practice by incorporating a fiber polarization controller [63] in one of the fiber arms or downlead (between laser and directional coupler), or by using waveplates in the launching system between laser and fiber.

In phase-stepped versions of the ESPI technique, it is convenient to use an optical-fiber phase modulator, based on straining part of the reference fiber using a piezoelectric element. A disadvantage of the fiber system, especially when used in harsh environments or where there are large time intervals between frames, is unwanted phase modulation arising chiefly from temperature changes in the fibers. However, it is straightforward to compensate for such effects [64]. Figure 12.19 shows a reference photodetector at the otherwise unused directional coupler arm in the reverse direction to the propagation of the main beams. The reference detector sees light reflected from the distal faces of both object and reference fibers. Accordingly, the two fibers behave like a fiber Michelson interferometer. Consequently, the

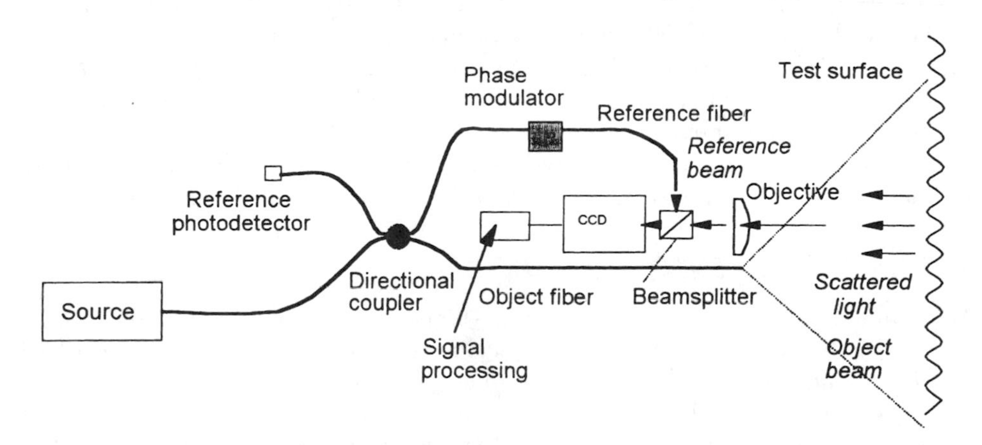

Figure 12.19 An example of the arrangement of a fiber-optic ESPI system.

intensity at the reference detector reveals phase changes between object and reference arms. A simple servo loop feeding back to the phase modulator thus serves to lock the relative phase of the object and reference beams to a fixed value for arbitrary periods.

12.3.1.6 Static Deformation Measurement

Of all the applications of ESPI, the measurement of static or slowly varying displacement is the most direct. The most common application is in determining the response of engineering components or structures under load, although it has also been exploited in architectural and civil engineering applications, and even for assessing deterioration in works of sculpture and art [65–67].

As a simple example, Figure 12.20 shows the deformation produced by bolt torque in the main bearing assembly of a prototype car engine [68]. The ESPI fringes were processed using phase-stepping with a four-step algorithm (12.38) to produce a direct measure of displacement, represented by the gray-scale plot shown. In common

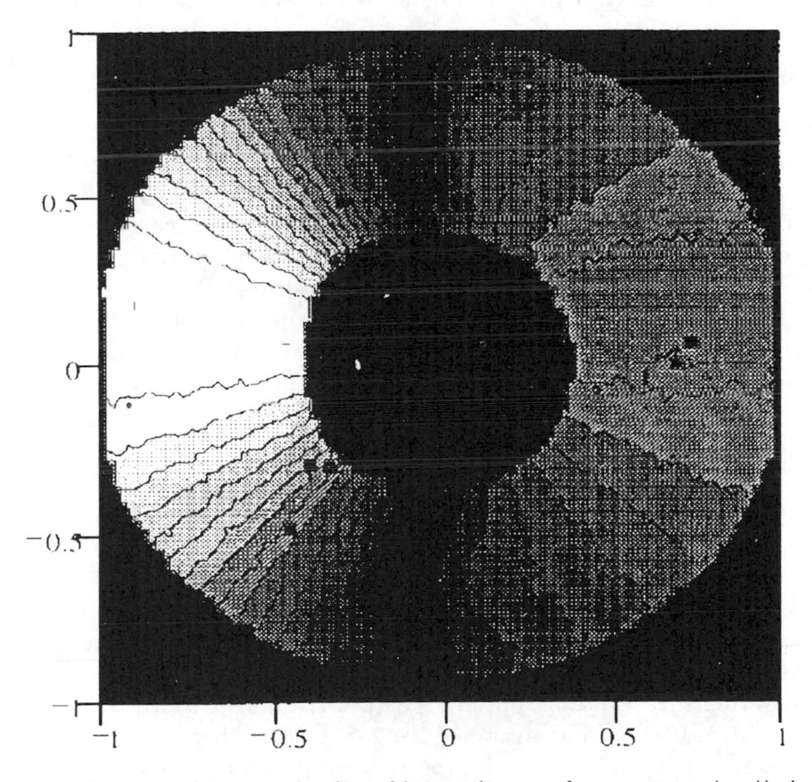

Figure 12.20 Deformation of the internal surface of the main bearing of a car engine induced by bolt torque, measured by phased-stepped ESPI. (*Source:* [68].)

with many similar experiments, the objective was to compare the observed displacements with the predictions of finite element models, to check for regions of anomalous displacement, and hence to assess the success of the design.

To convert the out-of-plane displacements into local in-plane surface strains requires a double differentiation of the displacement with respect to in-plane position. Such numerical differentiation is inherently noisy. Hence, in some situations it is desirable to use an optical technique to yield the spatial derivative of out-of-plane displacement—this is the technique of *speckle shearing*, or *shearography*, described in Section 12.3.2.

12.3.1.7 Vibration Amplitude Measurement

Section 12.3.1.3 described the process of time-averaged vibration measurements using ESPI. The technique provides a rapid and economic alternative to the use of multiplexed contacting accelerometers for the identification of vibration modes in structures. The test object is mounted such that it is as unconstrained as possible to avoid modifying significantly its vibrational behavior. The object is excited with a suitable actuator-typically an electromagnetic "shaker" for investigations in the audiofrequency range. In some cases, noncontact excitation is achieved by stimulating resonances in the object using sound waves. In all cases, harmonic excitation is used in order to avoid simultaneous stimulation of several vibration modes, which would lead to fringe patterns too complex to interpret. The excitation frequency is incremented, and at each frequency a set of speckle fringes is acquired. In practice, fringes are only observed at resonant frequencies of the test object, because at other frequencies the amplitude of vibration is too small to be observed.

The speckle fringes can be formed by first acquiring a reference frame with the unexcited object, and then at each excitation frequency subtracting the frame obtained from the reference frame. However, practical difficulties can arise because of unwanted whole-body motion of the test object, which is slow in comparison with the frame rate. An alternative is to form the speckle correlation fringes by sequential subtraction: no static reference frame is taken and instead consecutive frames obtained at a particular excitation frequency are subtracted. By imposing a reference beam phase shift of between the frames, time-averaged fringes similar to those obtained with a static reference frame are observed. The reference beam phase shift is readily produced using a phase modulator as described in Section 12.3.1.5. For objects showing significant whole-body motion on the timescale of the frame rate, it is not necessary to introduce a deliberate reference beam phase shift between frames, although the resulting fringes are of lower quality.

Recovering the vibration amplitude is not straightforward. It was shown in Section 12.3.1.3 that the time-averaged fringes are described by a Bessel function, where the argument of the function is proportional to the measured component of the vibration amplitude. Hence the simple phase-stepping routine applicable to the cosinusoidal fringes obtained in the measurement of static deformation is not appropriate. It is possible to produce cosinusoidal fringes in vibration measurement by using pulsed or stroboscopic illumination, as described below in Section 12.3.1.9.

Simple time-averaged vibration measurement using ESPI is valuable in rapidly identifying the vibration modes of a test object and assessing the approximate amplitudes of the antinodes. An example is shown in Figure 12.21. The test object is a car engine cylinder block, excited using an electromagnetic shaker. Figure 12.21 shows the time-averaged speckle fringes for a vibration mode. Three antinodes are visible as dark areas, with dark fringes contouring the zeros of the Bessel function; the nodes of vibration (zero amplitude) are revealed as bright fringes.

12.3.1.8 Vibration Measurement Using Heterodyned ESPI

Simple time-averaged vibration measurement has two limitations. Firstly, the visibility of the Bessel fringes declines with their order, thus limiting the maximum vibration amplitude that can be recovered before the fringe visibility falls below the noise threshold. Secondly, the time-averaging process intrinsically removes the vibrational

Figure 12.21 Vibration amplitude map of engine cylinder block (Rover Group Ltd.).

phase information, which would otherwise be of great importance in identifying modes of vibration and in assessing the acoustic energy that they will radiate. For example, for the practical case shown in Figure 12.21, the motion of the test surface is quite different for the cases where the two outer antinodes are either in phase or out of phase with the motion of the inner antinode.

A complete measurement of the phase of vibration is possible using pulsed illumination techniques, of which a stroboscopic version is described in Section 12.3.1.9. However, normal modes of vibration exhibit antinodes which are either in phase or exactly out of phase (by π rad) with each other. Under these circumstances, the phase information can be derived from a heterodyne technique, which is also capable of extending the dynamic range of the amplitude measurement.

The heterodyne technique involves modulating the optical phase of the reference beam in sympathy with the motion of the surface. For example, we have seen that the signal phase varies as

$$\phi_d = \frac{4\pi d_n}{\lambda} = \frac{4\pi a}{\lambda} \sin(\omega t + \alpha) \tag{12.39}$$

Let us suppose that the reference beam is phase-modulated harmonically at the same frequency as the vibration, so that it is described by

$$\phi_r = \frac{4\pi a_r}{\lambda} \sin \omega t \tag{12.40}$$

and the net phase of one of the speckles is

$$\Delta\phi = \phi_d - \phi_r = \frac{4\pi}{\lambda} [a \sin(\omega t + \alpha) - a_r \sin \omega t] \tag{12.41}$$

or

$$\Delta\phi = \frac{4\pi}{\lambda} (a^2 + a_r^2 - 2aa_r \cos \alpha) \sin \omega t \tag{12.42}$$

thus for points on the surface vibrating with the same amplitude and phase as the phase modulation of the reference beam, the net phase is zero and a bright fringe is observed. For other points on the surface, the intensity difference is given (following (12.36)) by

$$\Delta I \propto J_o(\Delta\phi) \tag{12.43}$$

An illustration of the effectiveness of heterodyne ESPI is shown in Figure 12.22 [61]. Figure 12.22(a) shows a normal time-averaged fringe map of a vibration mode

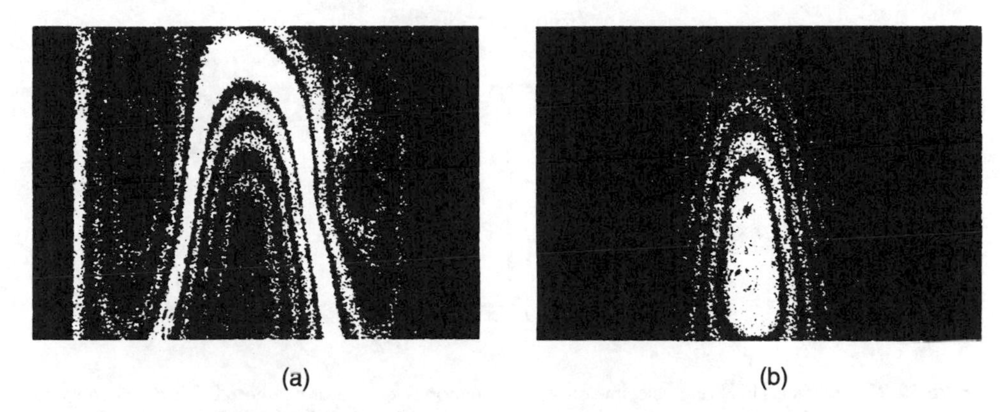

Figure 12.22 (a) Time-averaged bessel fringe map of a surface vibrating at 700 Hz and (b) with reference beam heterodyned by velocity signal from probe spot (dark spot in central fringe). (*Source:* [61].)

with three antinodes whose relative phase is indeterminate. Figure 12.22(b) shows the same mode, but with the reference beam heterodyned to match the motion of the central antinode, which thus appears bright. By contrast, the outer antinodes now appear dark, confirming that they are in antiphase with the central antinode.

Practical application of heterodyned ESPI requires considerable operator intervention and skill to match the amplitude, frequency, and phase of the phase modulation applied to the reference arm to match the motion of the chosen reference point. The procedure may be automated by using a single-point laser vibrometer (see Section 12.2.1.3.3) to generate an electrical signal proportional to the motion of the probed point on the test surface, which is then applied to the phase modulator used in the reference arm of the ESPI system. The matching condition is then satisfied simply by adjusting the gain of the driver of the phase modulator. An example of this technique using a separate fiber-optic vibrometer and ESPI system is described by Valera and others [73]. It is straightforward to traverse the vibrometer probe spot (which defines the reference point on the test surface), and hence to map out the phases of the vibrational modes. A further advantage gained by phase-locking the ESPI system to a vibrometer is that the vibrometer also follows any unwanted out-of-plane whole-body motion of the test surface and applies a compensating phase modulation to the reference arm of the ESPI system. The range of compensation is limited by the onset of speckle decorrelation [69], and no compensation is provided for in-plane motion.

The requirement for gain adjustment between the vibrometer output and the reference beam phase modulator in the ESPI system is avoided in the compound-fiber interferometer shown in Figure 12.23, which combines the functions of the vibrometer and ESPI system (which have a common reference arm) [70].

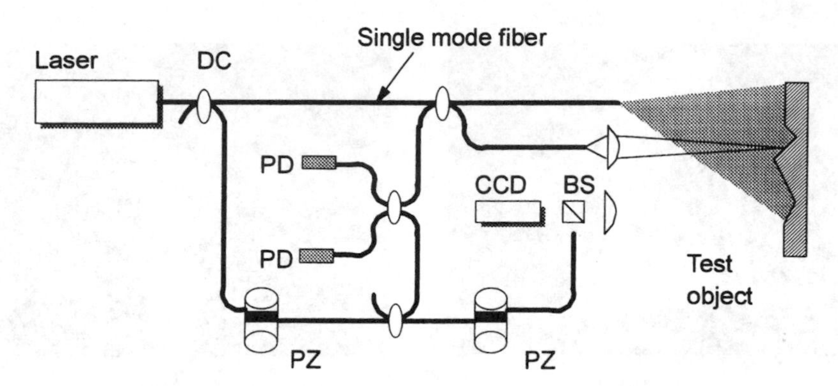

Figure 12.23 Combined ESPI and velocimeter with a common reference arm, where DC is directional coupler, PZ is fiber phase modulator, and PD is photodetector. (*Source:* [70].)

12.3.1.9 ESPI With Pulsed Illumination

While time-averaged techniques for vibration analysis are effective, especially when heterodyne methods are used, they retain some disadvantages. For simple harmonic vibrations, the Bessel function fringes cannot be accurately analyzed using the phase-stepping technique of Section 12.3.1.4; for composite modes, the phase of the vibration cannot be found given that it can attain values other than 0 or π radians; and for motion comprising a range of frequencies, no fringes are formed at all. For these reasons, techniques involving modulation of the source amplitude have been developed.

The conceptually simplest modulated source technique involves the use of double-pulsed illumination. At its most straightforward, the first pulse of illumination is used to store a frame describing the test surface in some initial state, and the second pulse at some later time is given by the interval between the pulses. The pulse interval is arranged to be much less than the period of any vibration frequencies present. Hence, the motion of the surface in some time interval is revealed. It is evident that the complete motion of the surface cannot be found from just a single pulse pair. However, in the case of periodic motion it is possible to repeat the experiment with successively incremented pulse intervals, and hence to map the complete motion. Nevertheless, the basic technique using a single pair of pulses is valuable in the study of transient events.

Practical difficulties arise with the double-pulsed technique. The framing rate of practical cameras tends to be long in comparison with the period of motion of the test surface in most situations, setting a lower limit to the pulse interval. To minimize the limitation, it is usual to arrange for the first pulse to arrive as near to the end of one frame as possible and the second close to the beginning of the next frame period. Even under these circumstances, the minimum practical interval tends to be in the ms range.

For shorter pulse intervals, it is possible to allow both pulses to arrive in the same frame. The speckle fields corresponding to the two deformation states thus add in intensity during that frame. Under these circumstances, speckle fringes are still formed, although the contrast of these *additive fringes* is much lower than for the usual subtractive ones.

Illumination sources for double-pulsed ESPI are usually Q-switched and frequency-doubled Nd:YAG lasers at a wavelength of 532 nm, with ns pulse durations and pulse energies in the 100-mJ range (although ruby lasers with similar energy levels have also often been used). The consequently high instantaneous power levels are beyond the capabilities of optical-fiber beam delivery systems, although pulse energies in the range of tens of mJ have been successfully delivered using large-core multimode fibers and fiber bundles for the technique of particle image velocimetry [71]. Consequently, for single-mode fiber systems a development of pulsed-ESPI using stroboscopic illumination has been demonstrated to yield the cosine type fringes from which the phase of the vibration can be derived for a harmonically vibrating surface.

The technique involves illuminating the test surface with a train of pulses, each short in comparison with the vibration period. A reference frame is first acquired with the surface at rest. The surface is then set into vibration. From (12.31), the difference between a frame taken with the object in motion and the reference frame is given by

$$\Delta I' = \frac{1}{2} I_o^2 V^2 (1 - \cos \phi_d) \tag{12.44}$$

where

$$\phi_d = \frac{4\pi a}{\lambda} \sin(\omega N \tau + \alpha) \tag{12.45}$$

where a is the amplitude of the vibration, ω is the circular frequency, τ is the period, α is the vibration phase, and N is an integer. The value of ϕ_d is recovered by phase-stepping. For example, by acquiring four frames with a $\pi/2$ optical phase modulation introduced into the reference arm between each frame, as explained in Section 12.3.1.4. Thus ϕ_d is found where

$$\phi_d = \frac{4\pi a}{\lambda} \sin \alpha \tag{12.46}$$

The phase of the illuminating pulse train is now changed, so that the pulses occur at times

$$t = \left(N + \frac{1}{4}\right)\tau \tag{12.47}$$

and the new value of ϕ_d is found, denoted by ϕ'_d, where we see that

$$\phi'_d = \frac{4\pi a}{\lambda} \cos \alpha \qquad (12.48)$$

Thus the phase of the vibration is found from

$$\tan \alpha = \frac{\phi_d}{\phi'_d} \qquad (12.49)$$

and the amplitude is found from

$$\phi_d^2 + \phi_d'^2 = \left(\frac{4\pi a}{\lambda}\right)^2 \qquad (12.50)$$

In practical systems, the necessary synchronization signals for generating the stroboscopic pulse train can be derived using a point vibrometer [72]. In a further development of the technique, the need for a separate reference frame can be avoided by sequential subtractions of frames while the test surface is vibrating [73].

By repeating the measurements for steadily incremented driving frequencies, a complete modal analysis of the test object can be carried out. As an example of a practical application, Figure 12.24 shows results obtained from a test object comprising a vehicle wheel and tire excited at their base, thus imitating the driving loads experienced in normal usage. The high levels of damping lead to the formation of composite vibration modes whose phase may take any value. Hence, in order to predict the likely radiation of acoustic noise from the structure, measurements of phase as well as vibration amplitude are essential. Figure 12.24 shows results for just one driving frequency as gray-scale representations of the vibration amplitude and the phase.

12.3.1.10 Two-Wavelength Contouring

In principle, ESPI can be used as a means for measuring the shape of test surfaces: ESPI fringes contour the surface with a contour spacing of $\lambda/2$. However, such a contour interval is small in comparison with the surface roughness, and no useful information is gained. In a development of the technique, two closely spaced wavelengths are used to illuminate the surface, either simultaneously or sequentially. Then, by analogy with classical two-wavelength interferometry (see Section 12.2.1.2), the contour spacing is increased to $\lambda_{eff}/2$, where

$$\lambda_{eff} = \frac{\lambda_1 \lambda_2}{\lambda_1 - \lambda_2} \qquad (12.51)$$

where λ_1 and λ_2 are the wavelengths of illumination.

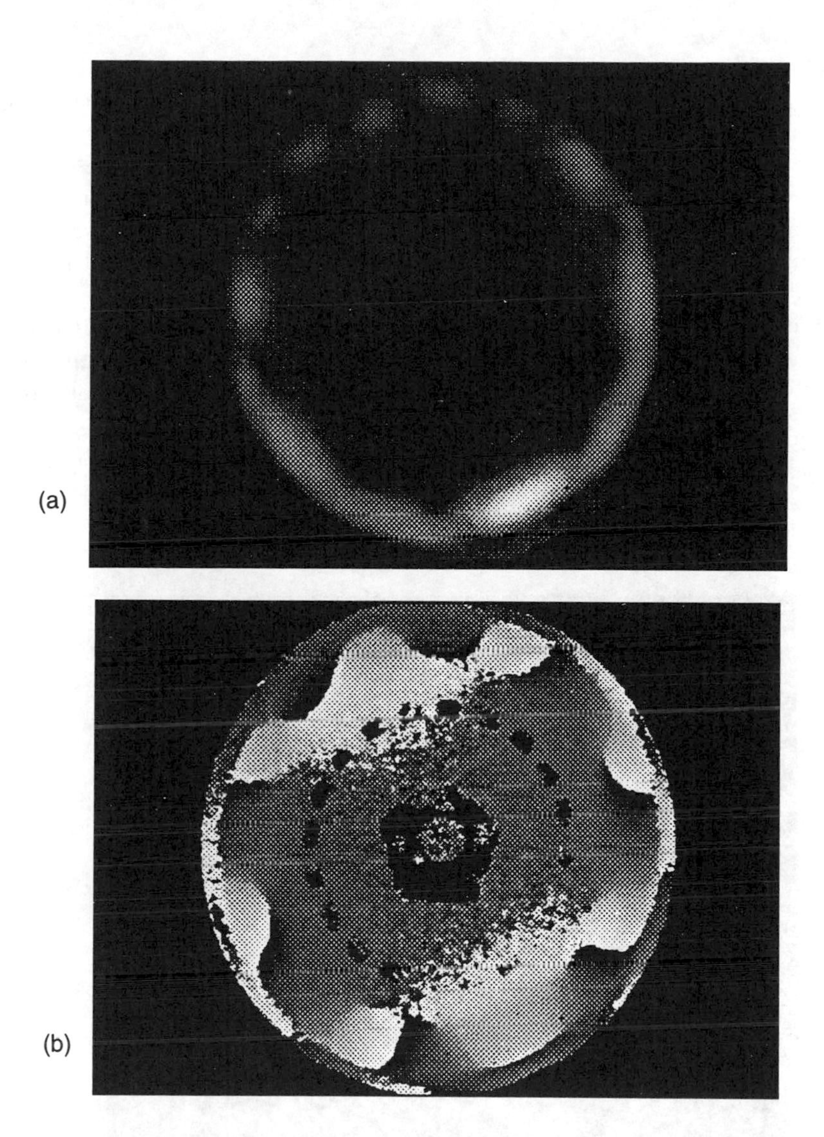

Figure 12.24 Vehicle wheel and tyre excited at their base; grey scale maps of (a) amplitude and (b) phase.

Two-wavelength contouring has often been implemented in optical-fiber ESPI systems using diode lasers as the source, given the wide choice of suitable wavelengths. Two versions of the technique have been used, using either two lasers or a single modulated laser. In two-laser systems, wavelength spacings of a few nm giving contour intervals of hundreds of μm are typically used. For objects with greater surface relief, larger contour intervals are appropriate, achievable by reducing the wavelength separation with progressively more stringent demands on the wavelength stability of the sources.

For very large contour intervals, it is more appropriate to use a single laser, and to synthesize the effect of two wavelengths by modulating its optical frequency—via its injection current, in the case of a diode laser [74]. Contour intervals in the sub-mm range are obtained in this way, with larger modulation amplitudes achievable by deliberately inducing longitudinal mode-hopping in the diode.

With either single- or dual-laser techniques, the phase is demodulated in the same way as for standard ESPI, typically by phase-stepping. As an example of the technique, Figure 12.25 shows speckle contours generated on the surface of a gas turbine blade, from which its shape is derived.

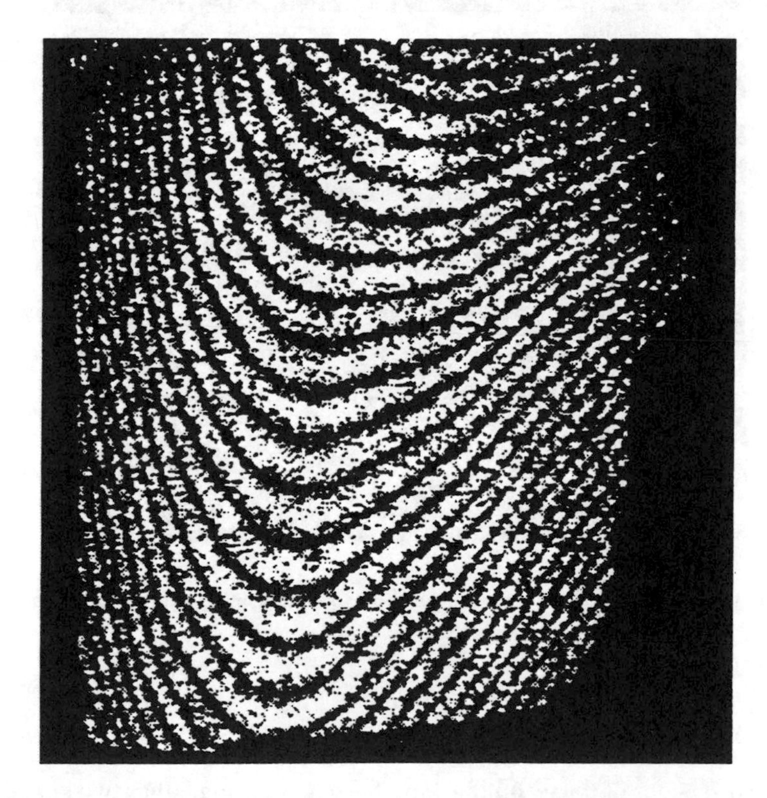

Figure 12.25 Two wavelength contour fringes (interval 0.9 mm) of a turbine blade surface. (*Source:* Atcha et al. 1991.)

12.3.2 Shearography

An important area of application of ESPI is for the detection of defects and in strain measurement. Defects can often be observed by anomalous changes in surface slope occurring when the object is loaded, thus demanding the measurement of the spatial derivative of the out-of-plane deformation. Similarly, in-surface strains are proportional to a double spatial differentiation of the out-of-plane displacement [75]. Numerical differentiation of the already noisy speckle fringes is undesirable, and it is for this reason that speckle-shearing interferometry—or *shearography*—has been developed, which performs the required spatial differentiation optically.

A speckle-shearing interferometer has a similar arrangement to standard ESPI, shown in Figure 12.19, except that there is no explicit reference beam. Instead, a single beam is used to illuminate the test surface. The surface is imaged through an optical arrangement that produces two *sheared* images of the object, which overlap on the face of the detector. Suppose that the shear is in the x direction such that points on the object with coordinates (x, y) in the first image coincide with those with coordinates $(x + \Delta x, y)$ in the second image at the same point on the detector, where the object is assumed to lie in the x, y plane.

The process of generating a speckle-shearing fringe map is similar to that used in standard ESPI. First, a speckle field of the object in its undeformed state is stored. The object is then deformed, a second field captured, and the fields are then correlated-normally by subtraction. The resulting phase map reveals an optical phase change arising from the deformation $\phi_{ds.}$ For simplicity, we shall assume normal illumination and viewing (i.e., in the z direction). Hence, in the limit of small Δz, by analogy with (12.34)

$$\phi_{ds} = \frac{4\pi}{\lambda} \Delta x \frac{\partial}{\partial x} \Delta z \qquad (12.52)$$

where $\Delta z(x,y)$ is the position-dependent out-of-plane deformation. As with standard ESPI, phase-stepping is a common technique for recovering the value of ϕ_{ds}.

Many arrangements have been used to generate a shear in the receiving optics, including optical wedges [76], a split receiver lens, narrow-angle beamsplitters [77], and various interferometer arrangements.

Figure 12.26 shows two configurations used in optical-fiber shearing interferometers. The configuration in Figure 12.26(a) uses a Twyman-Green interferometer with a tilted mirror to produce the shearing. The configuration in Figure 12.26(b) uses a Wollaston polarizing beamsplitter: the prism is placed between relay optics so that by translating it in the direction of the optical axis, the degree of shear can be controlled.

Phase-stepping is feasible by mechanical translation of the optical components-for example, by translating one of the mirrors in the arrangement of Figure 12.26(a)

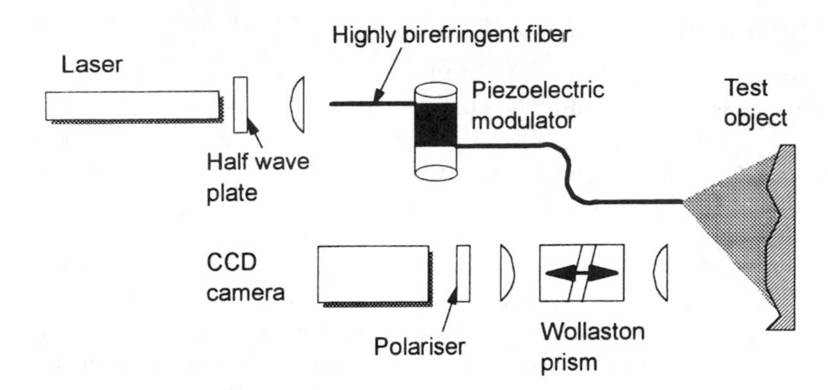

Figure 12.26 Fiber-based speckle shearing interferometer. W is Wollaston prism. (*Source:* [77].)

in a direction normal to its surface. However, phase-stepping by mechanical translation brings the risk of nonuniform phase shifts across the field of view, loss of alignment, and poor calibration of the size of the phase-step. Various techniques have been devised for phase-stepping in fiber-optic speckle-shearing interferometers without translation of bulk-optic components. For example, in the arrangement of Figure 12.26(a), the Twyman-Green interferometer used to shear the images is unbalanced, so that the phase can be stepped by modulation of the wavelength of the diode laser source used for illumination.

For the arrangement of Figure 12.26(b), phase-stepping can be achieved by modulating the state of polarization of the illuminating beam, under the assumption that the state of polarization is preserved on scattering from the test surface. A highly birefringent fiber is used in the illumination system, with its polarization eigenaxes aligned with those of the Wollaston prism used in the receiver optics, assumed to be vertical and horizontal. Hence, the vertical state of polarization corresponds to one of the sheared images, and the horizontal state to the other. Therefore, changing the relative phase between the horizontal and vertical states of polarization in the birefringent optical fiber produces a phase shift between the sheared images, as required. The differential phase modulation in the fiber is readily produced by straining it using a piezoelectric element [77].

12.3.3 Projected Fringe Shape Measurement

Speckle interferometry techniques are very effective for wavelength-scale measurements, with maximum feasible sensitivities on the nm scale. However, for large displacements (beyond the μm scale), the speckle fields lose correlation, and no useful information can be extracted. For these larger scale displacements, related techniques based on the use of *structured light* can be used to measure shape and deformation.

A laser (or, sometimes, an incoherent light source) illuminates the test object with a well-characterized intensity distribution, often in the form of a regular grid. By using an array detector or camera to view the object, its shape can be inferred from the appearance of the grid [7].

Consider the arrangement shown in Figure 12.27. By means of an interferometer, a set of plane-parallel fringes is projected onto a test object (rather as in a "Young's slits" experiment). Alternatively, the fringes can be generated using an aperture mask. The image of the fringes on the object is viewed using an off-axis camera. Imagine first that the test object is replaced by a reference plane normal to the optical axis of the camera. The camera sees a set of vertical (out of the plane of the diagram) interference fringes. The phase of the interference increases smoothly across the object in the horizontal direction, with an increase in phase of 2π radians per fringe. Suppose now that the test object replaces the reference plane. The phase at a point on the object now depends not only on the horizontal position, but also on the height of that point relative to the reference plane. Hence, by measuring the phase of the interference as a function of position, the height of the test object can be mapped.

The test-object when illuminated by plane-parallel interference fringes of unit visibility has a shape describable by its local height, z, relative to a plane (see Figure 12.27). The intensity in the image plane is thus given by

$$I \propto 1 + \cos 2\pi \left[\frac{x \cos \theta}{d} + \phi \right] \tag{12.53}$$

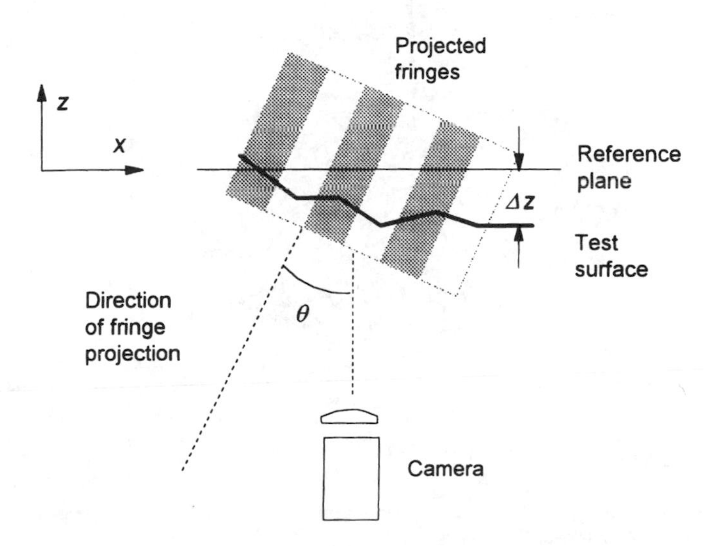

Figure 12.27 Optical arrangement for projected fringe shape measurement.

where d is the fringe spacing normal to the direction of projection, θ is the projection angle, and x is the dimension in the plane of the object normal to the fringes. The shape of the object can then be found, because $\phi = (z \sin \theta)/d$.

An example of the use of structured light techniques for the measurement of large deformations is shown in Figure 12.28, in which the technique was used to measure the wind-induced distortion of the roof of a prototype convertible soft-top motor car for validation of the design supplied by an external company [41]. The measurement was based on projecting interference fringes onto the soft-top motor car, with the complete vehicle in a wind tunnel. With projected fringe contouring, the experimental apparatus could easily be situated outside the tunnel, with only optical access to the test object required.

The apparatus is shown in Figure 12.29. The light source was a diode-pumped Nd:YAG laser coupled to optical fibers and divided into two fibers at a directional coupler. The two output fiber ends are close together and give diverging spherical wavefronts, producing approximately plane-parallel fringes in the far field. These fringes are viewed on the test object by an off-axis television camera. The recorded intensities are then converted to phase-by-phase stepping using the modulator shown and, ultimately, to displacement.

Figure 12.28 shows one representation of a subset of the results of the experiment. The deformations at a windspeed of 80 mph are shown on a gray scale. The

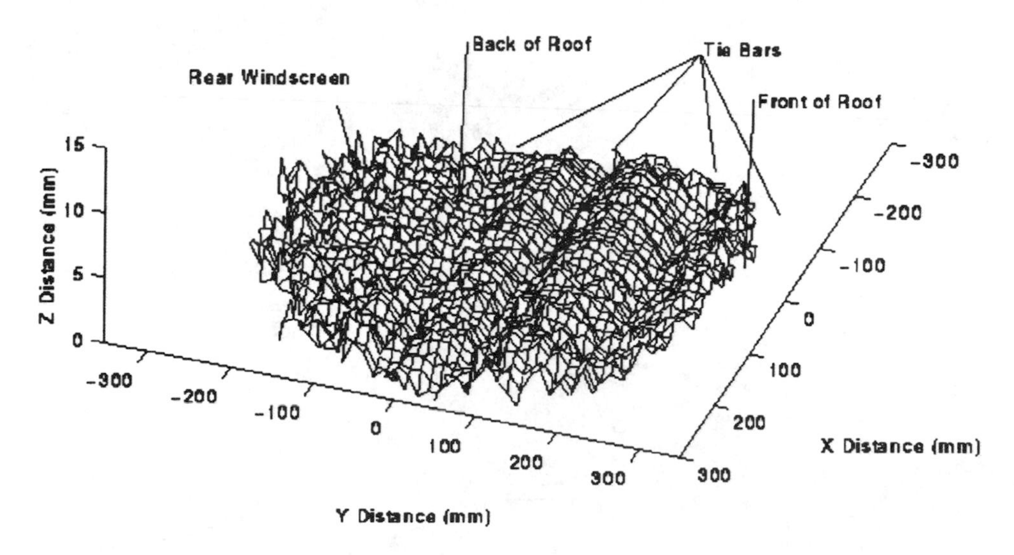

Figure 12.28 Deformation of vehicle soft-top in wind tunnel, measured by the fringe protection system of Figure 12.29 (Rover Group Ltd.).

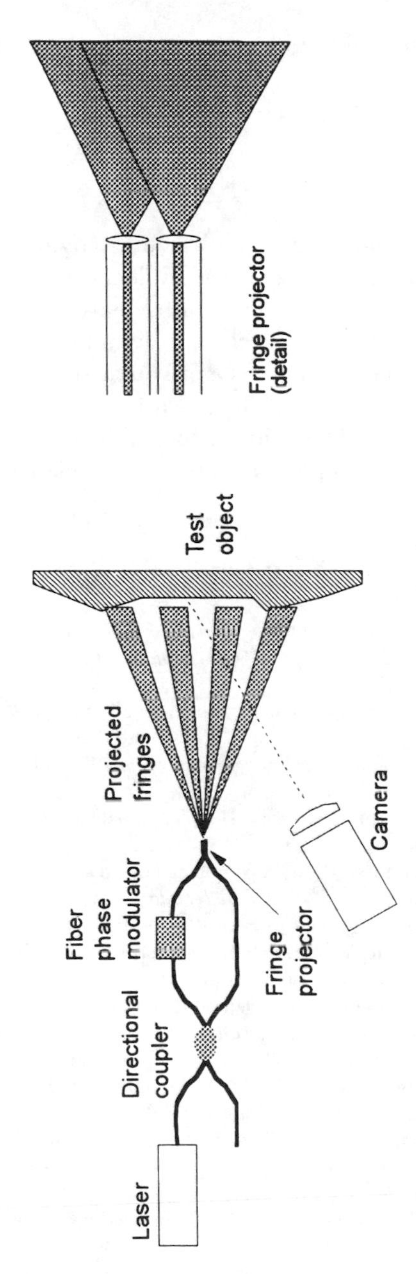

Figure 12.29 (a) Optical fiber system for shape measurement by fringe protection and (b) detail of the fringe protector.

positions of the reinforcements to the soft-top are seen as the positions of minimum deformation.

12.4 CONCLUSIONS

Fiber- and bulk-optic interferometers are established as engineering test tools, with commercial products very well established for laser velocimetry and vibrometry. Full-field fiber interferometers for shape and vibration measurement are approaching commercial reality. For demanding research applications (e.g., in aerodynamics design) fiber sensor development continues to be driven by the need for measurements not feasible using conventional techniques. In all cases, it is the availability of fiber technology that has made the optical technique acceptable for the application, steadily removing the requirement for the operator to possess specialist optical skills.

References

[1] Williams, D. C. (Ed.), *Optical Methods in Engineering Metrology*, Chapman & Hall, London, 1993.
[2] Lading, L., "Principles of laser anemometry," in *Optical Diagnostics for Flow Processes*, Lading, L., et al., (Ed.), Plenum, New York, 1994.
[3] Hinsch, K. D., "Three-dimensional particle velocimetry," *Meas. Sci. Technol.*, 6, 1995, pp. 742–753.
[4] Meyers, J. F., "Development of Doppler global velocimetry as a flow diagnostics tool," *Meas. Sci. Technol.*, 6, 1995, pp. 769–783.
[5] Chan, V. S. S., et al., "Iodine absorption filters for Doppler global velocimetry," *Meas. Sci. Technol.*, 6, 1995, pp. 784–794.
[6] Shan, Q., Chen, C. M., and Dewhurst, R. J., "A conjugate optical confocal Fabry-Pérot interferometer for enhanced ultrasound detection," *Meas. Sci. Technol.*, 6, 1995, pp. 921–928.
[7] Gasvik, K. J., *Optical Metrology*, John Wiley & Sons, Chichester, 1987.
[8] Rogers, A. J., "Essential Optics," in *Optical Fiber Sensors: Principles and Components*, Dakin, J. P., and Culshaw, B., (Eds.), Artech House, Boston, 1988.
[9] Kersey, A. D., Jackson, D. A., and Corke, M., "Demodulation scheme for fibre interferometric sensors employing laser frequency switching," *Electron. Lett.*, 19, 1983, pp. 102–103.
[10] Rashleigh, S. C., "Polarimetric sensors: exploiting the axial stress in high birefringence fibres," *1st Int. Conf. on Optical Fibre Sensors*, Conf. Pub., IEEE, London, 221, 1983, pp. 210–213.
[11] Koo, K. P., Tveten, A. B., and Dandridge, A., "Passive stabilisation scheme for fiber interferometers using 3 x 3 fiber directional couplers," *Appl. Phys. Lett.*, 41, 1982, pp. 616–618.
[12] Jones, J. D. C., "Monomode fibre optic sensors," in *Optical Methods in Engineering Metrology*, Williams, D. C. (ed.), Chapman & Hall, London, 1993.
[13] Leilabady, P. A., Jones, J. D. C., and Jackson, D. A., "Combined interferometric-polarimetric fibre optic sensor capable of remote operation," *Opt. Comm.*, 57, 1986, pp. 77–80.
[14] Vengsarkar, A. M., et al., "Fiberoptic dual-technique sensor for simultaneous measurement of strain and temperature," *J Lightwave Tech.*, 12, 1994, pp. 170–177.
[15] Gill, P., "Laser interferometry for precision engineering metrology," in *Optical Methods in Engineering Metrology*, Williams, D. C., (Ed.), Chapman & Hall, London, 1993.

[16] Wang, D. N., et al., "The optimized wavelength combinations of two broadband sources for white light interferometry," *J. Lightwave Technol.*, 12, 1994, pp. 909–916.
[17] Weir, K., Palmer, A. W., and Grattan, K. T. V., "Accurate measurement of small displacement using optical techniques," *Int. J. Optoelectronics*, 9, 1994, pp. 449–455.
[18] Hand, D. P., et al., "Profile measurement of optically rough surfaces by fibre optic interferometry," *Optics Letters*, 18, 1993, pp. 1361–1363.
[19] Caber, P. J., "Interferometric profiler for rough surfaces," *Applied Optics*, 32, 1993, pp. 3438–3441.
[20] Dyott, R. B., "The fiber optic Doppler anemometer," *IEEE J. Microwaves Opt. Acoust.*, 2, 1978, p. 13.
[21] Jentink, H. W., et al., "Small laser Doppler velocimeter based on the self-mixing effect in a diode-laser," *Applied Optics*, 27, 1988, pp. 379–385.
[22] James, S. W., et al., "Fibre optic based reference beam laser Doppler velocimetry," *Optics Commun.*, 119, 1995, pp. 460–464.
[23] Laming, R. I., et al., "Fiber optic vibration probe," *Electron. Lett.*, 22, 1986, pp. 167–168.
[24] Hand, D. P., et al., "Extrinsic Michelson interferometric fibre optic sensor with bend insensitive downlead," *Opt. Comm.*, 97, 1993, pp. 295–300.
[25] Lewin, A. C., Kersey, A. D., and Jackson, D. A., "Non-contact surface vibration analysis using a monomode fibre optic interferometer incorporating an open air path," *J. Phys. E: Sci. Instrum.*, 18, 1985, pp. 604–608.
[26] Nokes, M. A., Hill, B. C., and Barelli, A. E., "Fiber optic heterodyne interferometer for vibration measurements in biological systems," *Rev. Sci. Instr.*, 49, 1978, pp. 722–728.
[27] Livesley, D. M., et al., "Possible use of advanced optics in LDA particle sizing," *Proc. 3rd Int. Conf. on Laser Anemometry - Advances and Applications*, Swansea, 1989, pp. 32.1–32.8.
[28] Hutton, P. H., "Acoustic emission for continuous monitoring of light-water reactor systems: a status review," *Materials Evaluation*, 46, 1988, pp. 241–246.
[29] Wadley, H. N. G., and Mehrabian, R., "Acoustic emission for materials processing: a review," *Mat. Sci. and Eng.*, 65, 1984, pp. 245–263.
[30] Bruinsma, A. J. A., and Jongeling, T. J. M., "Some other applications for fiber optic sensors," in *Optical Fiber Sensors: Systems and Applications*, Culshaw, B., and Dakin., J. P., (Eds.), Artech House, Boston, 1989.
[31] McBride, R., et al., "Detection of acoustic emission in cutting processes by fibre optic interferometry," *Meas. Sci. Technol.*, 4, 1993, pp. 1122–1128.
[32] Farahi, F., and Jackson, D. A., "A fibre optic interferometric system for surface profiling," *Rev. Sci. Instrum.*, 61, 1990, pp. 753–755.
[33] Halliwell, N. A., and Eastwood, P. G., "The laser torsional vibrometer," *J. Sound and Vibration*, 101, 1985, p. 446.
[34] Liu, T Y., Berwick, M., and Jackson, D. A., "Novel fiber-optic torsional vibrometers," *Rev. Sci. Instrum.*, 63, 1992, pp. 2164–2169.
[35] Pannell, C. N., Jones, J.D.C., and Jackson, D. A., "The effect of environmental acoustic noise on optical fibre based velocity and vibration sensor systems," *Meas. Sci. Technol.*, 5, 1994, pp. 412–417.
[36] Knuhtsen, J., Olldag, E., and Buchhave, P., "Fibre-optic laser Doppler anemometer with Bragg frequency shift utilising polarisation-preserving single-mode fibre," *J. Phys. E: Sci. Instrum.*, 15, 1982, pp. 1188–1191.
[37] Chan, R. K. Y., Jones, J. D. C., and Jackson, D. A., "A compact all-optical fiber Doppler-difference laser velocimeter," *Optica Acta*, 32, 1985, pp. 241–246.
[38] Strunck, V., et al., "New Technologies for Laser Anemometers," in *Optical Diagnostics for Flow Processes*, Lading, L., et al., (Ed.), Plenum, New York, 1994.
[39] Jones, J. D. C., et al., "Miniature solid-state directional laser Doppler velocimeter," *Electron. Lett.*, 18, 1982, pp. 967–969.

[40] Boutier, A., "Three dimensional laser velocimetry systems," *Proc. 2nd Int. Conf. on Laser Anemometry - Advances and Applications*, Strathclyde, 1987, pp. 101–118.

[41] Jones, J. D. C., et al., "Fibre optic modulators for laser velocimetry," *Proc. Int. Conf. on Laser Anemometry - Advances and Applications*, Manchester, 1985, pp. 375–386.

[42] Dopheide, D., Strunck, V., and Pfeifer, H. J., "Miniaturized multi-component laser Doppler anemometers using high-frequency pulsed diode lasers and new electronic signal acquisition systems," *Exp. in Fluids*, 9, 1990, pp. 309–316.

[43] Jackson, D. A., and Paul, D. M., "Measurement of hypersonic velocities and turbulence by direct spectral analysis of Doppler shifted laser light," *Phys. Lett.*, 32A, 1970, pp. 77–78.

[44] Harvey, D., McBride, R., and Jones, J. D. C., "Fibre optic Sagnac interferometer based velocimeter," *Meas Sci Technol.*, 3, 1992, pp. 1077–1083.

[45] Knudsen, S., and Bløtekjaer, "An ultrasonic fiber-optic hydrophone incorporating a push-pull transducer in a Sagnac interferometer," *J. Lightwave Technol.*, 12, 1994, pp. 1696–1700.

[46] Dakin, J. P., "Distributed Optical Fiber Sensor Systems," in *Optical Fiber Sensors: Systems and Applications Vol. 2*, Culshaw, B., and Dakin., J. P., (Eds.), Artech House, Boston, 1989.

[47] Carolan, T. A., et al., "Fiber optic Sagnac interferometer for non-contact structural monitoring in power plant applications," in *Sensors & Their Applications VII*, Agousti., A. T., (Ed.), IoP Publishing, Bristol, 1995, pp. 172–177.

[48] Pannell, C. N., et al., "Fibre optic transit velocimetry using laser diode sources," *Electron. Lett.*, 24, 1988, pp. 525–526.

[49] Boutier, A., and Lefèvre, J., "Mosaic laser velocimeter for large facilities," *Meas. Sci. Technol.*, 6, 1995, pp. 1705–1716.

[50] Doorly, J. E., and Oldfield, M. L. G., "The theory of advanced multi-layer thin-film heat transfer gauges," *Int. J. Heat & Mass Transfer*, 30, 1987, pp. 1159–1168.

[51] Kidd, S. R., et al., "Wind tunnel evaluation of novel interferometric optical fibre heat transfer gauges," *Meas. Sci. Technol.*, 4, 1993, pp. 362–368.

[52] Kidd, S. R., et al., "Interferometric fibre sensors for measurement of surface heat transfer rates on turbine blades," *Optics & Lasers Eng.*, 16, 1992, pp. 207–221.

[53] Kidd, S. R., Barton, J. S., and Jones, J. D. C., "Demonstration of optical fibre probes for high bandwidth thermal measurements in turbomachinery," *J. Lightwave Technol.*, 13, 1995, pp. 1335–1339.

[54] Sadkowski, R., Lee, C. E., and Taylor, H. F., "Multiplexed interferometric fiber-optic sensors with digital signal processing," *Applied Optics*, 34, 1995, pp. 5861–5866.

[55] Parker, R. J., "Industrial applications of holographic interferometry," in *Optical Methods in Engineering Metrology*, Williams, D. C., (Ed.), Chapman & Hall, London, 1993.

[56] Jones, R., and Wykes, C., *Holographic and Speckle Interferometry*, Cambridge University Press, Cambridge, 1989.

[57] Takeda, M., Ina, H., and Kobayashi, S., "Fourier-transform method of fringe-pattern analysis for computer-based topography and interferometry," *J. Opt. Soc. Am.*, 72, 1982, pp. 156–160.

[58] Davies, J. C., and Buckberry, C. H., "Television holography and its applications," in *Optical Methods in Engineering Metrology* Williams, D. C., (Ed.), Chapman & Hall, London, 1993.

[59] Løkberg, O. J., and Krakhella, K., "Electronic speckle pattern interferometry using optical fibers," *Opt. Comm.*, 38, 1981, pp. 155–158.

[60] Davies, J. C., et al., "Development and application of a fibre optic electronic speckle pattern interferometer," *SPIE Proc.*, 863, 1987, pp. 194–207.

[61] Valera Robles, J. D., Harvey, D., and Jones, J.D.C., "Automatic heterodyning in fibre optic speckle pattern interferometry using laser velocimetry," *Optical Eng.*, 31, 1992, pp. 1646–1653.

[62] Atcha, H., and Tatam, R. P., "Heterodyning of fiber optic electronic speckle pattern interferometers using laser diode wavelength modulation," *Meas. Sci. Technol.*, 5, 1994, pp. 704–709.

[63] Lefevre, H. C., "Single-mode fibre fractional wave devices and polarisation controllers," *Electron. Lett.*, 16, 1980, pp. 778–780.

[64] Mercer, C. R., and Beheim, G., "Fiber optic phase stepping system for interferometry," *Appl. Optics*, 30, 1991, pp. 729–734.
[65] Gülker, G., et al., "Deformation mapping and surface inspection of historical monuments," *Optics and Lasers in Eng.*, 24, 1996, pp. 183–213.
[66] Steinbichler, H., and Gehring, G., "TV-holography and holographic interferometry: industrial applications," *Optics and Lasers in Eng.*, 24, 1996, pp. 111–127.
[67] Takemoto, S., "Holography and electronic speckle pattern interferometry in geophysics," *Optics and Lasers in Eng.*, 24, 1996, pp. 145–160.
[68] Jones, J. D. C., et al., "Lasers measure up for the car industry," *Physics World*, 8, 1995, pp. 33–38.
[69] Takai, N., Iwai, T., and Asakura, T., "Correlation distance of dynamic speckles," *Appl. Optics*, 22, 1983, pp. 170–177.
[70] Valera, J. D., Doval, A. F., and Jones, J. D. C., "Combined fibre optic laser velocimeter and electronic speckle pattern interferometer with a common reference beam," *Meas. Sci. Technol.*, 4, 1993, pp. 578–582.
[71] Anderson, D. J., et al., "An optical fiber delivery system for pulsed laser particle image velocimetry illumination," *Meas. Sci. Technol.*, 6, 1995, pp. 809–814.
[72] Anderson, D. J., Valera, J. D., and Jones, J. D. C., "Electronic speckle pattern interferometry using diode laser stroboscopic illumination," *Meas. Sci. Technol.*, 4, 1993, pp. 982–987.
[73] Valera, J. D., Doval, A. F., and Jones, J. D. C., "Vibration phase measurement by fibre optic electronic speckle pattern interferometry., ESPI, with stroboscopic illumination," *Proc. 9th Int. Conf. on Optical Fibre Sensors*, Firenze, 1992, pp. 381–384.
[74] Tatam, R. P., et al., "Holographic surface contouring using wavelength modulation of laser diodes," *Optics and Laser Technol.*, 22, 1990, pp. 317–321.
[75] Hung, Y. Y., and Durelli, A. J., "Simultaneous measurement of three displacement derivatives using a multiple image-shearing interferometer camera," *J. Strain Analysis*, 14, 1979, pp. 81–88.
[76] Hung, Y. Y., "Shearography: a new optical method for strain measurement and nondestructive testing," *Optical Eng.*, 21, 1982, pp. 391–395.
[77] Valera, J. D., and Jones, J. D. C., "Phase stepping in fiber-based speckle shearing interferometry," *Optics Lett.*, 19, 1994, pp. 1161–1163.

Chapter 13

Sensors in Industrial Systems

John W. Berthold
Babcock and Wilcox, U.S.A.

13.1 INTRODUCTION

This chapter describes several fiber-optic sensor systems developed for and applied to special industrial and aerospace applications. The aim is not to be comprehensive, in terms of either the breadth of industrial needs or the possible fiber-sensor technologies that might fill these needs, but to consider applications typical of those found in industry. The fiber-sensor systems developed illustrate the flexibility and adaptability of the technology, both to special one-off systems, and to systems having potential for high-rate unit production.

The chapter begins with a brief comparative case study of strain measurement, using Bragg grating, wavelength-modulated sensors and fiber-microbend, intensity-modulated sensors. The intent of this section is to discuss the tradeoffs, advantages, and disadvantages of each of these types of intrinsic fiber sensors, to help a designer choose which approach might be better for his or her intended application, and to provide a basis for evaluating alternative fiber-strain sensors.

The subsequent four sections discuss fiber-optic sensors developed for particular industrial applications. Three of the sensor systems are advanced prototypes that have been in use for some time. In Section 13.3 an extrinsic fiber sensor is described that collects and analyzes thermal radiation at multiple points across a moving web. Described in Section 13.4 is an extrinsic fiber sensor that measures the concentration of water in a flowing water-gas mixture. Section 13.5 reviews a feasibility test of a remote Raman spectroscopic probe for detecting corrosion inside pipes. In Section 13.6, an advanced extrinsic sensor is described. This was developed for the chemical process industry to measure opacity in a production flowline.

Section 13.7 summarizes the design, lab test, and flight test of a fiber-optic pressure transducer for an aerospace application that may eventually lead to large-scale production. Finally, Section 13.8 includes a discussion on marketing issues and recommendations for commercialization of fiber sensor products in the coming years.

13.2 STRAIN SENSOR CASE STUDY—FIBER-OPTIC BRAGG GRATING AND MICROBEND STRAIN SENSORS

13.2.1 Background

Strain gauges are important for their ability to measure mechanical strain in materials and for their use as the primary sensors in many other transducer types (e.g., pressure, flow, level, etc.). A wide variety of existing methods are used, or are being developed, for measuring strain, and in many situations there is a need to make contact strain measurements at high temperature.

Strain in a material is simply its fractional increase in physical length when stressed. In the linear elastic regime, strain ($\Delta\ell/\ell$) is equal to the ratio of the applied stress divided by Young's modulus of the material. Although noncontact strain measurement methods are available, and have been used for high-temperature applications, they are not particularly appropriate for long-term measurements in hostile environments. The degradation of the surface of components to be measured can make determination of strain very difficult using noncontact methods. Often, the components are covered by insulation during operation, and they are located in regions where "line of sight" interrogation is difficult, if not impossible. Furthermore, many of the noncontact methods are affected by the temperature, opacity, and turbulence of the intervening atmosphere.

The most common contact strain gauge available for high-temperature application is the electrical resistance gauge. Resistance gauges are available in a variety of forms and have been in use since the 1940s. Resistance strain gauges are commonly used at moderate temperatures (up to 700°F) for long-term monitoring, and at even higher temperatures for short-term or dynamic measurements. "Bonded" resistance strain gauges can be used up to 500°F in some applications. They are devices with relatively high mechanical compliance. "Weldable" resistance gauges usually have low compliance, but gain stiffness from their mounting package.

Resistance strain gauge technology is highly developed, and the gauges exhibit excellent sensitivity. They are the most widely used of any gauge type. For high-temperature applications, special (very costly) calibration efforts are required to improve the marginal reliability of static strain measurement data obtained with resistance strain gauges. A simple single-wire gauge will alter its resistance due to thermal expansion and due to changes in sensitivity with temperature. Both these may vary in a nonlinear manner and depend on how the gauge is mounted. The most difficult aspect of resistance gauge performance to address is the unique gauge output versus the temperature (often termed "apparent strain") and gauge output stability (drift) characteristics of each gauge. Hysteresis due to temperature cycling can also be problematic. Many researchers continue to investigate new materials and forms of this technology. At present, no resistance strain gauge, neither commercial or research, is considered adequate for measurement of long-term static strain at temperatures of 550 to 600°C. It is unlikely that a "step change" in the resistance strain gauge technology will take place in the near future.

The capacitance strain gauge has been developed specifically for high-temperature applications in order to combat the problem of drift. Commercial versions are generally not suitable for dynamic measurements above 100 Hz and capacitance gauges tend to exhibit significant hysteresis versus temperature. In addition, these gauges require significant care in calibration and installation. Nonetheless, they are stable at high temperatures under monotonic load conditions. The significant zero shift, which has been reported to occur during temperature cycling, limits their usefulness for obtaining absolute static strain. Their low drift performance, however, allows capacitance gauges to be used to measure creep strain changes at steady state.

R&D efforts in fiber-optic sensors have been underway since the late 1970s and various sensors have been developed where the optical intensity, phase, polarization, or wavelength may be modulated by the physical parameter being measured. A number of researchers have developed fiber-optic strain gauges using a wide variety of optical sensor concepts. These include Bragg grating [1], distributed microbend [2], extrinsic Fabry-Perot [3], and intrinsic Fabry-Perot [4]. All of these fiber strain sensors have the potential to operate in high-temperature environments, provided the fiber is suitably protected. For purposes of this case study, two fiber-optic sensing methods are discussed for strain measurement at a point: Bragg grating and microbend sensors.

Bragg grating sensors are based on the reflection and interference of multiple light beams traveling through the fiber, each being reflected by a small refractive index discontinuity in the grating (see Chapter 2) The sensitivity of these sensors to strain is primarily related to the consequential axial strain imposed on the fiber core region. The sensor requires a strong, stable bond to the test structure (workpiece) to allow the structure's strain to be transmitted to axial strain in the fiber.

The microbend sensor converts structural strain to a periodic microbending of the optical fiber. The microbending results in intensity modulation of transmitted light, which can be monitored (in forward, backward, or bidirectional mode) to measure strain. The microbend sensor does not require that the actual fiber itself be bonded to the test structure. We shall now examine the sensor types in more detail.

13.2.2 Bragg Grating Sensors

This sensor, which was developed by United Technologies [1], is shown in Figure 13.1. See also Chapter 2 for more details. Essentially permanent holographic gratings can be "written" into many types of commercial single-mode optical fiber. Fibers with germania-doped cores are well-suited for this process. The gratings were

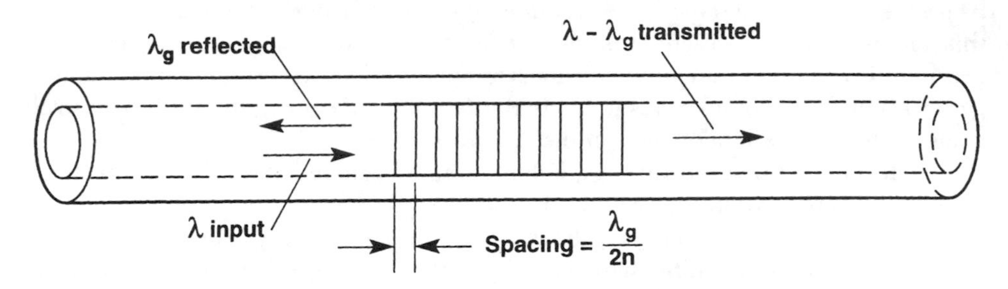

Figure 13.1 Diagram of fiber grating reflector and the Bragg condition for reflection. (*Source:* Morey, W. W., et al., *Proc. SPIE,* Vol. 1586, 1991, pp. 216–220. Reprinted with permission.)

first made by interfering two converging ultraviolet laser beams, incident from the side of the fiber [1]. The interference fringes in the germania-doped core produce a periodic modulation of the refractive index. The refractive index changes as the ultraviolet light modifies the oxygen vacancy defect absorption band. This method requires removal of the buffer coating for the exposure and then recoating. Changes up to 3×10^{-4} in the core refractive index can be produced. In the configuration shown in Figure 13.1, the grating reflects optical signals within the light-guiding core at a wavelength that is twice the optical spacing between high-index and low-index regions (grating spacing). This wavelength-matching condition is shown in Figure 13.1 and is called the Bragg condition.

A transmission spectrum of a typical grating is shown in Figure 13.2. High reflectivity near 100% has been demonstrated, and grating lengths may range from 1.5 mm to 15 mm. The longer the grating, the narrower is the spectral bandwidth. The grating spacing, and hence the Bragg wavelength, is controlled by the ultraviolet light wavelength and the angle between the incident ultraviolet beams. Reflection gratings have been produced having Bragg wavelengths ranging from the visible out to well beyond the communications wavelength of 1,550 nm in the near infrared.

When a section of fiber containing a grating is subjected to axial strain or to temperature change, the grating spacing and the refractive index change. Both affect the Bragg wavelength. In this fashion, the gratings act as sensors to detect strain and/or temperature. To separate the responses to temperature and strain, two gratings may be located either nearby or overwritten in the same fiber. If one grating can be mechanically isolated from the strain field, then it will sense temperature only, and the other (mechanically bonded) grating will sense both temperature and strain.

When the Bragg gratings are heated, the Bragg wavelength changes. The changes are caused mainly by the change in refractive index with temperature and, to a much lesser extent, the linear thermal expansion of the optical fiber [1]. A typical measure change in Bragg wavelength with temperature is $\delta\lambda/\lambda = 8.2 \times 10^{-6}$/°C. At a Bragg wavelength of 1,300 nm, a typical 0.01-nm/°C shift in the reflection peak is observed. The strain sensitivity of a fiber Bragg grating results primarily from the

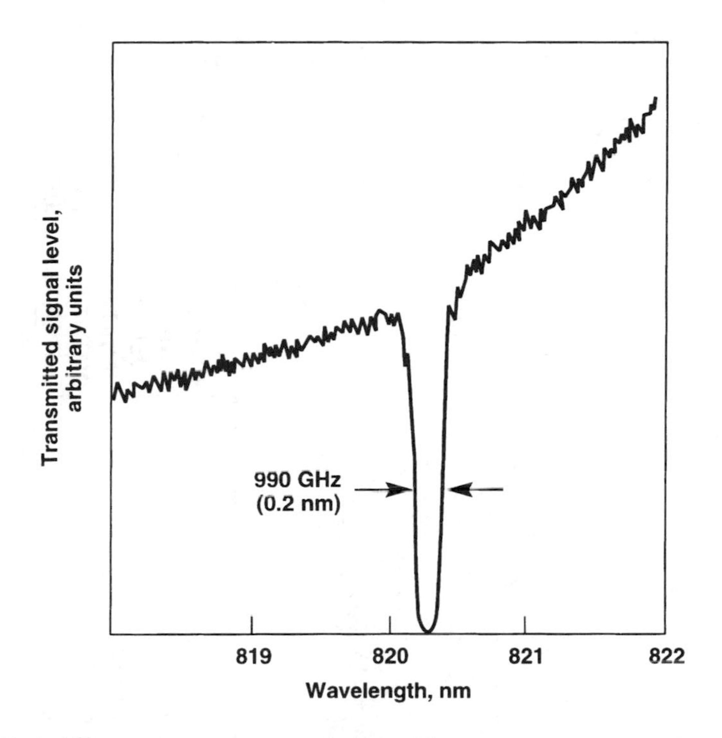

Figure 13.2 Typical fiber grating transmission spectrum. (*Source:* Morey, W. W., et al., *Proc. SPIE*, Vol 1586, 1991, pp. 216–220. Reprinted with permission.)

change in grating spacing when the fiber is stretched. However, a change in the axially polarized refractive index, caused by the stretching, also results from the photoelastic effect. For a fiber Bragg grating placed in tension or compression, the change in Bragg wavelength per unit wavelength is typically 74% of the strain; that is,

$$\delta\lambda/\lambda = 0.74\ \Delta\ell/\ell = 0.74\epsilon$$

This measured result agrees well with the result for silica fibers [1].

If it is desired to apply Bragg grating sensors to strain measurement at high temperatures, physical changes in the grating characteristics may occur as the time/temperature exposure increases. The thermal durability of gratings under these conditions is discussed in detail in Chapter 2.

13.2.3 Microbend Strain Gauge

The microbend sensor configured for strain sensing is shown schematically in Figure 13.3. This sensor was originally demonstrated by Fields and Cole [5]. A

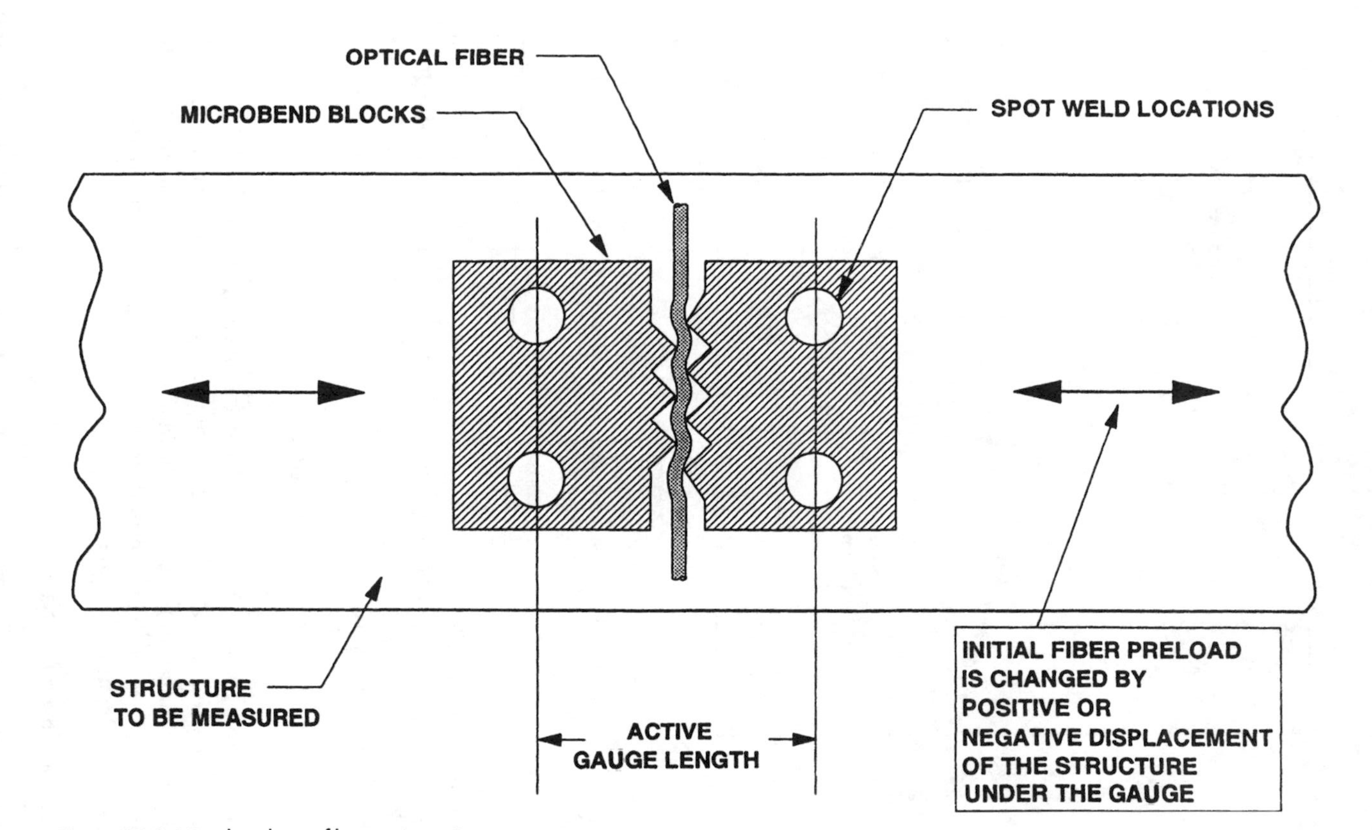

Figure 13.3 Microbend-type fiber-optic strain gauge.

multimode step index fiber is clamped between a pair of corrugated plates (tooth blocks), each welded to the workpiece. When strain is applied perpendicular to the fiber axis, the plates move in the plane of the workpiece and the fiber is spatially deformed into a sinusoidal shape. The deformation induces light amplitude changes in the fiber, which depend on applied strain. Thus, light passing through the fiber is attenuated accordingly. The light-loss sensitivity (scale factor) of the microbend sensor may be optimized by matching the sinusoidal deformation period to the difference between the propagation constants of adjacent high-order modes in the optical fiber [5].

The change in light intensity versus displacement of the microbend sensor corrugated plates (tooth blocks) is shown in Figure 13.4. Note that the displacement

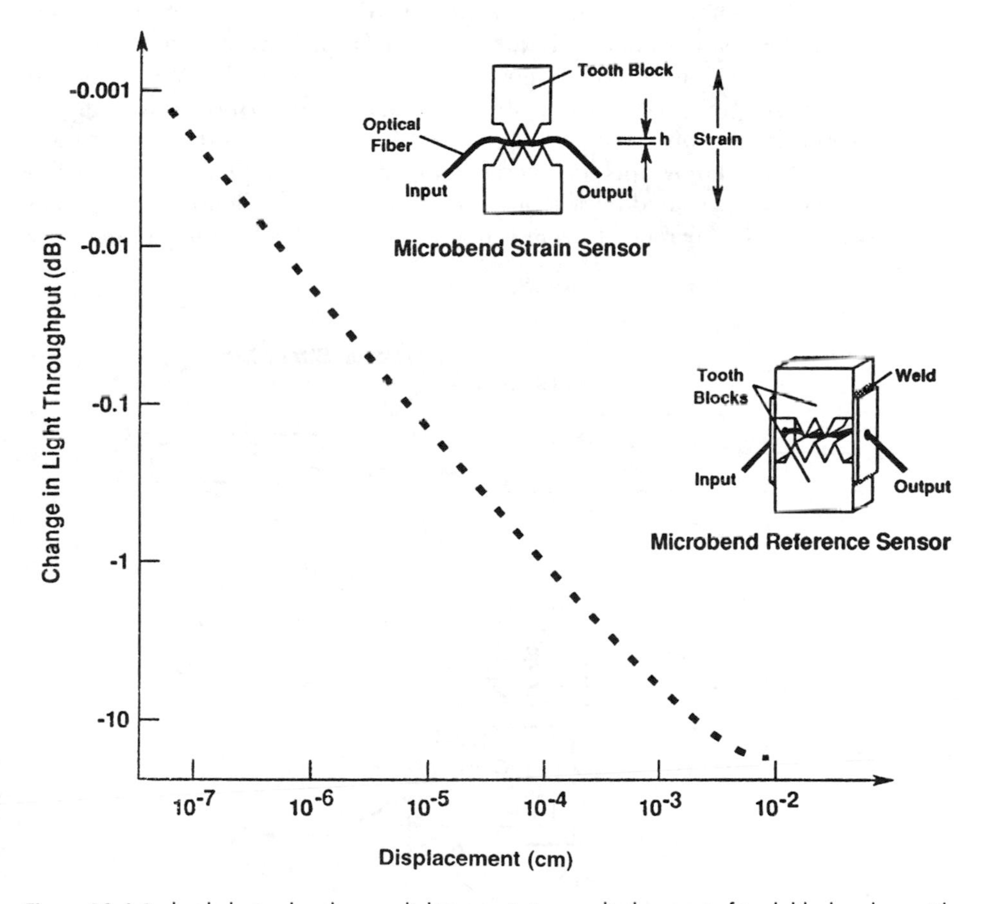

Figure 13.4 Preloaded microbend sensor light transmission vs. displacement of tooth blocks, along with an illustration of two microbend sensors-one configured for strain measurement and the second configured as a temperature sensor reference.

resolution is about 10^{-8} inch (in a 1-Hz bandwidth) and ranges up to 10^{-3} inch. The disadvantage of this sensor is that the scale factor depends on the modal power distribution at the measurement point, although in many applications this effect may be calibrated out.

In the simplest form, the change in light intensity from the microbend sensor vs. applied strain may be monitored from the amplified output from a photodiode; but to compensate for other bending-induced loss in the interconnecting fiber cable as well as random intensity changes from the connector insertion loss, self-referencing electronics may be used to compare the microbend sensor light intensity signal with a reference light intensity signal.

Instrumentation used to read out the microbend strain gauge is shown in Figure 13.5. The components used are conventional, inexpensive optoelectronics. The light source is a light-emitting diode (LED) and the detector is a silicon dual photodetector. With this configuration, a dc voltage output is proportional to the amount of light detected, and voltage can be acquired/measured by normal analog-to-digital data acquisition equipment. Detailed calibrations have been performed to compare microbend strain gauges with standard resistance foil strain gauges [6].

Figure 13.6 is a photograph of a microbend strain gauge welded to an alloy 600 tubular test specimen prepared for a series of mechanical tests. Figure 13.7 is a plot of strain output for the microbend strain gauge versus axial load applied to the alloy 600 tube.

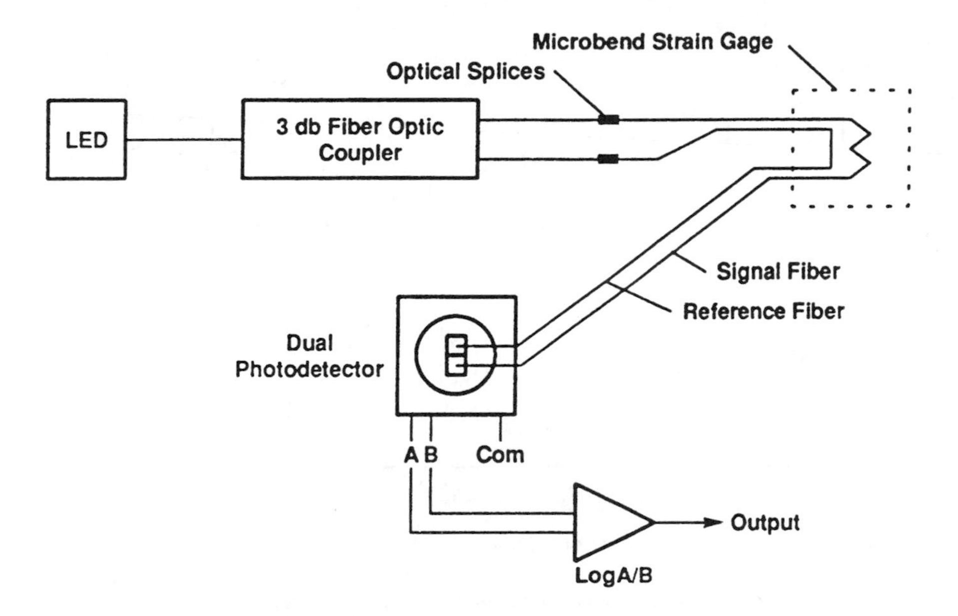

Figure 13.5 Microbend strain gauge instrumentation for single-point strain measurement.

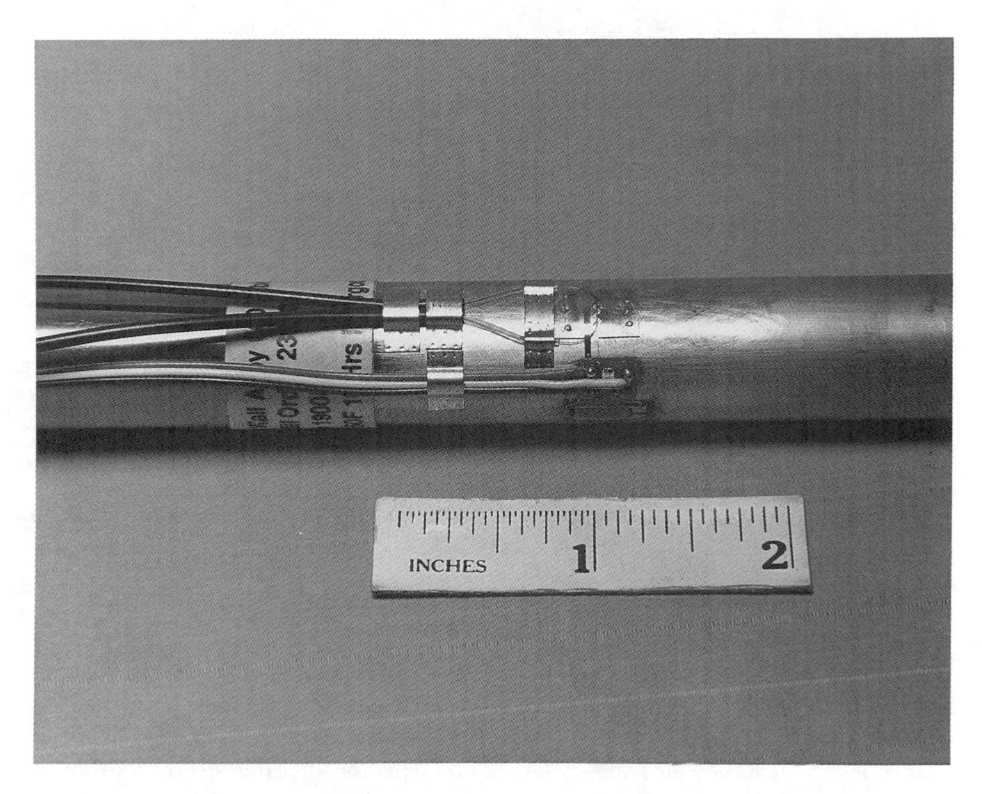

Figure 13.6 Photograph of microbend fiber-optic strain gauge (upper) and resistance foil strain gauge (lower) installed on tube.

13.2.4 Problems/Tradeoffs

13.2.4.1 Referencing Errors

Since Bragg grating sensors provide a measurable change in reflected wavelength with applied strain, it is generally not necessary to provide a light source intensity reference for Bragg grating sensors.

In contrast, the much simpler microbend sensors modulate the light intensity transmitted by the optical fiber, so microbend sensors must be desensitized to any other perturbations that could change light intensity. One way to provide a light source intensity reference for microbend sensors is to use a second "reference" optical fiber. Thus, in addition to the sensing fiber sandwiched between the plates, an unclamped reference fiber (space domain referencing) is also installed with approximately the same length in the hot zone. The reference fiber provides compensation for any mechanical perturbations in the sensing fiber or any variations in the intensity of

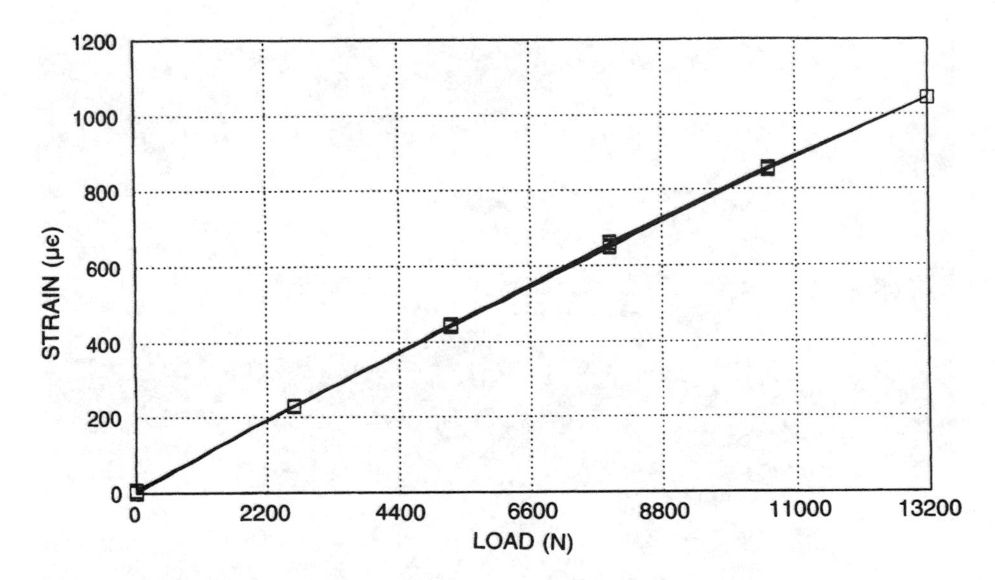

Figure 13.7 Strain versus axial load (three cycles) for microbend fiber-optic strain gauge.

the light source. This configuration is not ideal because perturbations on the reference fiber and microbend sensor fiber may not always result in equal light power changes in both. To avoid this problem, it is usually necessary to enclose the fiber leads in stiff metal tubes and to splice the leads to source and detector fiber pigtails rather than use connectors.

13.2.4.2 Temperature Compensation and Thermal Strain

Thermal expansion of the workpiece and temperature-dependent material characteristics of the gauge can introduce errors into the output of any strain gauge, including fiber-optic strain gauges. Methods are required for compensation of thermally induced strain (apparent strain) and gauge sensitivity changes with temperature. The fiber-optic compensation methods must be compatible with the signal-processing methods used to demodulate the fiber-optic sensor signals and convert from optical to electrical signals. In the case of Bragg grating sensors, temperature and strain affect the grating spacing similarly. Two nearby gratings in the same fiber may be used to provide signals proportional to both variables so that one sensor measures strain and temperature and the other temperature only. In this situation, the sensor measuring temperature only must be isolated from the strain field. See [7] for a detailed discussion of thermal compensation methods for Bragg grating sensors.

Temperature affects microbend sensors in two ways. First, the thermal expansion of the workpiece and the microbend tooth blocks will cause the tooth-block spacing to change with temperature. Second, the refractive indices of the fiber core and cladding depend on temperature, so the modal power distribution in a multimode fiber microbend sensor can change with temperature. Since both these effects are repeatable, the simplest method is to correct for them by measuring sensor temperature with a thermocouple or thermistor.

13.2.4.3 Signal Conditioning

Long-term measurements to detect small creep rates require excellent dc stability. Therefore, dc- rather than ac-coupled electronics must be used. Good dc stability puts extreme drift requirements on fiber-optic sensors, signal processing electronics, and the interconnecting fiber-optic cable. Thus, thermally induced drift in the electronics must be compensated or eliminated. These requirements apply equally to Bragg grating and microbend sensors.

13.2.4.4 Buffer Coating

With microbend sensors, the tooth-blocks push into the buffer coating, so it is important to understand the mechanical, physical, and chemical interactions among the tooth-block material, the buffer coating material, and the glass optical fiber. Empirical data has been gathered, much of it proprietary. The important issues with any buffer coating material are whether or not it introduces sensor zero shift and scale factor shift. All buffer coating materials introduce these errors to some extent, but the errors can be made negligible by doing a cyclic shakedown of the microbend sensor over the displacement range of the tooth blocks. It is desirable that fiber buffer coatings to be used in microbend sensors have characteristics such as wear resistance, mechanical stability, and good light absorption.

13.2.4.5 Packaging

Strain gauges must be reliably bonded to the workpiece. In general, the most reliable bond is a weld. Since glass optical fibers cannot be welded to metal, fiber-optic Bragg grating strain gauges must be packaged with a high-strength bond between the optical fiber and a metal package envelope. Then this envelope can, in turn, be welded or otherwise attached to the workpiece.

Since, with a microbend strain gauge, the optical fiber is sandwiched between two corrugated end plates, these plates may be made from the same material as the test component. Fabricating the gauge in this manner minimizes thermal output changes due to differences in thermal expansion between the gauge and test component. The fiber can be preloaded (bias compression) between the two tooth blocks prior to or during installation to provide tensile and compressive measurement capability and to set the fiber-bend amplitude to produce the sensitivity and linearity desired.

In summary, the requirements for temperature compensation, apparent strain compensation, and packaging entail tradeoffs and limit the applicability of different fiber sensors such as Bragg grating and microbend to strain measurement.

13.3 MULTIPOINT TEMPERATURE MONITOR

13.3.1 Application

Various industrial processes involve the transport of materials in the form of webs (sheets) that are moved and/or controlled by a series of rollers. In many cases, the temperature of the web and rollers is an important process variable. The multipoint temperature monitor is a fiber-optic-based radiation pyrometer designed and built to monitor the temperature distribution across a moving web up to 4m wide.

The multipoint temperature monitor (MTM) employs an array of up to 160 pick-ups. The outputs are delivered by optical fibers and optically multiplexed onto a 16-channel germanium photodiode array. The pick-ups are deployed linearly across the web on 2.5-cm centers.

13.3.2 Design

The design objective was to develop a system for continuous, online measurement of the temperature distribution across a moving web or roller. The specific target application that initiated this development involved temperatures in the range of 120 to 180°C, across webs up to 4m wide. Prior to this development, infrared cameras, mounted on a rail system and mechanically scanned across the width of the web, were used to monitor the temperature distribution [8]. The environment, however, was hostile to the scanning mechanism and continual maintenance problems initiated a search for an alternative method to obtain the temperature profile across the web.

With the MTM, the scanning mechanism is replaced with a stationary fiber-optic head that has no moving parts and that deploys a series of pick-ups across the width of the web. Each pick-up is an optical fiber aligned with its axis perpendicular to the direction of web motion. The system combines optical multiplexing with electronic multiplexing to provide monitoring of up to 160 points across the 4m web.

13.3.3 System

The three major subassemblies shown in Figure 13.8 include the pick-up assembly, the optical-fiber delivery cable, and electro-optics/computer. The pick-up assembly is rigidly mounted in a fixed position above the web. Radiation collected at the individual pick-up points is delivered via optical fibers to the electro-optics cabinet where the optical signals are converted to electrical signals. The electrical signals are then sent to a plug-in A/D board in the computer where they are digitized and converted to temperature readings.

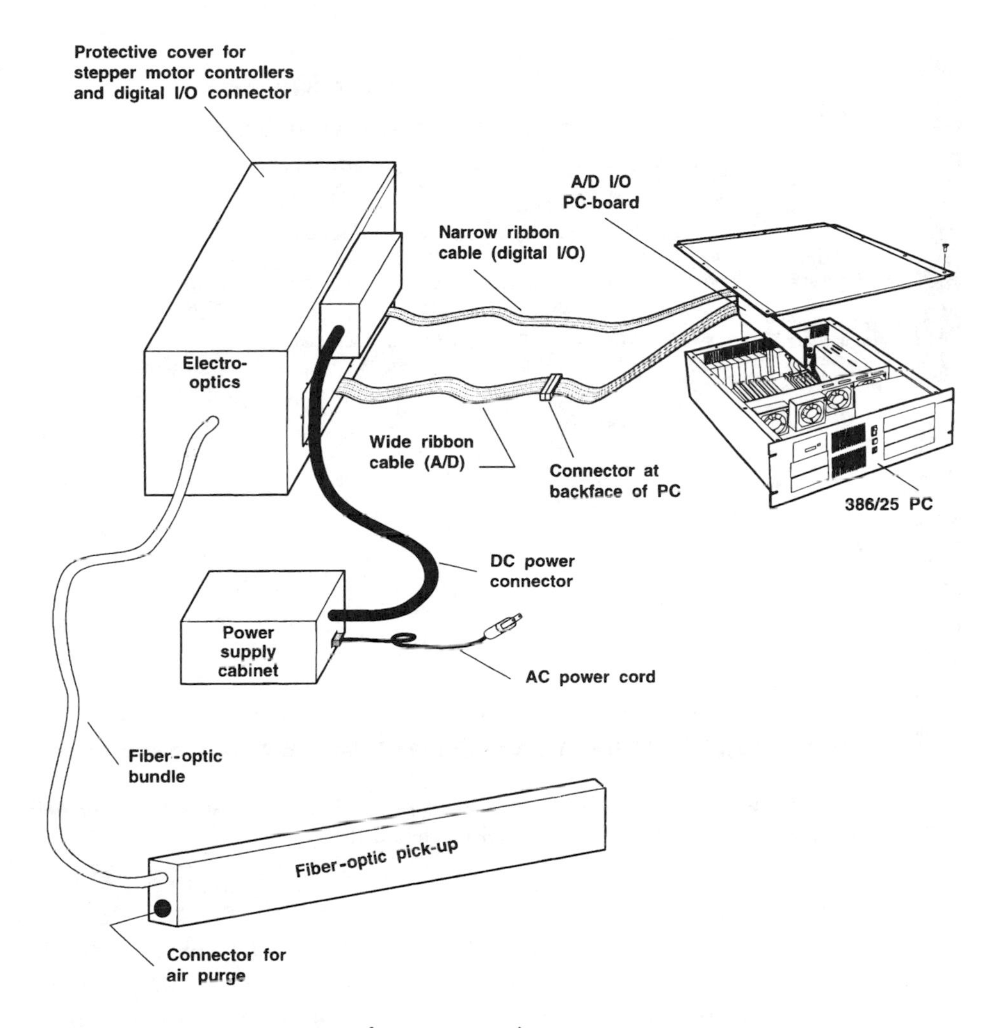

Figure 13.8 Component arrangement for prototype multipoint temperature monitor.

Details of the pick-up assembly are shown in Figure 13.9. Each pick-up consists of a 5-mm diameter ball lens coupled into a single strand of low OH, optical fiber with a core diameter of 400 μm. The pick-ups are deployed in a linear array in an air purged housing. The purging is designed so that most of the air passing through the upper plate exits to the side rather than through the holes in the lower plate. This allows sufficient air sweep to keep the retainer plate clean without perturbing the temperature of the web with the air exiting the lower plate.

The major components housed in the electro-optics cabinet are shown in

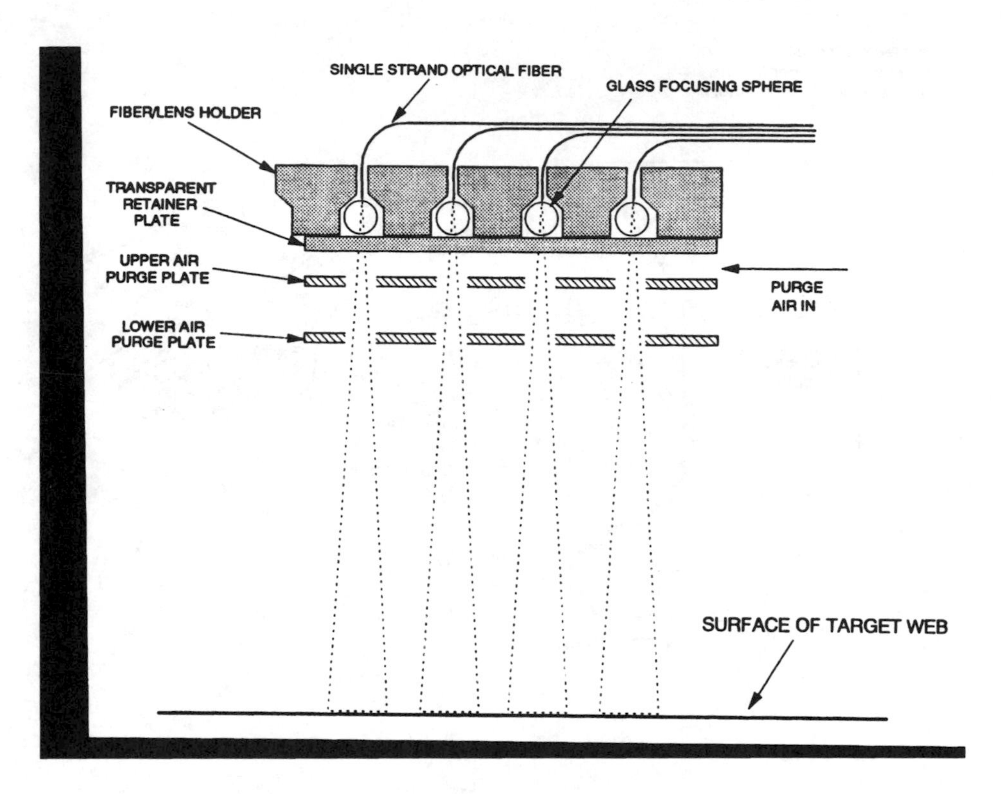

Figure 13.9 Detail of pick-up assembly for multipoint temperature monitor. (*Source:* Jeffers, L. A., *Proc. SPIE,* Vol. 2072, 1993, pp. 53–62. Reprinted with permission.)

Figure 13.10. The optical fibers are arranged in parallel rows of 16 fibers each. The prototype unit was limited to 32 fibers arranged in two rows, but the design allows future expansion up to 160 fibers in 10 rows. The fibers within each 16-fiber row are mounted on 1-mm centers so that the relay lens, which has a magnification of 1, images a single fiber on each of the 1 mm by 1 mm photodiodes in the detector array. The detector is a 16-channel, parallel output, germanium photodiode array. Each of the 16 detector signals is individually amplified and the resultant signals sent to a computer plug-in board, where they are electronically multiplexed and digitized.

The 16-channel electronic multiplexing is supplemented by the optical multiplexing provided by a stepper-motor-driven translation stage onto which the fiber bundle is mounted. Under control of the software, the stepper-motor moves the first row of 16 optical fibers into position to be focused onto the detector. After the signals from the fibers in the first row have been sampled, the motor moves the second row into position to be imaged onto the detector. Subsequent rows are moved into position

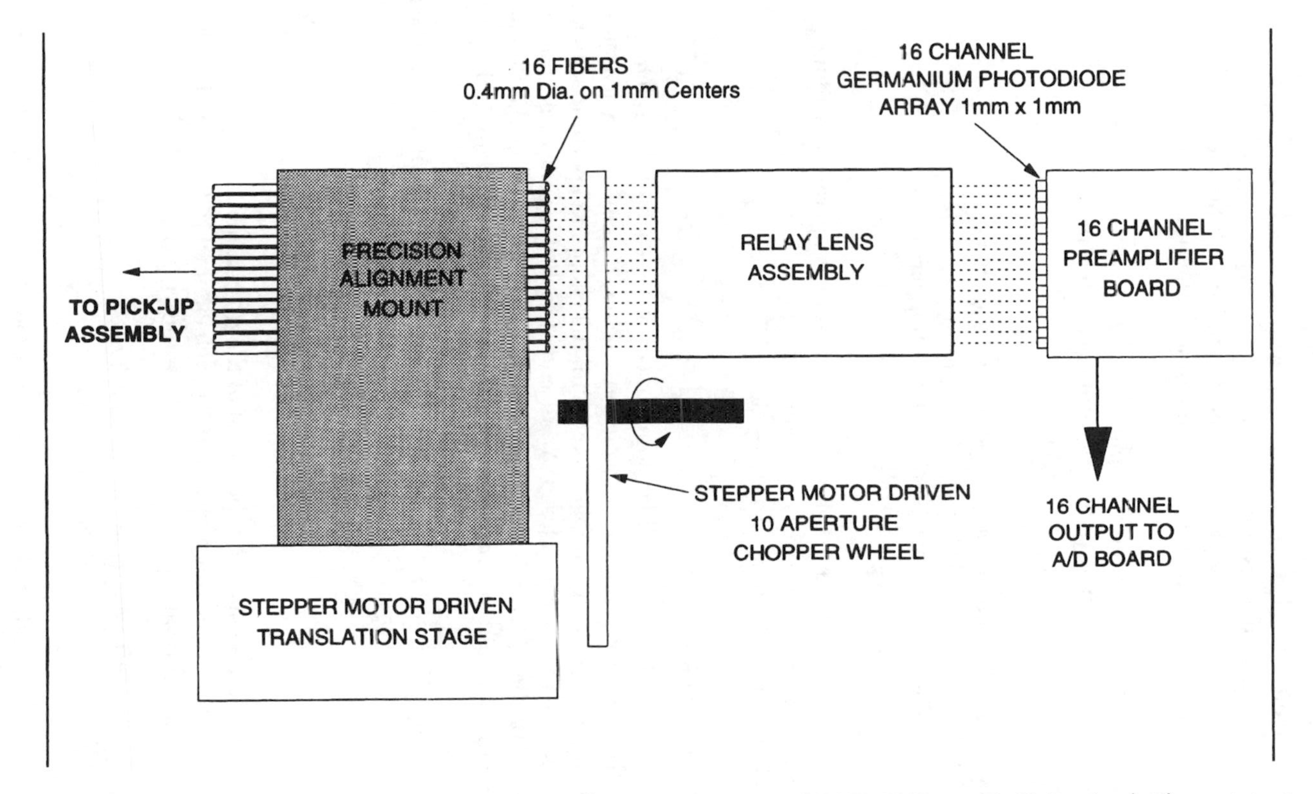

Figure 13.10 Electro-optics cabinet components. (*Source:* Jeffers, L. A., *Proc. SPIE,* Vol. 2072, 1993, pp. 53–62. Reprinted with permission.)

for readout in sequence until the final row has been sampled. At that point, the process repeats, starting at the first row.

A stepper-motor-driven chopper wheel, located in front of the fiber bundle, provides chopping of the optical signal in synchronism with the subsequent digitizing of the detector signals.

The detector signals are input to a plug-in A/D board located in the system computer. The 386/25 computer, running custom software, provides the following functions:

- Allows operator selection of various operating parameters;
- Controls the stepper motor for optical multiplexing;
- Controls the stepper-motor driving the optical chopper;
- Digitizes the detector signals in synchronism with the chopping;
- Calculates temperatures;
- Maintains a 32-channel bar graph display of current temperature readings;
- Stores a data file upon keyboard command;
- Allows calibration of the temperature readouts.

13.3.4 Signal Processing

The optical signals are chopped at 60 Hz simultaneously on each of the 16 channels in a given row. The detector signals are sampled at 2,400 Hz/channel in synchronism with the chopping to provide 20 samples taken with the chopping aperture open and 20 samples taken with the aperture closed. At the end of each chopping cycle, the sum of the readings with the aperture closed is subtracted from the sum of the 20 open readings to provide a resultant proportional to the radiant power delivered by the fiber.

Mathematically, the resultant signal, $R(j)$, for the jth channel can be expressed as

$$R(j) = \sum_{i=1}^{20} I_i(j) - \sum_{i=21}^{40} I_i(j)$$

where $I_i(j)$ is the ith sample taken on the signal of the jth channel.

The time required to generate a resultant for each channel is designated the loop time. For the 32-channel prototype system, the loop time t_{loop} is

$$t_{\text{loop}} = \frac{40 \text{ steps/channel} \times 32 \text{ channels}}{2400 \text{ steps/second}} = 0.5 \text{ sec.}$$

The temperatures can be updated as fast as once per loop (i.e., at 2 Hz). To improve the signal to noise ratio (SNR), the integration time can be specified in terms of the number of loops to be averaged for each update.

The temperature, $T(j)$, of the jth channel is calculated using the expression

$$T(j) = B[\ln S(j) - \ln A(j)] \tag{13.1}$$

where $S(j)$ is the measured time-averaged signal for the jth channel, and $A(j)$ and B are constants. Equation (13.1) is an approximation, valid to within 2°C over the range from 100 to 180°C. It is derived from the Planck blackbody function and the following simplifying assumptions: 1) the spectral transmission efficiency of the optical system is a constant over the relatively narrow range of effective wavelengths and 2) the spectral responsivity of the detector is given by $R = .4\lambda$ up to the cutoff wavelength, λ_{max}, and zero above the cutoff.

The constant B in (13.1) depends only on λ_{max} and is the same for all channels. The effects of target emissivity, detector efficiency, electronic gain, and transmission of lenses, windows, and fibers, are all lumped into the term $A(j)$ in (13.1). The system may be periodically standardized (calibrated) with a thermocouple, which provides an independent measurement of the web material temperature. This reference temperature is entered into software to obtain an *emissivity correction factor*, as discussed further in the next section.

13.3.5 Calibration and Test

The temperature calculated from (13.1) is a linear function of the natural log of the measured signal, which simplifies the calibration. The constant B is the same for all channels and remains constant with time, being a function only of the cutoff wavelength. Initial testing gave a value of B = 17.73°C, corresponding to λ_{max} = 1.68 μm. All of the factors that vary from channel to channel (and that tend to drift with time) are lumped in the term $A(j)$, which appears conveniently as an offset.

The software provides two calibration modes. The primary calibration mode (mode 1) is used to determine individual values of $A(j)$ for each channel to compensate for variations in alignment, transmission, and so forth. Mode 1 calibration involves storing a data file while viewing a source whose temperature at each channel position is independently measured. The known temperatures, $T(j)$, along with the measurand responses, $S(j)$, are then used in (13.1) to calculate the calibration constants, $A(j)$.

Mode 2 calibration provides a simple means for adjusting the system in the field, hence allowing for webs of differing emissivities. In this mode, the temperature of the target material needs to be independently measured at only one channel position. The software then determines the fractional change in A required to match the reference temperature reading. The same "emissivity correction factor" is then automatically applied to each of the $A(j)$ constants.

A source, consisting of a 1.2m long electrically heated metal bar with an array of 18 thermocouples located along the length was constructed for use in validation testing of the prototype. The same test program consisted of first calibrating the system and then storing a series of data files, taken for various nominal temperature settings of the source. At each temperature, files were stored for several different integration times.

Figure 13.11 shows the accuracy of the prototype measurements as a function of temperature. For each channel, the difference was calculated between the prototype measurements and the corresponding thermocouple measurements. The average (over all 32 channels) of the difference is plotted in Figure 13.11. It can be seen that the prototype measurements deviate from the thermocouple readings at both the high and low end of the range. At the high end, the difference increases because the temperature calculation is based on a linear approximation. The error at 190°C, expected as a result of ignoring the higher order terms, is 5°C; in practice, a 4°C deviation was measured. At the low end of the range, the errors result from nonlinearity and because the SNR degrades rapidly below 120°C.

13.3.6 Summary

A 32-channel prototype remote thermometer system has achieved a noise-limited precision of ±1°C over the range 140°C to 200°C. The noise increased to ±5°C at 100°C.

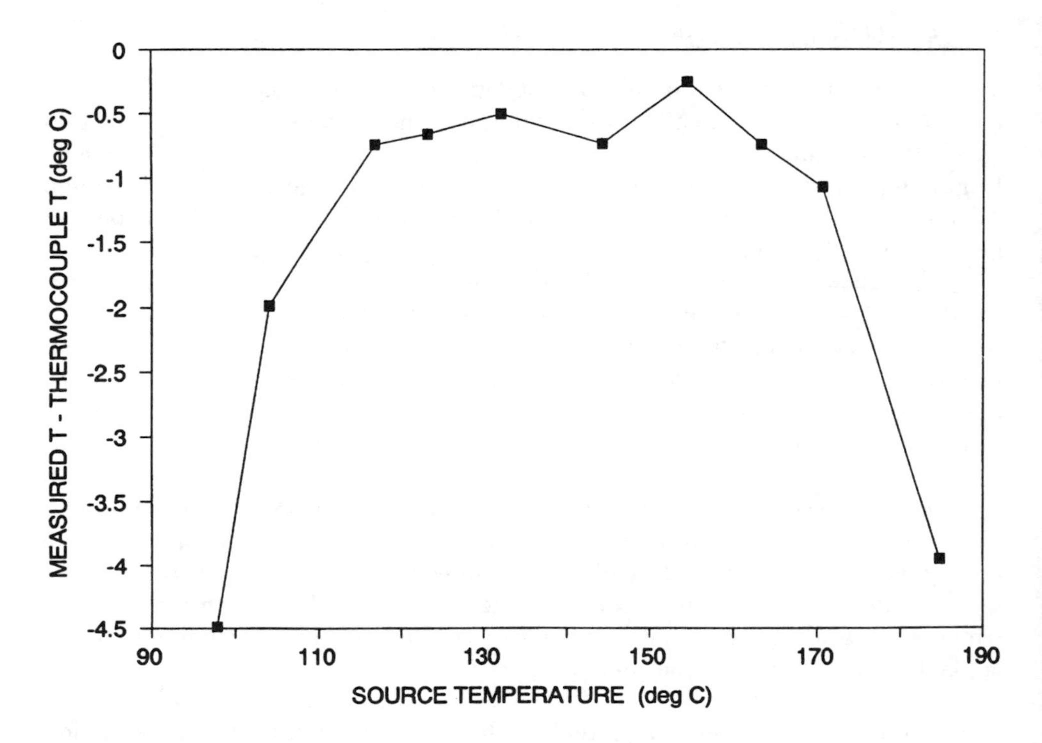

Figure 13.11 Accuracy vs. temperature. (*Source:* Jeffers, L. A., *Proc. SPIE*, Vol. 2072, 1993, pp. 53–62. Reprinted with permission.)

The 32-channel prototype was designed for operation in the range from 120°C to 180°C. Extension of the range on the high end can be achieved through the software by modifying the equation used to convert response to temperature. Extension of the range at the low temperature end will require the use of detectors with a longer cut-off wavelength.

13.4 VOID FRACTION SENSOR

13.4.1 Application

In a water-tube-type boiler, water and steam are present as a two-phase mixture at high temperature and pressure. If the water flow into the boiler is somehow interrupted, the effect of the unchanged heat input to uncirculated water in the heat exchanger tubing can become excessive. This results in "film boiling" at the inside surface of the tubing walls, formation of large steam bubbles within the heat exchanger, and potential catastrophic failure of the heat exchanger from excessive heating. It is extremely important to know the rate at which these events evolve after loss of water flow. However, many variables affect the rates, which makes modeling difficult. To support improvements to refine the models, full-scale simulated tests can be performed using sensors to detect the prevailing conditions.

The formation of steam bubbles in a heat exchanger can be quantified during the onset of film boiling in one of the water flow channels. The accepted definition of void fraction is the ratio of the volume of gas present to the sum of the volumes of gas and water. Thus, the void fraction can range from the zero (or fully flooded condition) to the 100% (steam only) condition.

The fiber-optic method described in this section measures void fraction in a small (near-cylindrical) volume along a line of sight across the flow channel. The bubbles cause light deflection away from the detector. To achieve a good approximation to the volumetric quantity of interest, time averaging of the bubble flow across the line of sight must be performed. This approach has some inherent error, particularly with high-void fraction, because the spatial distribution of voids varies with the flow regime [9,10].

The advantages of fiber optics for void fraction measurement are listed below, in comparison to the traditional electrical conductivity method that has been used in the past:

- Direct relationship to void fraction;
- Non-intrusive;
- High bandwidth;
- Insensitive to:
    - Surface tension effects;
    - Temperature;

- Water conductivity;
- Electrical noise pickup.

All of these advantages result in better repeatability and reliability than previously obtained with conductivity probes.

It is well known that the amount of light scattered in a given direction by small particles or discontinuities, such as bubbles in a liquid, is a very sensitive function of particle size. This sensitivity, along with the large number of bubbles present in two-phase flows, might lead to the conclusion that attempts to make a light transmission measurement to correlate with void fraction would be overwhelmed by scattered light. However, it has been demonstrated that although the effects of scattering are about 100 times greater than the magnitude of absorption, the scattering effects can be very nearly compensated for by making measurements at two nearby wavelengths [11]. With this approach, the measured concentration of liquid water gives very repeatable correlation with void fraction.

13.4.2 Design

The design criteria for probes for measurement of void fraction were defined as shown in Table 13.1 for the particular flow channel test geometry (see Figure 13.12) and anticipated operating environment.

These criteria were used to define the probe design and materials, to define the tests to demonstrate proof of principle, and to characterize the completed probes.

13.4.3 System

The operation of the fiber-optic void fraction sensor is based on the measurement of the differences in infrared transmission loss in liquid water as a result of the presence of less absorbing bubbles of gas or vapor. A schematic of the probe concept is shown

Table 13.1
Design Parameters for Void Fraction Probe

Test Geometry	*Operating Environment*
Void fraction range	0 to 1
Flow channel gaps	3.5 mm to 7.5 mm
Probe OD (max)	3.2 mm
Maximum temperature	150°C
Maximum pressure (P_{sat} at T_{max})	360 kPa
Maximum velocity	12 m/s^{-1}
Data rate/minimum bandwidth	100 Hz/30 Hz

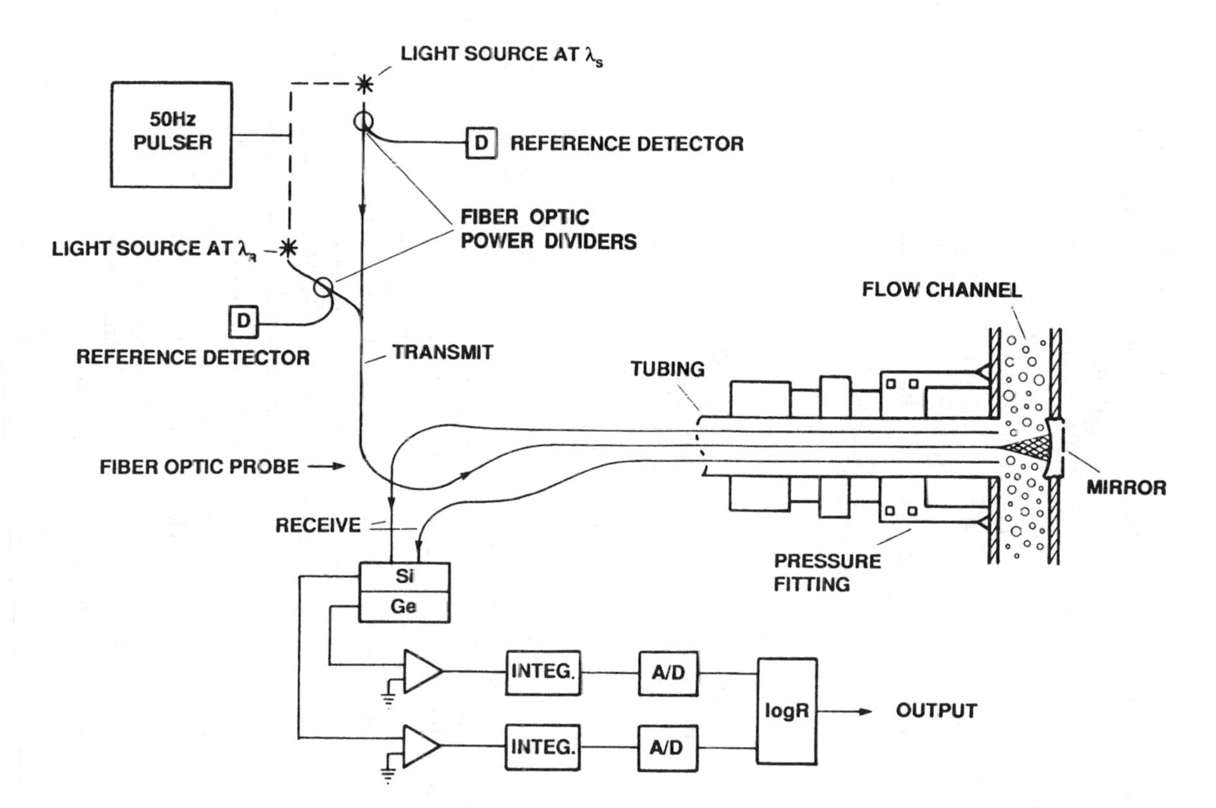

Figure 13.12 Diagram of fiber-optic void fraction sensor used for proof of principle tests. (*Source:* Berthold, J. W., et al., *Flow Meas. Instrum.*, Vol. 5, 1994, pp. 3–13. Reprinted with permission from Butterworth-Heinemann journals.)

in Figure 13.12. A light source is used that has a center wavelength arranged to overlap one of the infrared absorption bands of liquid water. The light is launched and transmitted to the flow channel through a multimode optical fiber. The light input fiber and several receiving fibers terminate flush with the wall of the flow channel in which the measurement is to be made. On the opposite wall, a mirror is positioned to reflect infrared light from the input fiber back to the receiving fibers. At the output of the receiving fibers, light attenuation is measured to determine the concentration of liquid water in the flow channel. The attenuation, as a function of concentration, obeys a Lambert-Beer's law relationship (see Chapter 7 by Dakin):

$$I = I_o(e^{-axc})$$

where

I = received light intensity;
I_o = transmitted light intensity;
a = absorption coefficient of water at the light wavelength;
x = optical path length (twice the flow channel width);
c = concentration of water in the optical path.

To obtain good range and resolution with optical path lengths of about 1/4 in to 3/4 in, the wavelength of the light is selected so that the attenuation coefficient for water is in the range of 0.5 to 3 cm^{-1}. Attenuation coefficients with this range of values are found in the wavelength range of 1.0 to 1.3 μm, where water vapor has virtually no attenuation. To provide compensation for changes in the transmitted light intensity (for example, due to a build-up of deposits on the fiber ends, changes in the reflectivity of the mirror surface, changes in the fiber transmission, or scattering losses), a second light source at a wavelength that is not absorbed by either water or water vapor is used. Wavelengths shorter than 900 nm have suitably low absorption coefficients. Any of the potential error sources listed above attenuate the second reference wavelength and the measurement wavelength in a similar manner. As a result, the ratio of the intensities at the two wavelengths is (to first-order) independent of all changes in light intensity, except for desirable ones caused by water absorption at the measuring wavelength. The wavelengths chosen for the sources are 830 nm for the reference wavelength and 1,300 nm for the measuring wavelength.

The fiber-optic probe (of 3.2-mm outer diameter), built for testing in two-phase flow, uses a silica core fiber with high OH (hydroxyl radical) fluoride-doped cladding and polymide buffer coating; the core, clad, and buffer diameters were 400, 480, and 505 μm, respectively. The fiber bundle in the probe has a single launch fiber in the center surrounded by a ring of seven receive fibers and was just over 1m long. Figure 13.13 is a drawing of the prototype probe. Light-emitting diodes were selected as the sources for the measuring and reference infrared wavelengths. The intensities were measured using a silicon/germanium two-color photodetector.

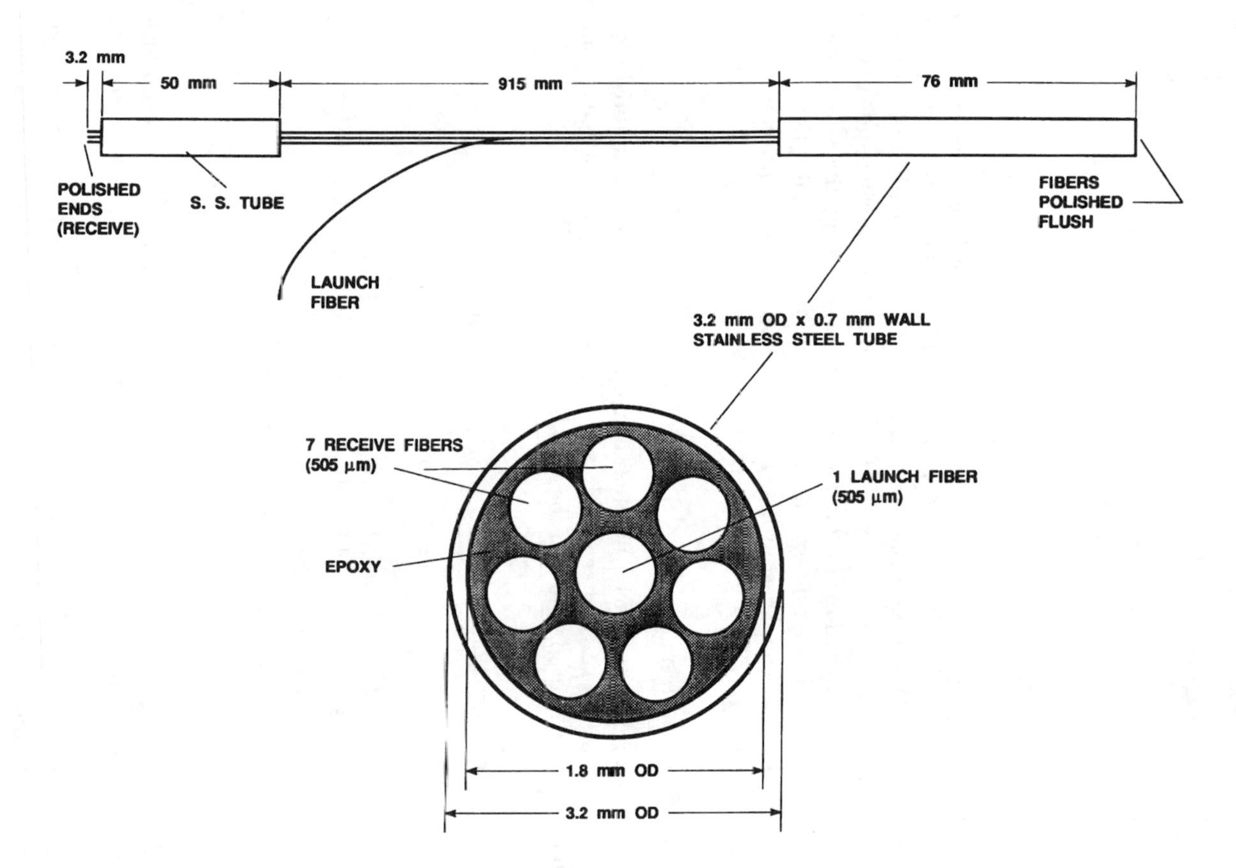

Figure 13.13 Diagram of fiber-optic probe geometry for fiber-optic void fraction sensor. (*Source:* Berthold, J. W., et al., *Flow Meas. Instrum.*, Vol. 5, 1994, pp. 3–13. Reprinted with permission from Butterworth-Heinemann journals.)

13.4.4 Signal Processing

The signal processing software solves for mean water concentration (c) in the volume hence, by subtraction, the void fraction (α):

$$\alpha = 1 - c = 1 - (1/-ax)\,(\ln(I/I_o))$$

where I and I_o are both measured intensities at 1,300 nm, I with an unknown void fraction and I_o with a void fraction of 1 (no water, and hence no absorption), and x and a are the constant absorption coefficient and optical path length, respectively. The intensity signals must also be corrected for losses due to scattering and deposits on the probe or mirror, variations in the dark current, source brightness with time, differences in the absolute brightnesses of the two sources, and initial differences in the optical path, such as coupler-splitting ratio (and excess loss), connector losses, mirror reflectivity, and wavelength-dependent fiber loss.

13.4.5 Calibration and Test

To determine the probe output as a function of void fraction, several hundred two-phase flow conditions were evaluated in a flow test rig. The measured voltage outputs from each detector were used to obtain the void fraction, α, as follows:

$$\alpha = 1 - [1/(-ax)]\ln[(V_{1300}/V_{835})/(V_{1300}/V_{835})_{air}]$$

where

a	= known (or calibrated) absorption coefficient (cm^{-1});
x	= optical path length (cm);
(V_{1300}/V_{835})	= voltage ratio at unknown α;
$(V_{1300}/V_{835})_{air}$	= voltage ratio at α = 1.

The optical path length used was twice the channel depth. The value for the absorption coefficient was determined from the voltage ratios in air and in water, and was found to be 1.48 cm^{-1}.

Figure 13.14 demonstrates that under identical flow velocity (4 m/sec) and geometrical conditions, the probe repeatability measured from four separate calibrations runs (A, B, C, D) is excellent. The "scatter" in the data results from the combined errors of the fiber-optic probe output and the difficulty of setting repeatable flow velocity in the test rig. Larger variations in response occur when the liquid-phase flow velocity is allowed to vary over its entire range from 1 m/sec to 5 m/sec. The physical basis for the nonlinearity in curve shape in Figure 13.14 is believed to be a com-

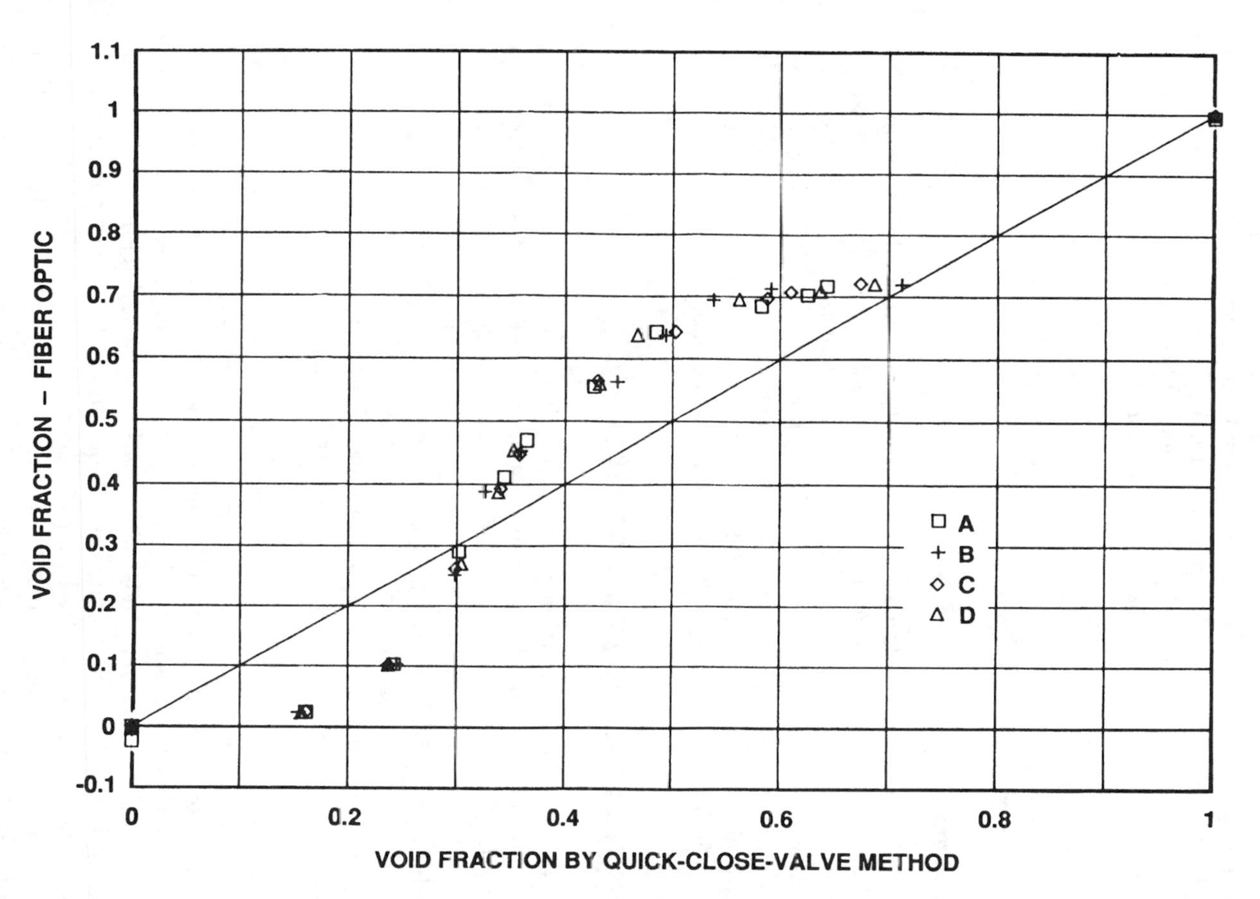

Figure 13.14 Repeatability of probe No. 2 output vs. void fraction. Each data set is a separate calibration with flow velocity held constant. (*Source:* Berthold, J. W., et al., *Flow Meas. Instrum.*, Vol. 5, 1994, pp. 3–13. Reprinted with permission from Butterworth-Heinemann journals.)

bination of hydraulic (void distribution variations) and optical (geometric and wavelength) phenomena. The nonlinear calibration data for each of the probe channels was fitted to a fourth-order, least squares polynomial. Thus, while the output of the fiber-optic void fraction system electronics was a nonlinear function of void fraction, the overall system output was made linear. The nonlinearity results in increased measurement uncertainty when the slope of the unprocessed calibration curve is small (0 to 0.2; 0.5 to 0.7 void fraction) but the system has sufficient resolution to provide useful results over the void fraction range of 0 to 1.

13.4.6 Summary

An improved method for void fraction measurement has been developed around fiber-optic extrinsic sensors. A three-probe, fiber-optic void fraction sensing system was used for five months to gather data during loss of coolant tests. The water environment in the vicinity of the probe tip was 400-kPa pressure and 130°C. For best accuracy, the probes must be calibrated in a flow channel having a geometry similar to that where the probes are to be used. The calibration should be performed over the expected range of liquid and gas phase velocities and flow regimes.

13.5 RAMAN SPECTROSCOPY VIA OPTICAL FIBER

Remote, fiber-optic-based Raman spectroscopy has many potential applications in environmental monitoring [12,13] and in industry [14]. (See also Chapter 7 by Dakin.) In the power boiler industry alone, applications include:

- Online monitoring of water chemistry;
- Online monitoring of corrosion and deposits;
- *In situ* inspection of steam generators during outages;
- Autoclave studies of high-temperature corrosion chemistry.

In this section, the results of a feasibility study are summarized. This study was conducted using laboratory equipment originally developed for fluorescence spectroscopy. Although far from optimum for Raman work, it nevertheless served the purpose of establishing the feasibility of remotely (through an optical fiber) obtaining Raman spectra of several compounds of interest. It also identified problems that must be addressed in the implementation of the technique; and provided background for planning the next phase of the development.

As discussed in Chapter 7, Raman spectra are the result of inelastic collisions between incident photons and molecules in the sample. As in the more familiar infrared spectroscopy, the analytic capability is based on the fact that the spectral distribution of the Raman light is characteristic of the molecular species involved.

Some of the advantages of Raman over conventional infrared are as follows:

- Measurements of transitions that have energies more representative of those of infrared photons can be made in the visible region of the spectrum, which opens the possibility of remote measurement through optical fibers using laser diode sources.
- Measurements can be made on solid, liquid, or gaseous samples.
- The signal is directly proportional to the concentration.

The main disadvantages of Raman are related to the fact that the effect is so weak. Raman cross-sections are typically 10 orders of magnitude smaller than cross-sections for absorption and 2 to 4 orders of magnitude smaller than cross-sections for Rayleigh scattering. Furthermore, the low signal level is subject to being swamped by undesirable fluorescence in the sample, the fiber, or even in impurities within the fiber.

13.5.1 Design

The use of optical-fiber probes provides several advantages, which include 1) the ability to make measurements in otherwise inaccessible regions, such as the interiors of steam generators in power boilers; 2) the ability to make measurements across a pressure boundary (autoclave for example); and 3) a remote measurement capability, which includes the possibility of multiplexing a large number of distributed measurement points into a single, centrally located analyzer.

The extreme weakness of the Raman effect poses special problems when using fibers. Most important of these is the generation of fluorescence and Raman lines in the comparatively long path length in the optical-fiber core material. Nevertheless, several successful applications of fiber-optic Raman spectroscopy have been reported [15,16].

13.5.2 Test Apparatus

Figure 13.15 is a block diagram showing the major components of the apparatus used for the feasibility tests. Excitation was provided by an argon ion laser operating at 488 nm. The average power delivered by the fiber to the sample was 100 mW. A small, 1/8 meter focal-length monochromator was used between the excitation laser and the fiber. The purpose of this monochromator prefilter was to prevent light in the nonlasing plasma lines from the laser discharge from entering the fiber and being ultimately transmitted to the detector as noise.

The excitation fiber was a 20m long, 600-μm core silica fiber (3M Company Part No. FG-600-UAT). This fiber is designed for optimum transmission in the blue and ultraviolet end of the spectrum. Typical attenuation at 400 nm is given by the manufacturer as .04 dB/m, so the loss in the fiber was minimal.

The pick-up consisted of a bundle of seven silica-core fibers, with a core diameter of 365 μm, arranged in a circular pattern at the pick-up end and in a linear array at the monochromator end. The linear array matched the rectangular geometry of the

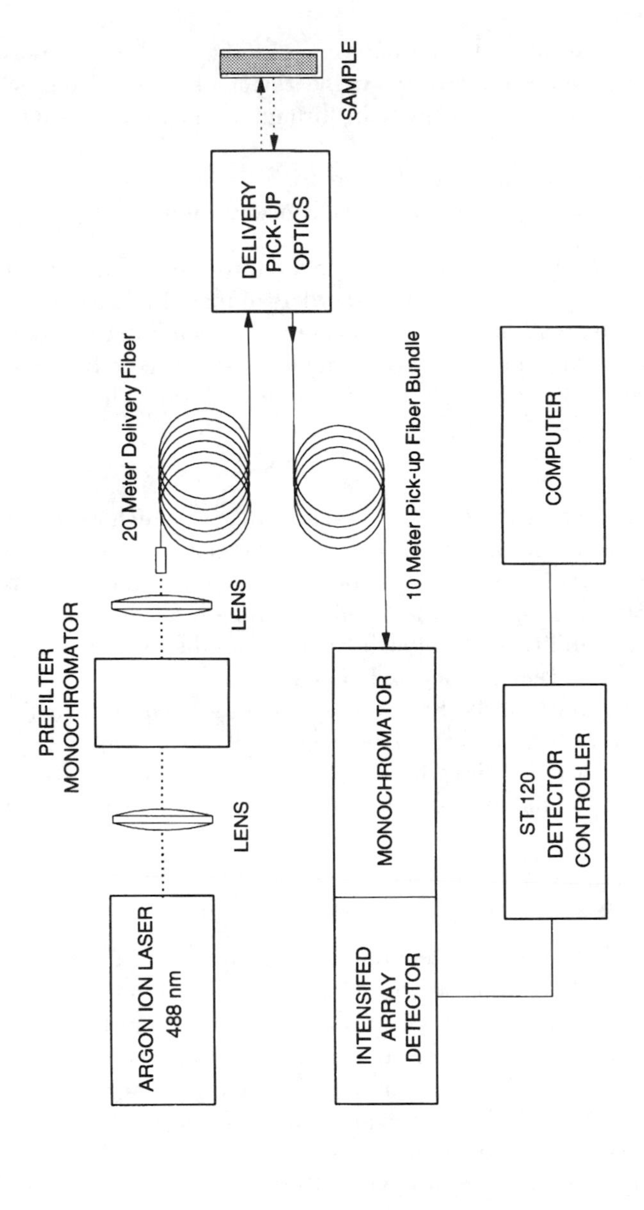

Figure 13.15 Block diagram of Raman chemical sensing apparatus. (*Source:* Jeffers, L. A. and Berthold, J. W., *Proc. NACE 12th Intl. Corrosion Congress*, Houston, Texas, 1993, p. 4313. Reprinted with permission.)

monochromator slit. The fiber bundle was designed for the study of fluorescence and was not very efficient for use in the Raman work. In the fluorescence work, spectral resolution is relatively unimportant and the monochromator slits can be opened to a width of 350 μm to pass nearly all of the light delivered by the fiber bundle. To give the desired resolution for the Raman spectra, the monochromator slits were narrowed to 40 μm, which resulted in about 14% delivery efficiency. The detector was a 1,024-element, intensified diode array that provides simultaneous parallel detection of all wavelengths.

Figure 13.16 shows the details of the delivery and pick-up optics assembly. The excitation light emerges from the delivery fiber and illuminates a small area on the surface of the sample. Lens 1 is spaced by its focal length from the sample surface, so that the reflected and scattered light collected by lens 1 is collimated and passed through the holographic notch filter. The filter has an optical density 4 at the excitation wavelength, and the transmission rises very rapidly on either side of the excitation wavelength. This allows a very high degree of discrimination against the elastically scattered light, while passing the inelastically scattered Raman light even at quite low frequencies. The light that passes through the filter is focused by lens 2 onto the circular end of the seven-fiber pick-up bundle.

13.5.3 Test Results

Raman spectra were obtained on the following samples: Na_2HPO_4 (DSP) both 1% and saturated solutions, Na_3PO_4 (TSP) both 1% and saturated solutions, Fe_2O_3 as powder, and Fe_3O_4 as powder.

Figure 13.17 shows the raw spectra obtained for the 1% solution of DSP. The plot is in terms of the measured intensity as a function of the difference in frequency, $\Delta\nu$, between the measured light and the excitation light.

The measured optical power signal, M in watts, can be written as

$$M = \text{NEP} + \{R + F + \rho[I_o + R_f + F_f]\}T \qquad (13.2)$$

where

NEP is the noise equivalent power of the detector;
R is the Raman signal from the sample;
F is the fluorescent signal from the sample;
ρ is the reflectivity of the sample;
Io is the optical power of the excitation light;
R_f is the Raman signal generated in the fibers and other optical components;
F_f is the fluorescent signal generated in the fibers and other optical components;
T is the transmission of the pick-up optical system.

Each of the terms in the equation (except for NEP) is dependent on the Raman frequency shift $\Delta\nu$.

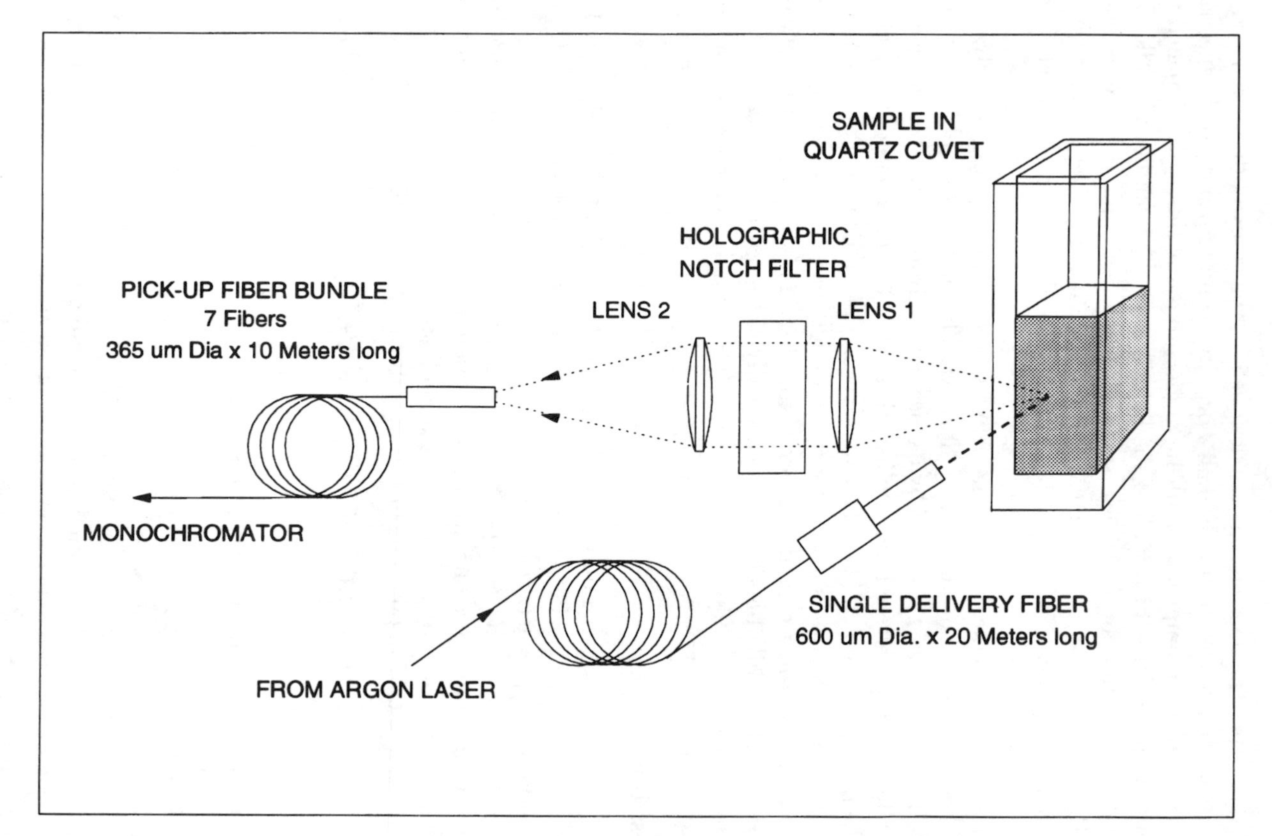

Figure 13.16 Delivery and pick-up optics for Raman probe. (*Source:* Jeffers, L. A. and Berthold, J. W., *Proc. NACE 12th Intl. Corrosion Congress*, Houston, Texas, 1993, p. 4313. Reprinted with permission.)

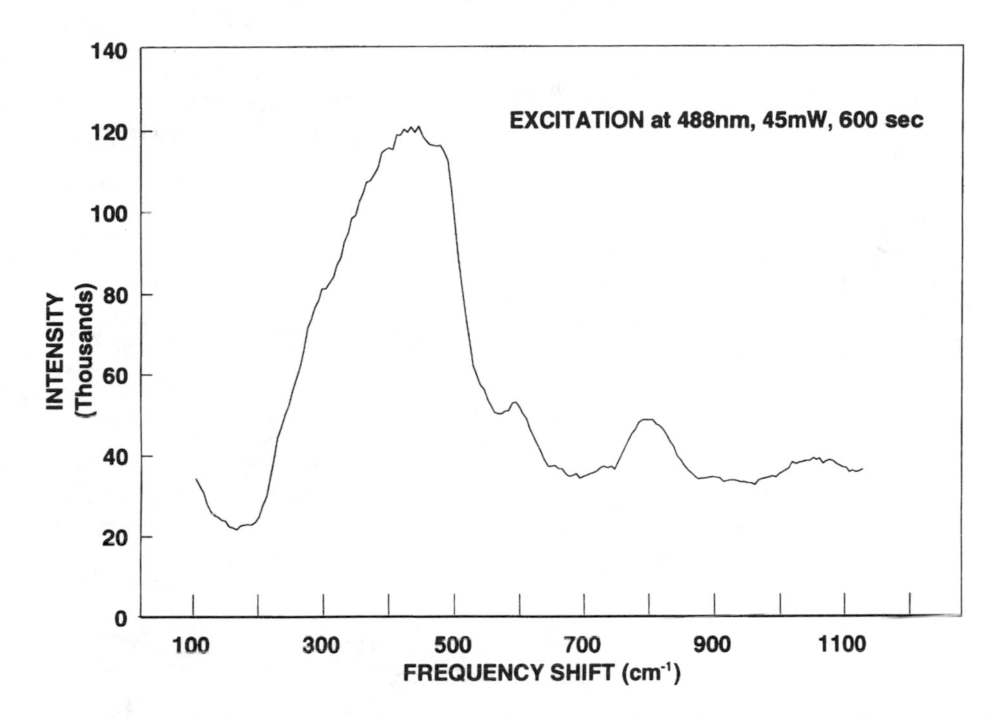

Figure 13.17 Uncorrected Raman spectrum of 1% disodium phosphate solution. (*Source:* Jeffers, L. A. and Berthold, J. W., *Proc. NACE 12th Intl. Corrosion Congress*, Houston, Texas, 1993, p. 4313. Reprinted with permission.)

Typically, I_o is many orders of magnitude greater than any of the other terms in (13.2). The holographic notch filter, however, has a transmission that is extremely low at the excitation wavelength and high everywhere else. As a result, the product I_oT is usually less than all of the other terms, except at wavelengths very close to the excitation wavelength (i.e., at small values of $\Delta\nu$.

All of the spectra look much the same as shown in Figure 13.17, because the sample independent terms R_f and F_f in (13.2) are about two orders of magnitude larger than the Raman signal from the sample. Although all of the spectra look superficially the same due to the dominance of the fiber-generated terms, there are differences that can be revealed by eliminating the "common mode" fiber contribution. Figure 13.18 shows the result of subtracting the measured spectrum (i.e. TSP) from that of DSP. The two positive peaks are Raman lines from the DSP, while the negative peak is from TSP. The values of frequency given in parentheses are published values for the peaks of these known Raman lines [17].

The background subtraction technique provides a very powerful tool for dealing with the fiber background problem and explains why the computer-linked

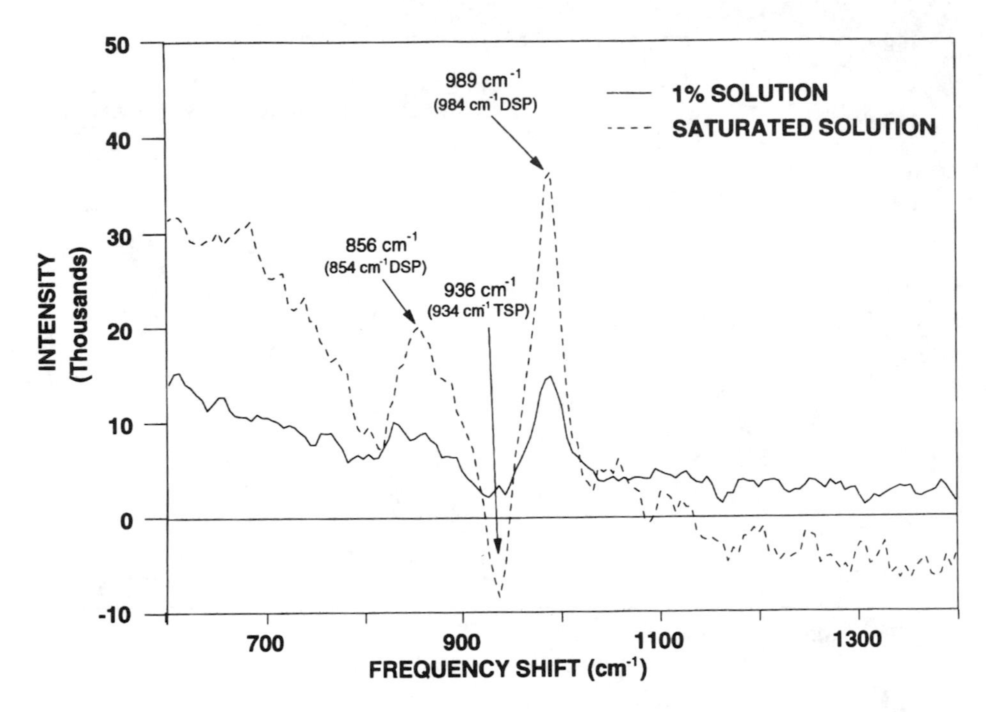

Figure 13.18 Difference Raman spectra, disodium phosphate - trisodium phosphate. (*Source*: Jeffers, L. A. and Berthold, J. W., *Proc. NACE 12th Intl. Corrosion Congress*, Houston, Texas, 1993, p. 4313. Reprinted with permission.)

detector array and digital signal processing are so vital to fiber-based Raman spectroscopy. The limitations of the technique can be appreciated by reference to (13.2). If there were no differences in the spectral reflection of the two samples, the fiber-related terms would cancel completely. In the case of the DSP and TSP samples, the cancellation was close enough to clearly reveal the Raman lines of both.

The spectra of the two iron oxides are closely spaced, and each spectrum was used for correcting the other, as shown in Figure 13.19 where several Raman lines were observed. The values in parentheses are those reported in the literature [18,19]. Values shown above the curve are for Fe_2O_3 lines, whereas those shown below the curve are for Fe_3O_4.

13.5.4 Summary

It is possible to collect Raman spectra of various compounds through 20m of optical fiber, even using a nonoptimized system. With the demonstrated detection capability

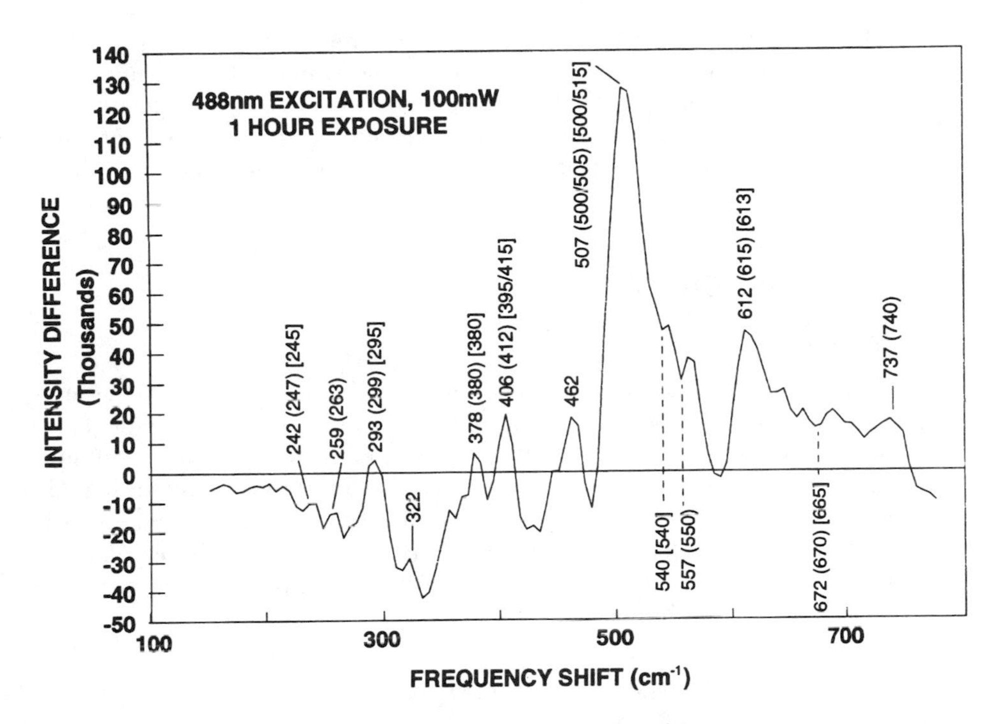

Figure 13.19 Difference Raman spectra, Fe_2O_3 - Fe_3O_4. (*Source:* Jeffers, L. A. and Berthold, J. W., *Proc. NACE 12th Intl. Corrosion Congress*, Houston, Texas, 1993, p. 4313. Reprinted with permission.)

for hematite, magnetite, and phosphates, an inspection system could be built that would be useful for identification of localized corrosion problem areas.

Using a previously-measured reference spectrum, measurements can be made in the presence of severe background noise arising from fluorescence and Raman generation within the fibers. Use of digital data acquisition and an array detector that allows simultaneous detection of the signal at all wavelengths is vital to the implementation of background correction. Even with background subtraction, uncertainties in the magnitude of the fiber background eventually set the limits of detectability when the improvements of signal averaging have been fully exploited to remove random noise.

Fiber fluorescence and attenuation loss are strongly wavelength-dependent-both increasing with decreasing wavelength. Also, photon energy and hence photon noise levels decrease at longer wavelengths, whereas silicon detector response improves out to 900 nm. This would suggest that a longer wavelength excitation source might be preferred. However, the intensity of the desired Raman signal is inversely proportional to the 4th power of the wavelength. The optimum wavelength will therefore be determined by a tradeoff between these factors. For very short delivery-fiber applications such as autoclave-corrosion studies, the 488-nm Argon line is ideal. At intermediate fiber lengths (20m or so) it may be advantageous to use the 514.5-nm argon line. For applications requiring delivery through 100m or more, or ones requiring compact optoelectronics, a near-infrared diode laser would be preferable.

13.6 FLUID OPACITY MONITOR

13.6.1 Application

Opacity monitors have many applications in industry. In one application, the measurement of opacity of gases in smoke stacks provides a measure of the amount of particles (smoke) in the gas. In another application, measurement of the opacity of high optical density liquids, such as paint, may be used to control the amount of ingredients added in the production process. For example, TiO_2 is a major ingredient in water-based paint, and due to its very high refractive index is optically very dense in high concentrations. Measurement of paint opacity during production ensures that the correct amount of TiO_2 has been added.

13.6.2 System Design and Signal Processing

The fluid opacity monitor (FOM) consists of a probe, light source, and connecting fiber optics. The probe contains photodetectors and electronics. The light source and fiber-optic cables are connected to the remote probe for insertion into the process pipeline. The FOM uses space division referencing to deliver two light signals to the sensor head from one light source. A diagram of the FOM is shown in Figure 13.20. A high-intensity incandescent light source transmits light to a power coupler, which divides the light between two optical-fiber bundles. One bundle delivers light to the sample channel and the other bundle to the reference channel in the sensor head. The ends of the bundles are axially aligned with two photodetectors in the sensor head. A cutaway view of the sealed sensor head assembly is shown in Figure 13.21. The photodetectors are connected to a log-ratio amplifier in the signal processor. An A/D converter is connected to the output of the amplifier and provides input to a computer for data processing, archiving, and display.

13.6.3 Calibration and Use

The opacity is defined by the following equation:

$$\text{Opacity} = \text{Log } I_{ref}/I_{sig} - \text{Offset} \tag{13.3}$$

where I_{ref} and I_{sig} are the photocurrents generated by incident light falling upon the photodetectors and the offset is the (dark) signal in the absence of light.

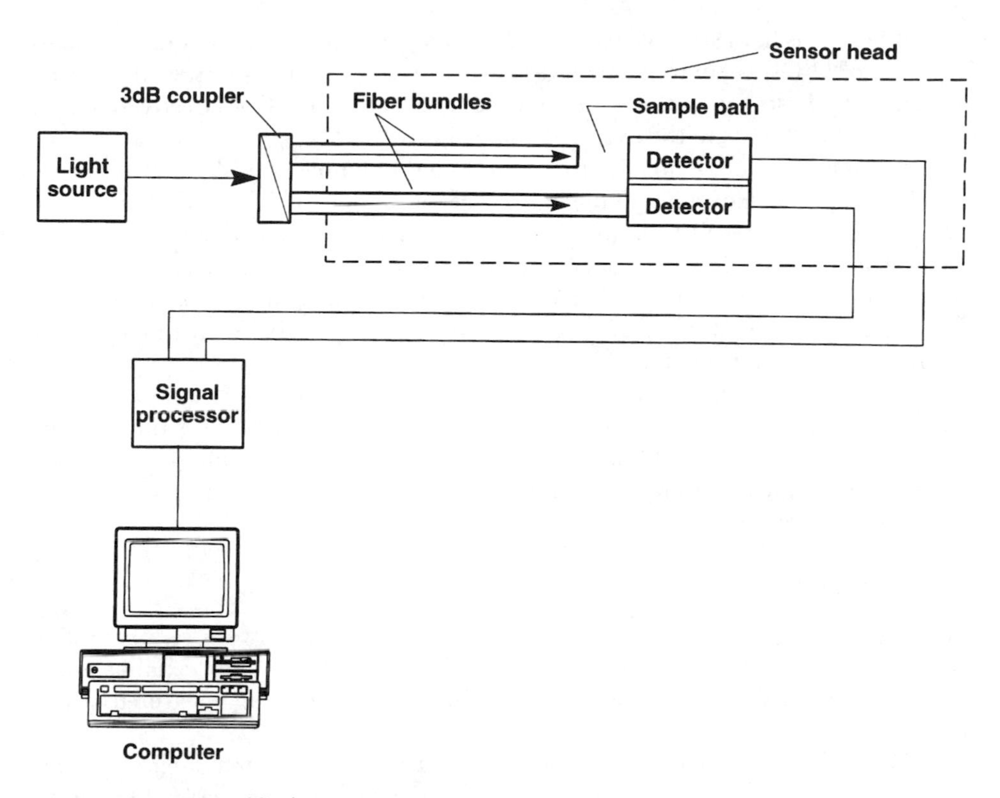

Figure 13.20 Diagram of fluid opacity monitor.

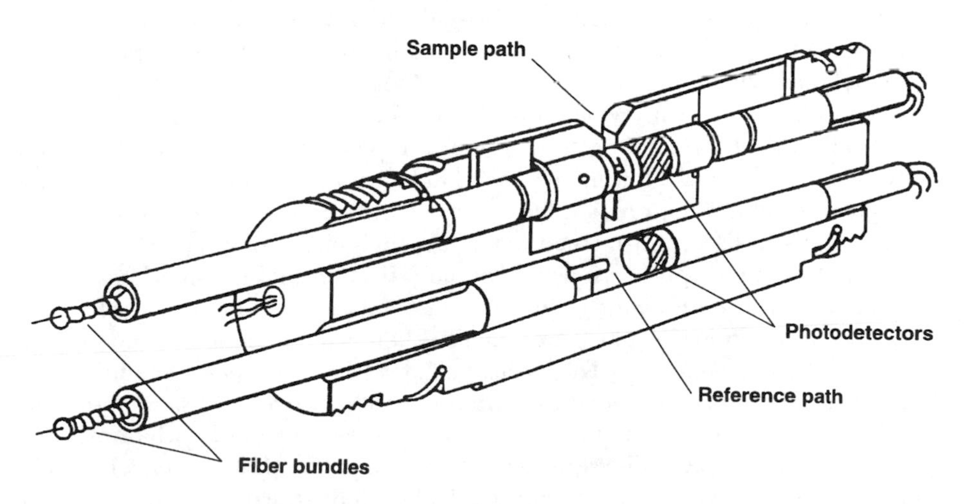

Figure 13.21 Cutaway view of sensor head assembly for opacity sensor.

When air is in the sample pathway, I_{ref} and I_{sig} are substantially equal and so the first term of (13.3) is approximately zero. An arbitrary offset is subtracted by means of a trimpot adjustment to bring the net result to exactly zero. This effectively sets the opacity reading for air to the expected value of zero.

When an opaque sample is introduced into the sample optical pathway, I_{sig} will drop by a value proportional to the opacity of the sample. If I_{sig} falls by an order of magnitude, the net result from (13.3) is an opacity unit of 1. Therefore, the opacity unit is defined as the difference (in orders of magnitude) of photocurrent between air (or some other chosen calibration fluid used as the reference) and the opaque sample. The FOM is capable of measuring opacity over a range of six (logarithmic) optical density units referenced to air.

Neutral-density filters (not shown) with nominal values of 1, 2, 3, and 4 that represent, in order, steps from 1→4 orders of magnitude reduction in light transmission may be used to verify proper operation. The neutral-density filters serve a dual purpose: first, they are useful as a calibration check, and second, as a standard for adjusting the zero offset.

13.6.4 Summary

The FOM satisfies a process industry need for a real-time, in-process sensor for fluid opacity that can operate repeatably over time, is easily maintainable, exhibits a wide measurement range, and permits a variety of opacity ranges to be measured using a single device.

The FOM constructed was capable of measuring opacity over a range of six units (decades) referenced to the opacity of air (zero-opacity units). An assembled unit is shown in Figure 13.22. Optical fibers were used to transmit light to the optical-sensing path.

The opacity sensor is configurable for spans of 1, 2, 4, or 6 opacity units (corresponding to 1,2, 4, or 6 decades of transmitted intensity) and 0 to 4 opacity units of offset by simple switch settings. This allows the same sensor to be used at many different processing points with widely variant nominal opacities. Calibration requires only zero adjustment (rather than normal zero and span adjustments). A value of zero opacity is assigned to air, allowing the sensor to be calibrated simply by purging the sample path of fluid. The provision for an adjustable zero offset permits compensation for changes in light transmission due to all envisioned error sources, including deposits or abrasion of optical windows, without affecting calibration or sensitivity.

Opacity sensor optics and signal processing are designed to be self-referenced with respect to variations in light source intensity, ambient temperature, and the condition of cables and connectors. The mechanical design of the probe simplifies sensor maintenance. The probe is shaped to permit a representative fluid sample to enter the optical path, while limiting the accumulation of deposits on the windows. Also, single-ended probe design allows the sample region of the probe to be withdrawn from the process pipe without compromising the fluid/pressure boundary. This design

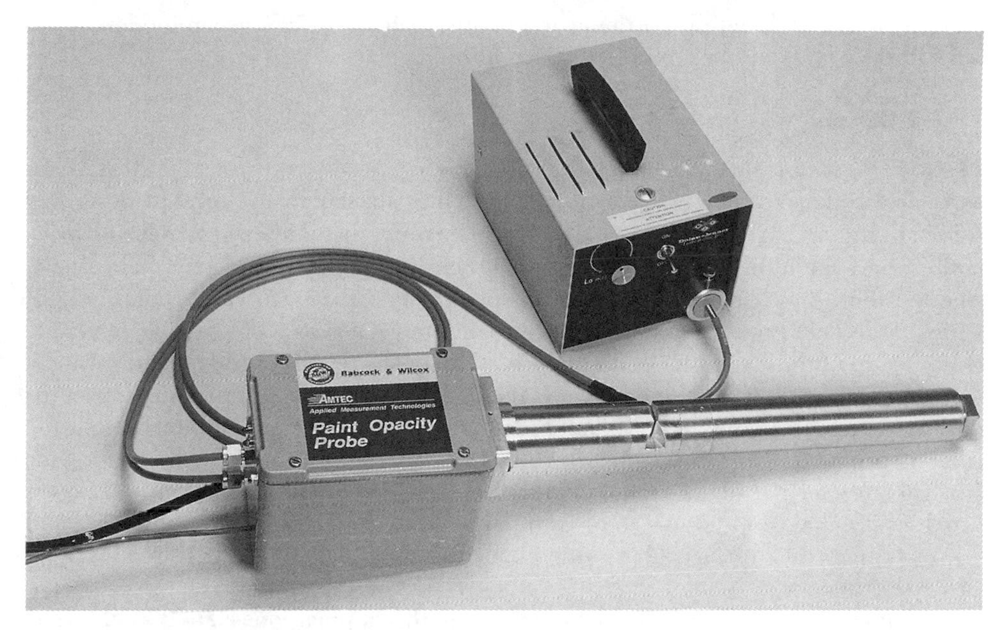

Figure 13.22 Photograph of fluid opacity monitor.

feature permits cleaning and offset adjustment without interrupting the manufacturing process.

13.7 AIRCRAFT TOTAL PRESSURE MEASUREMENT

13.7.1 Application

The National Aeronautics and Space Administration (NASA) Lewis Research Center has funded a program to flight test advanced fiber-optic sensor hardware in order to assess the usefulness of this equipment for future flight control systems. One device, which was built and flight-tested on the NASA program, was a temperature-compensated, self-referenced, fiber-optic pressure transducer for total pressure measurement. This transducer measures the sum of altitude-dependent background pressure and dynamic pressure from a Pitot tube, which is proportional to air speed. During flight tests on an FA-18 aircraft, the fiber-optic transducer and electronic reference transducer in the air-data computer were installed in the same pressure line from the Pitot tube. Simultaneous data from the two transducers was collected for comparison. The fiber-optic total pressure sensor (FOTPS) was located next to the electro-optic architecture (EOA) unit, which optically excites the FOTPS and receives an optical power signal proportional to pressure. The EOA can accommodate up to ten different types of fiber-optic sensors at a time.

13.7.2 Design

The FOTPS is similar in design to other fiber-optic pressure transducers [20–22]. An exploded and assembled cross-sectional drawing, similar to the present design, is shown in Figure 13.23. These transducers use the microbend fiber-optic sensing principle, where a multimode step index optical fiber is squeezed between a pair of corrugated tooth-blocks. One tooth-block is positioned at the center of a diaphragm, which deflects in proportion to the pressure difference across it. The deflection causes a closure of the tooth blocks and hence a change in the amplitude of the periodic distortion of the fiber squeezed between them. The sensor configuration is conceptually illustrated in the inset in Figure 13.23. Optical power transmitted through the fiber is attenuated in proportion to the amplitude of the mechanical distortions (microbends) of the fiber. Thus, measurement of the change in transmitted power provides a signal proportional to pressure applied to the diaphragm. The microbend sensor is an analog intrinsic fiber-optic intensity sensor, which is well-suited for pressure measurement.

The desired pressure measurement range in this application is 0 to 276 kPa. The need for absolute pressure reference requires that the region on the reverse side of the diaphragm (which contains the fiber and corrugations shown in Figure 13.23) should be evacuated and sealed. The diaphragm is sealed around its circumference by an electron-beam weld to the cap and body. The holes where the fiber enters the body are more simply sealed with vacuum epoxy. Finally, a plug is electron-beam welded to the body to hold a getter in place. The getter material absorbs outgassed and in-diffused gases such as hydrogen for extended time periods to maintain zero-pressure vacuum reference in the space behind the diaphragm where the microbend sensor is located.

13.7.3 Signal Conditioning

The electro-optic interface used for pressure transducer calibration is based on wavelength division multiplexing of optical signals [22,23]. It is similar to the EOA, but can only accommodate one sensor. The EOA, on the other hand, can accommodate numerous analog optical-intensity sensors, and digital optical-position sensors. Figure 13.24 illustrates how the pressure transducer is configured for wavelength self-referencing in both the calibration mode and in the operating mode with the EOA. Optical power is provided over a broad wavelength band at the sensor input. A wavelength demultiplexer (WDM) in the transducer head divides the input into two separate wavelength channels centered at λ_1 and λ_2. The channel centered at λ_1 is intensity-modulated by the microbend pressure sensor and the channel centered at λ_2 acts as an intensity and/or temperature reference. The device relies on the fact that

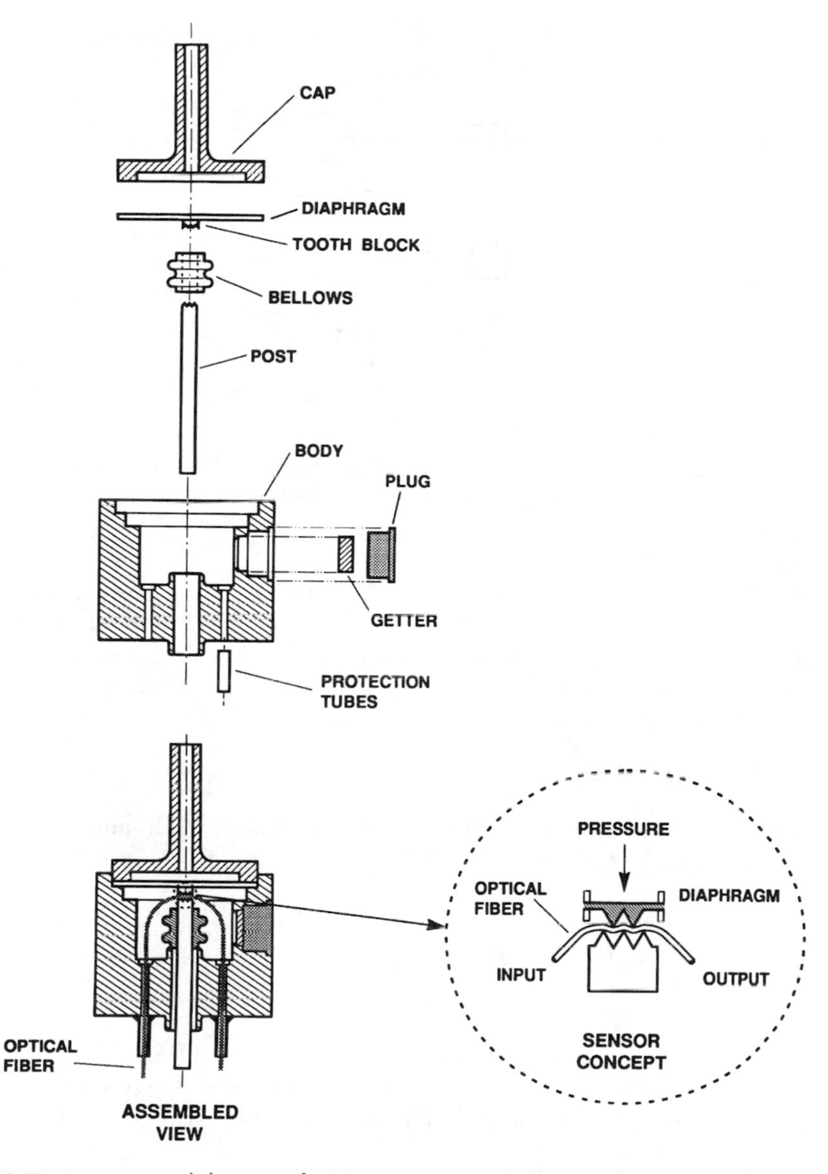

Figure 13.23 Cross-sectional diagram of FOTPS. (*Source:* Berthold, J. W., et al., *Proc. SPIE*, Vol. 2295, 1994, pp. 216–222. Reprinted with permission.)

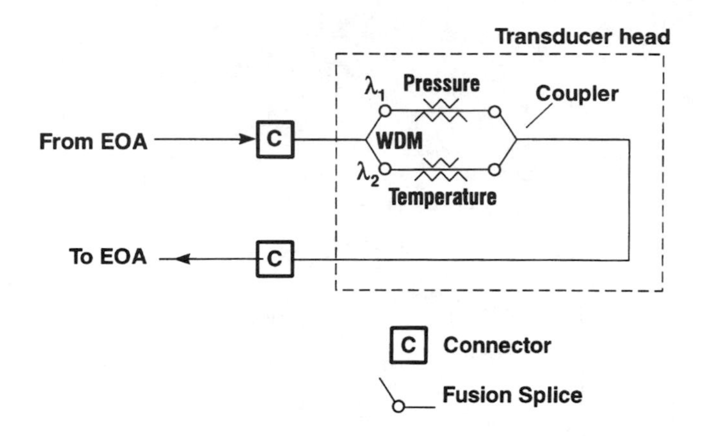

Figure 13.24 Configuration of FOTPS for wavelength self-referencing. (*Source:* Berthold, J. W., et al., *Proc. SPIE*, Vol. 2295, 1994, pp. 216–222. Reprinted with permission.)

light intensity in both channels is modulated in a similar manner by cable bends and connector losses. At the output, the two wavelength channels are recombined and the light signals return to the EOA receiver, where they are again wavelength demultiplexed, ratioed to correct for source brightness variations, and processed to extract a signal proportional to pressure.

13.7.4 Calibration and Test

The calibration results are presented in Figures 13.25 and 13.26 and show that the inherent repeatability (two sigma) is ±0.1% of range (range = 276 kPa = 40 psia = 81.5 inch Hg).

A calibration check was performed before flight tests when the FOTPS was integrated with the EOA. At that time, the following system characteristics were determined:

- The EOA adds additional nonlinearity that must be calibrated out.
- Standard deviation (one sigma repeatability) of the entire system (sensor and EOA) was measured to be 2.8 kPa (0.83 inch Hg), which is ±1% of range (range = 276 kPa).
- The calibration holds if the signal and reference optical power at λ_1 and λ_2 are held constant.

Several flight tests were performed in October through December 1993. A 27-sec snapshot of results from one of the flight tests on the FOTPS is shown in Figure 13.27. The raw data received from NASA were smoothed, using an alpha beta

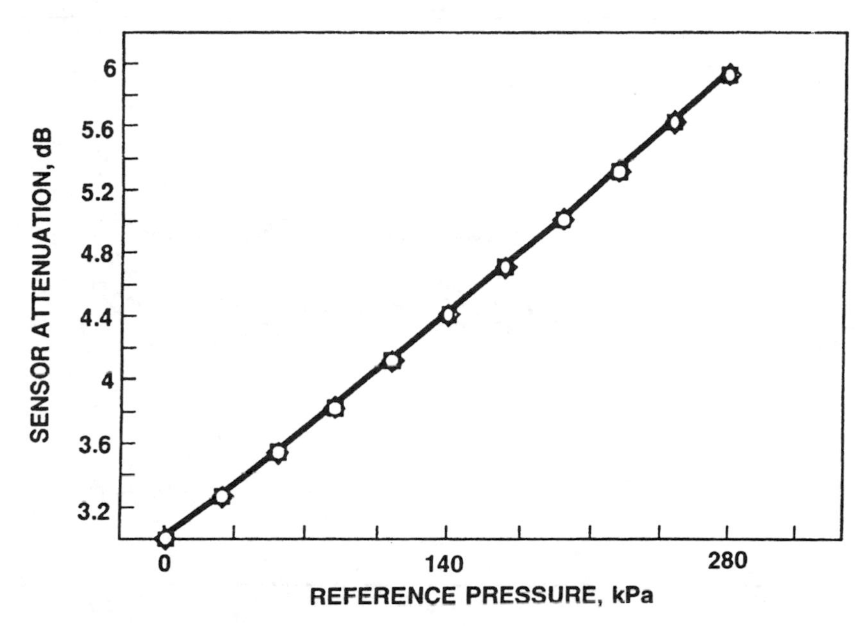

Figure 13.25 Fiber-optic total pressure transducer output vs. pressure measured during lab calibration. (*Source:* Berthold, J. W., et al., *Proc. SPIE*, Vol. 2295, 1994, pp. 216–222. Reprinted with permission.)

filter [24], and the smoothed data are presented in Figure 13.27. The data show that there are differences in offset and response time constant between the fiber-optic pressure transducer and reference pressure transducer in the air-data computer. The data also show that generally the two transducers track pressure changes together.

Subsequent analysis was performed on the entire data set from the early flight tests and the results for flight number 0524 are shown in Figures 13.28 and 13.29. Figure 13.28 is a plot of pressure and pressure difference versus time during the entire 3,300 sec (~1 hour) flight test. One hundred readings per second were recorded and stored by the data acquisition system on the aircraft. Note that the outputs from the reference pressure transducer in the air-data computer and the FOTPS are plotted and overlay almost exactly. The FOTPS data has been corrected for temperature changes that occurred in the compartment where the FOTPS was located.

Shown in Figure 13.29 is a scatter histogram of the pressure difference between readings from the FOTPS and reference transducer in the air-data computer. The full width at half-maximum of the error distribution is about ±3 kPa, which is the total combined error of both transducers. This result compares well with the ±2.8 kPa standard deviation measured during the calibration check of the FOTPS when integrated with the EOA.

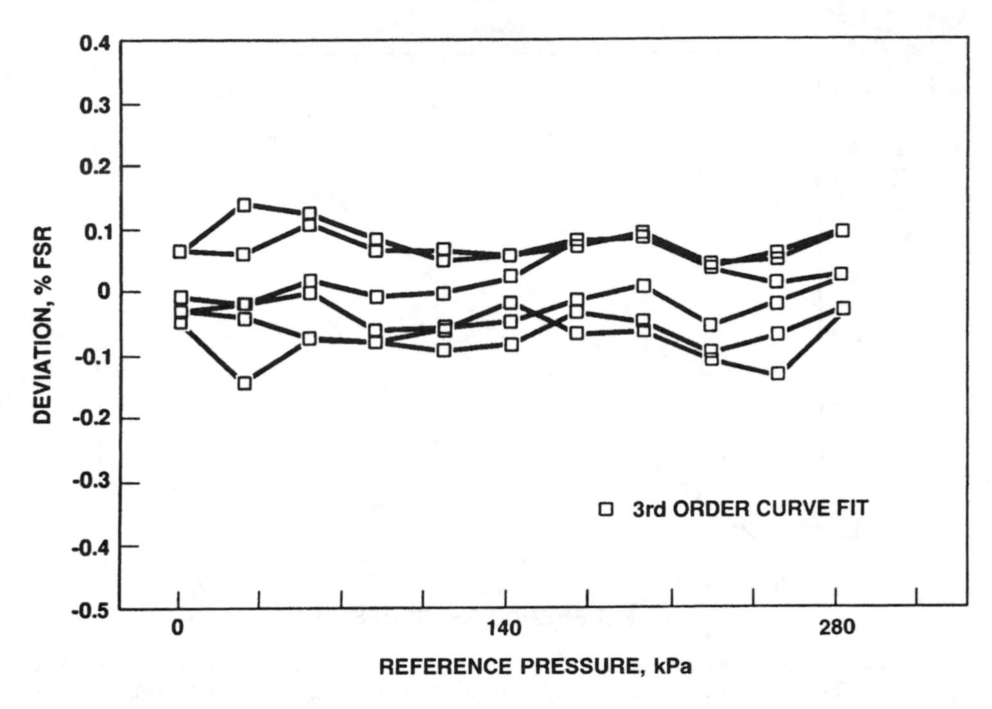

Figure 13.26 Deviation vs. pressure for fiber-optic total pressure transducer obtained from calibration data. (*Source:* Berthold, J. W., et al., *Proc. SPIE,* Vol. 2295, 1994, pp. 216–222. Reprinted with permission.)

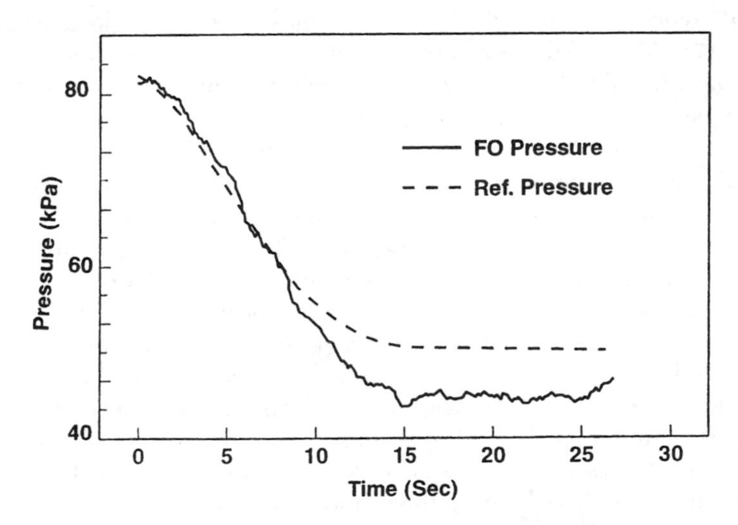

Figure 13.27 Measured pressure vs. time for fiber pressure sensor during loop maneuver—flight 0524. (*Source:* Berthold, J. W., et al., *Proc. SPIE,* Vol. 2295, 1994, pp. 216–222. Reprinted with permission.)

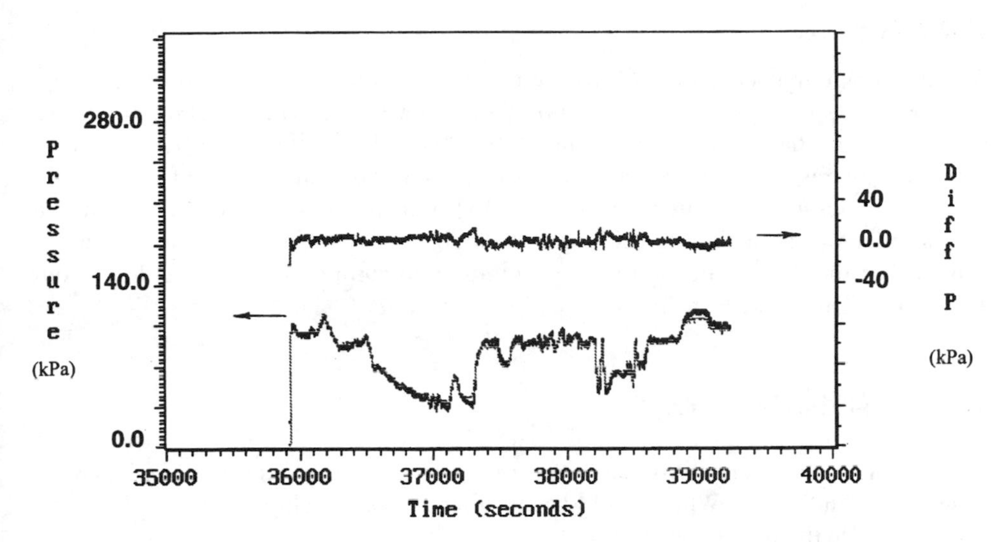

Figure 13.28 Plot of pressure and pressure difference vs. time during flight test 0524. After linear temperature correction of the FOTPS data, the outputs from the air-data computer and FOTPS are both plotted and overlay almost exactlly.

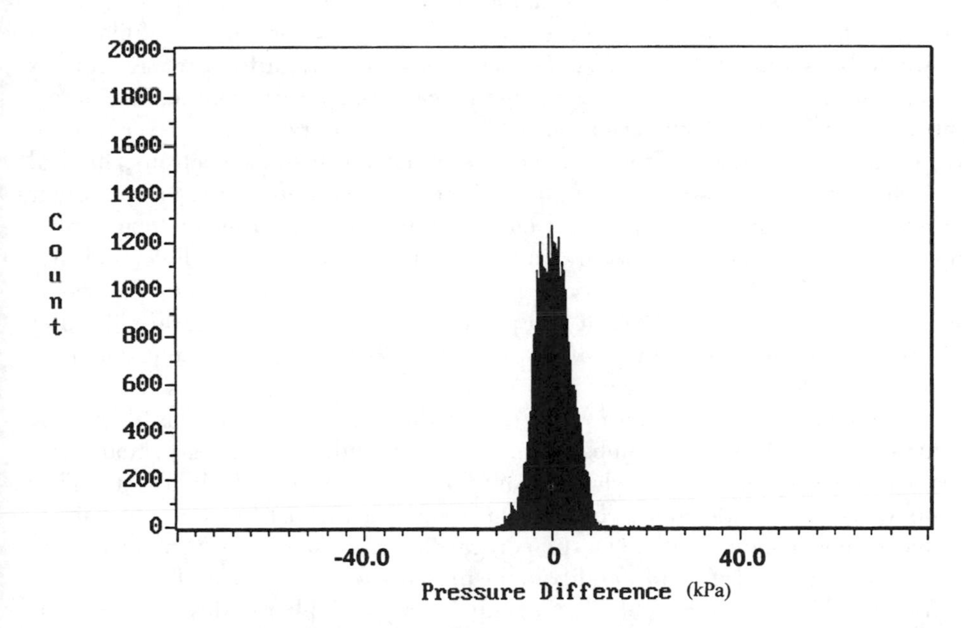

Figure 13.29 Error distribution vs. pressure difference for all data recorded by the air-data computer and FOTPS during flight 0524.

13.7.5 Summary

Results from a first series of flight tests on a FOTPS indicate reasonably good agreement with a reference pressure measurement from the aircraft air-data computer using an established sensor. Calibration of the FOTPS by itself in the laboratory indicated an inherent repeatability in the ±0.1% range, and calibration of the FOTPS with excitation and readout electronics (EOA) used for flight tests demonstrated a one-sigma repeatability in the ±1% of full range. Analysis of the flight test data has shown that the data from the FOTPS and air-data computer also agreed to about ±1%, in agreement with the preflight calibration.

13.8 KEY APPLICATION ISSUES

In this section, fiber sensors are discussed from a marketing perspective, and answers posed to the question, "What should we, as fiber-sensor developers, be doing to help commercialize the technology?"

13.8.1 The Customer First

The theme is that the successful application of fiber sensors will usually be customer-driven. In the early years, we thought that fiber sensors would displace existing technologies such as electrical strain gauges and that the commercial success of fiber sensors would be technology-driven. This scenario has, other than for a few exceptions such as distributed temperature sensing, not come to pass. For example, the success of fiber gyros is not based on performance alone, but rather performance versus cost. Gyro customers can choose from lower cost alternatives such as electromechanical and laser gyros. In this sense, fiber sensors are not technology-driven but rather market-driven, because customers have choices from other sensor technologies. Fiber sensors are an alternative technology and are in demand only where fiber is clearly the best choice for the job, both from a performance and a price standpoint. Some areas where fiber sensors have a unique performance advantage are in fiber hydrophones, distributed temperature sensing, and in a few selected chemical (optrode) sensors for specific chemicals.

Customer thinking has also evolved over the years regarding the fiber-optic products in general. A study published in 1991 [25] indicated that a primary customer concern was a lack of knowledge about fiber optics in general. High cost of fiber sensors was then considered a secondary concern. Over the last five years, the situation has changed significantly. Questions regarding fiber-sensor performance (e.g., accuracy, resolution, drift, and stability in the presence of environmental changes) are less frequent. Many customers have gained familiarity with fiber optics and fiber sensors through telecommunications and local area networks, and many have fiber-optic equipment installed and operating in production plants. They assume (correctly or not) that, as in the telecommunications industry, performance from fiber sensors will meet commercial specifications. Even though the scientific and engineering community realize that many technical challenges remain for fiber sensors, the major customer concern today is perceived to be their high cost.

It is important to distinguish between customer needs and desires. For example, when a customer asks, "Are fiber sensors better than those I am using now?," he or she is really asking a performance-related question. The answer obviously requires a response that compares fiber-sensor performance to existing technology. However, fiber sensors are an alternative technology and must have features that customers want, such as less frequent or autocalibration, small size, and a track record with proof of benefits. Even though we, the technology developers, try to concentrate on these desires, we must remember at the same time that customers really need performance characteristics, such as safety, reliability, repeatability, and low drift. Designing for these needs must come first.

Most customers do not want to be the first to buy any "risky" new technology; they want to know case histories and have proof of benefits from fiber sensors in their specific (niche) application. Fortunately, however, the world is full of niche applications, and maybe as much as 90% of the total sensors market is in niches, which is why so many small sensor companies thrive.

Cost is a multifaceted issue, but customers are now sufficiently educated to be generally concerned about the right issue-not the basic equipment price, but the total installed cost. This is where fiber sensors can have a big advantage. In the petrochemical industry, for example, where explosion hazards exist, fiber-optic installations do not require conduit, explosion-proof containers, or ground-fault protectors. These items represent a large potential savings in installed cost, which customers understand.

Customers do need proof of fiber-sensor benefits. Customers should expect fiber sensors to be safe and reliable. They should be able to "plug and play"—hook up fiber sensors to a data bus or signal line and operate them, just like they do with other instrumentation. In other words, whether or not instruments contain fiber sensors should be irrelevant to the user, because the instrument operation is totally transparent to the fiber optics contained inside. The customer should not need to do any extra work or spend additional money for such capability.

13.8.2 Reliability and Ease of Use

Important reliability issues that we must address for instruments containing fiber sensors include mean time to failure, how fiber sensors age and degrade with time (especially in hostile environments), how the sensors fail, whether or not they are fail-safe, how redundancy can help improve reliability, whether we should design for repair or replacement, and, when multiple units are sold to the customer, whether they should buy spare parts and learn to repair units themselves.

Instruments and equipment containing fiber sensors should be capable of communicating with the outside world. Units should, for example, be designed to be compatible with parallel/serial digital interfaces or the new field-bus architecture being introduced into the process control industry. All the engineering issues such as these must be addressed by fiber-sensor developers.

13.8.3 Cost

Consideration of cost issues for fiber sensors is not only very important, but also complex. Fiber sensors are often mechanically simple and easy to assemble and package, with fiber-optic delay coils often being the most mechanically complex component in a sensor system. The mechanical simplicity offers cost advantages compared to alternative technologies. However, this advantage is usually more than offset by the relatively high cost of electronic signal processing, which can often be complex. In many present applications, printed circuit board electronics assembly with low-level integration is used because of low product volume.

Since fiber sensors have not yet been accepted into the market place for high-volume applications, a good question to ask now is which fiber sensor technology has the best potential for low manufactured cost? From a manufacturing standpoint, the fiber-optic sensor winners will be those where the signal-processing functions can be provided with the most cost-effective digital signal processor and application-specific integrated circuit (ASIC) designs for high-volume applications. However, fiber-sensor technology is probably still several years away from this low-cost, high-volume breakthrough. Manufacturers will want to be certain that they have in place the right fiber-sensor technology market knowledge and strategy before investing in specialty ASIC design.

In addition to the direct costs for fiber-sensor hardware and their installation and spare parts, indirect costs are very important also. The indirect costs include maintenance, component sharing, and insurance savings. Many customers may discover they can get significant reductions in these indirect costs when using fiber-sensor hardware instead of conventional hardware.

13.8.4 Codes and Standards

The importance of codes and standards on the developing markets for fiber sensors cannot be overlooked. For example, before fiber sensors can be used on aircraft, they will need to undergo an extensive qualification program. The first step in this process was the NASA FOCSI program and the next step is the present ARPA FLASH program. ISO 9000 compatibility will be very important. Key organizations in codes and standards development are ASTM, NRC, API, ANSI (EIA), SAE, FDA, IEEE, ISA, and ASME.

To establish a new ASTM standard takes an average of two years, and many customers refuse to buy instrumentations for which standards do not exist. Before new standards can be written, new technologies must undergo a period of qualification, validation, and verification. In addition, standards preparation often takes an intense lobbying and promotion effort from engineers sitting on the subcommittees of the codes and standards organizations listed above. In the case of fiber-optic sensor hardware, the groundwork for this process is only just beginning.

13.8.5 Future Directions and Opportunities

Much more applications engineering is needed to address and follow through on the issues outlined in this section. Much of what is needed for fiber-sensor devices and measurement instruments has already been mentioned above (i.e., improve reliability, increase intervals between calibration, work on simple design and low parts count, design for manufacturability, meet international quality standards, and produce comparative performance data between fiber-optic sensors and the equivalent conventional hardware). We, as researchers, developers, and marketers, need to do this work. Customers will only buy when fiber sensors have these characteristics. The days of large "blue-sky" fiber-sensor research and development programs may be over. Customers now wish to see commercial products as a positive outcome from all of the earlier development expenditures.

13.9 SUMMARY

In this chapter, I have tried to provide a sense of breadth of possible applications for fiber-optic sensors, and the flexibility and adaptability of the technology. I have identified some of the key marketing and engineering issues that must be addressed before fiber-sensor technology is widely accepted and commercialized.

It is important to understand from an historical perspective that fiber-sensor technology is still relatively young when compared to the conventional sensor technologies such as pneumatics, electromagnetics, and electronics. Although major breakthroughs such as the Bragg grating sensor are perhaps not as likely as they were ten years ago, there remains much practical engineering work to be done to make incremental improvements to the many types of fiber sensors now in use. Attention to good design and engineering detail is essential if commercial success is to be achieved.

References

[1] Morey, W. W., et al., "Multiplexing fiber Bragg grating sensors," *Proc. SPIE*, 1586, 1991, pp. 216–220.

[2] Griffiths, R. W., and Nelson, G. W., "Recent and current developments in distributed fiber-optic sensing for structural monitoring," *Proc. SPIE*, 985, 1988, pp. 69–76.

[3] Tran, T. A., et al., "Extrinsic Fabry-Perot fiber-optic sensor for surface acoustic wave detection," *Proc. SPIE*, 1584, 1991, pp. 178–186.

[4] Lee, C. E., et al., "Optical fiber Fabry-Perot embedded sensor," *Optics Letters*, 14, 1989, p. 1225.
[5] Fields, J. N., and Cole, J. H., "Fiber microbend acoustic sensor," *Applied Optics*, 19, 1980, pp. 3265–3267.
[6] Berthold, J. W., and Reed, S. E., "Microbend fiber-optic strain gauge," U.S. Patent No. 5,020,379.
[7] Xu, M. G., et al., "Discrimination between strain and temperature effects using dual-wavelength fibre grating sensors," *Electronics Letters*, 30, 1994, pp. 1085–1087.
[8] Jeffers, L. A., "Multichannel fiber-optic temperature monitor," *Proc. SPIE*, 2072, 1993, pp. 53–62.
[9] Harms, A. A., and Laratta, F.A.R., "The dynamic-bias in radiation interrogation of two-phase flow," *Int. J. Heat Mass Transfer*, 16, 1973, pp. 1459–1464.
[10] Laratta, F. A. R., and Harms, A. A., "A reduced formula for the dynamic-bias in radiation interrogation of two-phase flow," *Int. J. Heat Mass Transfer*, 17, 1974, p. 464.
[11] Berthold, J. W., et al., "Fibre optic sensor systems for void fraction measurement in aqueous two-phase fluids," *Flow Measurement Instrumentation*, 5, 1994, pp. 3–13.
[12] Henderson-Kinney, A., and Kenny, J. E., "Spectroscopy in the field," *Spectroscopy*, 10, (7), 1995, pp. 32–38.
[13] Lieberman, S. H., et al., "Subsurface screening of petroleum hydrocarbons in solids, via laser induced fluorometry over optical fibers, with a cone penetrometer system," *Proc. SPIE*, 1716, 1992.
[14] Jeffers, L. A., and Berthold, J. W., "Remote monitoring of corrosion chemicals via fiber-optic Raman spectroscopy," *Proc. NACE 12th Int. Corrosion Congress*, Houston, Texas, 1993, p. 4313.
[15] Leugers, M. A., and McLachland, R. D., "Remote analysis by fiber-optic Raman spectroscopy," *Proc. SPIE*, 990, 1988, pp. 88–95.
[16] Williams, K. P. J., and Mason, S. M., "Fourier-transform Raman spectroscopy in an industrial environment," *Trends in Analytical Chemistry*, 9, (4), 1990, pp. 119–127.
[17] Nakamota, K., *Infrared Spectra of Inorganic and Coordination Compounds*, London, Wiley-Interscience, 1970.
[18] Thierry, D., et al., "In-situ Raman spectroscopy, combined with X-ray photoelectron spectroscopy and nuclear microanalysis, for studies of anodic corrosion film formation on Fe-Cr single crystals," *J. Electrochem Soc.*, 135, 1988, pp. 305–310.
[19] Ohtsuka, T., et al., "Raman spectroscopy of thin corrosion films on iron at 100 to 150C in air," *Corrosion-NACE 42*, (8), 1986, pp. 476–481.
[20] Berthold, J. W., et al., "Design and characterization of a high-temperature fiber-optic pressure transducer," *J. Lightwave Tech.*, LT-5, 1987, pp. 870–876.
[21] Berthold, J. W., "Field test results on fiber-optic pressure transmitter system," *Proc. SPIE*, 1584, 1991, pp. 39–47.
[22] Reed, S. E., et al., "Fiber-optic total pressure transducer for aircraft applications," *Proc. SPIE*, 2070, 1993, pp. 17–23.
[23] Berthold, J. W., et al., "Flight test results from FOCSI fiber-optic total pressure transducer," *Proc. SPIE*, 2295, 1994, pp. 216–222.
[24] Penoyer, R., "The alpha-beta filter," *The C Users Journal*, July 1993, p. 73.
[25] Vitale, A., "Fiber sensors: tech transfer needed," *Photonics Spectra*, Jan. 1991, p. 126.

Chapter 14

Distributed Sensors: Recent Developments

Tsuneo Horiguchi
NTT, Japan

Alan Rogers,
King's College, England

W. Craig Michie, George Stewart,
Brian Culshaw
University of Strathclyde, Scotland

14.1 INTRODUCTION

The capability for long-range distributed sensing is unique to optical-fiber technology: a distributed sensor returns a value of the measurand of interest as a function of linear position along the fiber length. The only contact between the point to be measured and the observation area is the optical fiber. The read-out is usually via some form or other of time-domain reflectometer (Figure 14.1) from which the signal is processed to produce a value of the parameter of interest as a function of linear position. The time-domain reflectometer gives a value of the intensity of the returned signal as a function of time unless the signal is mixed with a local oscillator, thereby enabling it to derive phase and/or frequency information.

The signal that is returned to the OTDR can be at the same frequency (a linear system) as the interrogating light or at a different frequency (nonlinear systems). The former have the advantage of simplicity but restrict flexibility since they are only capable of detecting modulations in returned intensity. The latter can be extremely complex, but offer substantial flexibility in implementation. The first generation of

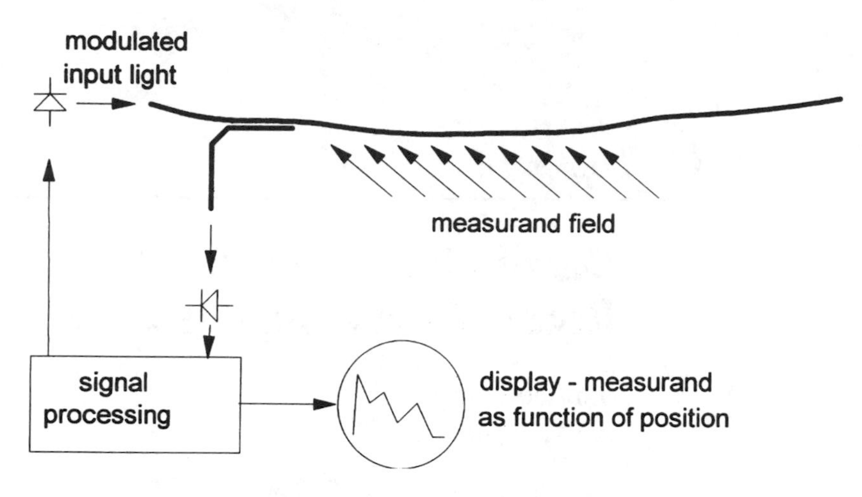

Figure 14.1 The essential elements of a distributed optical-fiber sensor. The detector can operate at the interrogating optical frequency or otherwise for linear and nonlinear systems, respectively.

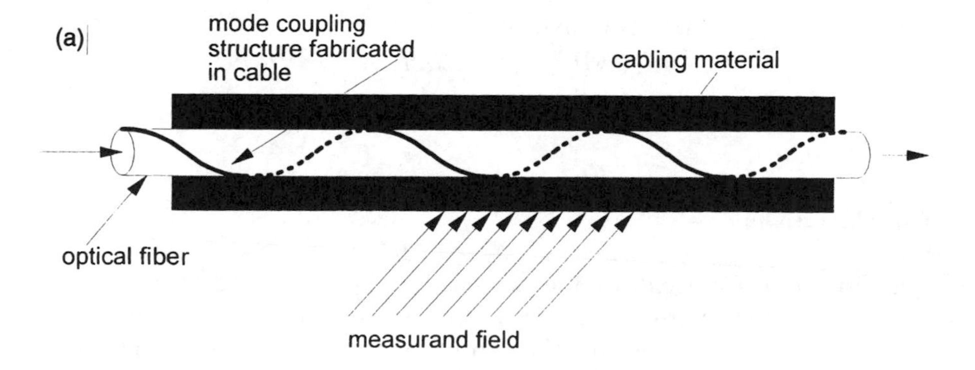

Figure 14.2 (a) General principles of microbend cables for sensing and (b) representative OTDR trace for microbend sensor (high, low, etc. refer to value of imposed measurand).

practical distributed optical-fiber sensors uses amplitude information to encode the measurand, but in both linear and nonlinear modes.

Microbend sensors (Figure 14.2) are useful loss modulation devices. The basic transduction mechanism is the depth of the microbend introduced by the measurand via some form of helical constraining member. The sensors are extremely simple—a great benefit-and using appropriate microbend to measurand transformers can

(b)

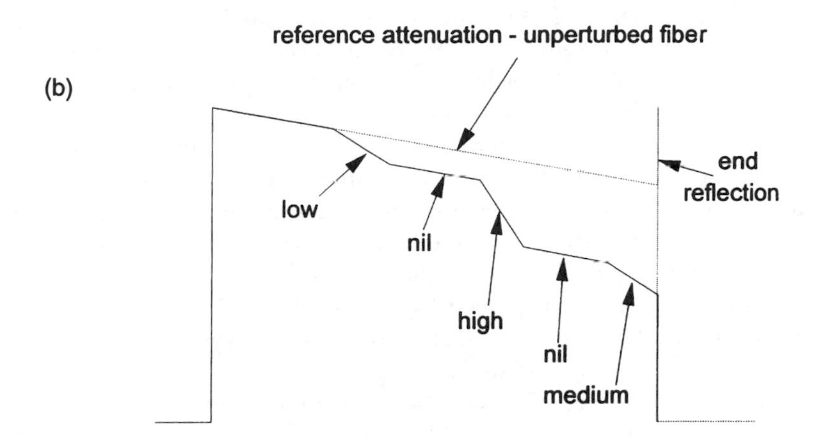

Figure 14.2 (continued)

address a surprising range of parameters. Currently available commercial systems are capable of measuring strain and strain distribution and of detecting hot spots. Future generations of microbend systems are featured in Section 14.4 and include a range of simple and effective chemical measuring systems. Evanescent wave absorption spectroscopy is also a distributed loss-inducing mechanism and through appropriate addressing techniques can be configured as a useful distributed sensor. The potential offered by these systems is also briefly mentioned in Section 14.4. Loss-based distributed sensors do, however, have a very fundamental limitation. Since the measurand does induce loss, then the power extracted by the imposition of the measurand at a particular point automatically implies that less power is available to measure at points beyond this lossy region. Further, the loss mechanism itself usually introduces some loss even in the absence of a measurand, so the systems tend to be limited in range. With typical commercial OTDR interrogation units, with dynamic ranges of 20 to 30 dB and resolutions in the region of 0.01 dB, loss-based distributed sensors can operate at ranges of up to a few kilometers with up to 1,000 or thereabouts effective interrogated sections along the fiber. Analog measurements to 10% accuracy or thereabouts are, in principle, feasible, but the concept is usually best suited to operation as a distributed alarm, especially since the sensitivity is a function of position due to increasing loss along the fiber and is also a function of the actual measurand field applied.

Nonlinear measurements are fundamentally different in nature. The loss in the probing signal is due to conversion to different optical frequencies, and this conversion process is—very broadly speaking—independent of the applied measurand (with the notable exception of some Brillouin scatter processes), so that the measurand field does not automatically influence the measurement sensitivity. The Raman distributed temperature sensor was described in Volume Two, and is the most basic of the nonlinear sensing mechanisms. The temperature-dependence of the ratio of the powers in the upper and lower (anti-Stokes and Stokes) sidebands of the Raman scatter signal is the basic physical property of the phenomena and provides a temperature-sensing system with ranges of tens of kilometers and resolutions on the order of degrees centigrade in ranges on the order of a meter. Such performance would be unthinkable in any loss-inducing transduction mechanism. The Brillouin scattering process promises similar achievements for strain and/or temperature measurements. In Section 14.2, Tsuneo Horiguchi describes the impressive achievements for the distributed strain measuring system capable of monitoring strain in optical-fiber cables in excess of 100 km in length with resolution of tens of microstrain over interrogation lengths of several meters. Of course, these nonlinear systems all require sensitive detection at optical wavelengths displaced from the original interrogation wavelength. In the case of stimulated Brillouin scatter, the detection system must measure the frequency difference between the interrogating light and the backscattered light to within fractions of a megahertz. The systems are therefore quite complex and can be expensive. However, nonlinear phenomena do offer relatively loss free and stable measurement systems, capable of operating over distances that cannot be even approached by other technologies. The Brillouin and Raman scatter processes (Figure 14.3) are essentially similar, corresponding to interactions between the interrogating light and optical and acoustic phonons, respectively.

The optical Kerr effect is another potentially useful nonlinear phenomena. Kerr

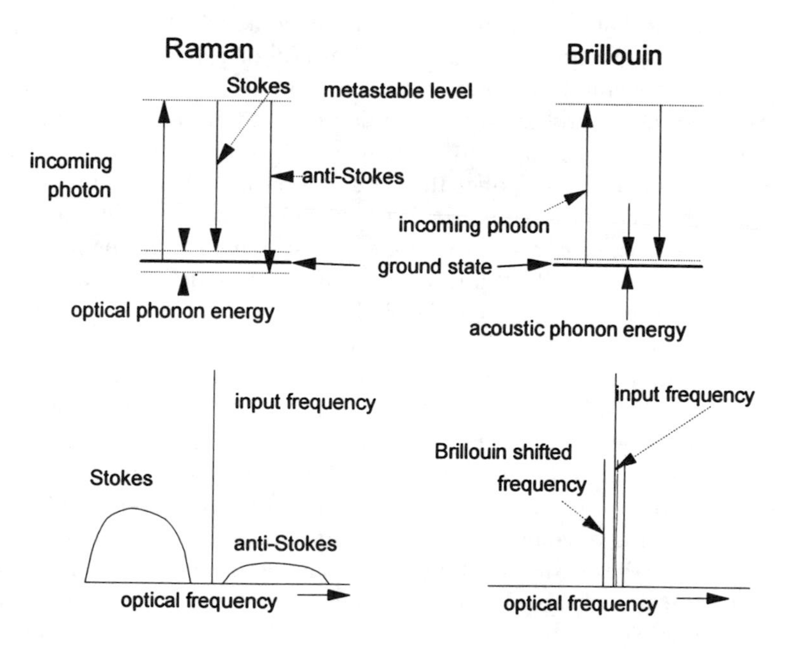

Figure 14.3 Raman and Brillouin scattering processes (these are exploited in some forms of optical-fiber sensors).

nonlinearities are observed in all transmission media, and even though silica has relatively weak Kerr coefficients, the extremely long interaction length that is possible in silica means that significant Kerr phenomena are frequently observed in optical-fiber systems. Some even argue that silica is the most nonlinear of optical media, due to its very low loss and, therefore, potentially very long interaction length. In optical-fiber gyroscopes, the Kerr effect introduces distinct nonreciprocal, nonlinearities that are compensated by the use of incoherent sources. In optical-fiber transmission systems, the combination of Kerr nonlinearities and anomalous dispersion explain the propagation of narrow pulse solitons over extremely long distances. In distributed sensing, the Kerr effect offers a number of interesting possibilities, which are described in Section 14.3 by Alan Rogers. The Kerr coefficient is itself a function of temperature and strain, so that the level of any nonlinear interaction can be related to these external physical parameters. However, it is as a nonlinear mixing process that the Kerr effect probably offers the most potential, where frequency offset pump and probe signals are coupled through measurand-modulated mechanisms, which result in a measurand-dependent frequency difference. The physics of these processes is interesting, but the viability as a measurement technique in a practical situation has yet to be proven.

There are other mechanisms, too. Variations in Rayleigh scatter can be used as a simple indicator of temperature. Fluorescent dyes can be introduced into the cladding of optical fibers. Porous systems can be used to measure the distribution of liquids or gases. However, we feel that we have, in the following pages, covered the principal emerging approaches to distributed sensing and presented an authoritative account of the underlying mechanisms by well-known experts in the field. Of the evolving optical-fiber sensing techniques, distributed measurements arguably offer the most potential and, with a few very specialized exceptions, are the least explored. There will no doubt be substantial contributions to come from the ideas described here and from techniques that are still to be discovered.

14.2 BRILLOUIN SCATTERING FOR MEASURING STRAIN AND TEMPERATURE

14.2.1 Introduction

There are three important types of scattering in optical fibers that can be exploited in distributed sensing. They are Rayleigh, Raman, and Brillouin scattering. Rayleigh scattering is elastically scattered light (i.e., it is at the same wavelength as the incident light) and accounts for most of the backscattered light. Raman and Brillouin scattering is light that is inelastically scattered by interaction with optical and acoustic phonons, respectively. In terms of power they are about 30-dB and 20-dB weaker, respectively, than Rayleigh scattering in silica fibers. Rayleigh scattering in liquid core fibers and Raman scattering in conventional silica fibers show large variations in their scattering coefficients with temperature, and have been advantageously used in distributed thermometry [1]. Raman-distributed thermometry systems, in particular, have been marketed by several manufacturers because highly reliable and relatively cheap silica communications fibers can be used as the sensing fibers. However, until comparatively recently there had been little research on distributed sensors using Brillouin scattering. In 1989, it was reported that the Brillouin frequency shift in silica fibers varies greatly with strain and temperature [2,3]. Since then, considerable attention has been paid to exploiting Brillouin scattering for distributed sensing. This is for the following reasons. First, strain is a very important parameter in relation to the reliability of the fibers themselves, in addition to its use for monitoring the integrity of large structures. Secondly, unlike the Raman technique, Brillouin frequency shift measurement does not require calibration of the optical-fiber loss. Furthermore, the Brillouin technique offers the potential for long-distance measurement. This is because the Brillouin frequency shift is small (about 10 GHz) and therefore the minimum loss wavelength region of the 1.55-mm band can be used. This should also be compared with the Raman technique, where the pump, Stokes, and anti-Stokes light wavelengths cannot all fall simultaneously in this wavelength region.

The following subsections describe the basic sensing mechanism based on Brillouin scattering, present detailed system configurations, and outline the main applications.

14.2.2 Sensing Mechanism

The Brillouin effect converts a small fraction of the incident light to scattered light with a shifted frequency. The spectrum of the scattered light is characterized by three parameters, as shown in Figure 14.4: the Brillouin frequency shift, ν_B, the Brillouin linewidth, $\Delta\nu_B$, and the Brillouin scattering coefficient, η_B.

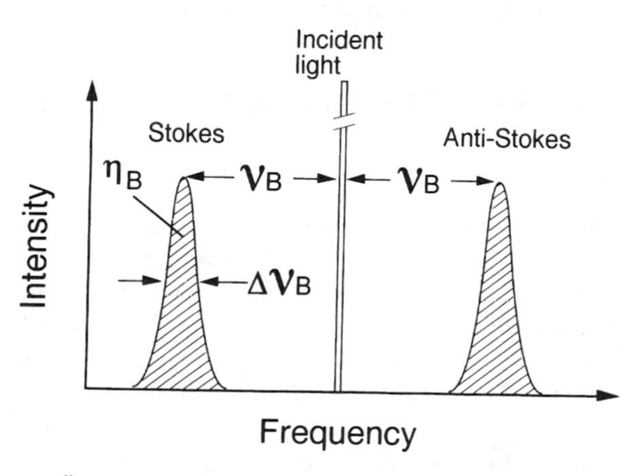

Figure 14.4 Schematic Brillouin spectra.

The Brillouin frequency shift, ν_B, is maximized when the light is scattered in a backward direction. This is the case for Brillouin scattering in optical fibers. Then, ν_B is given by [4]

$$\nu_B = 2nV_a/\lambda \tag{14.1}$$

where n is the fiber-core refractive index, V_a is the velocity of sound, and λ is the light wavelength. When we adopt typical values of $n = 1.46$ and $V_a = 5945$ m/s for silica glass and assume an operational wavelength $\lambda = 1.32$ μm, then $\nu_B = 13.2$ GHz. If the physical parameters cause n, V_a, or both to vary, the Brillouin frequency shift varies according to (14.1). Horiguchi and others [2] and Culverhouse and others [3] have proposed using the variation in the Brillouin frequency shift of single-mode fibers for distributed strain or temperature sensing. It has been confirmed from measured and calculated results that the major source of the variation in the Brillouin frequency shift is the change in the acoustic velocity. Careful measurements have revealed that, at a wavelength of 1.32 μm, the strain and temperature coefficients are 0.58 MHz/10 με and 1.2 MHz/K for a silica-based single-mode communication fiber, as shown in Figure 14.5 [5]. It is seen in the figure that the Brillouin frequency shift varies linearly with strain and temperature over wide ranges. Jacketing the fiber

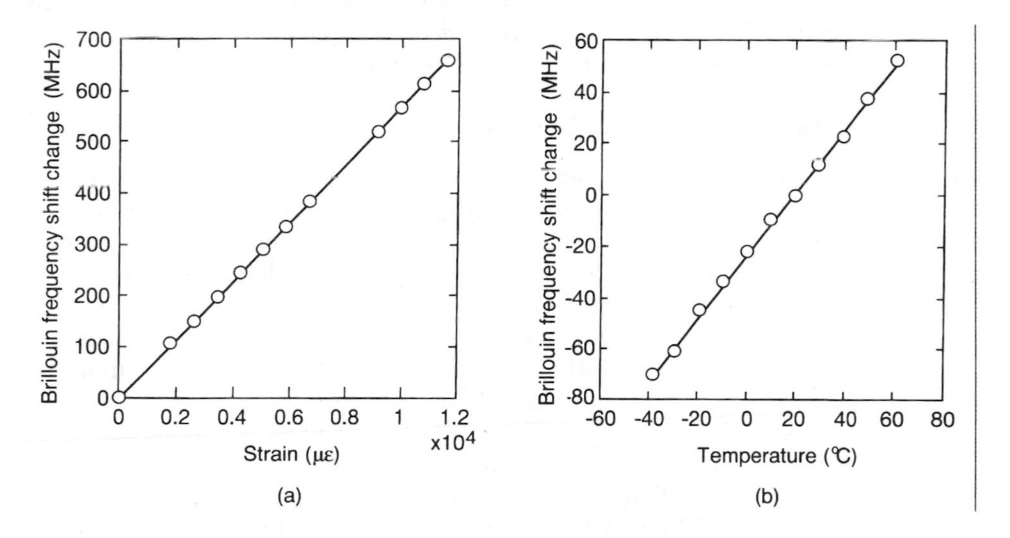

Figure 14.5 (a) Strain- and (b) temperature-dependence of Brillouin frequency shift (*After:* [5]).

increases the temperature sensitivity [6] because the thermal strain caused by the jacketing contributes significantly to the increase in the Brillouin frequency shift. A single-mode fiber coated with a 0.9-mm nylon jacket exhibits a temperature dependence of 3.7 MHz/K at around room temperature, which is three times that of a 250-μm acrylate-coated fiber. It is noted that calibration may be required to measure the absolute values of strain and temperature, since the Brillouin frequency shift also depends on the type and concentration of the material dopant in the silica fiber core.

Another important parameter for Brillouin scattering is the Brillouin linewidth, $\Delta\nu_B$, which is given by the inverse of the phonon lifetime. It has been shown that the linewidth does not change with strain, and has only a slight temperature-dependence of about −0.1 MHz/K at room temperature [7]. This value is one-tenth the dependence of the Brillouin frequency shift. This may rule out the use of the $\Delta\nu_B$ measurement for sensing. However, when the strain or temperature vary along the length of a sensing fiber over distances shorter than the spatial resolution of the Brillouin measurement system, the linewidth is broader than the intrinsic Brillouin linewidth. This increase in linewidth has been advantageously used to determine the magnitude of the fiber strain in bent cables [8] by assuming the type of strain variation, and to locate kinks in submarine optical cable [9] since the variation profiles of Brillouin scattering are known in advance except for those of variations due to strain magnitude.

Another parameter that characterizes Brillouin scattering is its scattering coefficient, η_B. Although the coefficient is also expected to change with strain and temperature, few studies have focused on exploiting any possible η_B dependence for distributed sensing.

In all the following subsections, we concentrate on a technique that uses ν_B rather than $\Delta\nu_B$ and η_B, to measure distributed strain and temperature, with the exception of a brief discussion of the possible use of η_B in conjunction with ν_B to distinguish between changes due to strain and temperature.

14.2.3 System Configurations and Operations

Techniques for measuring local changes in the Brillouin frequency shift along the length of a sensing fiber can be categorized as 1) an optical time-domain reflectometry (OTDR) method using coherent detection and 2) a method involving counterpropagating pulsed light and continuous light [10]. Both methods are used in conjunction with spectral analysis. The former method is called Brillouin optical time-domain reflectometry, using coherent detection, or simply BOTDR. The latter is called Brillouin optical time-domain analysis or BOTDA. The configurations for BOTDR and BOTDA are shown schematically in Figure 14.6(a,b).

With the BOTDR system, pulsed light is launched into the sensing fiber and spontaneous Brillouin backscattered light at a shifted frequency is detected. It is distinct from Rayleigh and Raman scattering because each type of scattering has a different frequency. The Brillouin scattering coefficient η_B, which is the ratio of the backscattered light power and the incident pulsed light power, is very low (~10^{-8} for

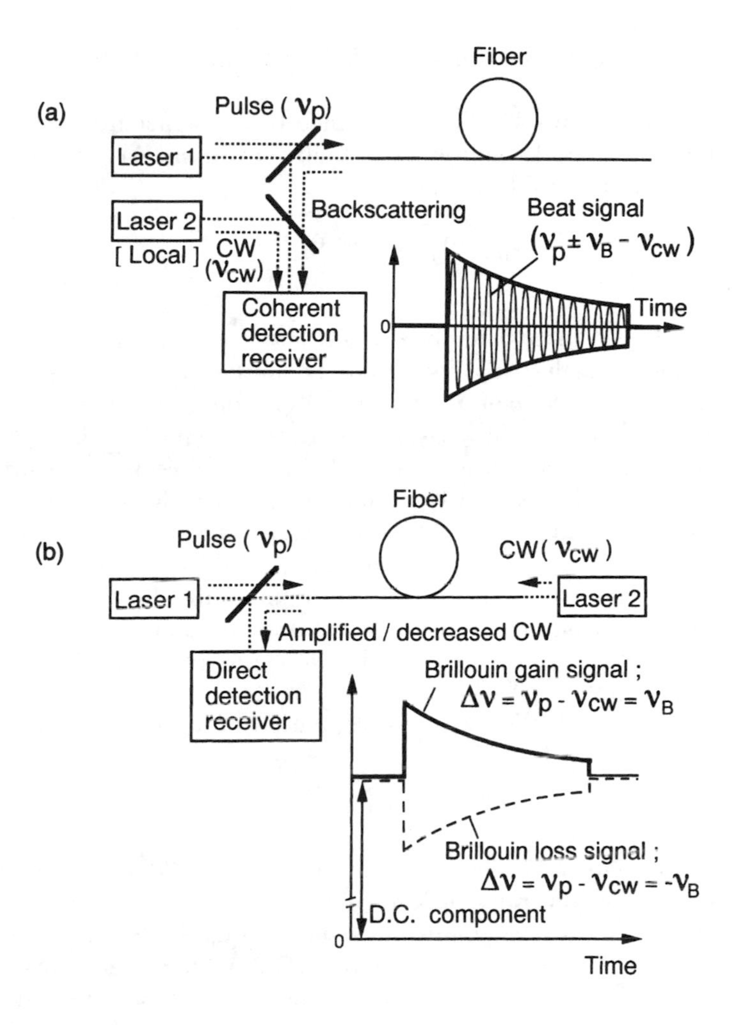

Figure 14.6 Configurations and signal waveforms of (a) BOTDR and (b) BOTDA.

100-ns pulsed light). So, a coherent detection method has commonly been used whose sensitivity is only limited by shot noise. This method has the advantage that the strong Rayleigh scattering is filtered out and the weak coherently detected Brillouin scattering is selected by an electrical rather than an optical filter. This also makes high-frequency resolution possible. The Brillouin spectrum can be measured either by sweeping the center frequency of the electrical bandpass filter to determine the heterodyne beat signal, or by varying the frequency difference between the incident pulsed light and the local oscillator light, while keeping the center frequency of the bandpass filter fixed.

Assuming the Brillouin spectrum is described by a Lorenzian function with a resonance frequency, ν_B, and a full width at half maximum, $\Delta\nu_B$, the minimum detectable change in ν_B is given by [10]

$$\delta\nu_B = \Delta\nu_B/\sqrt{2}(\mathrm{SNR})^{1/4} \tag{14.2}$$

where SNR is the electrical signal-to-noise ratio.[1] However, when light with a short pulse is used to achieve a high spatial resolution, the spectral width of the pulsed light may exceed $\Delta\nu_B$. Then, the width of the convoluted spectral distribution replaces $\Delta\nu_B$ in (14.2). The SNR required for the measurement can be calculated using (14.2) and the strain and temperature coefficients described above. The result is that strain and temperature accuracies of 100 $\mu\epsilon$ and 1K require respective SNR values of 25 dB and 52 dB, respectively, when $\Delta\nu_B$ = 35 MHz, which is a typical value for single-mode fibers. The use of curve fitting for the digitally sampled data of Brillouin spectrum reduces the strict SNR requirement for determining the Brillouin frequency shift [11,12]. This is because curve fitting using a large number of sampled data enhances the SNR in the same way as realized by the simple averaging of a large number of measurement runs.

As with conventional OTDR, the spatial resolution, δz, of BOTDR is determined by

$$\delta z = vW/2 \tag{14.3}$$

where v is the light velocity in the fiber and W is the incident width of the pulsed light.

BOTDR was first demonstrated by Kurashimà and others [5]. The configuration they used and their measurement results are shown in Figures 14.7 and 14.8. Two single-frequency Nd:YAG lasers, YAG1 and YAG2, with wavelengths around 1.32 mm, were used as a pulsed light source and an optical local oscillator. The pulsed light was obtained by gating with an acousto-optical modulator (AOM). The frequency difference between the two lasers $\Delta\nu$ was tuned around the Brillouin frequency shift, so that a narrow-bandwidth heterodyne receiver could be used. It is seen in Figure 14.8(a,b) that Brillouin scattered light from fibers 1 and 2 was clearly distinguished by tuning $\Delta\nu$ to the Brillouin frequency shift of each fiber. Here, the spatial resolution was 100 m, and the frequency accuracy was 3.6 MHz. The latter corresponded to strain and temperature accuracies of 60 $\mu\epsilon$ and 3K, respectively. This

[1]There is a minor difference between the factor $\sqrt{2}$ in (14.2) and equation (4) of [12], in that (14.2) was obtained by taking account of the presence of noise both at and around the center of the Brillouin spectrum, while equation (4) of [12] is only for the noise at its center.

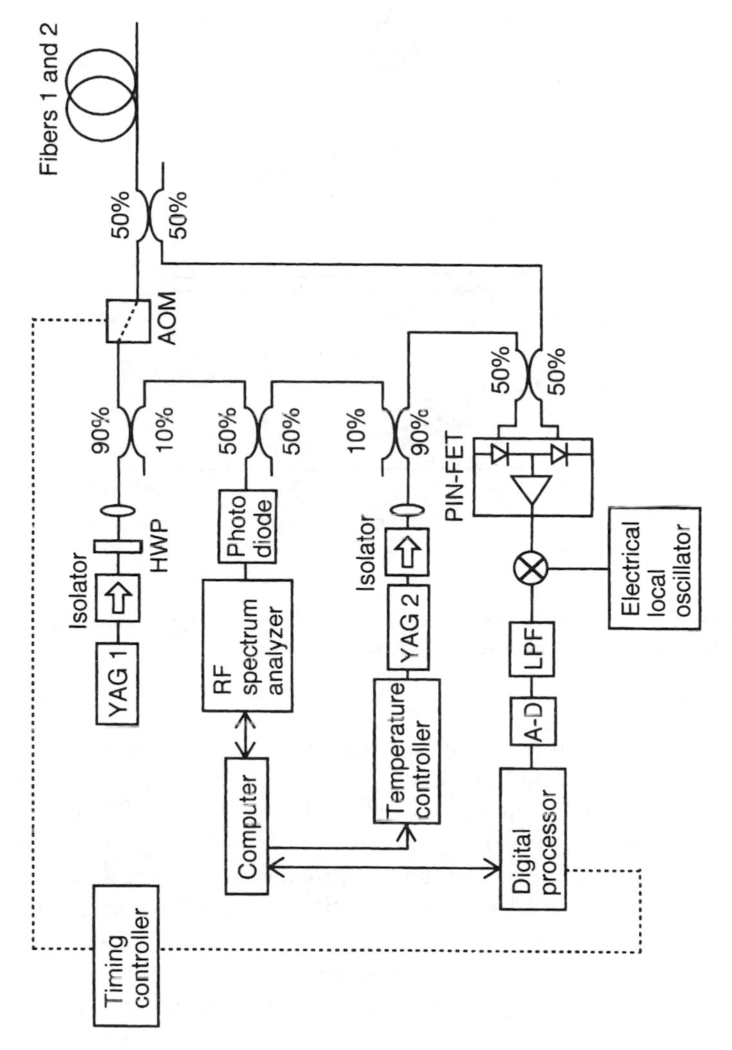

Figure 14.7 Experimental setup for BOTDR (*After:* [5]).

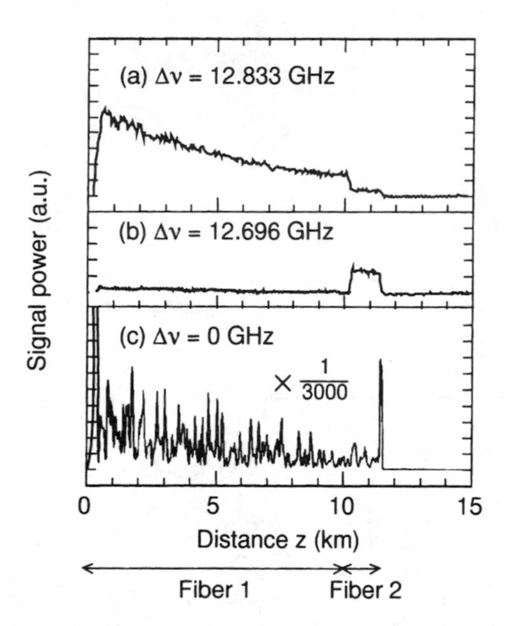

Figure 14.8 Backscattered signal traces: (a) Brillouin backscattering from fiber 1, (b) Brillouin backscattering from fiber 2, and (c) Rayleigh backscattering (*After:* [5]).

experiment also showed that Brillouin backscattered light does not suffer from "fading" noise [13], which is usually observed in Rayleigh backscattered light when a highly coherent light source is used, as shown in Figure 14.8(c). The reason for this reduction in fading noise with BOTDR is that the Brillouin backscattered light suffers random phase fluctuations due to spontaneous propagating density fluctuations in the fiber core. This beneficial feature enables BOTDR to provide accurate measurements.

For a BOTDA system, pulsed light at a frequency, ν_p, and a continuous wave (cw) light at a frequency, ν_{cw}, are counterpropagated in a sensing fiber. When the frequency difference $\Delta\nu$ ($= \nu_p - \nu_{cw}$) between the two optical signals is tuned to the Brillouin frequency shift, ν_B, and when the two guided beams interact, the continuous light is amplified by the pulsed light (the case where $\Delta\nu = -\nu_B$ will be discussed in Section 14.2.4). The BOTDA signal, resulting from the increase in the continuous light power, is detected as a function of time, as in OTDR. Therefore, local changes in the Brillouin frequency shift can be detected by measuring the changes in the continuous light power for various frequency differences, $\Delta\nu$, from the central Brillouin frequency shift peak for unperturbed fiber. When the fiber loss is ignored, the BOTDA signal power is given by

$$P_B = (G - 1)P_{cw} = \left(\exp\left[\frac{\gamma g P_p}{A} \frac{vW}{2} \right] - 1 \right) P_{cw} \tag{14.4}$$

where G is the Brillouin gain, g is the Brillouin gain coefficient, γ is the polarization factor, A is the effective core area, v is the light speed, W is the pulsed light width, and P_p and P_{cw} are the respective powers of the pulsed and the cw light, which interact with each other. When Brillouin amplification occurs in a linear regime (i.e., $G << 1$), (14.4) can be approximated as

$$P_B \sim \frac{\gamma g P_p}{A} \frac{vW}{2} P_{cw} \tag{14.5}$$

With parameters of $g = 2 \times 10^{-11}$ m/W, $\gamma = 0.5$ (this value will be discussed later), $A = 6 \times 10^{-11}$ m^2, $v = 2 \times 10^8$ m/s, $W = 100$ ns, and $P_{cw} = 0.1$ mW , we obtain a calculated BOTDA signal power of $2 \times 10^{-4} P_p$, which is about 100 times greater than the Rayleigh scattered light signal for the same pulsed light power. Then, Rayleigh scattering can conveniently be ignored. The measured Brillouin gain spectrum coincides with the spontaneous Brillouin scattering spectrum when $G << 1$. Then, the minimum detectable change in ν_B for BOTDA is given by (14.2), which is the same as that for BOTDR. The spatial resolution of BOTDA is also given by (14.3), because the Brillouin amplification occurs only during the interaction between the two light signals.

BOTDA was first used for distributed strain measurement by Horiguchi and others [14], using the arrangement shown in Figure 14.9. Here, the pulsed and continuous light sources were a Nd:YAG laser and a linewidth-narrowed laser diode, respectively, both oscillating around 1.32 μm. Measured results are shown in Figure 14.10. This figure shows that, in contrast to the tension-free fiber A, nonuniform strain occurred in fibers B–D, which were wound with various tensions around plastic drums. The spatial resolution was 100m and a strain resolution of 20 με was achieved.

Unlike the usual Rayleigh and Raman techniques, the Brillouin frequency shift technique uses a polarization-dependent scheme; a coherent detection technique for BOTDR and Brillouin amplification for BOTDA. The sensitivity of the coherent detection is maximum and zero, respectively, for coaligned and orthogonal polarization states between the backscattered light and the local oscillator light. The magnitude of the Brillouin amplification is maximum and zero for coaligned and orthogonal polarization states between the counterpropagating lights, respectively. It has been shown [15] that the degree of polarization of the Brillouin backscattering is 1/3 for long single-mode fibers, and that the polarization factor of Brillouin amplification varies between 1/3 and 2/3 for long interaction lengths if the polarization state is assumed to be distributed uniformly over the Poincare sphere. However, if a light source with a short pulse duration is used, as in BOTDR and BOTDA, this assumption is not valid, and hence the signal may drop to zero in the worst case. Therefore, polarization averaging is essential for the successful operation of both BOTDR and

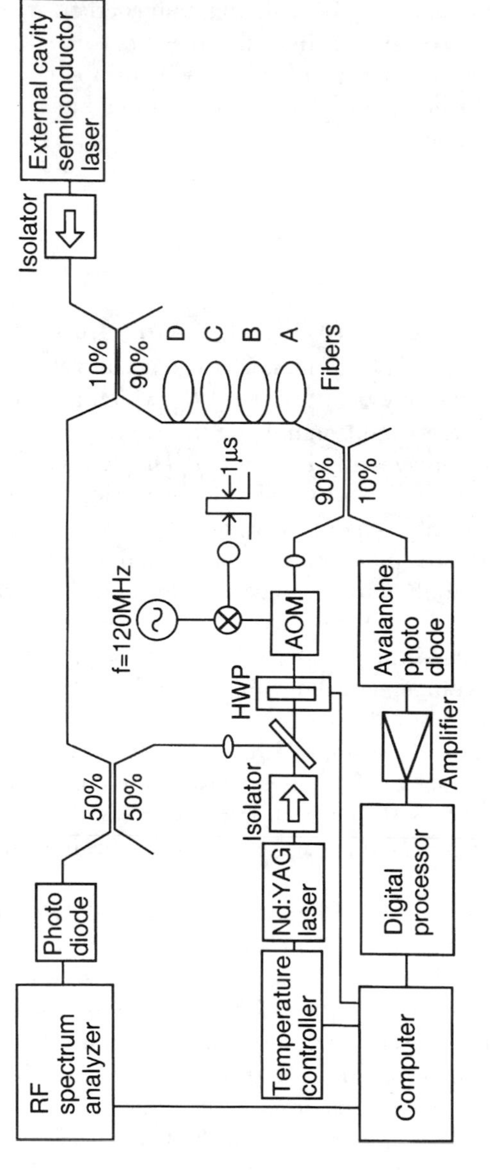

Figure 14.9 Experimental setup for BOTDA. (*After:* [14]. ©1990 IEEE.)

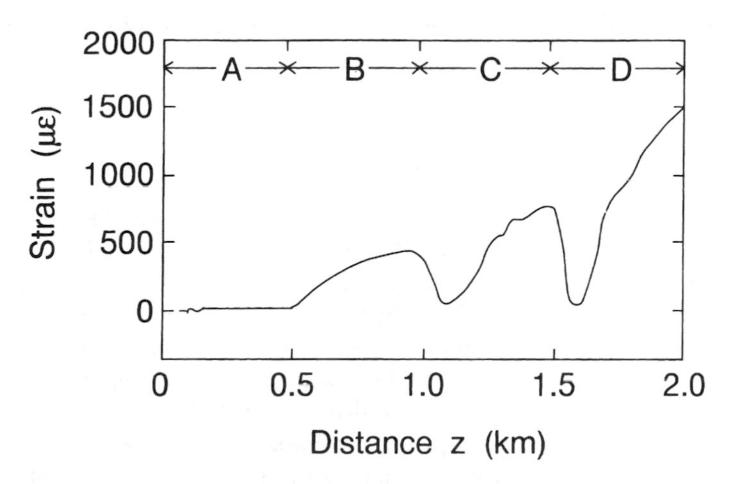

Figure 14.10 Strain distribution measured by BOTDA. (*After:* [14]. ©1990 IEEE.)

BOTDA, if weakly birefringent single-mode fibers are used for the sensing fiber. This polarization averaging was performed during the measurements in [5] and [14] by rotating a half-wave plate (HWP) (see the experimental setups in Figures 14.7 and 14.9). The expected average polarization factor for either coherent detection or Brillouin amplification was then 1/2.

14.2.4 Performance Improvements

Range performance has been improved by two methods. The first involves using a 1.55-μm wavelength source and using additional gain from erbium-doped fiber amplifiers [16]. The second consists of employing the Brillouin loss method [17] instead of the Brillouin gain method described in Section 14.2.3. With the first method, the probe light wavelength is changed from 1.3 μm to 1.55 μm, which is the lowest loss wavelength region for optical fibers, and the minimum value of the pulsed light power is boosted. The measurable fiber loss range has been extended to 25 dB, which corresponds to a fiber length of over 100 km. The spatial resolution and strain accuracies were 100m and 150 με. With the second method, the pulsed light frequency, ν_p, is set lower than the continuous light frequency, ν_{cw}, by a Brillouin frequency shift, ν_B, given by $\nu_B = \nu_{cw} - \nu_p$. Then, the continuous light experiences Brillouin loss, as shown by the dashed curve in Figure 14.6(b), whose magnitude is the BOTDA signal. This Brillouin loss signal is the same as the Brillouin gain signal, provided the total energy transferred between the two lights is insignificant compared to the incident pulsed light energy. Once this transferred energy becomes appreciable, the pulsed light is amplified in the Brillouin loss method, whereas it is depleted in the

Brillouin gain method. When the fiber is long, the magnitude of the amplification or the depletion, D, in the final section of the fiber is given by

$$D = 4.343 \frac{\gamma g P_{cw}(L)}{A\alpha} \text{ dB} \tag{14.6}$$

where $P_{cw}(L)$ is the incident cw light power and a is the loss coefficient of the fiber. It is (14.6) that yields the gain of a fiber Brillouin amplifier, which is experienced by the pulsed light when pumped by the cw light. For $P_{cw}(L)$ = 0.2 mW and $\alpha = 4.6 \times 10^{-5}\ \text{m}^{-1}$(0.2 dB/km), D becomes 3 dB. This different characteristic seems to favor the Brillouin loss method for very long distance measurement. Bao and others [17] measured the change in the Brillouin loss signal from 5m of fiber at a temperature of 40°C in an oven that was situated at a distance of 22 km along a 51-km sensing fiber. However, it seems to be impossible to evaluate the Brillouin spectrum in the final section of the sensing fiber accurately from the Brillouin loss signal without calibrating the incident pulsed light power. This is because the power of the amplified pulsed light entering the final section of the sensing fiber depends on $\Delta\nu$. The calibration method must therefore be studied in detail before this method becomes practical.

The spatial resolution of the Brillouin technique had been 5 to 100m, and this was limited by the use of a low-speed optical modulator such as an AOM. Recently, a spatial resolution of 1m with a frequency accuracy of 5 MHz has been achieved [18] over an 11-km fiber, using the combination of a fast electro-optical modulator (EOM) and an AOM for BOTDA. As the pulsed light width of 10 ns was shorter than the phonon lifetime of about 50 ns, the effective Brillouin gain was less than the steady-state gain [19]. In addition, the measured BOTDA signal spectrum for W = 10 ns broadened to about 80 MHz, as expected, which was double that for W = 100 ns. This broadening makes it increasingly difficult to achieve better spatial resolution, since, according to (14.2), a higher SNR is required. Calculations, however, predict that a spatial resolution of better than 1m will be possible in spite of these difficulties due to the spectrum broadening and the decrease in Brillouin gain.

It usually takes a few minutes to complete the measurement with the Brillouin technique because the frequency of the probe light must be scanned to find the maximum Brillouin signal and the signal must be averaged to improve the SNR. Very recently [20], a synchronously pumped fiber-ring Brillouin laser has been proposed to measure rapidly varying strain, because a large optical signal can be obtained from the output of the ring laser without scanning the laser frequency. A local change in the strain at a rate of about 1 Hz has been successfully measured by this technique.

Most of the Brillouin techniques described above have used two distinct lasers: one for the pulsed light of BOTDA and BOTDR, and the other for the counterpropagating continuous light of BOTDA or for the local oscillator light of BOTDR. However, it is not always easy to control the frequency difference between the two lasers accurately, and it may be impossible if the lasers exhibit mode-hopping. This is likely to result in a deterioration in the accuracy of the Brillouin frequency shift measurement, or an increase in the system cost. Several methods have been considered for solving this problem. A straightforward way to realize a single-laser measurement of the Brillouin frequency shift is to use a high-speed detector for the coherent detection [20]. On the other hand, a multimode laser, with a mode spacing near the Brillouin frequency shift, has been shown to be useful for measuring the Brillouin frequency shift even with a low-speed and inexpensive detector [21]. Such a laser can be used for BOTDR. The other way is to use a frequency translator such as a stimulated Brillouin scattering (SBS) generator [22], a high-speed $LiNbO_3$ optical modulator [23,24], or an optical ring containing an acousto-optical frequency shifter [25]. A fiber SBS generator produces a light that is downshifted in frequency by the amount of the Brillouin frequency shift of the fiber in the generator. This downshifted light and the original light have been used to create the pulsed light and the cw light, respectively, for measuring the Brillouin loss with the BOTDA scheme. However, this method has not been capable of accurately profiling the distribution of the Brillouin frequency shift due to the lack of a practical way of rapidly changing the frequency difference between the pulsed and the cw light. The SBS generator may also be employed in the BOTDR scheme. Another way to realize single-laser measurements is to use a high-speed LiNbO3 intensity electro-optic modulator (EOM), which modulates a light to produce frequency shifts near the Brillouin frequency shift [23]. The optical modulator can be also operated as an optical switch to produce a pulsed light. By using the first lower sideband of the modulated light, and a pulsed light with the original frequency, a single-laser BOTDA has been demonstrated that uses the Brillouin gain method, as shown in Figures 14.11 and 14.12. The most recent experiment has achieved a temperature resolution of 0.25K [24]. In [23,24], both counterpropagating light signals have been pulsed by the EOM to avoid depletion of the pump signal. Another way is to use an optical ring [25], which contains an acousto-optical frequency shifter and an optical amplifier, as shown in Figure 14.13. The optical ring can produce a pulsed light that undergoes a frequency shift of $\delta \sim 100$ MHz for each circulation around the ring. By compensating for the optical loss of the ring with the optical amplifier, the pulsed light can circulate around the ring a large number of times. As a result, the pulsed light frequency can be shifted by as much as the Brillouin frequency shift (about 10 GHz) without using high-frequency electrical and optical components. Furthermore, it is possible to tune the shifted frequency by slightly changing the acoustic frequency of the acousto-optic shifter. This type of optical ring has been successfully applied to single-laser BOTDR, even under conditions

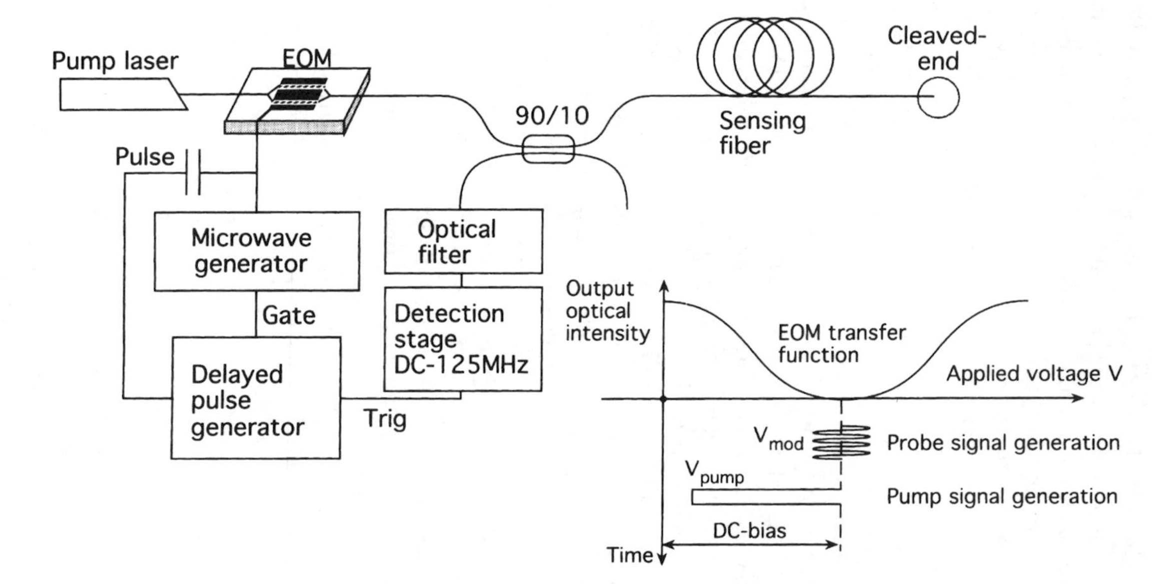

Figure 4.11 Experimental setup for single-laser BOTDA using EOM to generate both pump and probe pulsed signals. Both signals propagate back and forth along the sensing fiber via reflection at the far end (*After:* [23]).

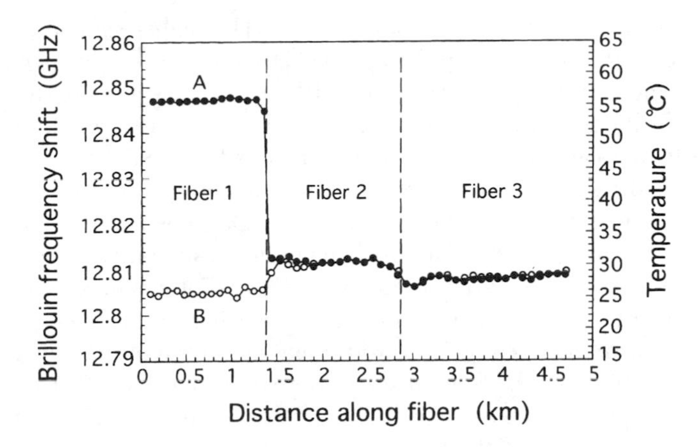

Figure 14.12 Distributed measurement of temperature using the setup of Figure 14.8: (a) fiber 1 placed in an oven at 55°C, fibers 2 and 3 at room temperature; (b) fibers 1, 2, and 3 at room temperature (*After:* [23]).

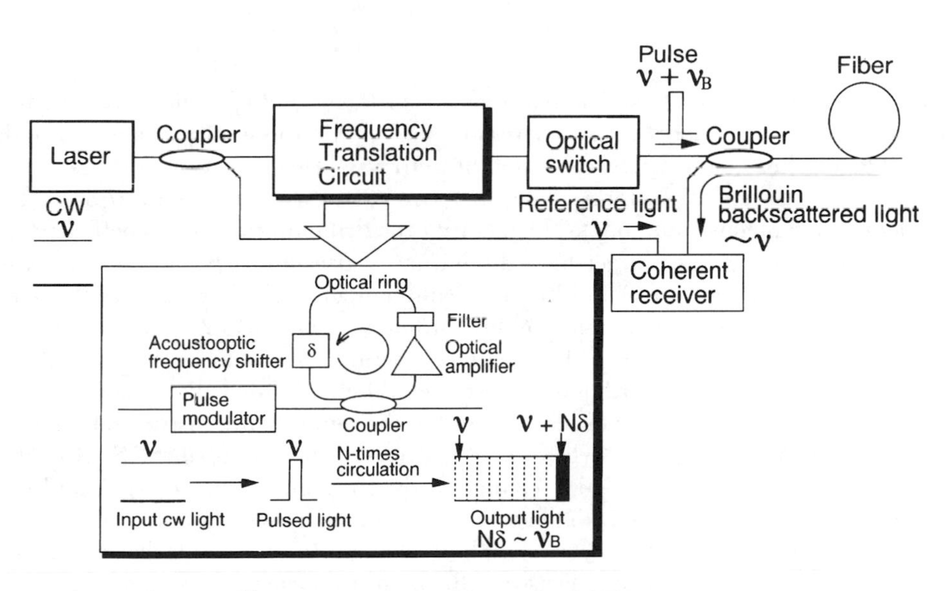

Figure 14.13 Schematic of frequency translation circuit used in BOTDR.

where the laser has exhibited mode hopping [25]. The high-speed optical modulator and the optical ring can be also applied to BOTDR and BOTDA, respectively.

The Brillouin frequency shift varies with strain as well as with temperature. Therefore the Brillouin technique has the potential to measure strain and temperature simultaneously, but can suffer from crosstalk from whichever is the undesired parameter. Bao and others [26] have realized the simultaneous measurement of both parameters by configuring a sensing fiber in such a way that one half detected both strain and temperature, while the other half detected temperature only. The development of a cable, in which some fibers are contained tightly and others loosely, to enable both strain and temperature sensing and temperature sensing, respectively, will make this simultaneous measurement easier. The additional use of the Raman technique [27] is another solution, because the Raman scattering coefficient has a large temperature-dependence and appears to have a negligible strain dependence. Another elegant way may be to use the changes in the Brillouin frequency shift ν_B and the changes in the Brillouin scattering coefficient η_B simultaneously. The relations describing the changes in Brillouin frequency shift, dnB, and Brillouin scattering coefficient, dhB, in response to changes in strain, de, and temperature, dT, can be expressed by a matrix equation:

$$\begin{bmatrix} \delta\nu_B \\ \delta\eta_B \end{bmatrix} = \begin{bmatrix} C_1 & C_2 \\ C_3 & C_4 \end{bmatrix} \begin{bmatrix} \delta\epsilon \\ \delta T \end{bmatrix} \tag{14.7}$$

where C_1 and C_2 are the strain and temperature coefficients of ν_B, which have already been described in Section 14.2.2, whereas C_3 and C_4 are those of η_B. Values of both $\delta\epsilon$ and δT can be determined by inverting the matrix in (14.7), provided $C_1C_4 \neq C_2C_3$ (i.e., provided the equations for ν_B and η_B in terms of strain and temperature are independent simultaneous equations). The fact that the Brillouin scattering coefficient η_B is one order of magnitude larger than the Raman scattering coefficient would favor this proposed technique for separating strain and temperature. Very recently, this scenario has been partly demonstrated by two different groups [28,29]. In [28], it has been clarified that η_B increases with absolute temperature. In addition, the ratio of the intensities of Rayleigh and Brillouin backscattered light (Landau-Placzek ratio) has been used to compensate for fiber loss to profile the temperature distribution. However, the fading noise in the Rayleigh backscattered light degraded the SNR of the measured temperature. In [29], the change in the Brillouin frequency shift and the change in the intensity of the Brillouin scattered light have been simultaneously measured, and it was found that, as expected, both changes increase with the increase in temperature. Unfortunately, the measured Brillouin frequency shifts were scattered, possibly due to the poor resolution of the Fabry-Perot filter, which was used to differentiate the Brillouin scattering from the Rayleigh scattering. Therefore, improvements are required in the accuracy of the simultaneous measurement of the Brillouin frequency shift and the intensity of the Brillouin scattered light. In addition, the unknown effect of strain on the intensity of Brillouin backscattered light must be measured to make the proposed method practical for the simultaneous measurement of strain and temperature.

14.2.5 Applications

Fiber strain is an important parameter as regards the reliability of optical-fiber cables because strain can cause degradation in fiber strength (stress corrosion), leading eventually to fiber failure. Therefore, Brillouin-distributed strain measurement has found many applications in the research and development of optical fibers and cables and their related technologies (see [10]). Reports have been published on the strain evaluation of optical-fiber cables, measurement of fiber strain in underground cables and in submarine optical cables before and after installation, the location of kinks in submarine optical cables, and sensing pressure caused by freezing of water in cable conduits. It has also been proposed that a BOTDA/BOTDR system be used for testing and monitoring the fiber strain in optical communication cables [30]. Brillouin-distributed strain and temperature-sensing techniques also have great potential for use in relation to employing fiber cables to monitor the integrity of large structures such as dams, nuclear plants, pipelines, and buildings.

14.2.6 Summary

A technique for measuring distributed strain and temperature using Brillouin scattering in optical fibers has been studied extensively. Two methods have been developed to produce prototype sensors: Brillouin coherent OTDR and the counterpropagating light method. The former method will be used in applications that require single-end measurement without access to the remote fiber end, while the latter will be used for long-distance measurement. Both methods have achieved levels of performance near their predicted values. Although the Brillouin technique is rather new and more sophisticated than other techniques, it has already found applications in the research and development of optical-fiber cables and their related technologies. This is mainly due to its unique capability for measuring distributed strain over long distances. The system cost will be reduced by employing simple configurations as achieved with single-laser BOTDA and BOTDR. Further investigations into ways of improving performance, separating sensitivity to strain and temperature, and installing a sensing cable (guidelines for which are described in the text) will make the Brillouin technique available for a wider range of practical applications.

14.3 DISTRIBUTED OPTICAL-FIBER SENSING USING THE OPTICAL KERR EFFECT

14.3.1 Introduction

Optical-fiber methods of distributed measurement sensing offer many important advantages for industrial use [31,32]. The fiber is a flexible, insulating, dielectric medium that can be readily installed in industrial plant without significant disturbance of the measurement environment. The range of measurands that are accessible to measurement by optical-fiber techniques is very large, since the propagation of light within an optical fiber is sensitive to a wide variety of physical influences external to it [33,34].

Optical-fiber distributed measurement sensing is a technique that utilizes the one-dimensional nature of the optical fiber as a distinct measurement feature. It is possible in principle to determine the value of a wanted measurand continuously as a function of position along length of a suitably configured optical fiber, with arbitrarily large spatial resolution. The normal temporal variation of the distribution is determined simultaneously.

Such a facility opens up an enormous number of possibilities for industrial application. For example, it would allow the spatial and temporal strain distributions in large critical structures such as multistory buildings, bridges, dams, aircraft, pressure vessels, electrical generators, and so forth to be monitored continuously. It would allow the temperature distributions in boilers, power transformers, power cables, aerofoils, and office blocks to be determined, and thus heat flows to be computed. Electrical and magnetic field distribution could be mapped in space so that electromagnetic design problems would be eased and sources of e/m interference would be quickly identifiable.

There really are two important definable reasons for requiring to obtain the information afforded by distributed optical-fiber measurement sensors. The first is that of providing continuous monitoring so as to obtain advance warning of any potentially damaging condition in the structure and thus allow alleviating action to be taken in good time. The second is that this spatial and temporal information allows a much deeper understanding of the behavior of large (or even relatively small) structures, with many implications for improvements in their basic design.

Conventional industrial measurement sensor technology does not provide this facility. When measurand distributions of any kind are vital in a given situation, the solution usually is to festoon the structure with a multitude of thermocouples, or strain gauges, or whatever. This then presents problems of multiplexing, logging, and calibration, and, in any case, relies on the choice of positions for each of the many sensors being the correct one—a choice that cannot properly be made without *a priori* knowledge of the very distribution one is seeking to measure. This "solution" is thus expensive, tedious, and usually broadly inadequate.

The optical fiber can be readily installed in an industrial plant (retrospectively, if necessary); produces minimal disturbance of the measurement environment; is cheap, passive, and electrically insulating; acts as its own telemetering channel; can easily be rearranged in accordance with acquired knowledge; and allows a choice of any or all measurement points along its length,within the limits of the spatial resolution interval. If such a technique can be made to work satisfactorily for a number of measurands, a new dimension appears in the field of industrial measurement.

A primary difficulty with DOFS systems (as with most measurement systems) is that of achieving good sensitivity together with high spatial resolution. The present aiming point of 0.1m spatial resolution (or better), with a measurement accuracy of 1%, is difficult to achieve simply because the optical interaction length of 0.1m allows relatively few photons to be modulated (per resolution interval) by the measurand field, unless optical powers are very high; and the power that can be propagated in an optical fiber is limited by the onset of nonlinear effects. There are two possible approaches to the solution of this problem: one is to use highly sensitive polarimetric/interferometric techniques; the other is actually to make use of those limiting, nonlinear effects. In addition, one can use the two simultaneously (i.e., nonlinear, polarimetric techniques). Three exploratory methods of this type are described in this section. They all utilize the same nonlinear effect: the optical Kerr effect.

14.3.2 Forward-Scatter Versus Backscatter Methods for DOFS

Most DOFS explorations to date have relied on backscatter arrangements [31,32,34]. In these, a pulse of light is launched into an optical fiber and is continuously Rayleigh backscattered as it propagates. Time resolution of the emerging backscattered light then reveals the spatial distribution of any external field that is capable of modulating, in a deterministic way, some characterizing property of the light, this latter being demodulated by the detector. Such a one-dimensional "lidar" arrangement has the clear advantage of simplicity, but it has a number of disadvantages.

The primary disadvantage is that Rayleigh backscatter is at a low level, of order 10^{-5} of the forward-traveling energy, per unit length of fiber. Moreover, since the fiber attenuation (in a propagation "window") will depend almost entirely on Rayleigh scatter, there will be an optimum length, for a given scatter coefficient, with regard to both sensitivity and resolution of the DOFS system.

Suppose, however, that there is no necessity to rely on backscattered light. Consider the forward-scatter schematic shown in Figure 14.14. In this illustration, a continuous wave (CW) source is launched at one end (F) of the fiber, into one of two possible modes: T and R (Figure 14.14(a)). These might be two propagation modes, two polarization modes, or even two separate cores within the same cladding. The important conditions that must obtained are that the two modes must be separately identifiable at the output from the fiber and that light coupling between them must both be

(a)

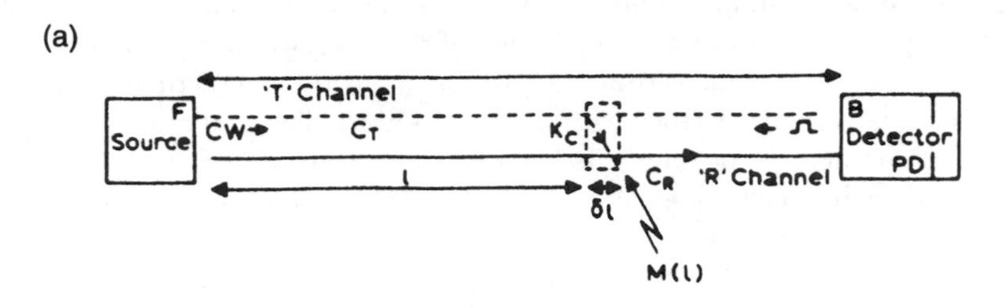

(b)

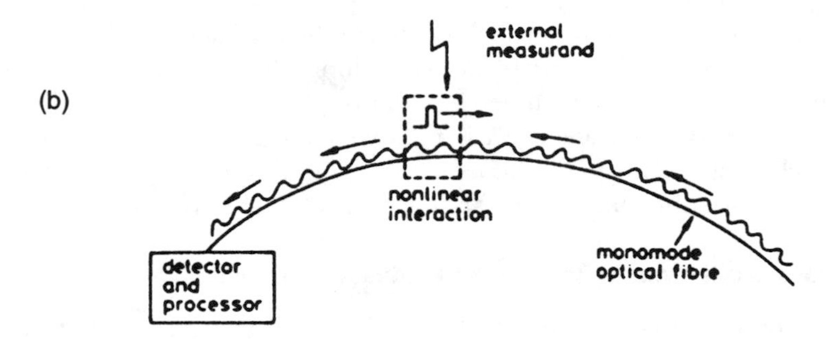

Figure 14.14 Forward-scatter DOFS: (a) schematic for F/S DOFS and (b) one form of operational forward scatter using a nonlinear interaction between a pulse and a counterpropagating CW.

possible and dependent in some way on the external field to be measured (measurand). The mode into which the CW is launched will be labeled the "filled" (in this case, T) mode while the other is the "free" (in this case, R) mode.

An optical pulse is now launched into the other end (B) of the fiber in such a way as to allow it, while propagating in the opposite direction to the CW, to couple the latter from the filled to the free mode. If, as was required above, the coupling mechanism is modulated by the measurand in a deterministic way, then, by time-resolving the CW power level emerging from end B in the free mode, one obtains the spatial distribution of the measurand field.

Figure 14.14(b) illustrates an operational arrangement where the pulse and the counterpropagating CW are coupled by means of a nonlinear interaction in the fiber at the point of overlap.

The important advantage that this method offers over the backscatter technique is that of increased received power level. There is no longer an optimum attenuation coefficient for a given fiber length since one is not relying on backscatter (although the mode-coupling requirement might impose other limitations on the minimum attenuation allowable). Clearly, a study of the methods by which forward-scatter DOFS might be implemented could well be worthwhile. Some such methods will be described in this section.

14.3.3 Broad Principles of Forward-Scatter DOFS

The first forward-scatter (F/S) method for DOFS to be studied relied on a Raman interaction between a counterpropagating pulse and a CW probe in a low-birefringence fiber [35]. In this, the CW probe received Raman gain from the pulse, and this gain depended on the relative polarization states of the probe and the pulse, these in turn being dependent upon the strain distribution along the fiber (Figure 14.15). As a demonstration of the basic principle of forward scatter DOFS, this was satisfactory, but it was not developed further since the signal processing

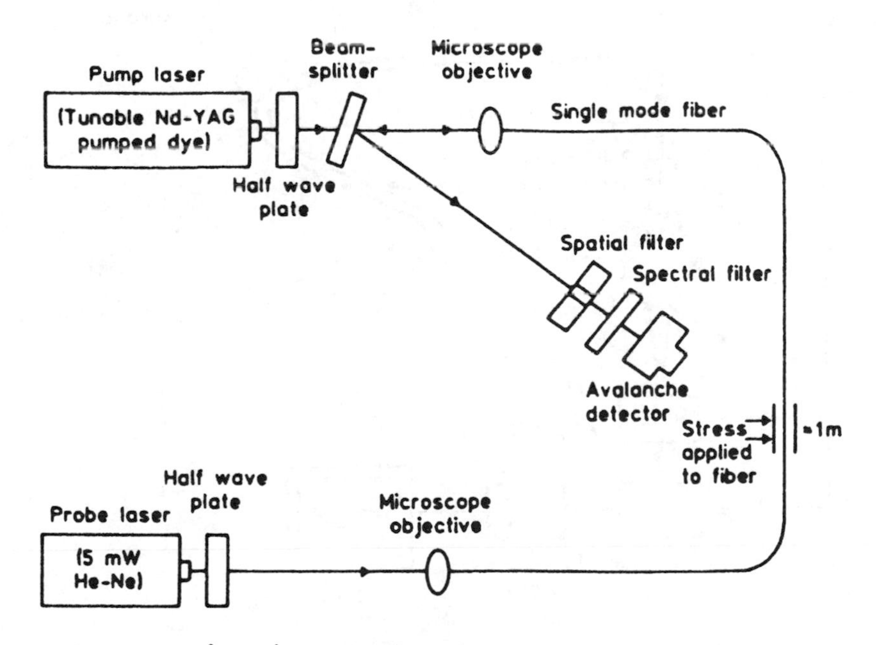

Figure 14.15 Raman-gain forward scatter DOFS.

required to determine the strain distribution was too complex for practical implementation. However, polarization methods are to be preferred since they are both very sensitive and independent (to first order) of fiber attenuation. High birefringence (hi-bi) fibers are now quite readily available. Such fibers possess a natural linear birefringence that allows good control over the propagation state of the propagating light over long distances. Attention recently has concentrated on the use of these fibers for DOFS.

The first attempt to use such fibers in DOFS involved a backscatter method [32,36,37] that was a variation on POTDR [38,39]. In this method, the propagating pulse, equally disposed in each of the two hi-bi eigenmodes, accumulated a linearly increasing phase difference between its two modal components as it propagated, and this was doubled for backscatter from any given point. By mixing the two backscattered eigenmode signals on emergence, a frequency was generated, and the time-dependence of this frequency mapped the distribution of birefringence along the fiber, corresponding to the pulse's passage along it (Figure 14.16). The method is

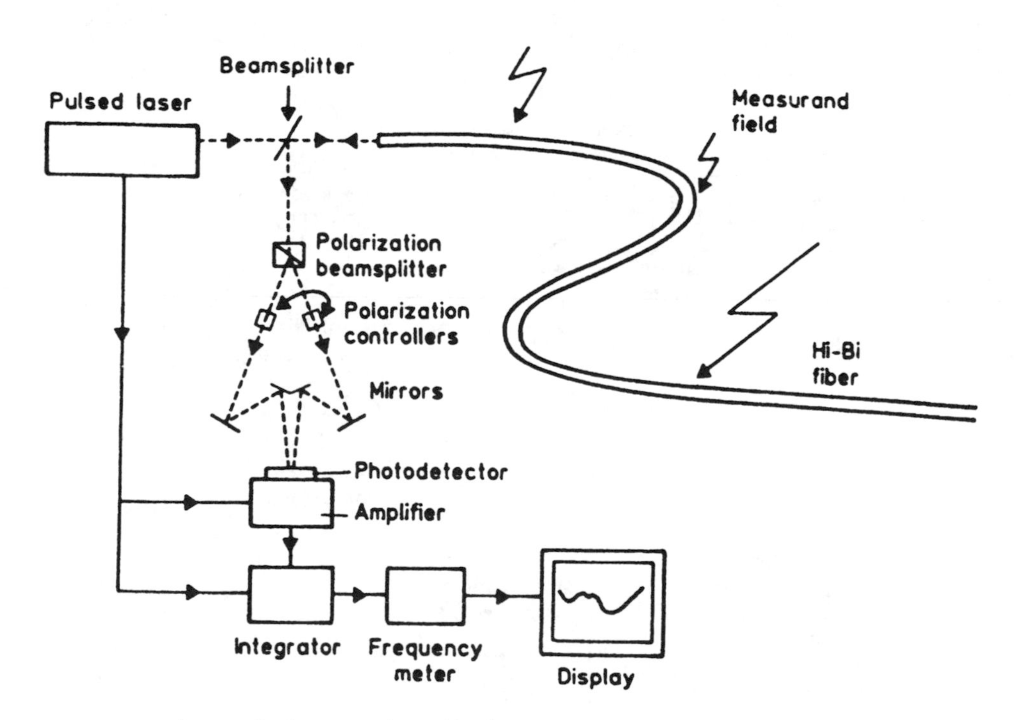

Figure 14.16 Schematic for frequency-derived backscatter DOFS.

known as frequency-derived DOFS. Any external field capable of modifying that birefringence was mapped correspondingly. While this method worked encouragingly well, it suffered still from that fact that it was a backscatter method, and thoughts turned towards attempts to embody the same, or a similar, idea in a forward-scatter arrangement.

The first thoughts along these lines involved the use of "resonant" coupling. If a pulse propagating in a hi-bi fiber could be arranged to impose a suitable periodic perturbation, with a spatial period equal to the hi-bi beat length (distance over which the differing polarization mode group velocities impose a phase difference of 2π between mode components), then this would lead to resonant coupling from one mode to the other of a counterpropagating CW probe. The birefringence could then be mapped via the coupling efficiency (level of coupled light) as a function of time.

The primary feature responsible for the polarization-holding property of hi-bi fiber is that of the difference in group velocities for the two eigenmodes. Any perturbation that acts to couple light between the modes therefore, in general, will couple components with phases that average the resulting amplitude to zero over one beat length. Since this latter is typically on the order of a few mm, only perturbations with spatial variations on this scale will cause significant coupling. Clearly, a periodic perturbation that has a spatial period equal to a beat length will couple optimally, since the points of maximum coupling will always lead to fully constructive interference. The most obvious coupling perturbation is a strain that, acting in a direction at a nonzero angle to the eigenaxes, will induce an extra birefringence and thus will effectively rotate the eigenaxes. The result is a coupling of each of the pre-existing eigencomponents into the new axes directions, a coupling that is not, in general, reversed when the perturbation ends.

Another, more readily controllable and thus more convenient coupling mechanism is via the Kerr effect. When an electric field is applied to an isotropic material (such as fused silica), linear birefringence is induced in the material, with the slow axis in the direction of the field and the fast axis orthogonal to it. The birefringence is given by

$$B_k = \lambda_0 bE^2 \tag{14.8}$$

where λ_0 is the vacuum wavelength, b the Kerr constant, and E the electric field.

Clearly, E could be the electric field of a propagating optical wave, in which case the phenomenon is called the *optical* Kerr effect.

The origins of the optical Kerr effect will now be discussed.

14.3.4 The Optical Kerr Effect

The nonlinear electric polarization, P, of an optical (dielectric) medium may be written as a power series in the applied electric field, E:

$$P(E) = \chi^{(1)}E + \chi^{(2)}E^2 + - - \chi^{(j)}E^j + - - \quad (14.9)$$

where the $\chi^{(j)}$ are the susceptibility coefficients, and they decrease rapidly in value with increasing j.

In an amorphous medium, such as the fused silica of which optical fibers are made, all even terms must be zero, since the value of $P(E)$ must simply reverse in sign when E is reversed (there being no "special" directions in an amorphous medium). Hence, to a very good approximation, the electric polarization for an optical fiber is given by

$$P(E) = \chi^{(1)}E + \chi^{(3)}E^3 \quad (14.10)$$

It is thus possible to define an "effective" susceptibility given by

$$\chi_{\text{eff}} = \frac{P(E)}{E} = \chi^{(1)} + \chi^{(3)}E^2 \quad (14.11)$$

Hence the effective refractive index of the medium is given by

$$n_{\text{eff}} = (1 + \chi_{\text{eff}})^{1/2} = 1 + \chi^{(1)}/_2 + \chi^{(3)}E^2/_2 \quad (14.12)$$

provided that $\chi^{(1)}, \chi^{(3)} E^2 << 1$.

Suppose now that E represents the electric field of an optical wave propagating in the medium. The intensity, I (power/unit area) of this wave will be proportional to E^2, so that, from (1), we may now write

$$n_{\text{eff}} = n_0 + n_2 I \quad (14.13)$$

where n_0 and n_2 are constants characteristic of the medium. Here, we now have an optical wave that is influencing the refractive index of the medium in which it is traveling, via its electric field strength.

This is the optical Kerr effect and it leads, amongst other things, to self-phase modulation (SPM), wherein the propagating light modulates its own phase via the intensity effect on the refractive index.

Since the action of the effect is to introduce a change in the refractive index for a particular direction of the optical electric field, it leads also to an electric-field-induced linear birefringence in the fiber, as mentioned in the previous section. It is this birefringence that is utilized in the DOFS methods that employ the optical Kerr effect.

Three such methods have been explored and these will now be described.

14.3.5 Frequency-Derived Forward-Scatter DOFS

Frequency-derived backscatter DOFS has been reported in the literature [32,36,37]. The forward-scatter version of this idea utilizes the optical Kerr effect [40].

Suppose that an intense optical pulse is launched into a hi-bi fiber, with equal power in each of the two eigenmodes. As the pulse propagates, the two components

(a)

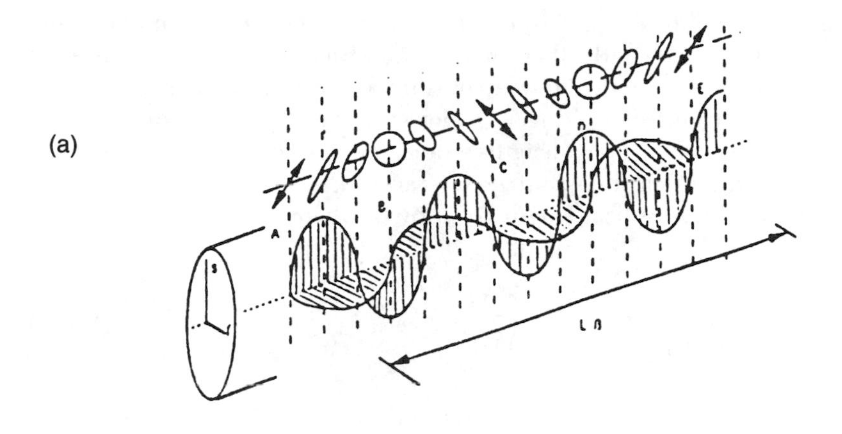

(b)

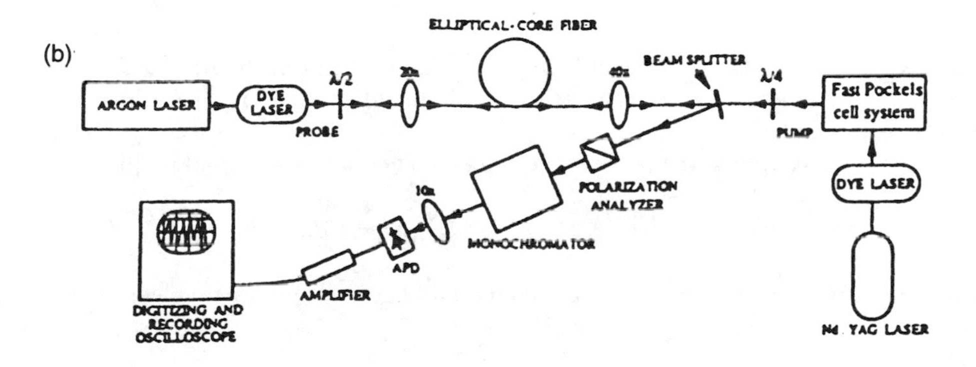

Figure 14.17 Optical Kerr effect DOFS: (a) polarization-state evolution down a hi-bi fiber and (b) optical Kerr effect DOFS.

will come into phase once per beat length and into antiphase at the intervening half-beat lengths (Figure 14.17(a)). The electric field of the resultant wave thus will maximize in a direction lying at ±45 deg to the intrinsic axes, every alternate half-beat length. The Kerr-effect-induced birefringence resulting from this field consequently will cause the resultant birefringence axes to rock around the intrinsic direction with a spatial period equal to the beat length.

Consider now a CW with the same wavelength as the pulse launched, again equally into the two eigenmodes, in the opposite direction to the pulse. When it encounters the pulse, the Kerr-induced rocking will produce optimal phase-matched coupling (since the rocking occurs naturally at the "resonant" frequency), as a result of which probe power will be coupled from one mode to the other. When pump and probe are at the same wavelength, there will be an exact correspondence in beat length that will maintain the coupled power constant as the pulse propagates down a fiber with uniformly distributed birefringence. However, if pump and probe are at different wavelengths, the differing beat lengths will lead to the condition that optimum coupling will occur only when the rocking axes can couple-in-phase probe components; that is, with an effective beat length given by L_e where

$$\frac{1}{L_e} = \frac{1}{L_1} - \frac{1}{L_2}; L_e = \frac{\lambda_1\lambda_2}{\Delta\lambda B_1} \quad (\Delta\lambda = \lambda_1 - \lambda_2) \tag{14.14}$$

From this we may conclude that the optimum coupling occurs at a "beat" frequency given by

$$\Delta f = cB_1 \cdot \frac{\Delta\lambda}{\lambda_1 \lambda_2} \tag{14.15}$$

where λ_1 is the wavelength of the pulse light, λ_2 that of the probe light, and L_1, L_2 the respective beat lengths; c is the velocity of light in the fiber and B_I is the intrinsic birefringence of the fiber.

For small coupling, the power coupling efficiency, k^2, is governed by [41]

$$k^2 = (\sigma^2/[\sigma^2 + (\theta/2)^2]\} \ \sin^2(L_{int}[\sigma^2 + (\theta/2)^2]^{1/2}] \tag{14.16}$$

where $\theta = (2\pi/L_e)$ is the dephasing parameter, and L_{int} is the interaction length. Also, the coupling coefficient σ is given by [10]

$$\sigma = \{\pi B_k/2\lambda_1\} \tag{14.17}$$

where B_k is the optical Kerr-induced birefringence. As the wavelength of the pump is shifted with respect to the probe, a phase mismatch will appear between the coupled wave elements along the fiber. As a result, the net coupling would be reduced compared to the case where both sources have the same wavelength, as expected from (14.16). The pump power cannot be raised indefinitely to compensate for this, because of a limitation on the peak pump power in the fiber due to the onset of spectral broadening effects (e.g., Raman, Brillouin), which would further reduce the coupling efficiency. A large wavelength shift between the pump and probe beams therefore cannot be achieved without paying a substantial penalty in terms of the coupling efficiency. Equation (14.15) shows that this, in turn, limits the tuning range of the derived probe frequency. To avoid this problem, the pump pulse should be short compared with the spatial beat length between the polarization states of the pump and probe beams, so as to avoid the "smoothing" effect caused by the probe pulse embracing many beat lengths (i.e. to keep L_{int}/L_e (in (14.16)) small). The experimental setup is illustrated in Figure 14.17(b). The fiber used was 24m long with an elliptical core (supplied by Andrew Corporation) and was single-mode at 633 nm, where it exhibited a beat length of 4 mm. The pump pulses were generated by a dye laser pumped by a Q-switched, frequency-doubled Nd: YAG laser. The probe was provided by a cw Argon-pumped dye laser that was counterpropagating with respect to the pump, and linearly polarized at a small angle to the fiber's birefringence axes, with a launched power of ~300 mW. In principle, the strongest probe signal should be provided when the eigenmodes are equally excited by a launched polarization set at 45 degrees to the fiber's principal axes. The signals would be then superimposed on a high mean-received power. However, an avalanche photodiode detector was used, and this rendered high mean signal levels undesirable. After reflection from a beamsplitter at near-normal incidence, the probe beam was analyzed with a Glan-Thompson analyzer. A monochromator was placed before the avalanche photodiode detector in order to block the large pulse predominantly produced by the front face reflection from the fiber. This pulse would have otherwise saturated the input amplifier. The pump and probe beams were set at various wavelength differences up to 2.4-nm apart, generating derived signal frequencies up to ~200 MHz. The signals were recorded and analyzed using a computer-interfaced digital storage oscilloscope, which was used to average the results from 256 pump pulses. The averaging process occupied approximately 10 seconds.

For the length of fiber used in the experiment, the spectrum began to broaden when the first stimulated Raman line was generated. This occurred at a pulse energy of about 36 nJ. The experimentally achieved coupling efficiency was about 1%. The variation of the derived frequency with wavelength offset was measured by the following method. The pump and probe were initially set at the same wavelength of 646 nm. Then, by tuning the pump wavelength with respect to the probe, the required wavelength shift was set. An example of a derived frequency signal averaged over 256 pump pulses is shown in Figure 14.18. The wavelength shift between the pump and probe in this measurement is 1.5 nm, and the corresponding derived frequency is 71 MHz. The measured signal to noise ratio was 19.3 dB. The experimentally observed and theoretically calculated relationships between the frequency and wavelength shift are shown in Figure 14.19. This figure shows the expected linear trend connecting the derived frequency with the wavelength shift. There was good agreement between the theoretical prediction represented by the line in the figure, and the experimental values represented by the points. The tunable frequency range offering

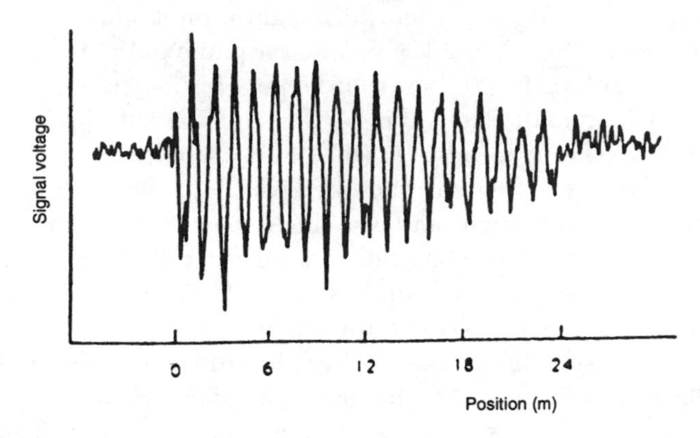

Figure 14.18 Derived frequency signal from Kerr-effect F/S DOFS.

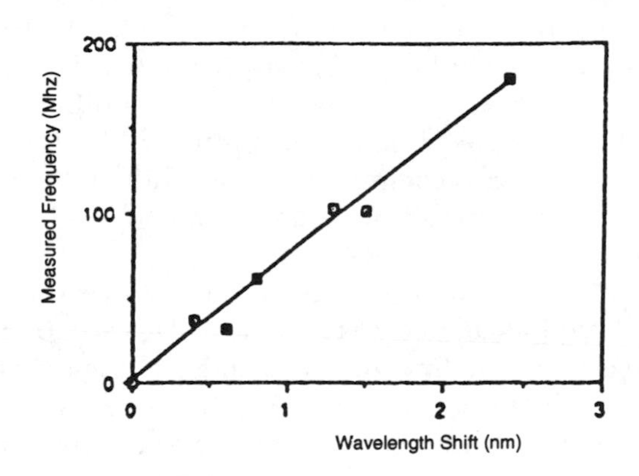

Figure 14.19 Derived frequency vs. wavelength difference between pump and probe.

strong signals was limited. This agrees closely with the theory given by (14.16), in that, for a 2-ns pump pulse, the first zero of the coupling response was expected to occur at a wavelength offset of ~3.2 nm, and from (14.15), the equivalent signal frequency would then be ~200 MHz.

Using the novel technique presented here, the spatial variation of the birefringence of a polarization-maintaining fiber can be measured remotely in a short time and, since the signal is in the form of a frequency, it is immune from the common error sources present in intensity-coded systems. Any change in the applied stress or ambient temperature will tend to change the local birefringence and will therefore be indicated by a shift in the frequency due to the external perturbation. Thus this technique may be applied to implement a distributed optical-fiber sensor. In this application, the tunability of the derived frequency could provide some extra flexibility in the system design. This method possesses the important advantage of providing a signal indication that is independent of signal level, and thus of capricious variations in fiber attenuation.

14.3.6 Polarization-State-Dependent Kerr Effect Forward-Scatter DOFS

In this second method, the emphasis presently is on the spatial location of a perturbation rather than on the measurement of its magnitude, although suitable processing is capable, in principle, of revealing the latter. In its present form, it is capable of providing good spatial resolution and rapid response for application to, for example, intrusion monitoring or vehicle location.

The optical arrangement employs a length of polarization-maintaining fiber carrying two counterpropagating beams (Figure 14.20 [42,43]). A CW probe beam is launched from one end of the fiber so as to excite equally the two eigenmodes, and the polarization state of this beam is detected at the far end of the fiber by means of a beamsplitter and analyzer oriented at 45 degrees to the birefringence axes. An intense pulsed pump beam is launched on one of the birefringence axes. This arrangement is similar to the well-known Kerr effect shutter system [44].

As in the Kerr shutter, the pump pulse causes a phase shift between the eigenmodes of the probe beam, leading to a change in the output polarization state of the probe. This is detected as a sharp change in the probe intensity passed by the analyzer when the pump is initially launched into the fiber.

If, now, a force acts at an angle to the axes along a section of fiber, coupling of the pump light to the other axis will occur, and the Kerr effect on the probe will thus be modified. The probe light itself will also experience mode-coupling, which will further modify the output polarization state. The actual change that occurs will depend *inter alia* on the states of polarization of the beams as they enter the perturbed region and thus, unless the birefringence perturbation is very small compared with the intrinsic birefringence, there will exist a mutual dependence of effects from different measurement locations, which only fairly complex signal processing would be able to resolve. However, it is clear that for any change in the direction of birefringence axes consequent upon the perturbation by a measurand, there will, in general, be a change in optical Kerr effect. A differentiated signal thus will, at least, indicate differential features of the measurand distribution, even though a fully quantified spatial distribution is more elusive.

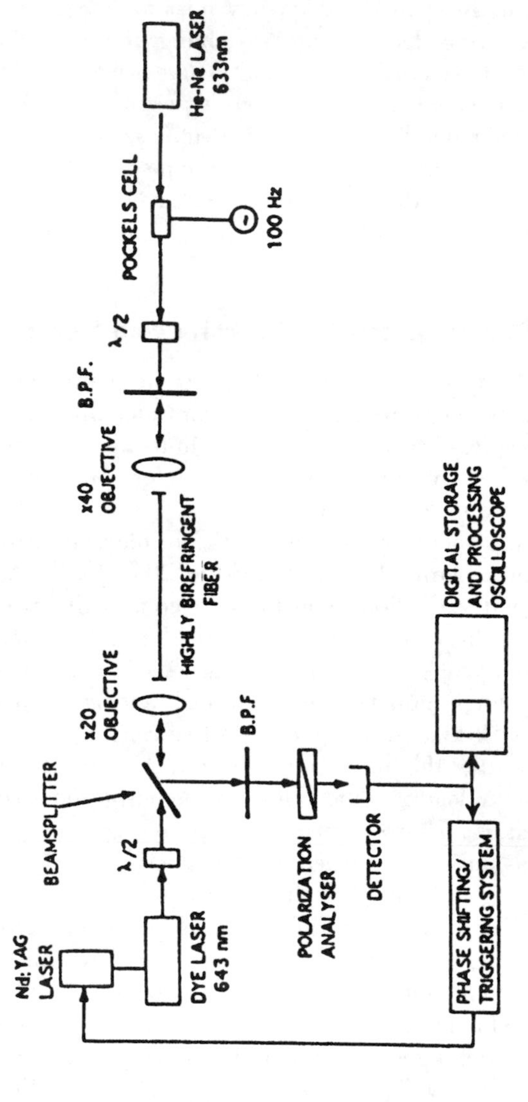

Figure 14.20 Experimental arrangement for polarization-state-dependent Kerr effect F/S DOFS.

The fiber used in the experiments [43] was a monomode high-birefringence fiber manufactured by Andrew Corporation, with a diameter of 67 μm, attenuation of 35 dB/km at 633 nm, and core-to-cladding refractive index difference $\Delta n = 0.032$. The length of the fiber was about 100m. The arrangement of the experiment is shown in Figure 14.20. Pump pulses (617 nm) of 8 ns (FWHM) duration were generated in a dye laser with a repetition rate of 50 Hz. These pump pulses were launched onto one of the birefringence axes of the fiber with the help of a half-wave plate and with a peak power of 3W measured at the output end of the fiber. The linearly polarized probe beam of wavelength 633 nm, from a He-Ne laser, was launched into the fiber at 45 deg to the birefringence axes. On emergence, the probe beam was directed by a beam splitter to the detector, via the polarization analyzer; its average power at the detector was about 25 μW. The He-Ne laser and the detector were protected from the pump light by use of bandpass filters at 633 nm. Force was exerted by pressing metal rods on the fiber. The received signals were recorded, averaged, and differentiated using the functions of the digital storage oscilloscope.

In the absence of any measurand-induced perturbation, the Kerr effect of the pulse is to modify the local value of birefringence as it propagates, a modification that is sensed by the probe beam as a phase shift between the eigenaxes, and that is, in principle, constant for the duration of the pulse's passage through the fiber. The effect of this phase shift on the optical signal passing through the analyzer is shown in Figure 14.21. In practice, the slow fall in the value of the phase shift is due to the attenuation of the pump pulse with distance along the fiber. Figure 14.21 also shows the effect of differentiating this signal with respect to distance.

Figure 14.22 shows the fluctuating analyzer signal when the fiber was perturbed at two points, and its differential with respect to distance. The points at which the weights were applied are clearly evident. Such a system, even as it stands, could be used as an intruder alarm or as an indication of anomalous disturbance of almost any kind.

A further development is shown in Figure 14.23. In order to guard against the possibility of the detector system possessing low sensitivity to the polarization change brought about by the perturbation (by virtue of the polarization bias comprised by the emerging state), polarization diversity detection is employed: three different detectors are employed, each with a different inserted bias, corresponding to an effective measurement of the Stokes parameters [45] for the emerging polarization state. The processing then takes the modules of each detected signal and adds them to produce a final output, as shown in Figure 14.24.

Of course, it is true that if the perturbing force acts along one of the eigenaxes, no rotation of the axes occurs and there is no resultant polarization perturbation. Hence, either the direction of the perturbing force must be shown (e.g., a vertical

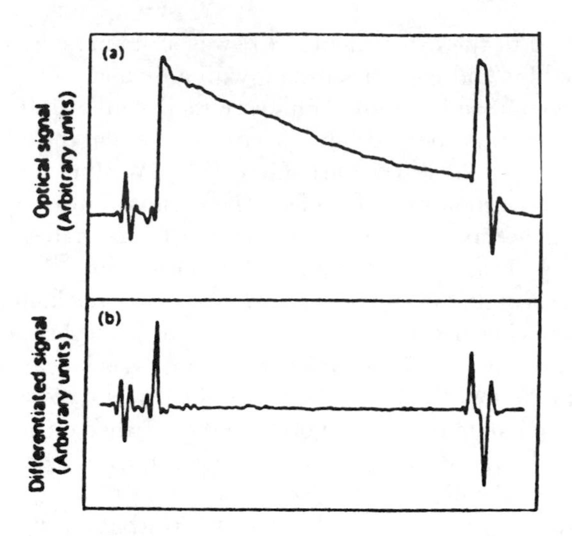

Figure 14.21 Kerr effect phase shift.

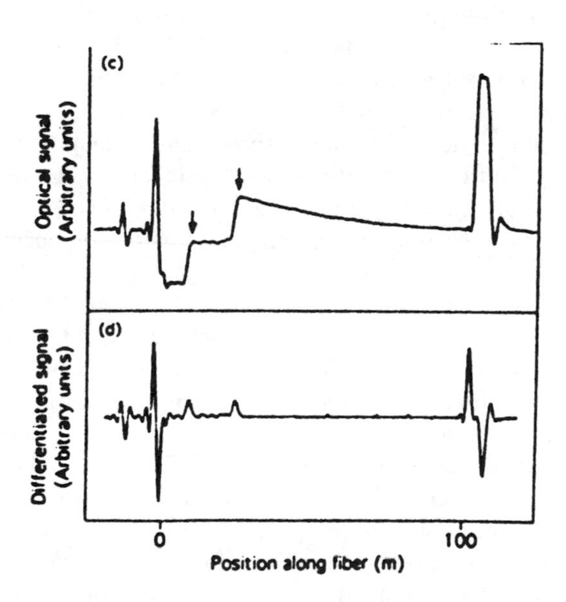

Figure 14.22 Effect of stress perturbations on Kerr effect phase shift.

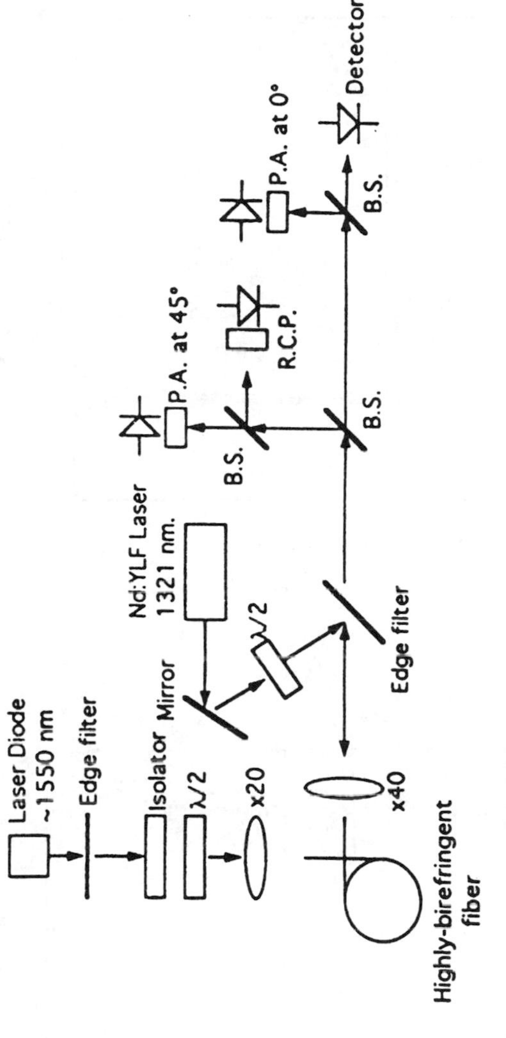

Figure 14.23 Experimental arrangement for polarization-diversity detection.

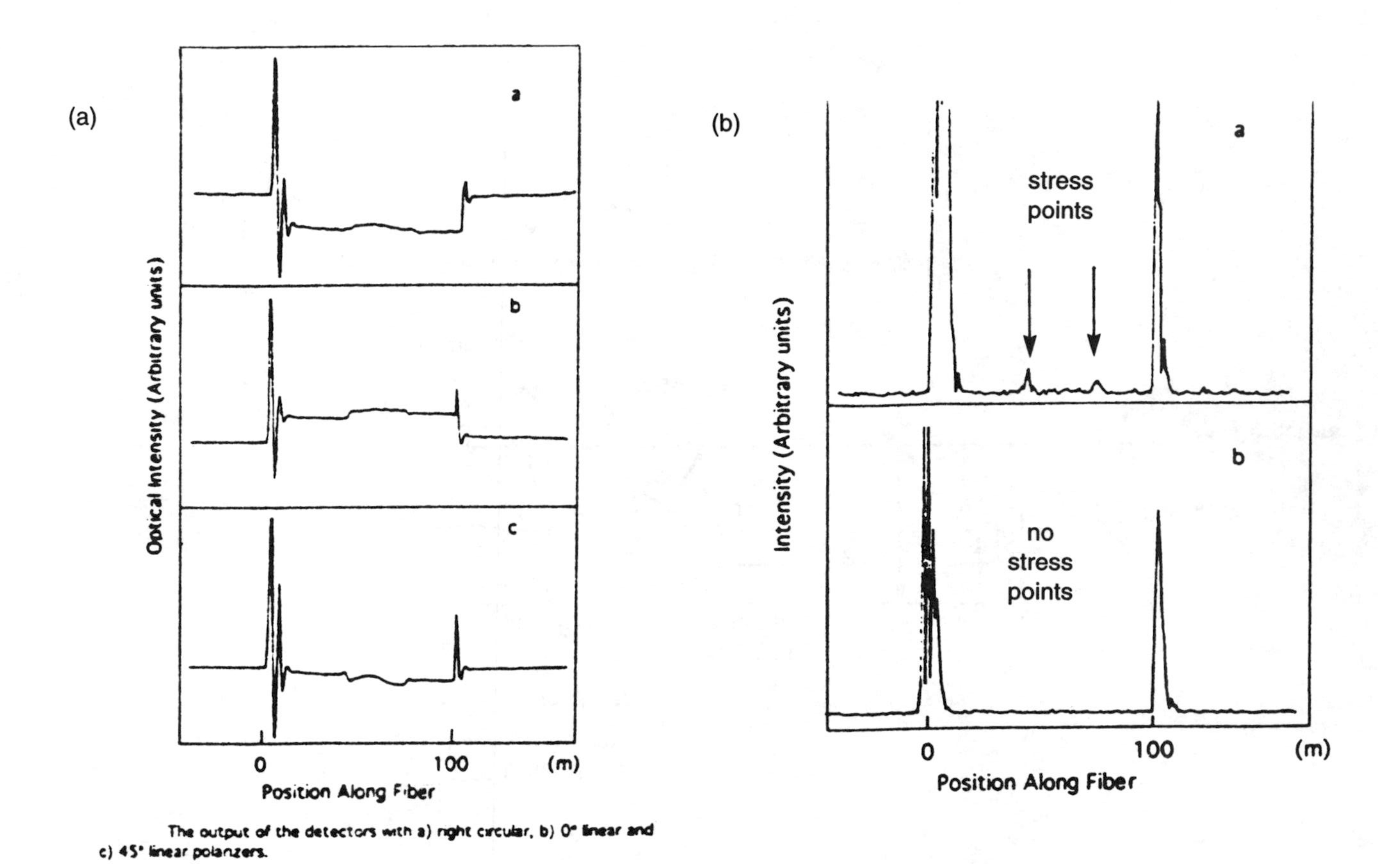

Figure 14.24 Perturbational signal for polarization-diversity detection: (a) individual detector outputs and (b) summed-moduli signals.

weight) or two fibers may be used: these should run closely parallel and be orientated with their birefringence axes at 45 degrees to each other.

Thus is demonstrated the use of the optical Kerr effect to determine the locations of discrete mode-coupling points spaced along a polarization-maintaining fiber. Differentiation of the received signal with respect to time provides a simple way to reduce confusing interactions when multiple coupling points are present. Further work on this method is continuing, with a view to developing practical real-time processing and theoretical analysis of the potentialities of the technique.

14.3.7 DOFS by Pulse Walk-Off in Four-Photon Mixing

Our third and final method comprises a scheme for a DOFS system that makes use of the pulse walk-off effect between a pump and waves generated by four-photon mixing (FPM) [46]. This scheme has two distinctive features. These are 1) the application of birefringence phase-matched FPM, which allows a variety of measurands to be detected and 2) a copropagating rather than counterpropagating system, which leads to a very simple optical structure. Only the pump wave needs to be launched into the fiber, thus most potential difficulties that might appear in counterpropagating interactions can be avoided. Measurands such as temperature or strain, which modulate the fiber's birefringence, can be detected with good spatial resolution and high sensitivity. The spatial resolution of such a DOFS system is analyzed by including the effect of fiber birefringence in addition to chromatic dispersion. The useful operating length is calculated by considering the combined effect of group velocity dispersion (GVD) and self-phase modulation (SPM). Effects that might limit the performance of the system, such as stimulated Raman scattering (SRS), are also taken into account. Finally, an experimental arrangement will be considered.

Four-photon mixing in optical fibers is a well-known phenomenon [47]. It is a nonlinear effect, again mediated by $\chi^{(3)}$ (so that it is essentially another aspect of the optical Kerr effect). The process relies upon the fact that, for optical frequencies ω_p, ω_a, ω_s such that

$$2\omega_p = \omega_a + \omega_s$$

then $\chi^{(3)}$ allows the mutual generation processes:

$$\chi^{(3)}E_p^{\,2}E_a \rightarrow E_s$$
$$\chi^{(3)}E_p^{\,2}E_s \rightarrow E_a$$

where E_p, E_a, and E_s represent the optical electric fields at the pump (ω_p) and anti-Stokes (ω_a) and Stokes (ω_s) frequencies. The actual values of ω_a and ω_s for a given value of ω_p are determined by the phase-matching conditions (i.e., constructive interference). In a hi-bi fiber, the primary phase-matching condition is realized via the velocity difference between the polarization eigenmodes; constructive interference in the generation, by ω_p, of other frequencies can only occur when the velocity of the pump in one eigenmode is equal to that of ω_a, say, in the other eigenmode, via birefringence and GVD effects. Clearly then, the values of ω_a and ω_s will be dependent upon the local values of birefringence (GVD is not significantly influenced by external factors). It is this dependence that serves as the sensing mechanism in this case.

When a strong pump is launched into the fiber, and if a phase-matching condition is satisfied, significant amounts of energy will be transferred from the pump waves to a frequency-downshifted Stokes wave as well as an upshifted anti-Stokes wave. Both waves can be generated from spontaneous emission if no signal is input initially. By using birefringence phase-matching techniques [48], the frequency shift between the pump and the generated waves, $\Delta\Omega$, is determined by the fiber's birefringence and the pump wavelength.

In a fiber made of dispersive media, both Stokes and anti-Stokes waves will experience a walk-off effect with respect to the pump pulse. In visible and near-IR wavelengths, the Stokes wave will overtake the pump pulse and the anti-Stokes wave will be delayed. As the strong pump pulse travels down the fiber, both new waves are generated continuously, with frequencies decided by the local birefringence $B(z)$. Thus two temporally broadened and frequency-modulated pulses are produced. A component generated near the launching end of the fiber will arrive at the output end near the leading edge of the Stokes wave, and near the trailing edge of the anti-Stokes wave. If either or both waves are detected and analyzed at the output end, the spatial distribution of birefringence, and hence of a measurand that modulates the birefringence of the fiber, can be acquired.

Figure 14.25 shows this idea schematically. Stokes and anti-Stokes waves generated near the launching end of the fiber are shown with oblique shading, while those components generated near the middle of the fiber are shown with vertical shading; and, finally, components generated near the output end of the fiber are shown unshaded. The temporal dispersion is indicated by the migration of each part of the generated waves away from the center of the pump pulse, until the waves leave the fiber.

As shown in Figure 14.25, the spatial resolution of this type of DOFS is decided by the walk-off parameter. Considering a partially degenerate FPM case where the two pump photons are identical, the phase matching condition will be

$$2k_p - k_s - k_a = 0$$

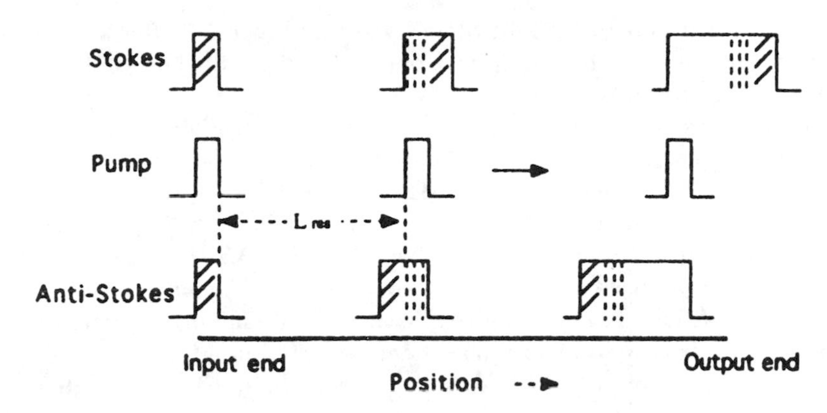

Figure 14.25 Schematic illustration of walk-off phenomenon for pump, Stokes, and anti-Stokes waves.

The spatial resolution distance L_{res} is decided by $L_{res} = T/(d_m + d_b)$ where d_m is the contribution of chromatic dispersion to the walk-off parameter:

$$d_m = \beta\Delta\Omega/2\pi c$$

and d_b is the birefringence contribution: $d_b = B/c$.

Here, T is the pump pulsewidth, β is the dispersion coefficient, $\Delta\Omega$ is the frequency shift, B is the fiber birefringence, c is the velocity of light in vacuum, and k_p, k_s, and k_a represent the propagation constants of the pump, Stokes, and anti-Stokes waves, respectively.

It is necessary to discuss the role of the birefringence contribution d_b to the walk-off parameter in a birefringence phase-matched FPM system. In the case of small frequency shift, where the pump wave is divided equally between the two polarization eigenmodes, the Stokes wave is polarized parallel to the slow axis of the fiber and the anti-Stokes wave parallel to the fast axis (this corresponds to Process II in [48]). For this situation, it can be shown that the walk-off effect due to chromatic dispersion will be canceled completely by the birefringence contribution for both the Stokes and the anti-Stokes waves (i.e. $d_m = -d_b$). This results in a very long interaction length, which is undesirable for sensing purposes. However, in the case of a large frequency shift, where the pump is polarized parallel to the slow axis only while both Stokes and anti-Stokes waves are polarized parallel to the fast axis (corresponding to Process V in [48]), it is found that the birefringence contribution cancels only a fraction of the walk-off in the anti-Stokes wave and increases the walk-off in the Stokes wave. In this case, d_m and d_b have opposite sign for the anti-Stokes wave and the same sign for the Stokes wave.

For example, considering the Stokes wave in a system adopting the latter phase-matching technique, if we use typical fiber parameters: $\lambda = 532$ nm, $\boldsymbol{\beta} = 2.3 \times 10^{-5}$ cm, $B = 6.4 \times 10^{-4}$, then we have $\Delta\Omega = 1.8 \times 10^{3}$ cm^{-1}, $d_m = 22.1$ ps/m, $d_b = 2.1$ ps/m. For a resolution distance $L_{res} \approx 1$m, we find $T \approx 25$ ps will be needed. If a temperature distribution is to be measured, taking $\Delta B/(B\Delta T) \sim 2.5 \times 10^{-3}$ °C (typical for strain-induced birefringent fiber), then the frequency difference corresponding to 1°C is found to be about 67.5 GHz or 0.064 nm in wavelength.

It is relatively easy to realize phase-matched FPM in a birefringent fiber. Following the calculation of [49], we find that, even with 1m interaction length and 10W pump power, the Stokes (or anti-Stokes) signal amplification factor is expected to be well over 10 in a single-mode fiber operating at 532 nm, which provides sufficient SNR for detection. For this arrangement, signal intensity is less important than the signal frequency.

A possible experimental arrangement is shown in Figure 14.26. A mode-locked laser can be used to provide short pump pulse, on the order of 20 ps, for launching into highly birefringent single-mode fiber. The pump pulses are launched into the slow axis of the fiber with the help of a half-wave plate. A bandpass filter should be used to block the pump pulses at the output end and to pass the Stokes or anti-Stokes signals. Time-resolved optical spectrum analysis is needed to measure the output wavelength versus time. There are commercialized fast detectors, such as multichannel streak cameras, available to fulfill this task. Finally, the time scale corresponds to the position within the fiber and the wavelengths of the recorded signals reflect the local birefringence.

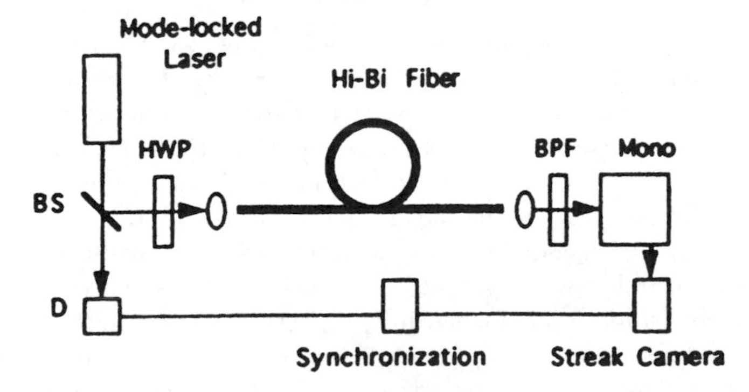

Figure 14.26 Four-photon mixing experimental arrangement (BPF: bandpass filter; HWP: half-wave plate; BS: beamsplitter; D: detector; M: monochromator).

This type of sensor clearly has advantages of simplicity, in that it requires only a pump laser. However, spectral analysis is needed for the detection of the signal, and very short duration pulses are required. Work is continuing on this system in an attempt to improve its appeal.

14.3.8 Summary

Distributed optical-fiber measurement systems offer an extra level of information gathering for large structures. The potential for applications in civil engineering, aerospace engineering, mining, petrochemical engineering, and in research diagnostics is immense.

In order to achieve high spatial resolution with good sensitivity and relatively simple signal processing, it is very fruitful to look towards systems that use forward-scatter techniques, polarimetric techniques, or both simultaneously. The optical Kerr effect is extremely useful in this regard.

This section has indicated some of the possible arrangements of this type that are under active investigation, and has illustrated the kind of system that might prove viable in certain application areas. There is continuing development of these ideas. It is clear that there is much work yet to be done before such methods can be shown to be viable for operational measurement. In particular, it is clear that great reliance will, of necessity, be placed upon the fabrication of special fibers for enhancement and control of the nonlinear optical interactions, for control of the polarization state, and for optimized interfacing with the measurand field. The potential payoffs in terms of a generic range of fully distributed optical-fiber sensors capable of providing for the majority of requirements in industrial measurement (in addition to allowing access to an extra level of information) are deemed such as to make the research investment conspicuously worthwhile.

14.3.9 Acknowledgments

Section 14.3 has provided a summary of work that has been done by a number of people, as is evidenced by the references in the text. In particular, the excellent efforts of the following are gratefully acknowledged: Dr. V. A. Handerek, Dr. L. C. G. Valente, Dr. I. Cokgor, Mr. S. U. Ahmed, Mr. J. Zhang, and Mr. F. Parvaneh.

14.4 NEW DISTRIBUTED CHEMICAL MEASUREMENT SYSTEMS

Distributed chemical measurement systems are a relatively recent addition to the optical-fiber sensors repertoire. A number of new sensing techniques have emerged, typically based on either distributed spectroscopy or on chemical to mechanical measurand transformers. This section examines examples of these system concepts.

14.4.1 Introduction

The ability to address a large number of electrically passive sensor locations and provide information on measurement parameters as a function of linear position is often seen as a key advantage that fiber-optic sensors have over other technologies. This capability enables complex sensor arrays to be constructed with the minimum of wiring difficulties. In recent years, much research effort has focused on developing the means to exploit these advantages using a diversity of techniques such as wavelength division multiplexing of Bragg gratings [50], optical time-domain reflectometry (OTDR) [51], Raman OTDR [52], and Brillouin OTDR [53]. The common feature of these measurement approaches is that they all address physical parameters such as temperature and strain. The following section will review recent methods that have been examined for the purpose of recovering spatially distributed information relating to chemical parameters.

14.4.2 Evanescent Wave Sensors

While a great deal of work has been carried out on discrete chemical sensors using optical fibers, the development of practical concepts for distributed measurement systems has been limited. Systems considered generally employ the evanescent field, allowing interaction with chemical species along extended lengths of fiber (see Chapter 3 on evanescent sensors). In principle, local changes in attenuation or index caused by the chemical species can be interrogated using OTDR to recover the information on a distributed basis, as illustrated in Figure 14.27. In practice, these changes can be extremely small, especially when attempting to produce gas absorption sensors in the near infra red (IR), and the most viable systems to date are based on the use of a cladding material impregnated with indicator dyes to mediate the sensing. Preliminary work on a quasi-distributed sensor of this type was first reported in 1989 by Kvasnik and McGrath [54]. They used a multimode, 200-mm core diameter, plastic-clad silica (PCS) multimode fiber for evanescent field interaction. Three sensor regions, each 25 cm in length, were formed in series and spaced at intervals of 5 or 10m in a 25m long fiber section. The sensor regions were formed by removing the plastic cladding and replacing it with a cellulose acetate film containing the immobilized pH-indicator dye, cresol red. Excitation of the dye at 540 nm was accomplished using 5-ns pulses from a nitrogen-pumped dye laser and changes in the time-resolved backscattered signal were monitored as the sensor regions were selectively exposed to acidic or alkaline vapors (1.8×10^{-4} mole/liter HCl or 5.6×10^{-3} mole/liter NH_3 in air). Qualitative changes in the backscattered signature were observed as the different sensor regions were exposed, but further work would be necessary to make a viable system.

More recently, Kharaz and Jones [55] have reported a similar type of

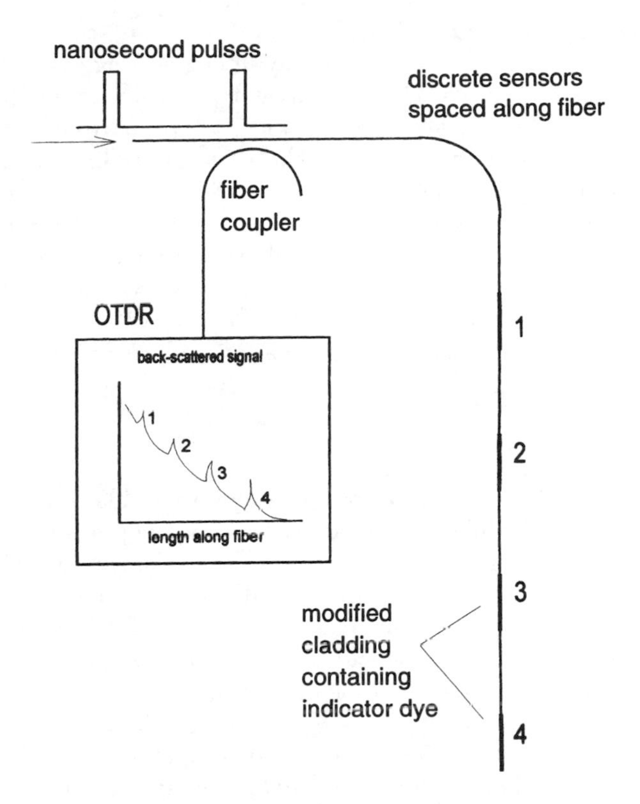

Figure 14.27 Distributed sensing system based on evanescent field absorption.

quasi-distributed system for relative humidity measurement. In this case, the sensor region was constructed by replacing a 50-mm length of cladding of a 600-mm core hard-clad silica (HCS) fiber with a gelatin film containing cobalt chloride. The cobalt chloride salt changes color (blue to pink) on exposure to water molecules, and the change in absorption at 670 nm was monitored. A second measurement at 850 nm, where the absorption is unaffected by relative humidity, was used as a reference. One important difference from the previous example is that the gelatin film index is higher than that of the silica core, so the interaction is much stronger than that via the evanescent field. For demonstration purposes, four sensors were formed in a length of fiber and spaced at intervals of 20m. An OTDR with two pulsed laser diode sources at 670 and 850 nm was used to interrogate the network. Tests were carried out over a 20 to 80% range of relative humidity, and attenuations, typically of about 1 to 3 dB for this range of RH, were measured at each sensor from the OTDR trace.

Despite the strong interest in optical chemical sensors based on evanescent techniques, there has been little activity in the way of engineering evaluations of sensor systems based on these concepts. The difficulty of reliably protecting the optical-fiber sensor, while allowing the optical field to interact with the surrounding environment in a stable and predictable manner, is one of the main causes. Interaction between the target chemical species and the light guided in the fiber core necessarily implies that the fiber itself is exposed to the surrounding environment. The fiber may then be vulnerable to the influence of OH ion ingress, which has a dramatic effect on its strength. The conflicting requirement of allowing the optical wave within the fiber to interact with the chemical parameters to be measured and ensuring that the fiber is protected from undesirable chemical attack presents a considerable technical challenge.

14.4.3 Microbend-Based Sensors

14.4.3.1 General Concept

Until recently, the use of mechanical transducers, or novel cable designs, which enable the target parameter to modulate the microbend loss in an optical fiber, had not been explored to any degree. This is perhaps surprising since changes in backscatter signal strength can readily be measured to within 0.01 dB and located (with centimeter resolution) using a wide range of commercially available OTDR instruments. This is therefore a convenient means by which existing measurement instruments can be used to produce a complete range of distributed measurement systems. The approach is particularly suited to instances where a simple on/off alarm indicator is all that is required. A recent development making use of this capability is a distributed water detection sensor that has been demonstrated to be effective in detecting water ingress in cables, which could extend to several kilometers in length [56].

14.4.3.2 Hydrogel Sensors: Principle of Operation

Hydrogels are a range of hydrophilic macromelecular polymers that swell when absorbing water, but do not dissolve in aqueous media [57]. They have been used extensively in biomedical and pharmaceutical applications where their ability to swell in the presence of target stimuli is used for controlled drug release devices [58]. The swelling action of the hydrogel is readily converted to a mechanical response in the form of force, pressure, or bending moment, and is the basis of the detection process in the microbend-based water sensor.

The water sensor, depicted schematically in Figure 14.28, converts the swelling action of the hydrogel to microbend losses. Hydrogel is deposited onto a central supporting carrier and is held in contact with an optical fiber by a helically wound Kevlar thread. In the presence of water, the gel swells and squeezes the fiber against the Kevlar winding. Light passing through the fiber at this point will therefore suffer microbend losses, which can be used to identify the presence or absence of water at

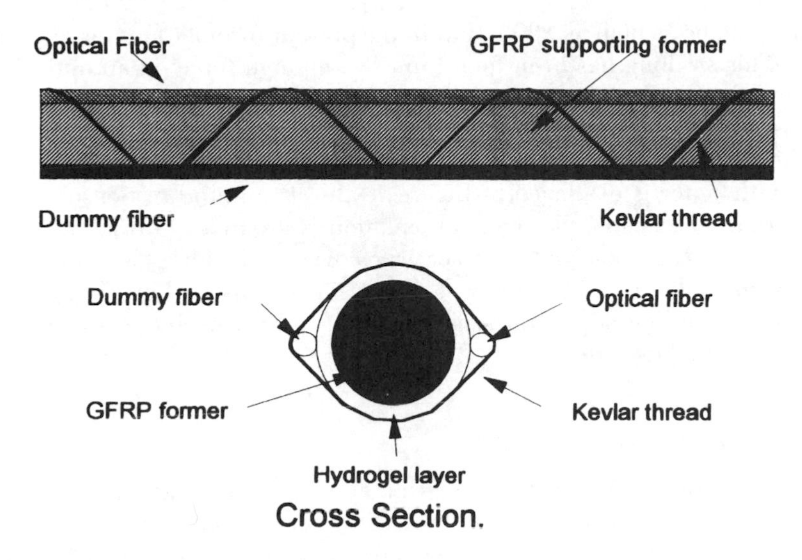

Figure 14.28 Schematic diagram of a hydrogel-based sensor.

this location. Resonant power transfer between radiative and guided modes within the fiber takes place most strongly when the Kevlar winding pitch Δ_m is given by

$$\Delta_m = \frac{2\pi a}{\sqrt{2\delta}}$$

where a is the core radius and δ is the maximum refractive index difference between the core and the cladding [59]. Standard acrylate-coated graded-index multimode fiber, with core and cladding diameters of 62.5 μm and 125 μm, respectively, displays a resonance peak when the microbend spacing is approximately 1 mm. In the sensor cables constructed here, the Kevlar thread introduced a microbend perturbation with a 2-mm spatial periodicity. The hydrogel used was one of a family of poly(ethylene oxides) PEO formed from the polymerization of ethylene oxide monomers [60].

14.4.3.3 Sensor Characterization

The sensor performance can be quantified by measuring the change in signal loss per unit sensor length when the target measurand is present. Once the spatial period of the microbend is established, this loss value is a function of the tension in the wrap and the gel thickness. Hydrogel polymers absorb different amounts of water according to their composition and consequently swell to different degrees. The volumetric expansion can be as high as 200%, but in the present formulation it is approximately 35%. This swelling has been found to exert enough force on an optical fiber to induce losses of 110 dB/km with a layer of gel [56], which is less than 20 μm in thickness.

The evolution of the signal loss as a function of time after the sensor comes in contact with water is displayed in Figure 14.29. The initial response is extremely rapid; within 30 seconds, the signal attenuation is around 60 dB/km. The swelling process is completely reversible; the sensor recovers to its base loss state as it dries. Generally, in a laboratory environment the sensor will dry within 15 minutes. However, under humid conditions, the sensor can take up to several hours or even days to recover to its low-loss state.

14.4.3.4 Distributed Water Detection

The sensor has been developed as a distributed monitor for water ingress that operates as an alarm to indicate when (and where) water has penetrated into a particular piece of apparatus. Increases in the attenuation gradient of the Rayleigh backscatter signal are used to identify wet regions. In the example displayed in Figure 14.30, the sensor shows two distinct sections of high loss situated between the positions of 9m to 15m and from 21m to 26m along its 30m length. In practice, detection of wetted lengths of sensor smaller than 1m in extent might be required. With the present sensor

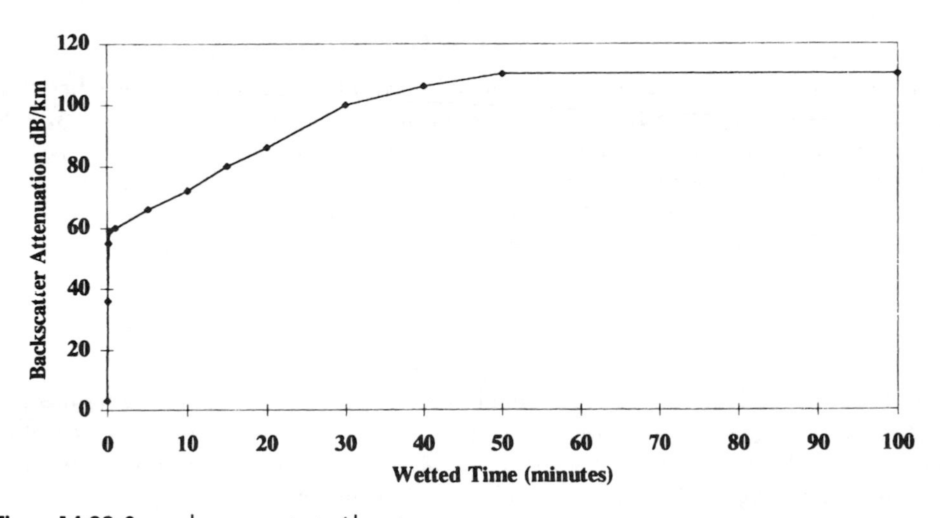

Figure 14.29 Sensor loss on contact with water.

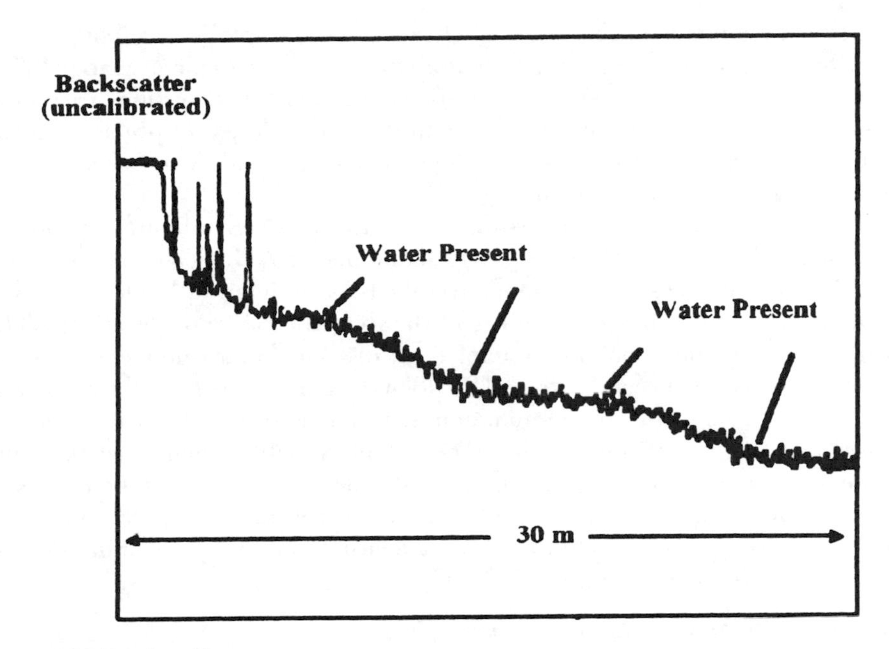

Figure 14.30 Distributed water ingress measurement.

design this can be readily achieved without difficulty. A thin layer of hydrogel produces a loss of around 100 dB/km, or 0.1 dB/m when wet. An OTDR unit with a loss resolution of 0.01 dB can detect variations with a low probability of false alarm provided the loss is greater than 0.05 dB; therefore, 0.5m spatial resolution could be readily obtained. Improving the resolution further is possible by increasing the coating thickness or the winding tension, but this is achieved at the expense of operational range, which is dictated by the dynamic range of the OTDR unit.

14.4.3.5 Measurement of pH Value and Other Chemical Parameters

A powerful feature of this sensor is its simplicity of construction and generic design, enabling it to be readily adapted to detect a number of chemical species without modifying the basic sensing process. The active component in the sensor is the hydrogel itself. These gels have been demonstrated to be sensitive to a number of parameters, including pH, amino compounds, ionic strength, photoirradiation, and temperature. Selection of an appropriate gel system allows the sensor to be adapted to suit a specific measurement requirement. Figure 14.31 shows the water uptake (swelling) characteristics of a recently developed gel system that is responsive to changes in pH [60]. This gel has the same host polymer as that used in the moisture detection experiments and is therefore immediately compatible with the sensor design to provide a sensor with a pH-triggered response. Other pH ranges can be covered by incorporating different acid (or basic) complexes into the gel.

With appropriate signal processing, the sensor can be configured to monitor analog quantities (such as pH value) provided that the gel swelling characteristics can be tuned to vary in a monotonic manner over the range of interest. While this may be possible in many circumstances, limits in the dynamic range of practical OTDR instruments restrict the capability to implement distributed measurements. Typically, OTDR units have between 10 dB and 20 dB of dynamic range from the ratio of the normal level of backscatter to the minimum detector sensitivity. Distributed analog measurements can rapidly deplete the system of all available signal, since the signal must be attenuated in order for the parameter under investigation to be measured. With the above OTDR parameters and a specified measurement precision of 1%, between 10 and 20 sensors could be simultaneously addressed (assuming that the OTDR can resolve signal changes of 0.01 dB).

14.4.3.6 Soil Water Content Measurements

The hydrogel sensor was designed to detect the presence (or absence) of water. In the applications envisaged, it was assumed that free water would be available, and the response of the sensor to different relative humidity (RH) conditions was not considered. However, the basic feature of the sensor is that the gel swells in accordance with the amount of water that it absorbs; therefore, in situations where the RH value is high but less than 100%, it is possible for the sensor loss to increase. Characterization of this kind of response is important not only from the viewpoint of sensor stability, but also because it opens up the possibility of a new range of sensors focusing on relative humidity value.

The equilibrium water uptake (EWU) of a hydrogel polymer is the standard means of determining the ability of the hydrogel to absorb water (and hence swell) in a particular environment and is defined as

$$\text{EWU} = \frac{\text{SW} - \text{DW}}{\text{DW}} \times 100$$

where SW is the swollen weight of the gel and DW is the dry weight. The variation of hydrogel EWU with relative humidity value, estimated using samples of bulk gel cast from the same host polymer as used in the sensor, is displayed in Figure 14.32. The gel samples were suspended above a series of saturated salt solutions in sealed

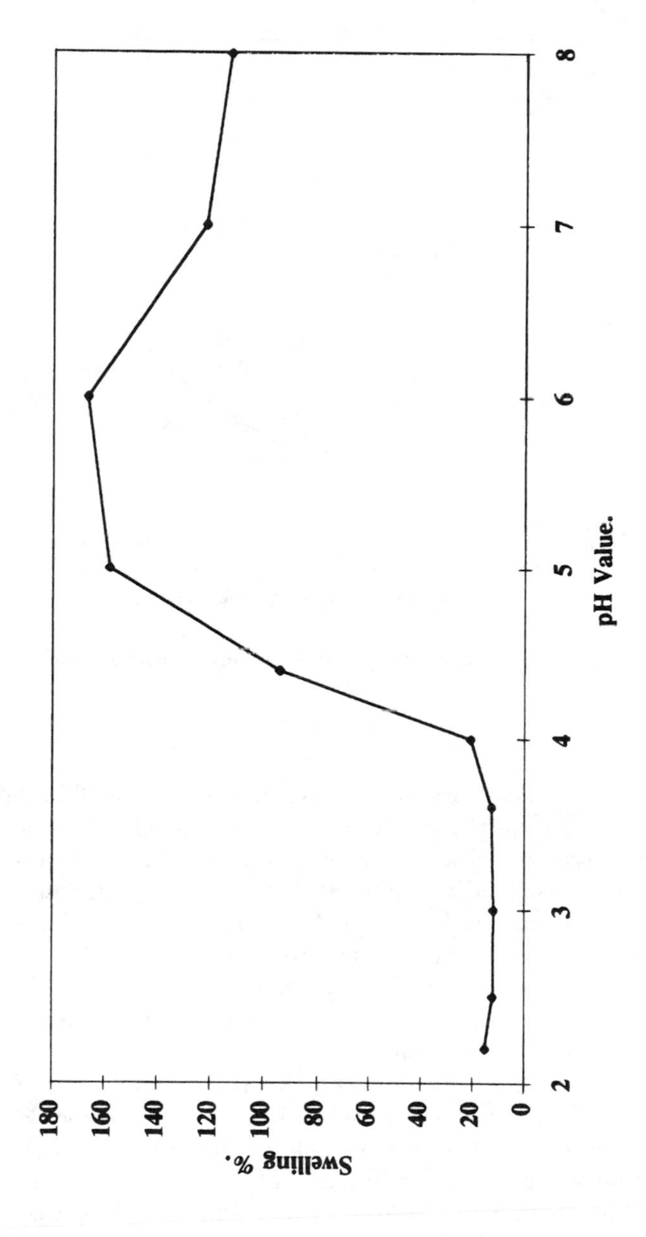

Figure 4.31 Swelling behavior of pH-sensitive gel.

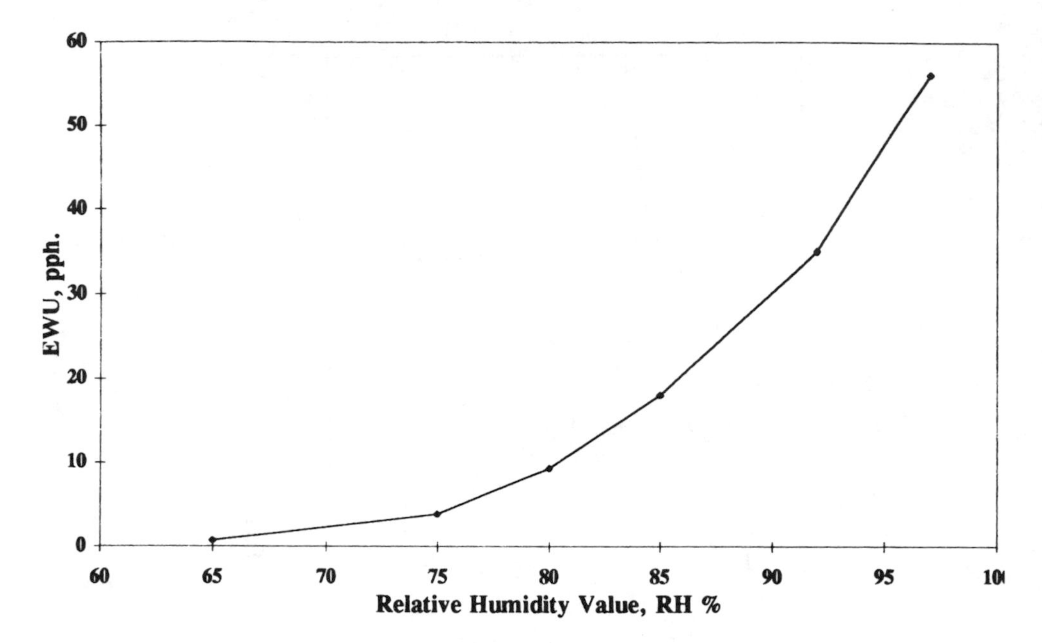

Figure 14.32 Equilibrium water uptake of hydrogel samples vs. relative humidity level.

containers. Selection of the appropriate salt solution allows the RH value of the air in the confined space adjacent to the solution to be controlled [61]. The samples were allowed 48 hours to reach their EWU value, and these values were recorded over RH values ranging from 65% ($NaNO_2$ solution) to 97% (K_2SO_4 solution).

The EWU can be observed to rise for RH values above 70%, reaching a maximum value of around 60 parts per hundred (pph) when the RH value reaches 97%. It should be noted that this is significantly lower than the EWU value for hydrogel samples immersed in distilled water when the gel takes in approximately 140 pph. The implication of this is that the sensor will be much less sensitive to changes in relative humidity value and that it will not respond at all at RH values of less 70%.

Investigation of the effect of relative humidity on the water sensor is more difficult since the sensor cable takes up more volume than the small hydrogel samples; therefore, the relative humidity is more difficult to control. However, preliminary experiments indicate that the conclusions reached above are broadly correct. The sensor attenuation is significantly lower, even at high RH values, than is observed for immersion in free volume (see Figure 14.33). However, it must be noted that significant experimental difficulties are encountered in attempting to regulate (or indeed

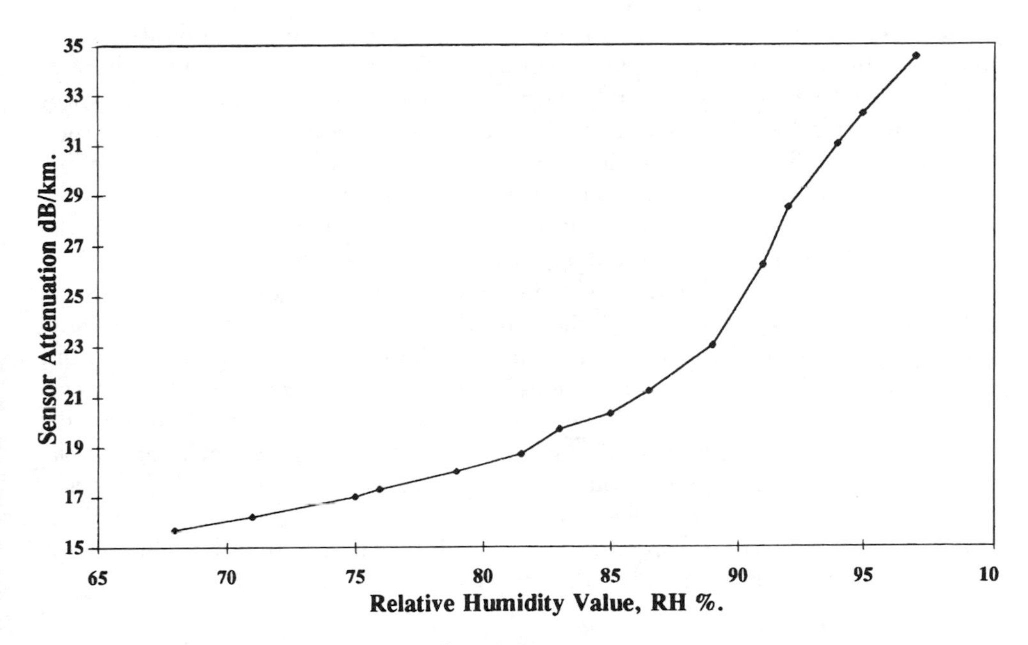

Figure 14.33 Sensor attenuation vs. relative humidity.

measure) the humidity of an environment at high RH values (> 95%). The data presented in Figure 14.33 can therefore only be used to provide a general indication of the sensor behavior.

Much interest in the use of this type of sensor to determine soil water content has emerged. In civil engineering, the soil water content is important since it has a strong influence on the ground load-bearing capability. In agriculture and soil science, the interest lies in determining relationships between soil water content and plant/crop growth. Although the sensor displays a much reduced sensitivity to changes in relative humidity, this need not necessarily be the case for soil water content measurements, where the water transport mechanisms are different, as is illustrated in the following.

The obvious route to assess the capability of the sensor to determine soil water content is to add water to dried soils and allow the mixture to reach equilibrium. However, this can be an extremely time-consuming process (taking several weeks), especially for large soil samples. To circumvent this difficulty, soil scientists generally use salt solutions of different concentrations to simulate the relative humidity level of soils [62]: the water uptake of plants can be directly related to the ionic content of water (concentrated solutions represent dry soil conditions). The EWU of bulk hydrogel polymer and the corresponding sensor attenuation over a range of NaCl solution concentrations [63] are displayed in Figures 14.34 and 14.35. The attenuation of the sensor can clearly be seen to fall as the salt concentration is increased (representing the drier soil condition). As the solution concentration is increased towards the 6-molar level, the signal attenuation approaches that of the dry sensor.

Clearly, the sensor displays a significant variation in signal attenuation over a range of soil moisture conditions. Full calibration of the sensor as a means of measuring soils moisture content on a spatially distributed manner is progressing, but early results confirm what the experiments outlined above suggest (i.e., that the sensor can detect differences in soil water content with a high degree of precision). Figure 14.36 shows the evolution of sensor attenuation over a period of several weeks following the addition of 10 cc of water to a 5-kg sample of oven dried soil. The steady rise in the sensor attenuation reflects the time required for the soil to reach an equilibrium water content. The subsequent decay indicates that during the course of the experiment, water was being lost to the external environment.

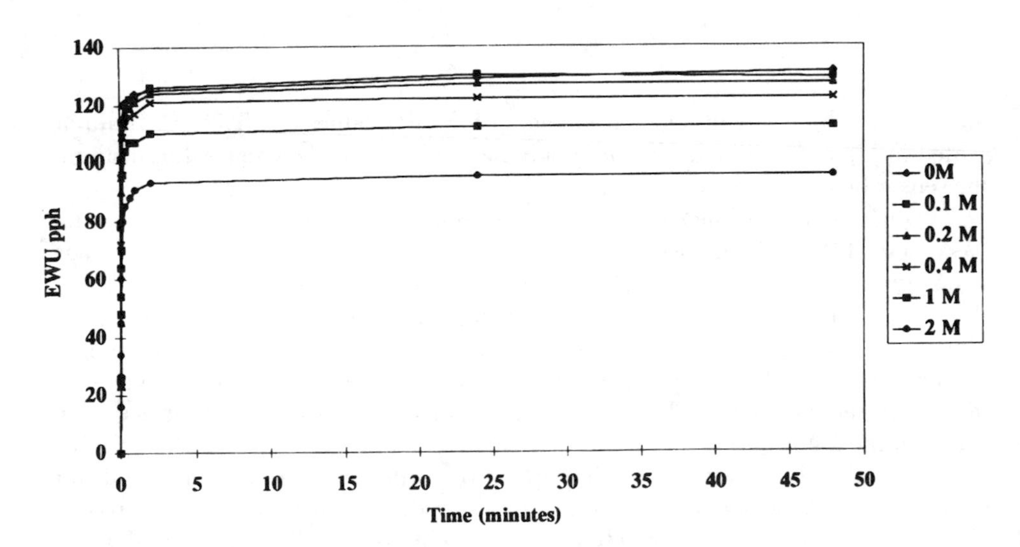

Figure 14.34 Equilibrium water uptake for hydrogel in NaCl solutions.

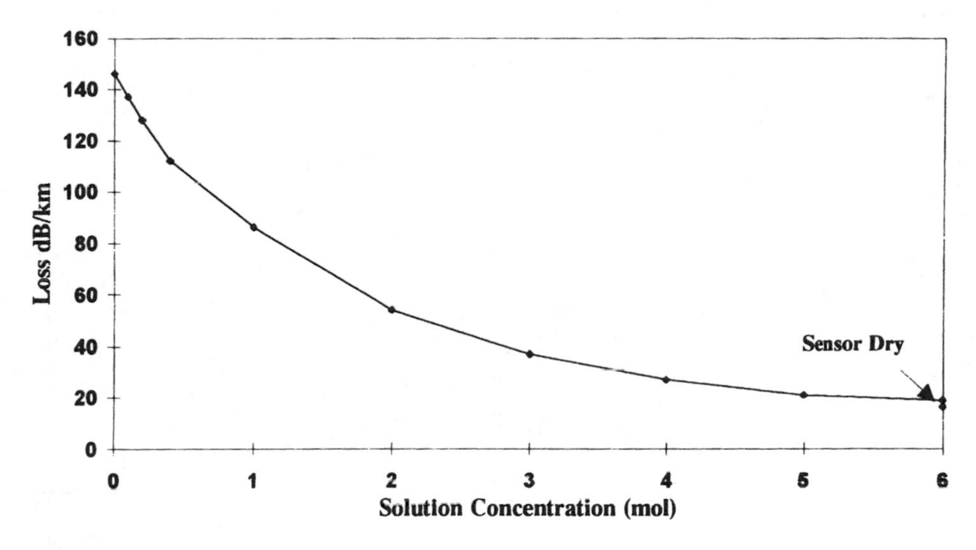

Figure 14.35 Sensor signal attenuation in NaCl solutions.

14.4.3.7 Application Trials: Detection of Grout Through Water Presence

The hydrogel sensor was developed as a water detector aimed at applications such as protection of telecommunications cables against water ingress. Additional applications have also emerged in areas that are less obvious. The following section outlines one such use.

Post tensioning is a strengthening method for concrete structures whereby steel reinforcing tendons are inserted into ducts and tensioned to apply a compressive load to the concrete. This ensures that the concrete, which has a low tensile strength, remains in compression under all the design loading conditions of the structure. The force each tendon exerts to achieve this aim is extremely high, and if a tendon begins to corrode there is a high risk of dramatic failure with a subsequent loss of support for the host structure. Therefore, to guard against this, the ducts that contain the supporting tendons are commonly filled with a cement-based grout that acts as a barrier against water and chemical ingress. One of the difficulties that a site engineer faces is determining when the duct is completely filled with grout. Voided regions can arise for a variety of reasons, and these areas can compromise the protection offered by the grout. The cement grout used to seal the ducts has a high water content, therefore it was decided to use the distributed water detector as a means of assessing the level of grout fill along the duct.

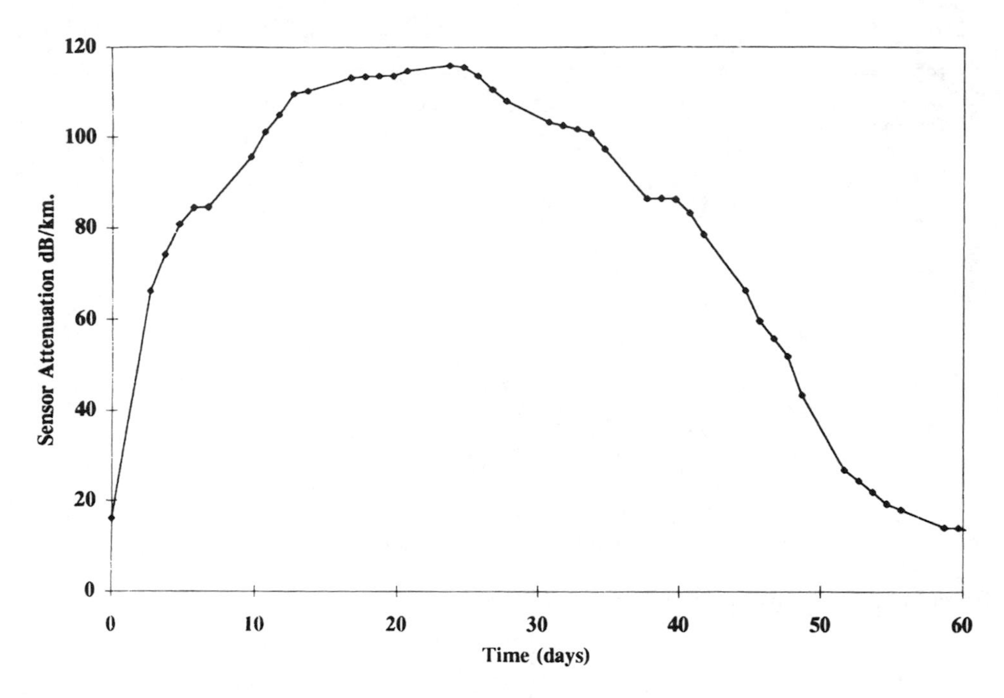

Figure 14.36 Evolution of sensor attenuation in soil: 10 cc of water added to 5 kg of dry soil.

To evaluate the potential of the sensor as a grout fill monitor, grouting trials were carried out on a scaffold frame assembly containing a 100-mm diameter tendon duct 17m in length. The anchorage units at each end of the duct had provision for 19 stranded tendons. Eighteen tendons were inserted into the duct and tensioned, and a spare hole was left for the hydrogel cable. A 17m length of cable was installed with the use of a draw line. Grouting was performed under normal working conditions and voided regions were then introduced into the duct by puncturing the duct wall and using compressed air to expel grout from selected places. The sensor was then monitored and the loss profile constructed [64]. This profile is displayed in Figure 14.37 along with data on void location as determined from visual inspection. Comparing the areas of low sensor loss and the location of poorly grouted sections along the duct length (Figure 14.37) shows clearly that a high degree of correlation exists between the poorly voided sections and the dry sensor locations. Detecting the water content of the grout therefore provides a powerful means of indicating how well the duct has been filled. All of the significantly voided regions (5, 6, 7, 8, 12) were detected. The amount of grout fill in each of these locations varied from around 30% to 95% of the area of the duct. In all cases, the grout covering of the sensor was inadequate to induce

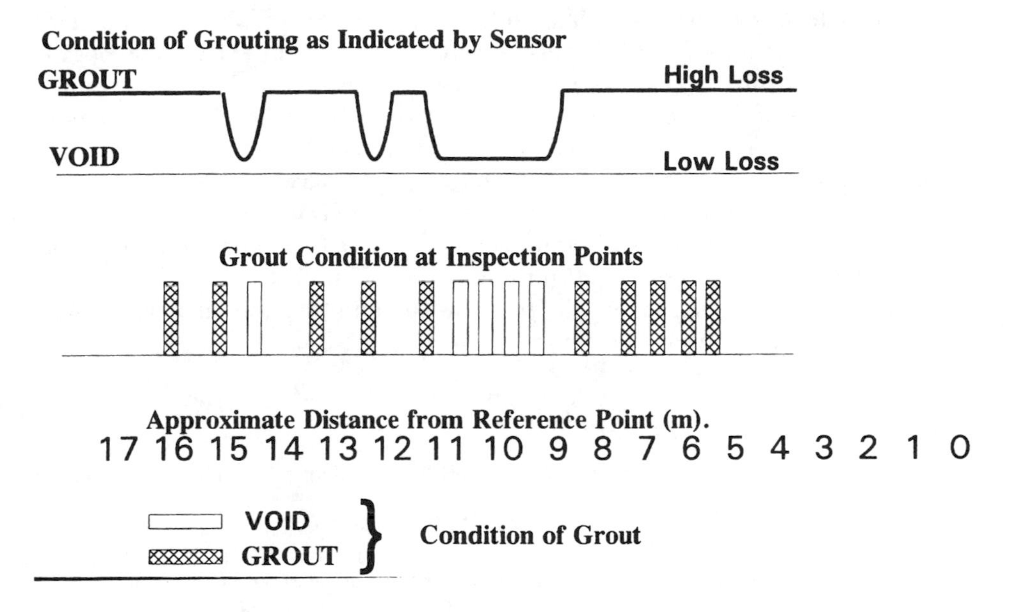

Figure 14.37 Estimation of tendon duct grout fill.

the magnitude of loss that would be consistent with a totally wetted section of sensor and hence a void was assumed to be present. However, it should also be noted that a voided region was indicated around section 10, where inspection of the duct revealed that the sensor was fully immersed in grout. The cause of this is not yet known.

14.4.4 Summary (Section 14.4)

Optical-fiber sensors for a variety of chemical species have been demonstrated for some time now, but development of distributed systems remains an area of active research. Evanescent field techniques offer, in principle, a convenient method for distributed sensing, but in practice a number of difficulties arise. The low percentage of total power that is carried by the evanescent field means that sensitivity is poor and direct absorption-based sensors using near-IR lines of gases are difficult to realize, especially in distributed form. The use of indicator dyes or chemical reagents to mediate the sensing is more promising for distributed systems, but here again the number of IR dyes currently available is limited (most dyes absorb in the blue or UV spectral regions). Development of new IR dyes will be a key factor in future chemical sensor systems. Rather than using evanescent field methods, quasi-distributed systems may be more viable using, for example, micro-optic gas cells inserted at intervals along a fiber-optic system. New sources such as DFB lasers tuned to specific absorption lines will also be important for future systems as well as advances in OTDR technology.

To date, difficulties associated with the protection of the fiber core have limited the investigations of the above distributed systems to laboratory trials. The simpler approach of using a chemically swellable material to modulate the loss in an optical fiber appears to offer a measurement solution that can be developed in a much shorter timescale, and preliminary trials in realistic field environments have been very encouraging. The drawback of this approach is that it is best suited to applications where an alarm signal (yes/no to a chemical presence) is all that is required. The sensor operates by introducing loss to the transmission and therefore analog measurements will always be restricted by dynamic range.

References

[1] Dakin, J. P., *Optical Fiber Sensors Vol. II.*, Culshaw, B. and Dakin, J. P. (Eds.), Boston, London: Artech House, 1989, pp. 575–598.

[2] Horiguchi, T., Kurashima, T., and Tateda, M., "Tensile strain dependence of Brillouin frequency shift in silica optical fibers," *IEEE Photonics Tech. Lett.*, 1, (5), May 1989, pp. 107–108.

[3] Culverhouse, D., et al., "Potential of stimulated Brillouin scattering as sensing mechanism for distributed temperature sensors," *Electron. Lett.*, 25, (14), July 1989, pp. 913–915.

[4] Cotter, D., "Stimulated Brillouin scattering in monomode optical fiber," *J. Opt. Commun.*, 4, (1), 1983, pp. 10–19.

[5] Kurashima, T., et al., "Brillouin optical-fiber time domain reflectometry," *IEICE Trans. Commun.*, E76-B, (4), April 1993, pp. 382–390.

[6] Kurashima, T., Horiguchi, T., and Tateda, M., "Thermal effects on the Brillouin frequency shift in jacketed optical silica fibers," *Appl. Opt.*; 29, (15), May 1990; pp. 2219–2222.

[7] Kurashima, T., Horiguchi, T., and Tateda, M., "Thermal effects of Brillouin gain spectra in single-mode fibers," *IEEE Photonics Tech. Lett.*, 2, (10), Oct. 1990, pp. 718–720.

[8] Horiguchi, T., et al., "Brillouin characterization of fiber strain in bent slot-type optical-fiber cables," *J. Lightwave Technol.*, 10, (9), Sept. 1992, pp. 1196–1201.

[9] Horiguchi, T., et al., "Locating kinks in submarine optical fiber cable," *Tech. Dig. Optical Fiber Measurement Conference*, Torino, Italy, Sept. 1993, pp. 77–80.

[10] Horiguchi, T., et al., "Development of a distributed sensing technique using Brillouin scattering," *J. Lightwave Technol.*, 13, (7), July 1995, pp. 1296–1302.

[11] Savitzky, A. and Golay, M. J. E., "Smoothing and differentiation of data by simplified least squares procedures," *Anal. Chem.*, 36, (8), July 1964, pp. 1627–1638.

[12] Dhliwayo, J., and Webb, D. J., "Temperature error analysis for a distributed temperature sensor based on stimulated Brillouin scattering," *Tech. Dig. Conference on Optical Fiber Sensors*, Sapporo, Japan, May 1996, pp. 554–557.

[13] Healy, P., "Fading in heterodyne OTDR," *Electron. Lett.*, 20, (1), Jan. 1984, pp. 443–444.

[14] Horiguchi, T., Kurashima, T., and Tateda, M., "A technique to measure distributed strain in optical fibers," *IEEE Photonics Tech. Lett.*, 2, (5), May 1990, pp. 352–354.

[15] van Deventer, M. O., and Boot, A. J., "Polarization properties of stimulated Brillouin scattering in single-mode fibers," *J. Lightwave Technol.*, 12, (4), April 1994, pp. 585–590.

[16] Kurashima, T., et al., "Large extension of dynamic range in distributed-fiber strain measurement using Brillouin spectroscopy," *Tech. Dig. Conference on Lasers and Electro-Optics*, Baltimore, MD, May 1991, paper CThN5.

[17] Bao, X., et al., "Experimental and theoretical studies on a distributed temperature sensor based on Brillouin scattering," *J. Lightwave Technol.*, 13, (7), July 1995, pp. 1340–1348.
[18] Horiguchi, T., Kurashima. T., and Koyamada, Y., "1 m spatial resolution measurement of distributed Brillouin frequency shift in single-mode fibers," *Tech. Dig. Symposium on Optical Fiber Measurements*, Boulder, CO, Sept. 1994, pp. 73–76.
[19] Pohl, D., and Kaiser, W., "Time-resolved investigations of stimulated Brillouin scattering in transparent and absorbing media : Determination of phonon lifetimes," *Phys. Rev. B*, 1, (1), Jan. 1970, pp. 31–43.
[20] Ogawa, O., Kamikatano, M., and Matsuzawa, T., "A new technique for measuring a time-varying optical fiber strain," *Tech. Dig. Conference on Optical Fiber Sensors*, Sapporo, Japan, May 1996, pp. 670–673.
[21] Czarske, J., and Müller, I., "Heterodyne detection technique using stimulated Brillouin scattering and a multimode laser," *Opt. Lett.*, 19, (19), Oct. 1994, pp. 1589–1591.
[22] Heron, N. A., et al., "Brillouin loss based distributed temperature sensor using a single source," *Tech. Dig. Conference on Optical Fiber Sensors*, Sapporo, Japan, May 1996, pp. 630–633.
[23] Niklès, M., Thévenaz, L., and Robert, P. A., "Simple distributed temperature sensor based on Brillouin gain spectrum analysis," *10th International Conference on Optical Fiber Sensors*, Culshaw, B., and Jones, J. D. C. (Eds.), *Proc. SPIE 2360*, Glasgow, Scotland, Oct. 1994, pp. 138–141.
[24] Niklès, M., Thévenaz, L., and Robert, P. A., "Simple distributed fiber sensor based on Brillouin gain spectrum analysis," *Opt. Lett.*, 21, (10), May 1996, pp. 758–760.
[25] Shimizu, K., et al., "Coherent self-heterodyne Brillouin OTDR for measurement of Brillouin frequency shift distribution in optical fibers," *J. Lightwave Technol.*, 12, (5), May 1994, pp. 730–736.
[26] Bao, X., Webb, D. J., and Jackson, D. A., "Combined distributed temperature and strain sensor based on Brillouin loss in an optical fiber," *Opt. Lett.*, 19, (2), Jan. 1994, pp. 141–143.
[27] Dakin, J. P., et al., "Distributed optical fiber Raman temperature sensor using a semiconductor light source and detector," *Electron. Lett.*, 21, (13), June 1985, pp. 569–570.
[28] Wait, P. C., and Newson, T. P., "Landau Placzek ratio applied to distributed fiber sensing," *Optics Comm.*, 122, (1), 1996, pp. 141–146.
[29] Parker, T. R., et al., "Simultaneous measurement of spontaneous Brillouin scattering amplitude and frequency as a function of temperature," *Tech. Dig. Conference on Optical Fiber Sensors*, Sapporo, Japan, May 1996, pp. 662–665.
[30] Sankawa, I., et al., "Optical fiber line surveillance system for preventive maintenance based on fiber strain and loss monitoring," *IEICE Trans. Commun.*, $F3E76-B, (4), April 1993, pp. 402-409.
[31] Rogers, A. J., "Distributed optical-fiber sensors" *J. Phys. D. Appl. Phys.*, 19, 1986.
[32] Rogers, A. J., "Distributed optical-fiber sensors for measurement of pressure, strain and temperature," *Physics Reports (Physics Lett.)*, 169, (2), 1988, pp. 99–143.
[33] Dakin, J. P., and Culshaw B., *Optical Fiber Sensors*, Norwood, MA: Artech House, 1988.
[34] Dakin, J. P., *The Distributed Fiber-Optic Sensing Handbook*, Springer Verlag, 1990.
[35] Farries, M. C., and Rogers, A. J., "Distributed sensing using stimulated Raman action in a monomode optical fiber," *Proc. 2nd Int. Conf. on Optical-Fiber Sensors*, Stuttgart, 1984, pp. 121–132.
[36] Handerek, V. A., Parvaneh, F., and Rogers, A. J., "Frequency-derived distributed optical-fiber sensing : signal frequency downshifting," *Elect. Lett.*, Feb. 1991, 27, (5), pp. 394–396.
[37] Rogers, A. J., and Handerek, V. A., "Frequency-derived distributed optical-fiber sensing : Rayleigh backscatter analysis," *Appl. Opt.*, 31, (21), 1992, pp. 4091-4095.
[38] Rogers, A. J., "Polarization-optical time domain reflectometry." *Appl. Opt.*, 21, (6), 1981, pp. 1060–1074.
[39] REF missing
[40] Parvaneh, F., et al., "Forward-scatter frequency-derived distributed optical-fiber sensing using the optical Kerr effect." *Elect. Lett.*, June 1992, 28, (12), pp. 1080–1082.

[41] Ahmed, S. U., et al., "Phase-matched polarization coupling in high-birefringence fibers via the optical Kerr effect," *Opt. Lett.*, 17, (9), 1992, pp. 643–645.

[42] Cokgor, I., Handerek, V. A., and Rogers, A. J., "Rapid mapping of transverse stresses along a highly-birefringent fiber," *SPIE Symposium*, Boston USA, Sept. 1993, paper 2071-18, 2071, pp. 221–228.

[43] Handerek, V. A., Rogers, A. J., and Cokgor, I., "Detection of localized polarization mode coupling using the optical Kerr effect," *Proc. 8th Int. Conf. on Optical Fiber Sensors (OFS 8)*, Monterey, CA, Jan. 1992, pp. 250–253.

[44] Dziedzic, S. M., Stolen, R., and Askhin, A., "Optical Kerr effect in long fibers," *Appl. Opt.*, 20, 1981, pp. 403–406.

[45] Born, M., and Wolf, E., *Principles of Optics*, New York, NY: Pergamon Press, 1959, Section 1-4.

[46] Zhang, J., Handerek, V. A., and Rogers, A. J., "Distributed optical-fiber sensing by four photon mixing pulse walk-off," *Opt. Lett.*, 20, (7), April 1, 1995, pp. 793–746.

[47] Stolen, R. H., Bjorkholm, J. E., and Ashkin, A., "Phase-matched three-wave mixing in silica fiber optical waveguides," *Appl. Phys. Lett.*, 24, pp. 308–1974.

[48] Jain, P. K., and Stenersen, K., "Phase-matched four-photon mixing processes in birefringent fibers," *Appl. Phys. B.*, 35, pp. 49–1984.

[49] Stolen, R. H., and Bjorkholm, J. E., "Parametric amplification and frequency conversion in optical fibers," *IEEE J. Quantum Electron.*, QE-18, p. 1062, 1982.

[50] Berkoff, T., and Kersey, A. D., "Eight element time division multiplexed fiber grating sensor array with an integrated optic wavelength discriminator," *Proc. 2nd European Conference on Smart Structures*, Glasgow 1994, pp. 350–353.

[51] Baronski, M. K., and Jensen, S. M., "A novel technique for investigating attenuation characteristics," *Appl. Optics*, 15, 1976, pp. 2112–2115.

[52] Dakin, J., et al., "Distributed optical fiber temperature sensor using a semi-conductor light source and detector," *Electronics Letters*, 21, 1985, pp. 569–570.

[53] Shimizu, K., Horiguchi, T., and Koyamda, Y., "Measurement of distributed strain and temperature in a branched optical fiber network using brillouin OTDR," *OFS 10*, pp. 142–145, 1994.

[54] Kvasnik, F., and McGrath, A. D., "Distributed chemical sensing utilising evanescent wave interactions," *Chemical, Biochemical and Environmental Sensors*, SPIE, 1172, 1989, pp. 75–82.

[55] Kharaz, A., and Jones, B., "A distributed fiber optic sensing system for humidity measurements," *Measurement and Control*, 28, (4), May 1995, pp. 101–3.

[56] Michie, W. C., et al., "Distributed sensor for water and pH measurement using fiber optics and swellable polymeric materials," *Optics Letters*, 20, (1), 1995, pp. 103–105.

[57] Ricka, J., and Tanaka, T., "Swelling of ionic gels: quantitative performance of the Donnan theory," *Macromolecules*, 17, (2), 1984, pp. 2916–2921.

[58] Peppas, N. A., *Hydrogels in Medicine and Pharmacy*, Vol. II, Polymers, Boca Raton, FL: CRC Press, 1987.

[59] Fields, J. N., "Attenuation of a parabolic index fiber with periodic bends," *Applied Physics Letters*, 36, 1979, pp. 779–801.

[60] Random Block Co-polymers, UK Patent Application. 9306887.2

[61] *The Pharmaceutical Handbook*, 19th Edition, Wade, A. (Ed.), London, UK: The Pharmaceutical Press, 1980.

[62] Lang, A. R. G., "Osmotic coefficients and water potentials of sodium chloride solutions from 0 to 40 C," *Aust. Journ. Chem.*, 1967, 20, pp. 2017–23.

[63] Hadjiloucas, S., et al., "Hydrogel based distributed fiber optic sensor for measuring soil salinity and soil water potentials," *IEEE Coll oquium, Progress in Fiber Optic Sensors and their Applications*, Nov. 1995, London.

[64] Michie, W. C., et al., "Optical fiber grout flow monitor for post tensioned reinforced tendon ducts," *Second European Conference on Smart Structures and Materials*, 1994, pp. 186–190.

Chapter 15

Multiplexing Techniques for Fiber-Optic Sensors

Alan D. Kersey
Naval Research Laboratory, U.S.A.

15.1 INTRODUCTION

Many of the applications for fiber-optic sensors involve the use of multiple sensors, configured to monitor either the same parameter at a number of different spatial locations or, possibly, a variety of different parameters at a single or multiple locations. In either case, the ability to multiplex sensors onto common interrogating fibers is a significant advantage of the technology, and one which has been explored for a wide range of applications. These application areas include high-sensitivity underwater acoustic sensor systems, industrial process control sensors, chemical sensing, and environmental and structural sensing. In each of these applications, the use of multiplexing techniques can be beneficial with regard to a number of system aspects including reduced component costs, lower fiber count in telemetry cables, ease of electro-optic interfacing, and overall system immunity to EMI. These advantages enhance the competitiveness of fiber sensors, making them more attractive for a range of applications.

This chapter discusses basic techniques used for multiplexing several "point" fiber sensors, including time, frequency, code, wavelength, and coherence-based addressing techniques. The systems described include simple serial arrays of sensors based on optical time-domain reflectometry (OTDR) processing concepts to more sophisticated interferometric fiber sensors.

15.2 QUASI-DISTRIBUTED NETWORKS

OTDR techniques [1–5] have been used in the development of a range of intrinsic distributed sensor systems. These systems utilize inherent scattering processes in optical fibers, including Rayleigh [6–10], Raman [11–13], and Brillouin [14–16] scattering to provide continuous "intrinsic" distributed sensing. The intrinsic Rayleigh scattering in a fiber is the basic OTDR interrogation mechanism used for analysis of loss in fiber links to derive losses in splices, connectors, and so forth. The capability of OTDR techniques to analyze small losses in a fiber can be used to form a quasi-distributed sensor system, where losses intentionally introduced into a fiber via a transducing element can be monitored.

15.2.1 Loss-Based Sensors

This type of serially concatenated system involves an extension of fully distributed fiber-sensing techniques to interrogate a finite number of discrete sensors. Figure 15.1 shows an implementation of a quasi-DFOS (QDFOS) system. Various sensing methods and addressing techniques have been used to implement quasi-distributed sensor systems. For example, modified fiber sections with sensitized optical properties can be spliced at intervals into a long fiber to provide enhanced localized variations in, for example, loss, backscatter intensity, or polarization. This is different from intrinsic distributed fiber sensing in that the measurand can be determined at a finite number of locations only, rather than continuously along the fiber path. Alternatively, discrete nonfiber sensor elements that vary in transmittance or reflectance with the measurand field can be incorporated into the fiber line.

Such an arrangement for distributed temperature sensing was demonstrated at a early stage in the development of QDFOS technology [6]. This system used ruby-glass-sensor elements, the attenuation of which increases with temperature for light of wavelength ~600 to 620 nm (absorption edge shifts at a rate ~1.2 Å/°C). OTDR-type interrogation of the system was used to determine the loss at each sensor element and a second wavelength, removed from the absorption edge, was used to provide a temperature-independent (reference) output. Other materials, such as semiconductors, are also suitable for this approach, as are fibers doped with certain elements (e.g., holmium, neodymium) [7,8].

The major limitation of this type of system, as with other OTDR loss-based systems, is the fact that the attenuation is accumulative; the light levels at the most distal sensor are very weak and also depend on the measurand at each sensor along the fiber [17]. This places demanding requirements on the dynamic range of the detection system and limits the number of sensors that can be used in a practical system.

15.2.2 Reflective Quasi-Distributed Sensors

An alternative form of quasi-distributed sensor employs partially reflective sensing elements, the reflectivity of which depends on the measurand strength. Each sensor element then reflects a pulse back to the optoelectronics, the peak power of which is a measure of the reflectivity at that sensor.

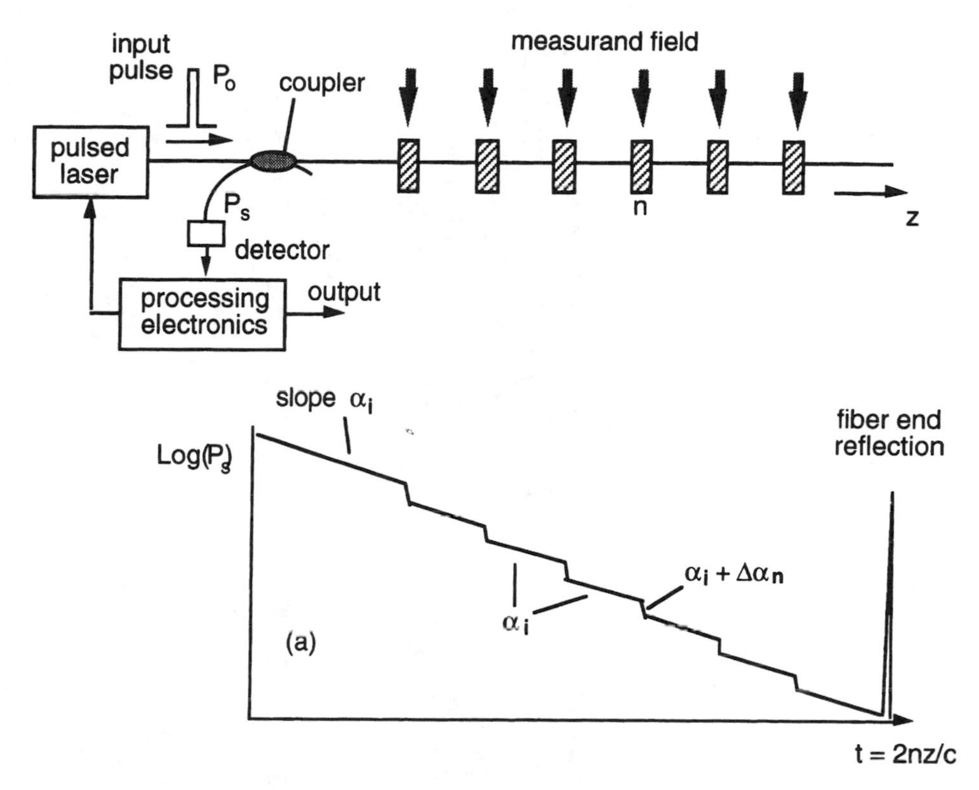

Figure 15.1 Quasi-distributed serial array using OTDR processing.

Figure 15.2 shows two examples of reflective transducers that have been reported [17–19]. In Figure 15.2(a), a bend in the fiber is used to cause a degree of coupling of the light from the core of the fiber, which is coupled via index-matching material to a retroreflective film. This reflective film returns a portion of the light back along the incident path, and thus is recoupled into the fiber in the reverse direction. The efficiency of this process is dependent on the distance between the fiber and the film (or more simply for an ON/OFF-type sensor, the presence or absence of the film), and thus the effective reflectivity of the element can be modulated. A second example is shown in Figure 15.2(b). Here, a fiber-to-fiber coupling element, similar to a mechanical splice, is used. The index of the material between the two fibers is chosen to be different to that of the fiber core, and to be temperature-dependent; as an example, index-matching oils, or oils such as mineral oil can be used. The index difference between the fiber and the oil produces a weak Fresnel reflection, the magnitude of which depends on temperature. As an example of the strength of reflector that can be formed, an index difference of 0.1 results in a reflectivity of 0.001 (0.1%). This weak reflector is still ~10-dB stronger than the "effective reflectivity" due to Rayleigh scattering in a 10m length of fiber. To avoid multiple reflection points with the sensor, one

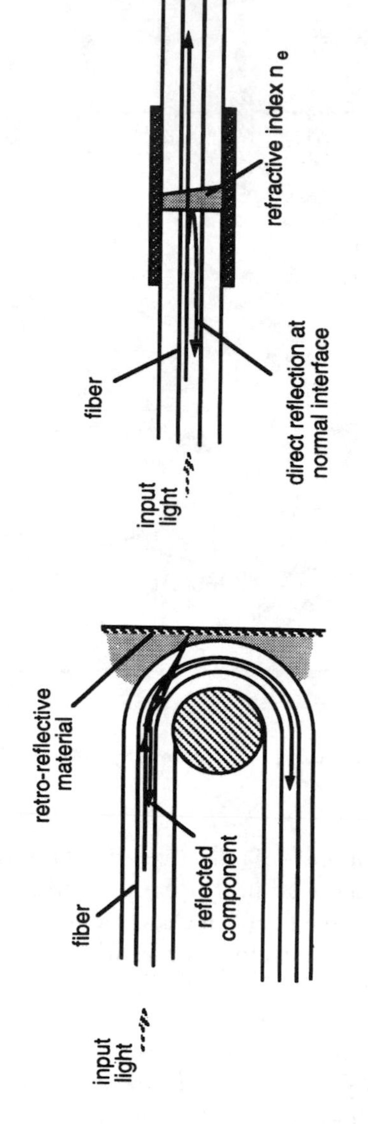

Figure 15.2 Reflectometric sensor elements.

of the fibers can be angle-cleaved, as shown, to suppress the Fresnel reflection at the second interface. This type of sensor has been used for temperature sensing of "hot spots" in electrical distribution systems.

15.3 FIBER BRAGG GRATING-BASED SENSORS

The area of quasi-distributed sensing has become a major area of research interest since the advent of the in-fiber Bragg grating. These intracore fiber Bragg grating (FBG) sensors have attracted considerable interest because of their intrinsic nature and wavelength-encoded operation, which has certain advantages over time-domain analysis approaches. Bragg gratings are photoinscribed into Ge-doped fiber by side-exposure to UV interference patterns, created either holographically [20] or by phase mask techniques [21]. The writing of gratings using point-by-point techniques have also been demonstrated [22], and techniques for single-pulse fabrication either offline [23–25] or on the fiber draw tower [26] have been reported. Bragg grating-based devices will prove to be useful in a variety of applications in both the fiber communications [27] and sensing fields [28,29], in particular in the area of advanced composite materials or "smart structures" where fibers can be embedded into the materials to allow real-time evaluation of load, strain, temperature, vibration, and so forth.

15.3.1 Grating Sensing Mechanism

Figure 15.3 shows the generic sensing concept using Bragg gratings: with broadband light incident on the grating, a narrowband component, or spectral slice is reflected back at the Bragg resonance wavelength, which is determined by

$$I_B = 2\,n\,L \tag{15.1}$$

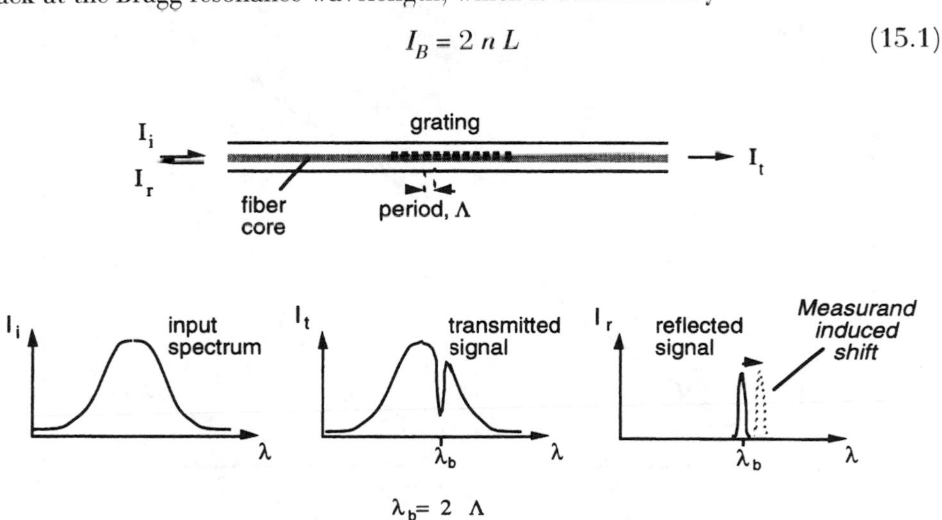

Figure 15.3 Basic Bragg grating-sensing mechanism.

where L is the grating pitch and n is the fiber index. Perturbation of the grating results in a shift in the grating pitch and/or a change in the fiber index. The primary measurand influence of interest in the sensor community is the strain-dependence of the Bragg wavelength. The strain response arises due to both the physical elongation of the sensor (and corresponding fractional change in grating pitch) and the change in fiber index due to photoelastic effects. The shift in Bragg wavelength with strain can be expressed using [29]

$$DI_B = (1 - p_e)\, I_B \cdot \epsilon \tag{15.2}$$

where ϵ is the applied strain and p_ϵ is an effective photoelastic coefficient term given by

$$p_\epsilon = (n^2/2)\{P_{12} - \mu(P_{11} + P_{12})\} \tag{15.3}$$

where the $P_{i,j}$ coefficients are the Pockel's coefficients of the strain optic tensor and μ is Poisson's ratio. The factor pe has a numerical value of ≈0.22. Figure 15.4 shows the measured strain response of a 1.3-μm FBG. The normalized strain response is found to be

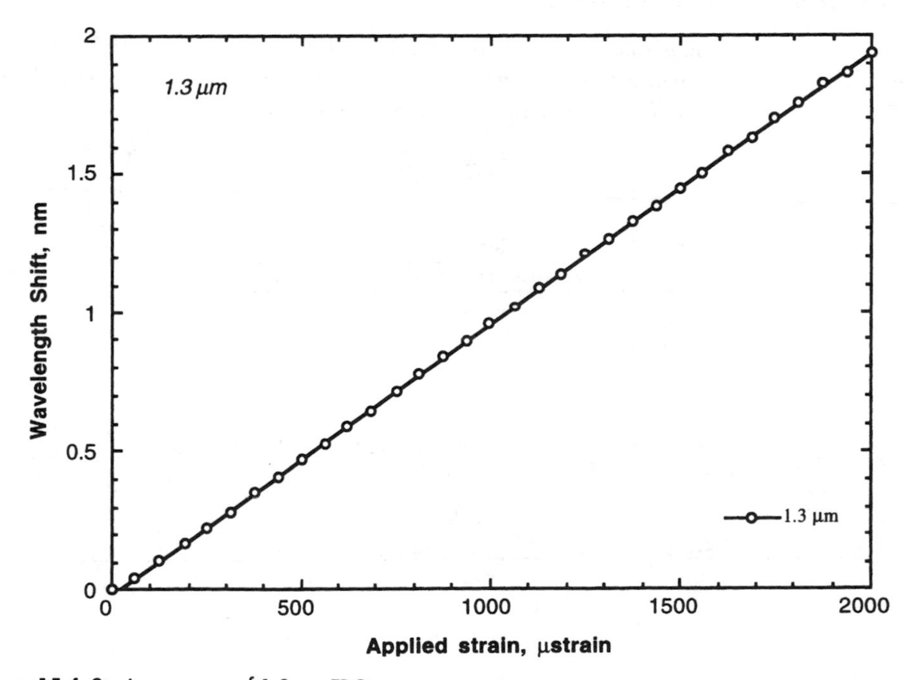

Figure 15.4 Strain response of 1.3-μm FBG.

$$\frac{1}{I_B} \cdot \frac{dI_B}{d_\epsilon} = 0.78 \times 10^{-6}\ \mu e^{-1} \tag{15.4}$$

This responsivity gives a "rule of thumb" measure of the grating-shift scale factor with strain of 1 nm per 1,000 µϵ at 1.3 µm.

The dependence of the Bragg wavelength on temperature arises due to two primary effects: the dependence of the index of refraction of the glass on temperature and the thermal expansion of the glass. In silica fibers, the former of these is the dominant effect, accounting for ~95% of the observed shift. The shift can be expressed as

$$DI_B = \{(dL/dT)/L + (dn/dT)/n\} \cdot I_B DT \tag{15.5}$$

The first part of this expression relates to the thermal expansion of the fiber, whereas the second term is due to the thermal dependence of the refractive index. Figure 15.5 shows the shift in wavelength of a 1.3-µm fiber Bragg grating with temperatures over the range 5 to 85°C. A normalized responsivity

$$\frac{1}{I_B} \cdot \frac{dI_B}{dT} = 6.67 \times 10^{-6}\ °C^{-1} \tag{15.6}$$

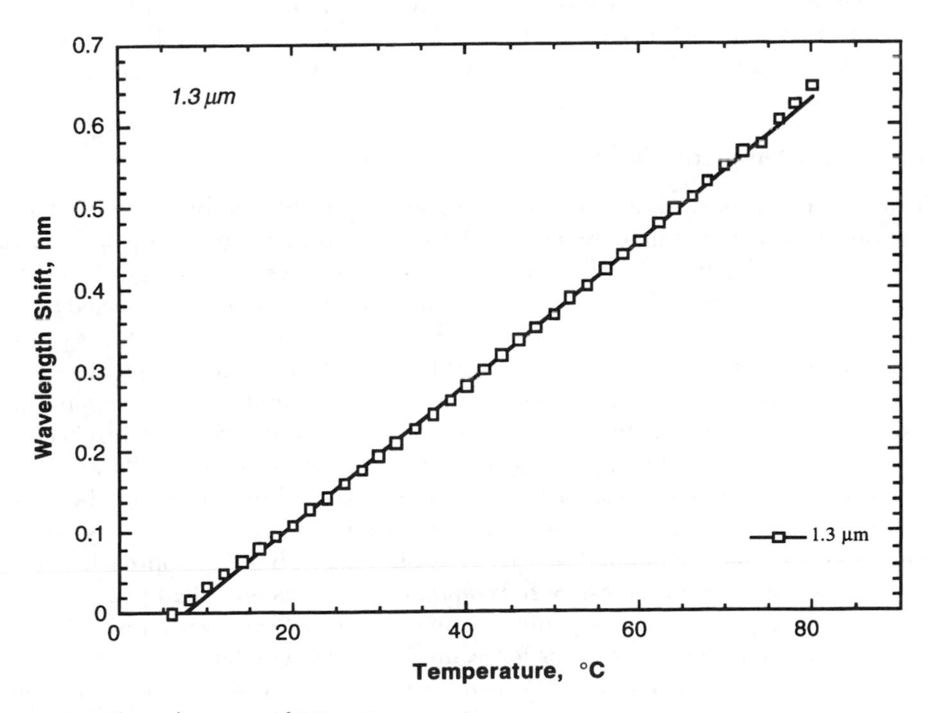

Figure 15.5 Thermal response of 1.3-µm Bragg grating.

is observed. A wavelength resolution of ~0.5 ppm, equivalent to a wavelength shift of 0.0001 nm at 1.3 μm, is required to resolve a temperature change of ~0.1°C. Although this wavelength resolution is easily achieved using laboratory instrumentation, the ability to resolve changes on this order using small, packaged electro-optics units, is more of a challenge.

The inherent wavelength-encoded output of Bragg gratings has a number of distinct advantages over other sensing schemes. One of the most important of these is that as the sensed information is encoded directly into wavelength, which is an absolute parameter; the output does not depend directly on the total light levels nor on losses in the connecting fibers and couplers or source power, provided these do not exhibit strong spectral variation Furthermore, the wavelength-encoded nature of the output also facilitates wavelength division multiplexing [30] by assigning each sensor to a different slice of the available source spectrum. This allows the distributed sensing of strain or temperature. The concept is illustrated in Figure 15.6. for three gratings, but can clearly be used for larger number of devices. The upper limit to the number of gratings that can be addressed in this way is a function of the source profile width. With current devices, it is possible to multiplex 20 or more devices along a single fiber path.

The key to a practical sensor system based on FBGs lies in the development of instrumentation capable of determining the relatively small shifts in Bragg wavelength of FBG elements (induced by strain or temperature changes in these sensor elements). This area has received significant attention lately, with a variety of approaches demonstrated.

15.3.2 FBG Interrogation Techniques

The most straightforward means of interrogation of an FBG sensor element is based on passive broadband illumination of the device, as is depicted in Figure 15.7. Here, a light source with a broadband spectrum covering the wavelength range of the FBG sensor is launched into to the system, and the narrowband component reflected by the FBG is directed to a wavelength detection system. Several options exist for measuring the wavelength of the optical signal reflected from a FBG element. These include the use of a simple miniaturized spectrometer, wavelength to amplitude conversion using passive optical filtering, tracking using a tunable filter, and interferometric detection.

Simple filtering techniques based on the use of broadband filters allow the shift in the FBG wavelength of the sensor element to be assessed by comparing the transmittance through the filter compared to a direct reference path [31], as shown in Figure 15.8. A relatively limited sensitivity is obtained using this approach, due to problems associated with the use of bulk-optic components and alignment stability. One means to improve on this sensitivity is to use a fiber device with a wavelength-dependent transfer function, such as for example, a fiber wavelength division multiplexing (WDM) coupler. Fused WDM couplers for 1,480/1,550 nm operation are commercially available, and can be custom-fabricated [32] to shift the operation

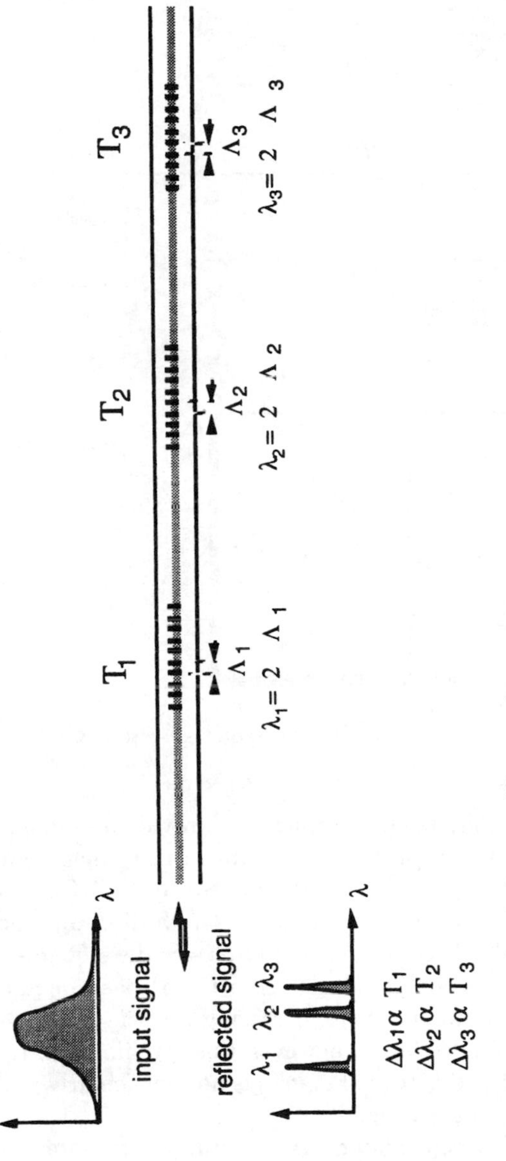

Figure 15.6 Wavelength-division addressing of Bragg gratings.

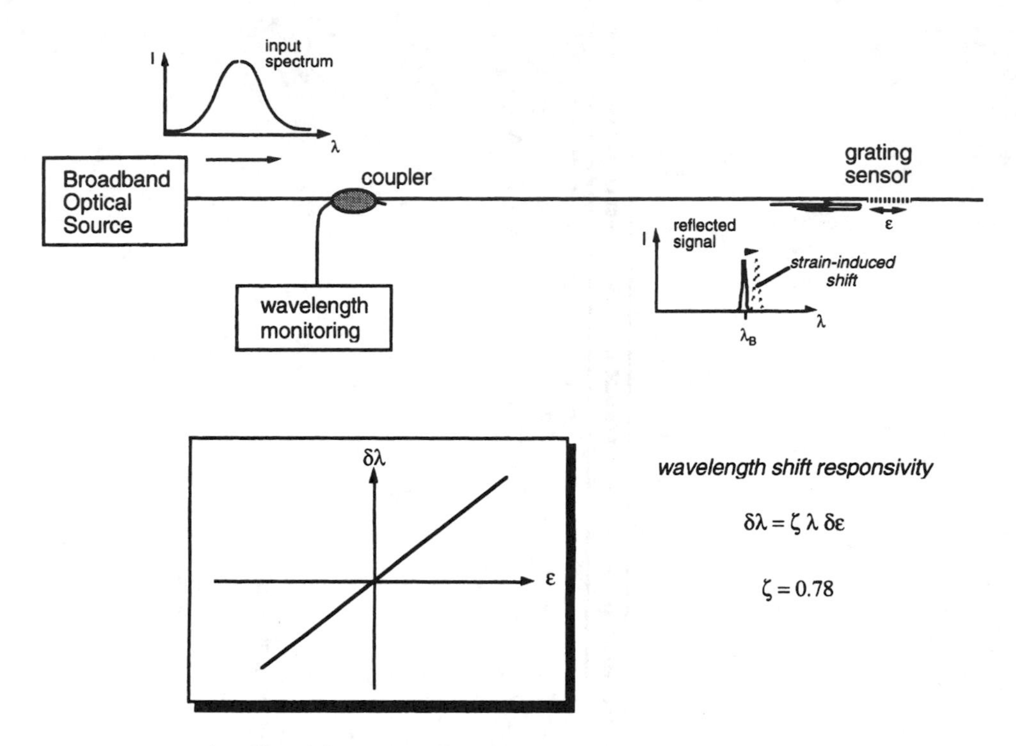

Figure 15.7 Basic "broadband illumination" fiber Bragg grating-sensor system.

wavelengths to 1,500/1,570. This coupler can provide a monotonic change in coupling ratio between two output fibers for an input optical signal over the entire optical spectrum of an erbium broadband source (~1,520 to 1,560 nm) and thus has a suitable transfer function for wavelength discrimination over this bandwidth. The use of an all-fiber arrangement improves the system stability, and a resolution ~ ±5 μstrain, corresponding to ~ ±0.5°C has been demonstrated using this approach [32]. An alternate means to increasing the sensitivity is to use a filter with a steeper cutoff (i.e., an edge filter); however, this can limit the dynamic range of the system. Although simple, this type of system is not particularly well-suited to provide high-resolution multipoint sensing.

One of the most attractive filter-based techniques for interrogating FBG sensors is based on the use of a tunable passband filter for tracking the FBG signal. Examples of this type of filter include Fabry-Perot filters [33], acousto-optic filters [34], and FBG-based filters [35]. A system based on a scanning Fabry-Perot filter is depicted in Figure 15.9. In this system, light reflected from an array of Bragg grating sensors is passed through a Fabry-Perot (FP) filter. This devices passes one narrowband wavelength component, depending on the spacing between the mirrors in the device.

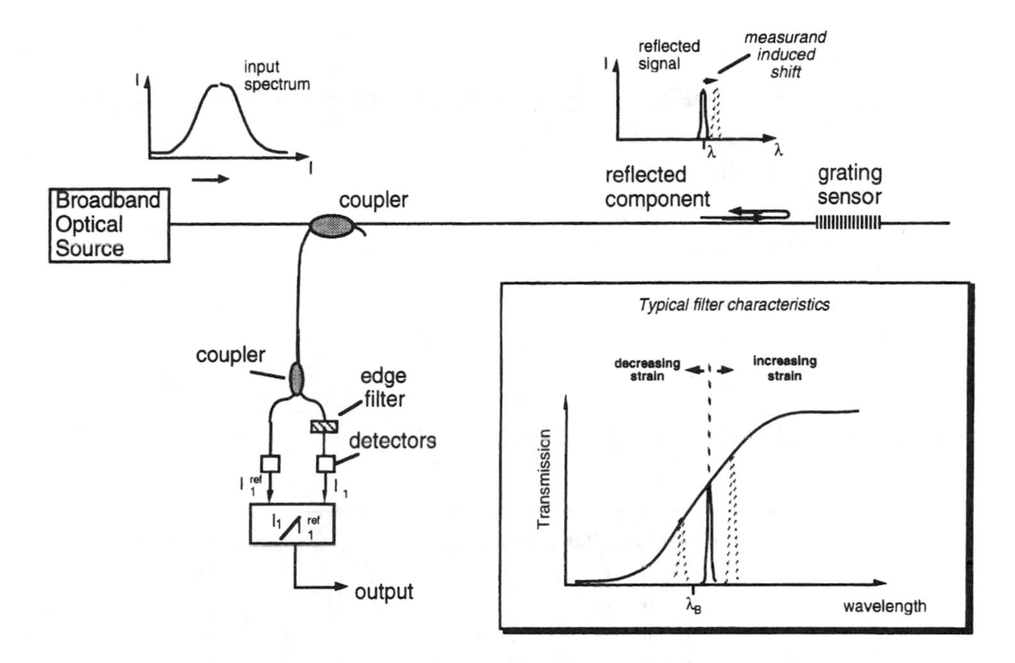

Figure 15.8 Basic filter approach to grating wavelength shift detection.

Electrical control of this mirror spacing via piezoelectric stacks allows for tuning of the filter passband wavelength. As the filter is tuned, the passband scans over the return signals from the gratings, and the wavelengths can be determined and recorded from the voltage applied to the filter as the return signals are detected. This technique provides a simple technique for sensing an array of gratings. Resolutions on the order of ~1 μstrain for strain sensing and ~0.2°C for temperature have been achieved. The advantage of this technique compared to the fixed-filter approach is that it is capable of analyzing the optical components reflected from a series of gratings. Figure 15.10 shows the optical transmission characteristics of a 12-FBG sensor array with grating spacing of 3 nm. The scanning FP filter approach has been used recently for the monitoring of up to 60 FBG sensors [36,37].

The other tunable filter approach that is also attracting some interest is the acousto-optical tunable filter (AOTF) approach [34]. This device can also be used in a scanning mode, and can be scanned faster than typical FP filters. Another possibility with the AOTF demodulator is the ability to track multiple grating lines by applying multiple RF signals to the device.

The interferometric detection technique [38] is also a form of filtering with a transfer function of the form $\{1 + \cos f\}$, with the phase term dependent on the input

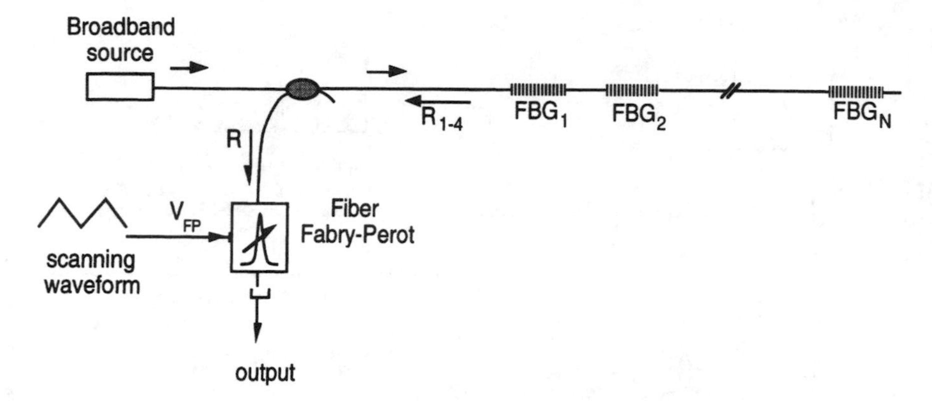

Figure 15.9 Scanning filter approach to grating wavelength shift detection.

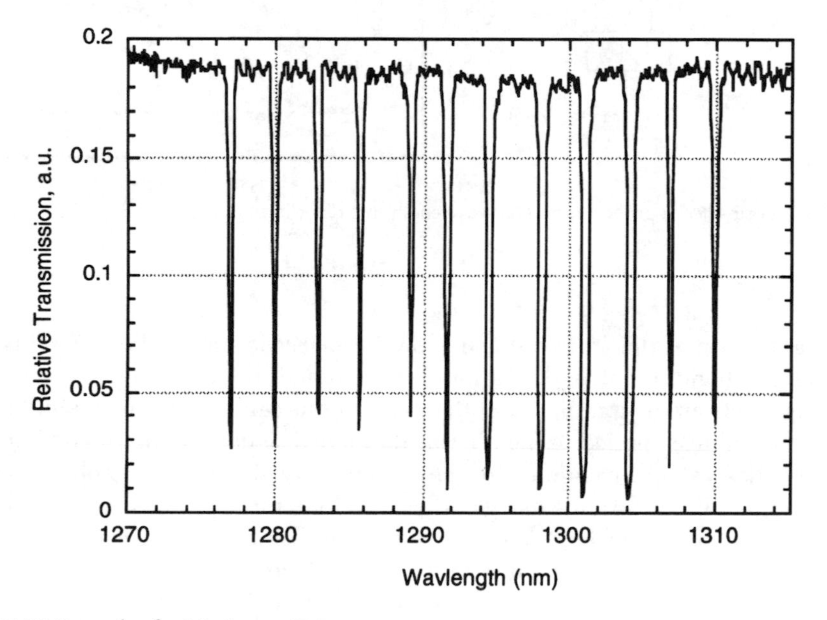

Figure 15.10 Example of a 12-element FBG array transmission spectrum.

wavelength. Figure 15.11 shows the general principle of this technique. Light reflected off a grating is directed through an interferometer that has unequal paths.

Due to the inherent wavelength-dependence of the phase of an unbalanced interferometer on the input wavelength, shifts in Bragg wavelength are converted into phase shifts. As the interferometer output can be modulated via control of the path imbalance between the interferometer arms, various phase-reading techniques can be applied to determine the FBG wavelength. This technique is extremely

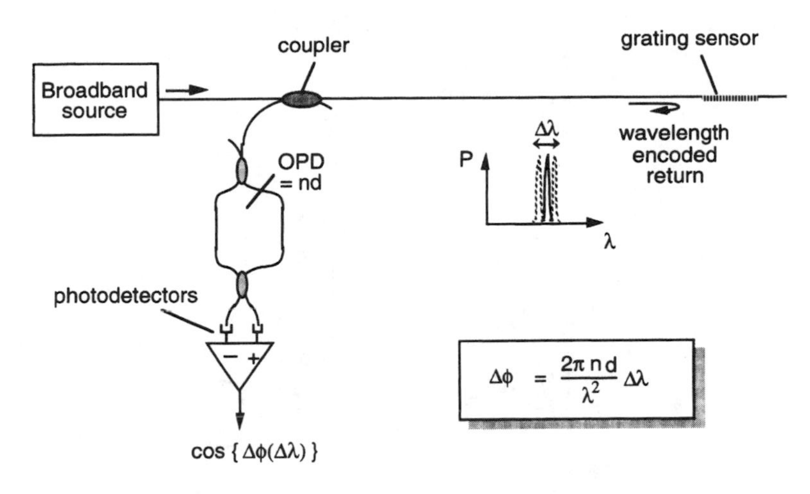

Figure 15.11 Interferometric approach to grating wavelength shift detection.

sensitive to weak dynamic Bragg wavelength shifts [38]. Experimentally, dynamic shifts in wavelength on the order of ~10^{-6} nm have been monitored during weak mechanical perturbation of the grating. This corresponds to a strain resolution of ~1 nstrain (dl/l ~ 10^{-9}) or an equivalent temperature variation of ~10^{-4} °C, illustrating the potential sensitivity of the approach. Due to random low-frequency phase shift in the interferometer, this approach is susceptible to drift and therefore, despite having superb resolution, it does not perform well for quasi-static measurements. Referencing techniques have, however, been demonstrated for strain [39] and temperature [40] monitoring.

Multiplexing using this interferometric technique has been demonstrated using two basic approaches: wavelength division addressing and time division addressing. In each case, a single readout interferometer can be used to process the returns from all the gratings simultaneously. Figure 15.12 shows a time division addressed array: Eight gratings have been multiplexed using this approach [41].

Direct spectroscopic analysis of the grating return signals is possible. One technique is based on the use of a conventional dispersive element (line grating, prism, etc.) and a linear array of photodetector elements, or charge-coupled device (CCD) array [42]. The high response of current state-of-the-art CCD arrays and their low noise allow simple miniature spectrometer systems to be constructed that can operate with very low power optical sources and provide reasonable resolution.

Another form of direct spectroscopic tool suitable for analyzing the return signals is Fourier analysis. In this case, the light from an array of grating sensors is fed to an interferometer in which one arm can be scanned to change the relative optical path lengths [43,44]. The concept is shown in the schematic in Figure 15.13. When

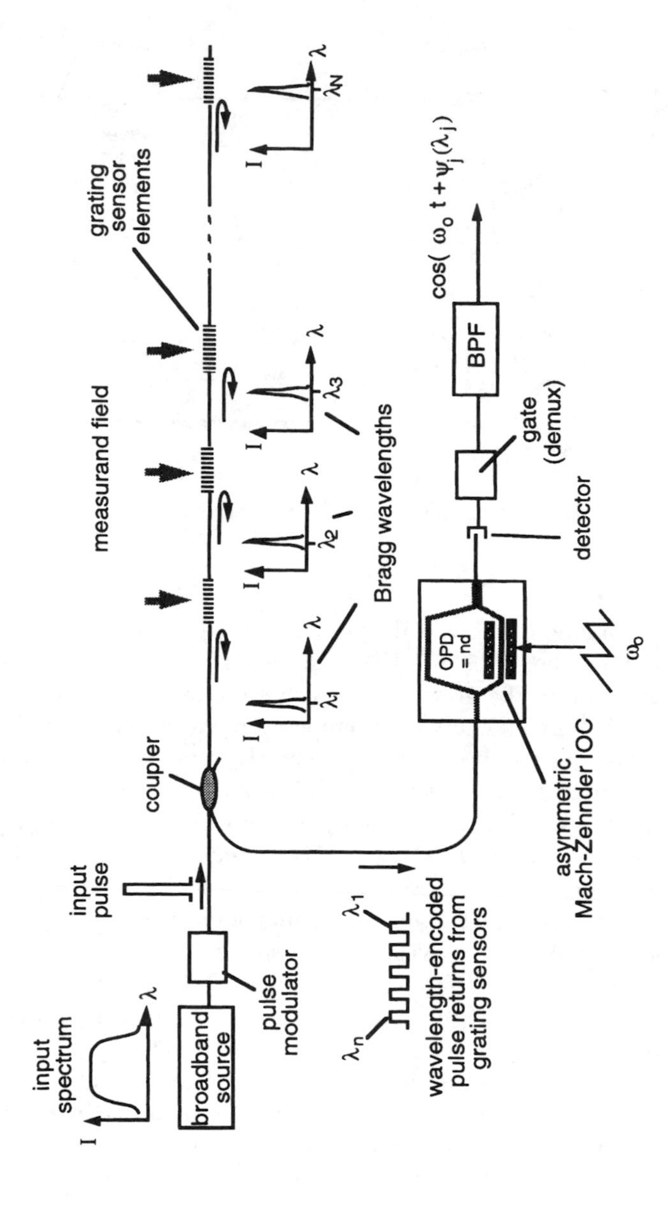

Figure 15.12 Muxed interferometric FBG system.

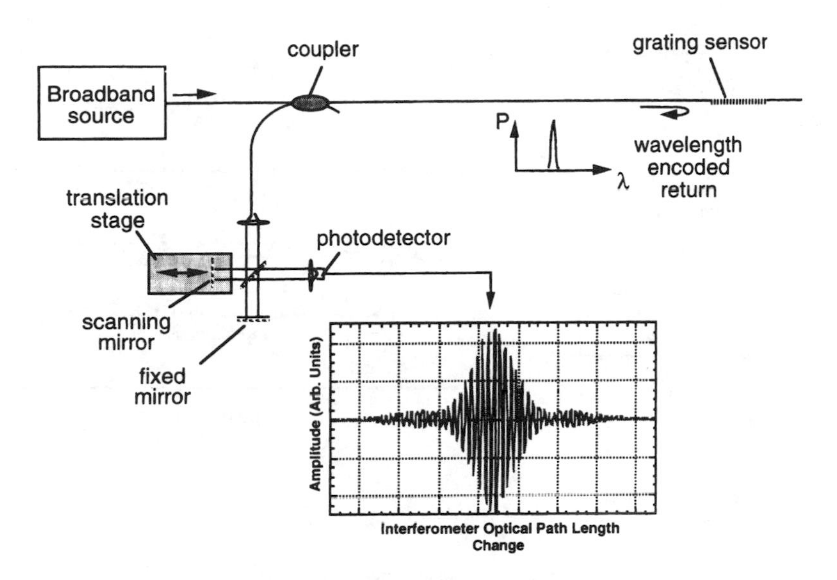

Figure 15.13 Fourier Transform approach and result.

the length of one arm is scanned, a beat signal between the optical components is generated at the detector as the path difference passes through zero. For a multiwavelength signal such as that produced by the grating array, the detector output comprises a series of discrete frequency components, each of which corresponds to a particular grating. These components may be determined by Fourier analysis of the output. Temperature or strain variations at a given grating modulate the corresponding frequency component, which are detected by monitoring shifts in the Fourier frequency components.

Due to the finite bandwidth of broadband optical sources, including superluminescent diodes (SLD), edge-emitting LEDs (ELEDs) and fiber superfluorescent (SFS) sources, a limited number of gratings can be multiplexed using direct WDM techniques A means to overcome this limitation is to adopt some form of TDM in conjunction with the inherent WDM capability of the grating sensors. Figure 15.14 shows a schematic of this concept using both TDM and WDM for addressing a large number of elements. This approach has been demonstrated using a 3×3 wavelength grating array [45]. The gratings used were low reflectivity (5 to 10%) and written at three nominal wavelengths in the 1.53- to 1.55-μm range. Three subarrays were connected in series with a delay fiber of 5m separating between each one. Detection of the wavelength shifts was accomplished by using a scanning Fabry-Perot filter, whereas the subarrays were discriminated by using pulsed operation of the source and delayed gated detection. Figure 15.15 shows the recorded outputs obtained for the total of nine sensors, with a strain signal applied to just one FBG element (sensor # 9). This

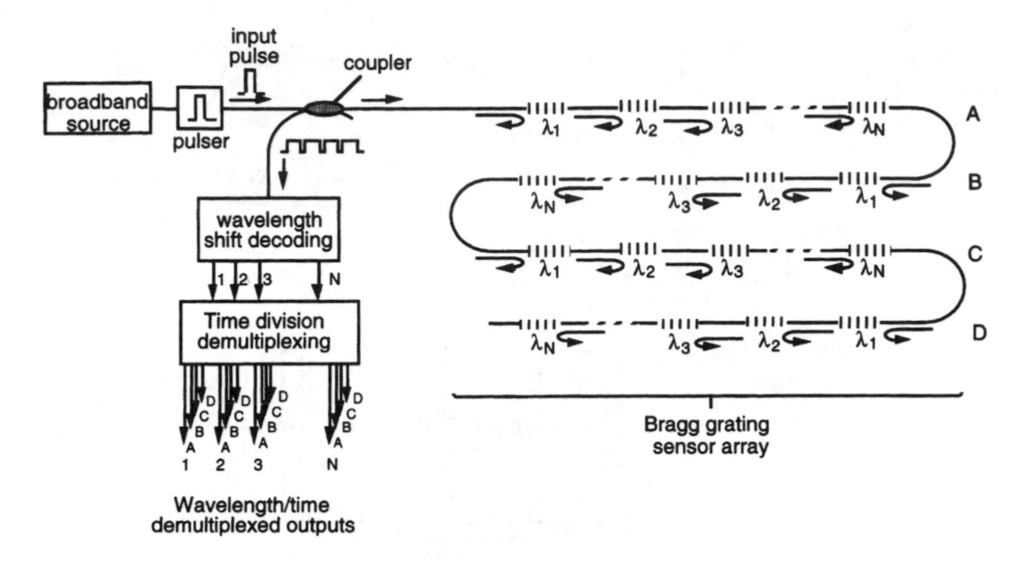

Figure 15.14 FBG array based on time and wavelength-division addressing.

type of signal processing may allow a large number of grating sensors to be interrogated in a serial array, which would be of interest in applications such as embedded sensor systems for smart structures. Applications include embedded distributed sensing in concrete and various composite structures [46–50].

15.3.2.1 Bragg Grating Laser Sensors

Fiber Bragg gratings are also ideal for use as spectrally narrowband reflectors for creating all-fiber cavities for fiber lasers. This area is attracting interest from both the communications research community (for the development of tunable single-frequency devices for WDM networks) [51,52], and in the sensing field (for strain, temperature, and very high resolution dynamic strain monitoring). Several variations on this concept have been reported over the past few years.

The basic form of a fiber Bragg grating laser utilizes two gratings with matched Bragg wavelength to create an in-fiber cavity. The use of a doped fiber section between the gratings (e.g., erbium-doped fiber) allows the system to be optically pumped to provide cavity gain and thus lasing. The device can be implemented in various ways and operated in either a single frequency or multimode fashion. In order to provide single-mode operation, a short cavity laser with highly spectrally selective elements in the cavity is required to effectively limit the bandwidth therein allowing only a single dominant mode to be supported. Fiber Bragg gratings are ideal for this purpose, as they provide high reflectivity over a very narrow spectral band (as low as 0.1 nm). Multimode operation of the cavity will occur if the cavity-mode spacing ($\delta_n = c/2nL$,

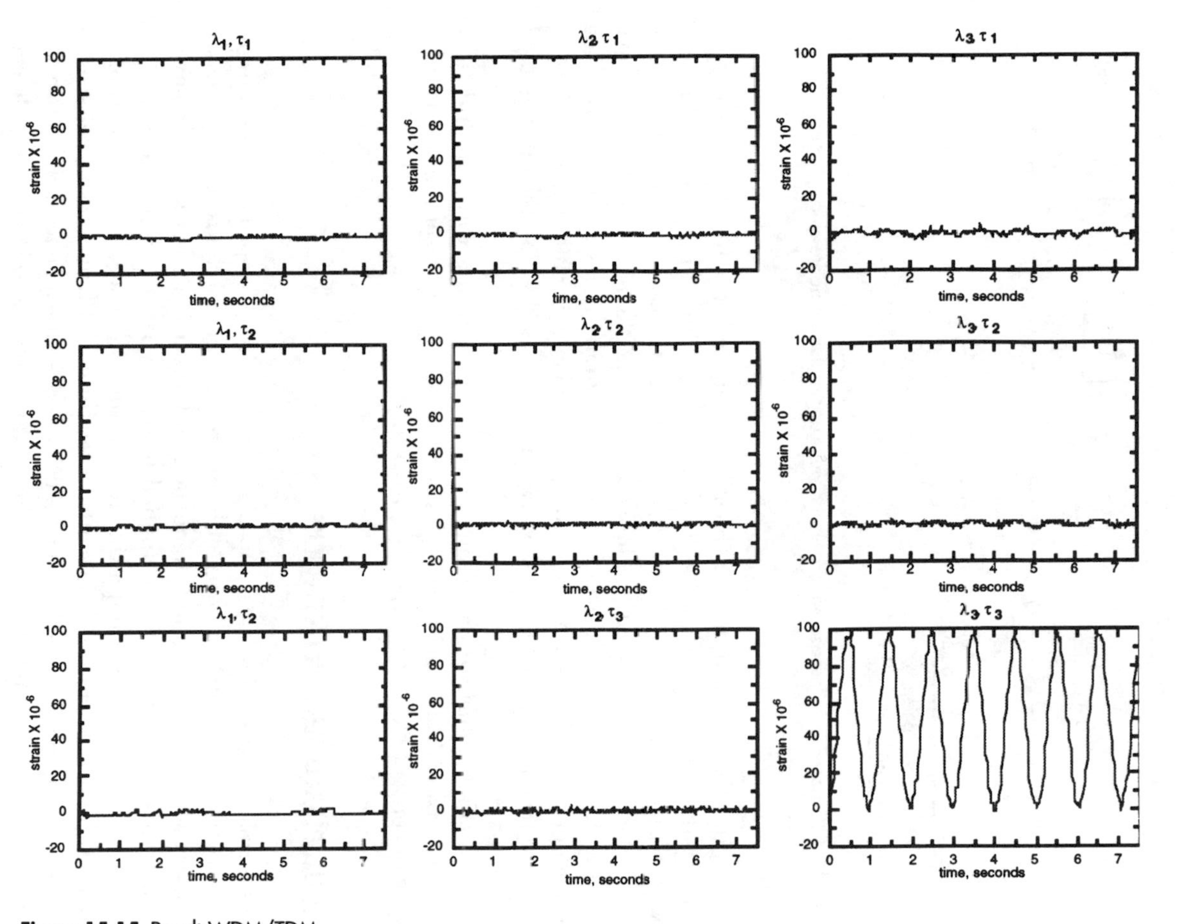

Figure 15.15 Result WDM/TDM.

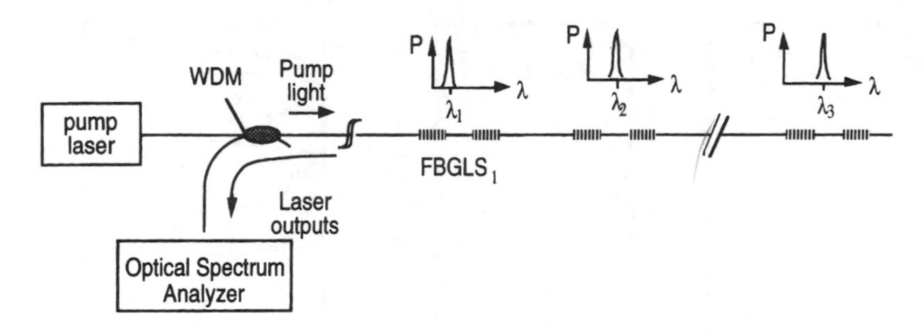

Figure 15.16 FBG-based fiber laser sensor array with wavelength-division addressing.

where L is the cavity length, n is the fiber index, and c is the free-space velocity of light) is significantly less than the bandwidth of the gratings. For single-mode operation with typical FBG bandwidths of ~0.1 to 0.2 nm, the cavity is required to be on the order of a few cm (2 to 5 cm) . This can necessitate the use of very highly doped fiber and high pump powers to achieve sufficient gain in the cavity to support lasing. Fibers codoped with Er/Yb, which provide much higher gain efficiencies, can be used to overcome this problem, allowing short cavities to be used while still allowing modest pump powers.

Figure 15.16 illustrates a schematic representation of a wavelength division multiplexed array of FBG laser sensors. A series of up to four laser sensors based on this type of concept have been reported [53–59].

15.4 MULTIPLEXED SENSOR NETWORKS

Various forms of networks or array topologies have been developed to support multisensor systems using both intensity and interferometric methods. In many cases-particularly when considering time, frequency, and wavelength division multiplexing-the same types of topology can be applied to both types of sensor systems. Other schemes such as coherence multiplexing rely on the coherence characteristics of the interrogating light, and thus are inherently suited for use with interferometric sensors.

The term "intensity-based sensor" is used to describe a generic class of sensors that depend on monitoring changes in some characteristic related to the detected intensity at the sensor output to derive information on the measurand of interest. Examples include sensors based on attenuation, reflectance, fluorescence signal, and modal modulation. A number of different types of branching networks have been investigated for use with intensity-based sensors, particularly those based on simple concepts such as attenuation. Sensors can be addressed using schemes based on optical analogs of conventional electronic TDM and FDM techniques or by using schemes devised for use in optical communications systems such as WDM.

Interferometric fiber sensors have been developed for a wide range of application areas where high sensitivity is required, including acoustic pressure, and magnetic and electric fields. A number of different multiplexing topologies have been devised and tested by research groups working in this field. Although much of the early work in this area demonstrated the multiplexing of small numbers of sensors, the work laid the foundation for later system demonstrations with arrays of up to 64 sensors reported using TDM and 48 sensors using FDM techniques.

15.4.1 Time Division Multiplexing

Nelson and others [60] proposed and demonstrated the first TDM sensor array for addressing a number of discrete reflective intensity-based sensors. These sensors were spaced at different distances from the source and detector such that a single pulse of appropriate duration at the input to the network produced a series of distinct pulses at the output. These pulses represent time samples of the sensor outputs interleaved in time sequence, as shown in Figure 15.17. The required duration of the input pulse is determined by the effective optical delay of the fiber connecting the sensor elements. Repetitive pulsing of the system allows each sensor to be addressed by simple time-selective gating of the detector. Network configurations for both transmissive and reflective intensity-based sensors have been described. Spillman and Lord also reported a self-referencing TDM intensity sensor network where recirculating fiber loops [61,62] provide a referencing capability for the sensors.

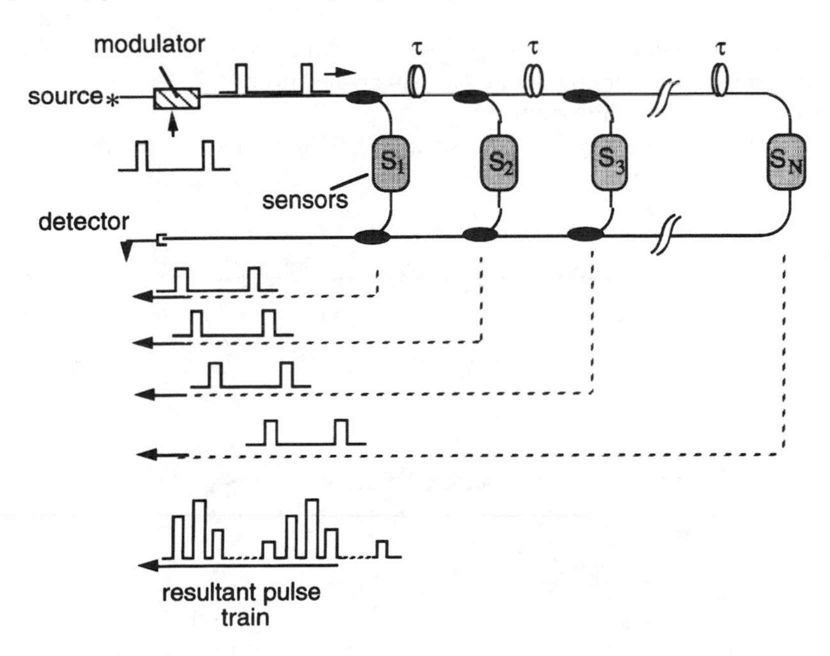

Figure 15.17 Basic time division multiplexed sensor array.

Considerable interest has been directed towards the experimental demonstration and evaluation of these types of TDM topologies [63–66] for use with interferometric sensors. The ladder array system of Figure 15.17 is readily used with interferometric sensors, provided a means for demodulating the interferometer outputs is also incorporated. In such an arrangement, the output pulses obtained from the array of interferometers represents a time sample of each of the interferometers at a given instant and is thus a convolution of the interferometric phases of interest and the nonlinear cosine transfer function of each interferometer. The required phase can be recovered from these outputs by utilizing one of several "passive" phase-carrier demodulation techniques that can be used if the sensor interferometers are configured as path "unbalanced" devices. This is typically accomplished by phase-modulation of the interferometers via laser frequency-modulation.

Subsequent work in TDM systems concentrated on alternative forms of ladder arrays that reduced the number of couplers required to implement the configurations [67,68]

Naturally, Mach-Zehnder arrays are not the only interferometer elements that can be used as sensors in theses topologies. Arrays based on Michelson interferometers, for instance, have been proposed and demonstrated [69]. Figure 15.18 shows two Michelson configurations based on discrete and nondiscrete interferometer elements. In the array shown in Figure 15.18(a), the Michelsons are formed as separate (discrete) interferometers and are optically addressed via the passive branching network of couplers and delay coil. In Figure 15.18(b), however, the Michelsons are formed within the network such that the networking couplers and delay coils also serve as the interferometer coupler and sensing coil. This reduces the number of components required to implement the array.

These Michelson arrays are also of interest from another standpoint, namely

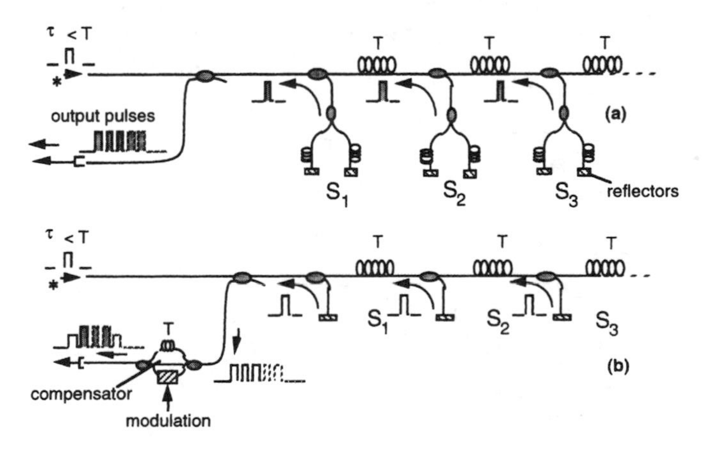

Figure 15.18 Time division multiplexed serial arrays based on Michelson interferometer sensor elements

polarization sensitivity. In fiber interferometric systems constructed using low-birefringence fibers, the random evolution of polarization in the fiber arms can lead to fading [70] of the interferometric signal. This issue has received much research attention, and a number of approaches to overcoming its effects have been reported [71–76]. In an array configuration, fading tends to occur randomly over time in an uncorrelated way at the various sensor outputs. The use of Michelson interferometers can allow for the use of a passive birefringence compensation scheme that eliminates polarization-induced fading.

The birefringence compensation method is based on use of the *orthoconjugate reflector* of Edge and Stewart [77], which consists of a 45-deg Faraday rotator followed by a plane mirror. For an optical beam that retraces its path in a fiber, Pistoni and Martinelli [78] have demonstrated that the insertion of a Faraday rotator and mirror (FRM) results in a state of polarization (SOP) at the exit that is orthogonal to the SOP at the input to the fiber. As depicted in Figure 15.19, when employed in a Michelson interferometer [79,80] the SOPs returned from each of the devices placed at the end of the interferometer arms will be orthogonal to the common-input SOP. Consequently the two output SOPs are aligned with each other and this ensures maximum fringe visibility is obtained at the interferometer output. A four-sensor array configuration based on the technique with miniaturized pigtailed FRMs as the reflectors based on the topology of Figure 15.18(b) has been built and tested [69]. The fringe visibility in this system was estimated to be > 0.95 (fading < 0.5 dB) and was maintained simultaneously for all the sensors and remained at this level even under deliberate manual birefringence perturbations of the fiber leads in the system.

The ladder configuration of Figure 15.17 is "return-coupled" in nature. One of the issues associated with this type of configuration is the need to equalize the optical

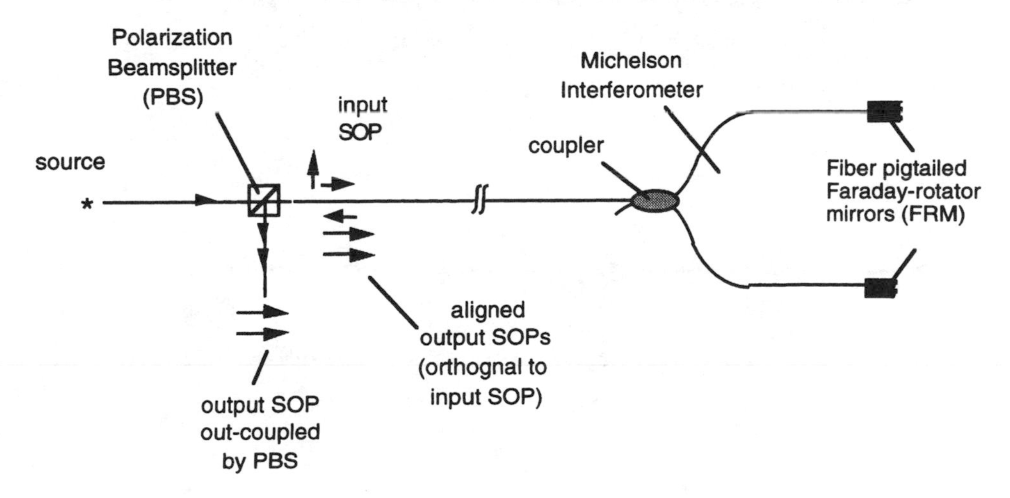

Figure 15.19 Polarization-independent Michelson interferometer configuration.

power returned from each sensor. This can be accomplished by appropriate choice of the splitting ratio of the couplers used to implement the sensor network. The coupler-splitting ratios (for the jth coupler in the ladder input and output fiber bus, see Figure 15.20(a)) can be easily calculated for an ideal lossless system: the coupling ratios are set according to

$$kj = 1/(N - j + 1) \tag{15.7}$$

However, when losses are taken into account, the more distal sensors experience higher net loss and thus return lower power levels. This can be overcome by retailoring the coupler-splitting ratios, but for large arrays the coupler ratios required can become very low, and tolerances on setting these can become an issue in the practical implementation of the array.

An alternative architecture utilizes a forward-coupled array ladder topology, shown in Figure 15.20(b). Here, the light is fed forward through each sensor rung such that each path accumulates approximately the same net losses. The optimum coupler-splitting ratios for this system are still the same as those defined by (15.7),

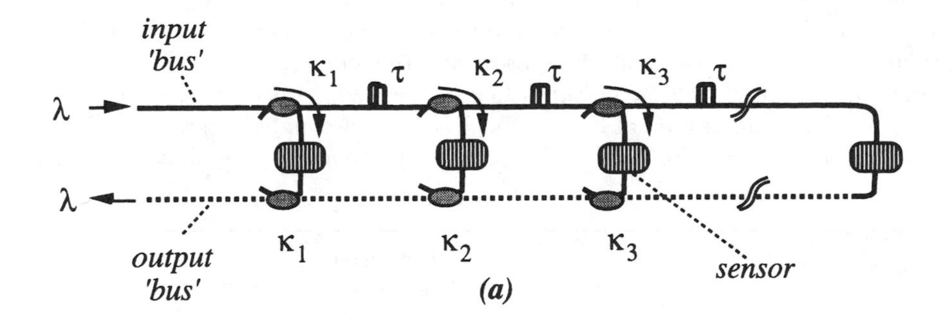

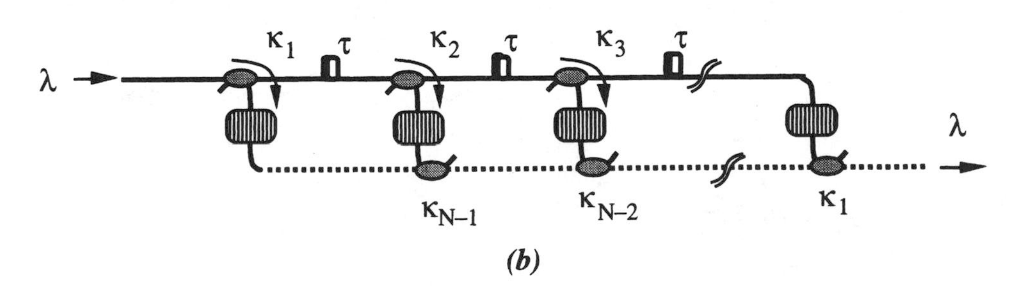

Figure 15.20 Forward-coupled array concept.

but now do not have to be adjusted to compensate for losses (within reasonable limits), as each sensor path through the system experiences approximately equal losses due to couplers, and splices. This array configuration was recently used to demonstrate the multiplexing of 64 sensors using TDM addressing in a two-tier forward-coupled array configuration [81].

The use of partial reflectors along a length of fiber allows for the multiplexing a of several interferometer elements in an inline configuration. The initial work in this area utilized inline partially reflective fiber splices, or in-fiber reflectors [82] between fiber-sensing coils, each of length L, to form inline an array of Fizeau or low-finesse Fabry-Perot [83] interferometric elements, which can be interrogated with pulsed operation of the source. Providing the width of the input pulse t is less than the round-trip delay ($T_d = 2L/nc$) between reflectors, a train of nonoverlapping pulses is formed at the output. An interferometric signal from each element in the array can be generated in time sequence by applying a double-input pulse interrogation signal, with the two input pulses separated by an interval T_d, such that at the output pulses reflected from consecutive partial reflectors in the array overlapped, thus mixing interferometrically. By setting the optical frequencies of the two input pulses to be different, the overlapping pulses at the array output produce beat frequencies. This concept, termed "differential delay heterodyne" interrogation, was the scheme originally used with this multiplexing arrangement. An alternative form of addressing utilizes a compensating interferometer and single-input pulse interrogation. In this approach [84], the separated pulses produced by each reflective splice with single-pulse interrogation are fed to a compensating interferometer, of delay (path length between the two arms) equal to the round-trip delay (T_d) between the reflectors (Figure 15.21). In this way, portions of light in adjacent pulses are forced to interfere, creating the interferometric outputs desired. An array of six acoustic sensors based on this approach has been field tested [85].

Due to the possibility of multiple pulse reflections between partial reflectors, the reflectometric topology gives rise to multiple pulse interactions, which leads to crosstalk. The magnitude of this crosstalk is proportional to the partial mirror

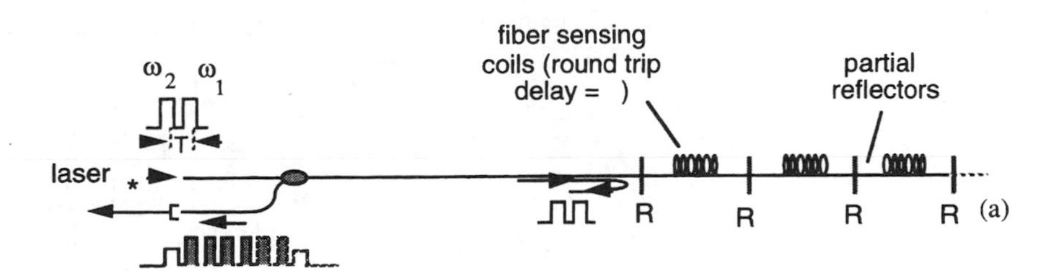

Figure 15.21 Time division multiplexed reflectometric serial array.

reflectivity [86], and thus low reflectivities are required to reduce the effect. The fabrication of low-reflectivity splices with consistent reflectance was originally problematic, but the advent of Bragg gratings has provided an ideal solution to this problem. As discussed later, the use of Bragg gratings also provides wavelength selectivity to the reflectors, which enables additional multiplexing capabilities.

An alternative form of time-domain-based interrogation of ladder arrays utilizes code modulation of the interrogating light compared to the normal pulse interrogation. Code-division-multiplexed (CDM), or spread spectrum, techniques [87] have been applied to a variety of communications applications, including optical-fiber systems [88]. This type of signal processing has also been previously investigated for optical time-domain reflectometry (OTDR)-based sensing [89]. When used with fiber sensor arrays [90], the interrogating laser source is modulated using a pseudo-random bit sequence (PRBS) of length $(2^m - 1)$ (maximal length sequence, or *m*-sequence), and correlation is used to provide synchronous detection to identify specific sensor positions. A delay equal to an integer multiple of the bit (or 'chip') period separate the sensors along the ladder structure. The received signals from the array are then encoded by delayed versions of the PRBS, and correlation techniques can be used to extract the individual signals.

Figure 15.22 diagrammatically represents the principle of operation of the CDM

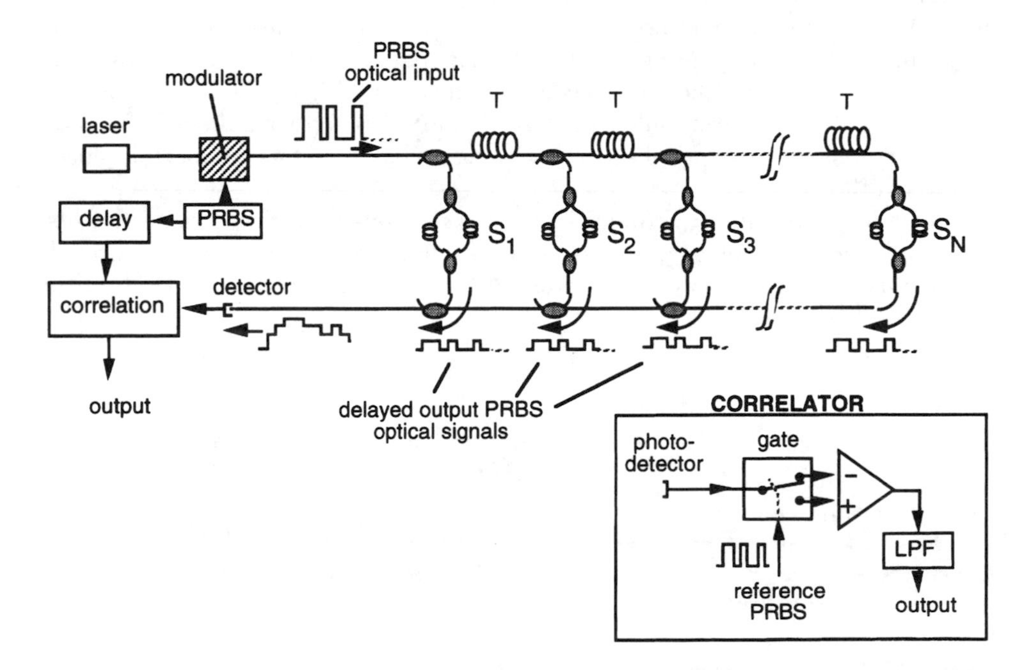

Figure 15.22 Code-division-multiplexed array using spread-spectrum techniques.

approach applied to an interferometric sensor array. The PRBS input optical signal is fed to each of the N sensors, delayed by a multiple, nj, of the bit period T, where j denotes a specific sensor ($1 \leq j \leq N$). The total output signal comprises the intensity sum of the overlapping delayed PRBS sequences (each modified by the appropriate sensor transfer function). This results in a complex up-down staircase-like function at the optical detector that can be decoded using synchronous correlation detection involving multiplication of the received signal with an appropriately delayed reference PRBS.

Although this method may provide advantages in terms of power budget for time division multiplexed systems, it would also seem to be limited by excess phase noise effects arising due to mixing of time-coincident pulses from different sensors and relatively high crosstalk between sensors. Recent work addressed these limitations of the technique using a detection/signal-processing approach that yields improved crosstalk and noise performance [91]. In this work, crosstalk levels lower than those expected from consideration of the code length were obtained using a mix of bipolar and unipolar codes, which produces an improvement in the channel/channel isolation. This arises due to the correlation function of a bipolar with a unipolar m-sequence PRBS, shown in Figure 15.23, which has a value $2^{(m-1)}$ for an aligned code, but is zero for any asynchronous alignment of the codes (this is in contrast to the conventional bipolar-bipolar autocorrelation that has a value of $(2^m - 1)$ for code

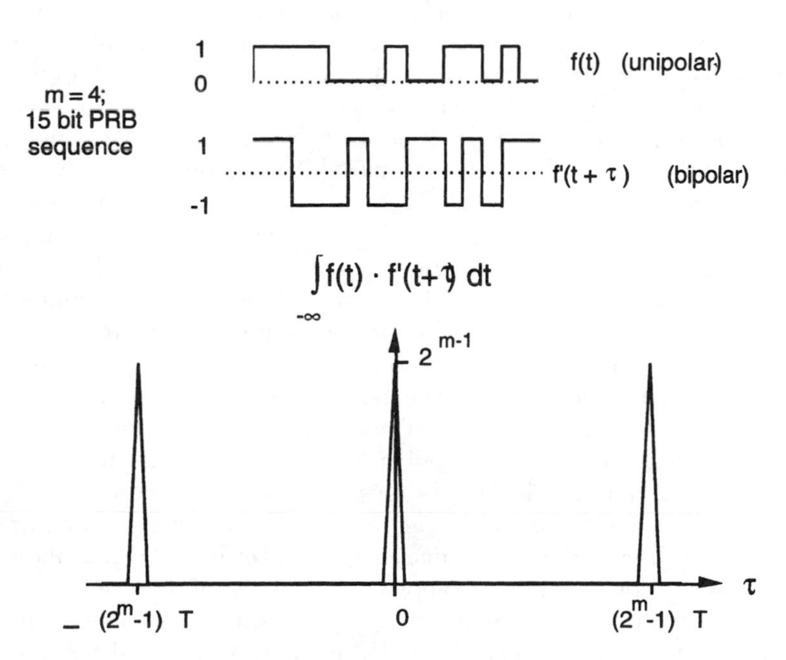

Figure 15.23 Correlation function between a bipolar and a uniploar PRBS of length 2m- 1, with $m = 5$.

alignment, but a value of −1 for asynchronous alignment) [91]. This feature ensures good crosstalk performance without the need to utilize excessively long PRBS codes; indeed, low crosstalk can be obtained providing the code length $(2^m - 1) \geq N$, where N is the number of sensors in the array (assuming a one-bit time delay between sensors).

15.4.2 Frequency Division Multiplexing

A number of novel concepts for frequency-domain-based multiplexing schemes [92,93] for intensity sensors have also been reported. An example of such is a radar-based frequency-modulated continuous wave (FMCW) technique to allow frequency division addressing with a network of intensity-based sensors [93]. In this case, a chirped RF intensity-modulated source is used to interrogate a number of simple reflective intensity sensors, and the detector output is electrically mixed with a reference chirp signal. The FMCW technique provides a ranging capability that enables the sensors, located at different spatial locations, to be discriminated. The processing produces a beat frequency associated with each sensor element, allowing frequency demultiplexing of the outputs, with the magnitude of the beat signal a measure of the attenuation (or reflectance) of the sensor.

One of the earliest approaches developed for the multiplexing of interferometric sensors was also based on the FMCW concept [94–96], but with an optical-frequency-chirped laser source. This scheme relies on the use of unbalanced interferometers arranged in a serial (see Figure 15.24) or parallel network illuminated by an optical-frequency-chirped source. Due to the inherent sensitivity of an unbalanced interferometer to input optical frequency, a beat frequency is generated at each sensor output, the period of which depends on the frequency excursion of the chirp, the chirp rate, and the interferometer optical-path difference (OPD). Assigning a different OPD to each interferometer allows the beat frequencies associated with each sensor element to be distinct, and thus separable using band filtering. The phase of the beat signal that is encoded with the interferometric phase of interest can then be recovered using phase-locked loop (PPL) or other phase-analysis techniques.

One major problem that arises with this type of multiplexing technique is cross-terms due to unwanted interferometric components arising differentially or additively between sensors, or "ghost" interferometers arising from connecting fiber paths in conjunction with the interferometers. These cross-terms lead to sensor-to-sensor interference, or crosstalk, which is a problem in most applications where the full capability of using interferometric sensors, in terms of the detection sensitivity and dynamic range, are important. These "stray" components can be minimized using certain topologies based on predetermined sets of OPDs, but cause significant design complexity for an array involving an appreciable number of sensor elements.

An alternate FDM approach utilizes the spatial and frequency-domain separation of sensor signals shown in Figures 15.25. In Figure 15.25(a), the outputs from N sensor elements, all powered from a common source, are spatially-multiplexed onto

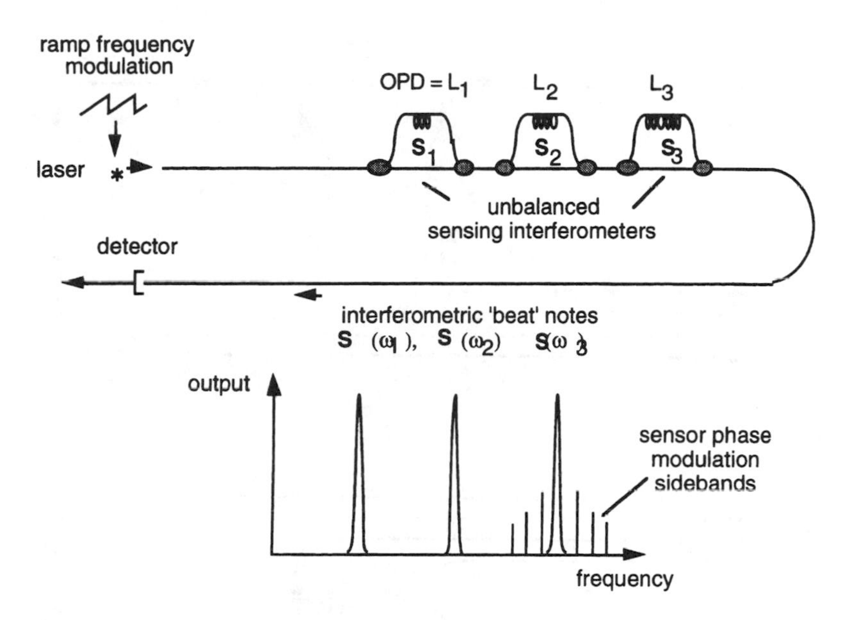

Figure 15.24 FMCW interferometric sensor multiplexing.

separate fibers. In Figure 15(b) the outputs from N sensors, which are independently illuminated by separate sources, are combined onto a single output fiber. Using phase-generated carrier (PGC) interrogation [97] with each laser operated at a different carrier frequency, the sensor outputs in Figure 15.25(b) can be separated using synchronous detection or band filtering. Combining these techniques allows a matrix-type array [88] to be configured with N input fibers and N output fibers, which contains $N \times N$ sensor elements, as schematically represented in Figure 15.26. This system is somewhat unique in that the operation of the remote PGC interrogation (demodulation) scheme automatically provides both the demodulation and demultiplexing functions, provided the sources are modulated at different carrier frequencies.

This type of array is the most highly developed topology demonstrated to date. An array comprising 48 acoustic sensors has been successfully demonstrated in a sea test under a joint NUSC/NRL advanced technology demonstration program in 1990 [99].

15.4.3 Transducer Multiplexing

The multiplexing techniques described in the previous sections involve the networking of many interferometric sensors, with each interferometer designed to detect a particular measurand at a particular location (or averaged over an extended spatial

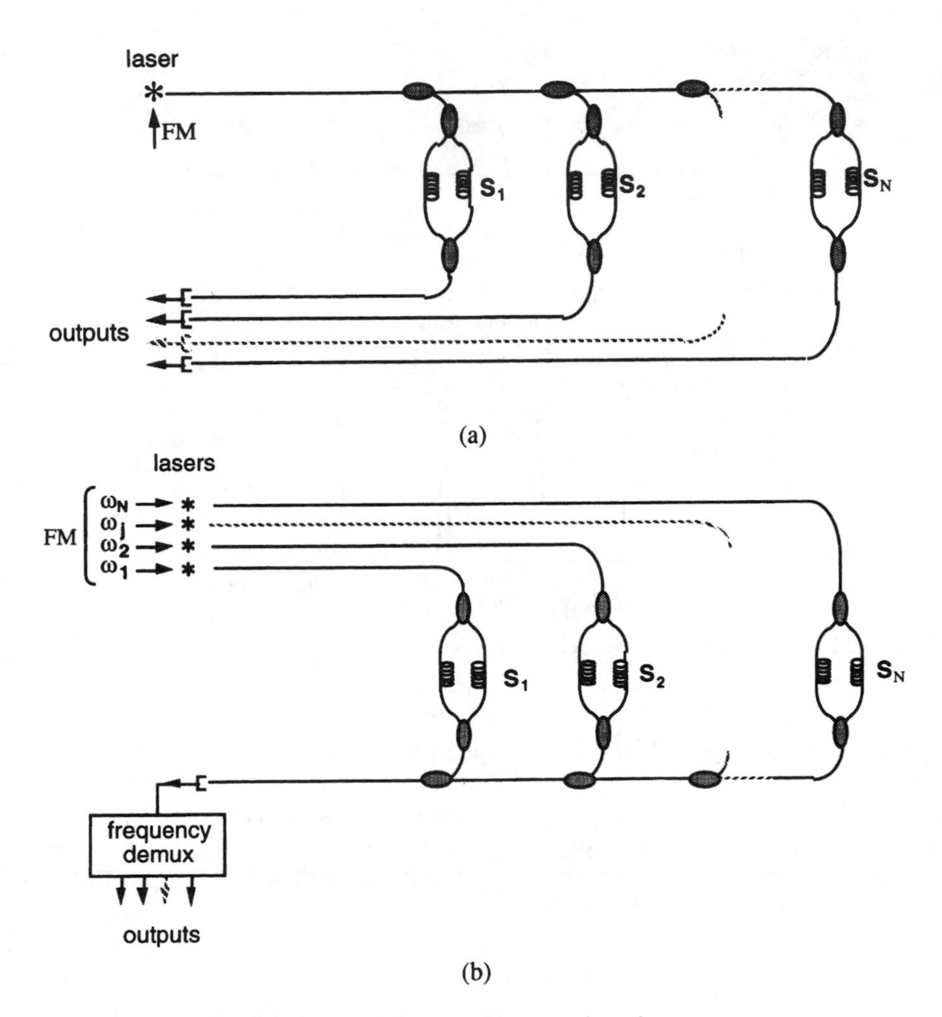

Figure 15.25 (a) Spatial and (b) frequency-domain addressing of interferometers.

aperture). It is also possible to multiplex the transducer elements within a single interferometric sensor using frequency division techniques. Several fiber transducers have been developed in which the strain imparted to a fiber in an interferometer is proportional to the square to the applied measurand field. Examples include magnetostrictive [100] and electrostrictive materials [101], and a displacement-sensing geometry [102] based on the lateral displacement of a fiber supported at two fixed points.

In general, the fiber strain in these types of sensors can be expressed as

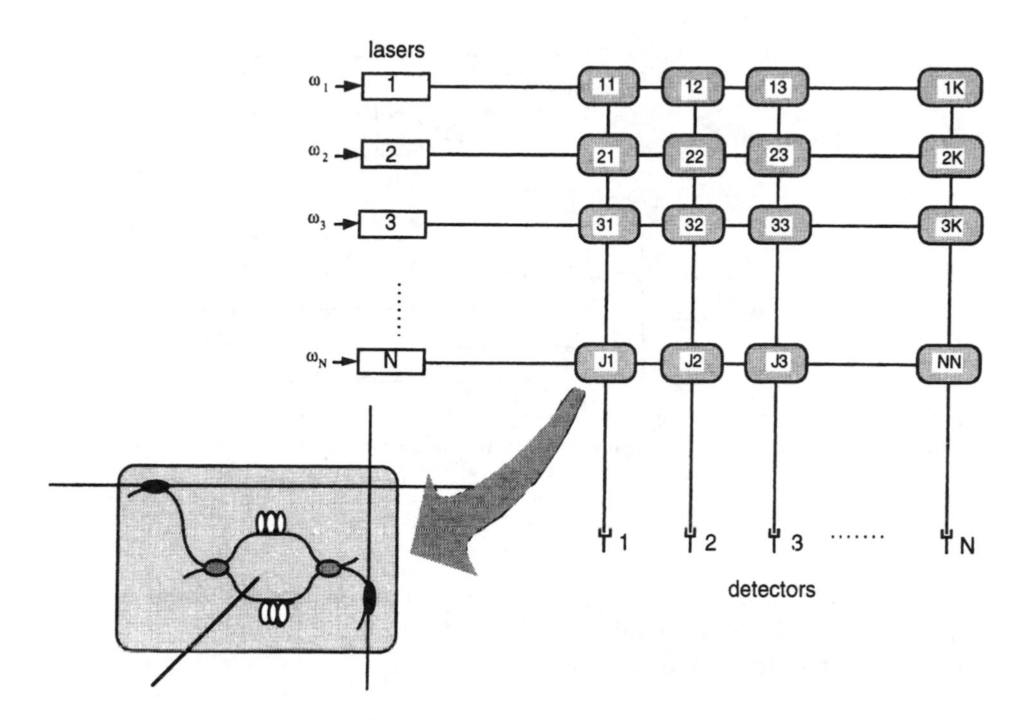

Figure 15.26 *N* × *N* matrix array configuration based on the spatial and frequency-domain multiplexing concepts of Figure 15.25.

$$\epsilon = CM^2 \tag{15.8}$$

where C is a constant that depends on the material parameters or the exact geometrical arrangement of the transducer, and M is the measurand field (i.e., H, E, or z in the cases of magnetic, electric fields, and displacement, respectively). If M comprises two components $M_o + DM \sin w_d t$ where M_o is proportional to the measurand field amplitude, and $DM \sin w_d t$, is a dither signal, the component of the strain induced in the fiber at the fundamental (w_d) of the dither is given by

$$\epsilon(w_d) = 2CDM\, M_o \tag{15.9}$$

which is linearly proportional to M_o and thus the measurand field of interest. This strain can be detected using a fiber interferometer, and a number of such nonlinear transduction elements can be incorporated in a single interferometer system by using different dither frequencies for each sensor, as shown schematically in Figure 15.27. Using this basic concept, the multiplexing of transducers for pressure,

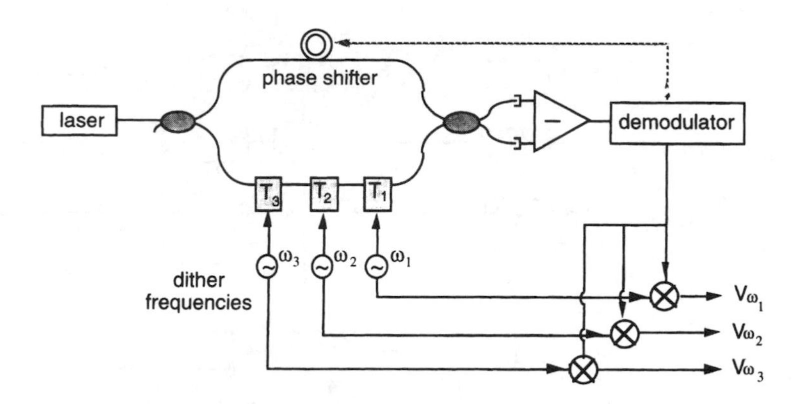

Figure 15.27 Interferometer configuration with transducer multiplexing.

displacement, and magnetic field [103] using a single interferometer has been demonstrated. Other measurands such as acceleration and remote optical dithering have been demonstrated [104].

15.4.4 Wavelength Division Multiplexing

Wavelength division multiplexing has emerged as a powerful technique for use in fiber communications systems and is being exploited in both local area and long-haul communications links [105]. Figure 15.28 shows the type of arrangement possible using WDM technology applied to sensor array systems.

The approach, which is applicable to both intensity and interferometric sensor types, is theoretically the most efficient technique possible, as all the light from a particular source can, in principle, be directed to a particular sensor element and then onto a corresponding photodetector with minimal excess loss. There have been, however, few demonstrations of this capability: the reason for this is due to the limited availability of wavelength-selective couplers (splitters and recombiners, e.g., $N \times N$ star and $1 \times N$ tree couplers), which are required to implement the technique. These types of wavelength-routers [106] are, however, becoming available for use in dense WDM fiber communications systems and can be expected to find their way into sensor systems in the near future.

An important development in this area is that of fiber Bragg gratings. These devices offer some of the most powerful capabilities for WDM of interferometric sensors. By using a modified reflectometric configuration of the type shown in Figure 15.21, very efficient WDM sensor systems can be implemented [30].

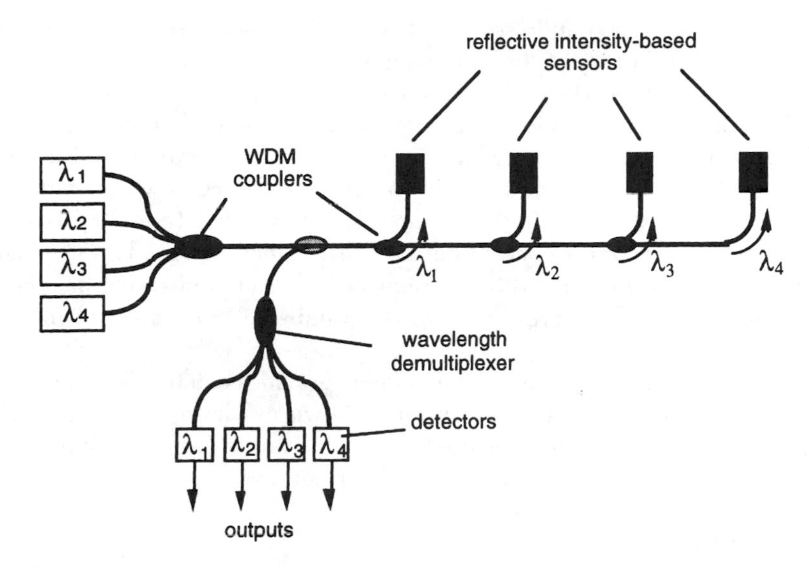

Figure 15.28 Wavelength division multiplexed sensor array.

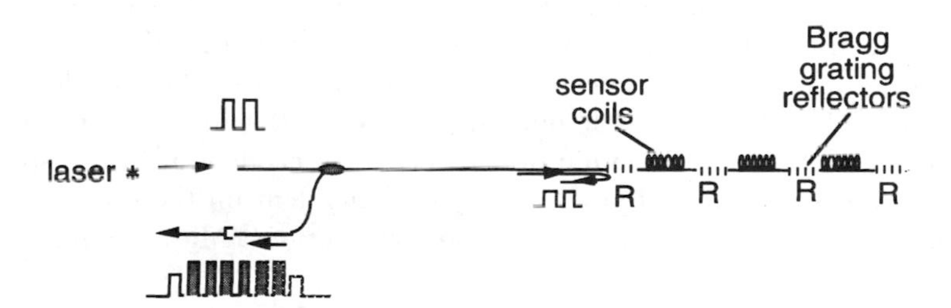

Figure 15.29 FBG reflectometric WDM.

Figure 15.29 shows an example topology. Alternatively, in-line Fabry-Perot (low-reflectivity) interferometers formed in-line in a fiber by using Bragg-wavelength matched pairs of gratings l_1 to l_n can be formed. This type of array can be interrogated using either a multiwavelength source or a tuned laser that can sequentially address each interferometer [107].

15.4.5 Hybrid TDM/WDM System

A means of improving the multiplexing capability of a network is to utilize a system based on a hybrid of addressing approaches. A possible means for this is combining time or frequency division addressing with WDM. As discussed in the above subsection, due to the complexity of the components required to implement on large-scale WDM-based sensor networks (i.e., components to selectively tap certain wavelengths from a fiber bus to sensors and recombine them onto a single output fiber), WDM arrangements have received somewhat limited experimental attention. Furthermore, as the crosstalk between sensors is determined by the degree of wavelength isolation that can be achieved with the WDM couplers (typically only ~15 to 20 dB), only very limited demonstration of the concept has been performed to date. However, combining WDM concepts with time [108] or frequency division addressing has the potential to allow a manyfold improvement in the number of multiplexed sensors in an array.

Again, the use of gratings in these systems provides enhanced systems capabilities: By combining features of both systems shown in Figures 15.21 and 15.29, a TDM/WDM system can be implemented in which a large number of interferometric sensors could be multiplexed along a single fiber path.

15.4.6 Coherence Multiplexing

The basic principle of the coherence multiplexing concept for interferometric sensors is shown in Figure 15.30. Here, light from a broadband source is coupled through a sensor interferometer, the OPD of which is greater than the coherence length of the source. Consequently, at the output of the interferometer, no interference is observed. The light is then coupled to a second interferometer, the path difference of which is closely matched to the sensor interferometer (to within the coherence length of the source) [109–111]. Four possible optical paths exist through this system, two of which are of equal length (to within the source coherence length). Consequently, an interference term is produced at the detector, dependent on the differential phase experienced by the light in both interferometer sections. Multiplexing using this

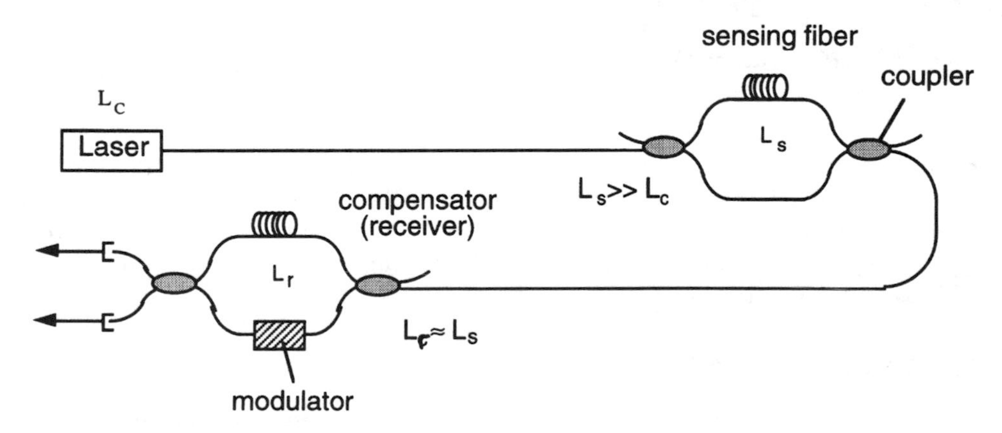

Figure 15.30 Basic principle of coherence addressing.

concept is accomplished using a configuration of the type shown in Figure 15.31. Here, sensor and compensator pairs of matched OPDs, DL_1, DL_2, ... DL_n, are coupled in series, and the output light passed through the sensor chain is fed to a bank of receiving "compensating" interferometers. An interference term associated with each interferometer pair is then produced at each of the corresponding detectors. Unfortunately, this approach is, in practice, extremely problematic due to the very large number of second-order interferometers that can result due to the multitude of optical path combinations that can occur. The number of these crosstalk components increases dramatically with the number of sensor elements. This results in increased noise and crosstalk between sensors, and the technique is thus not well-suited for high-sensitivity interferometric applications.

A two-element multiplexed temperature sensor system based on a wavelength-modulation scheme for monitoring interferometric OPD [112] and using a coherence-

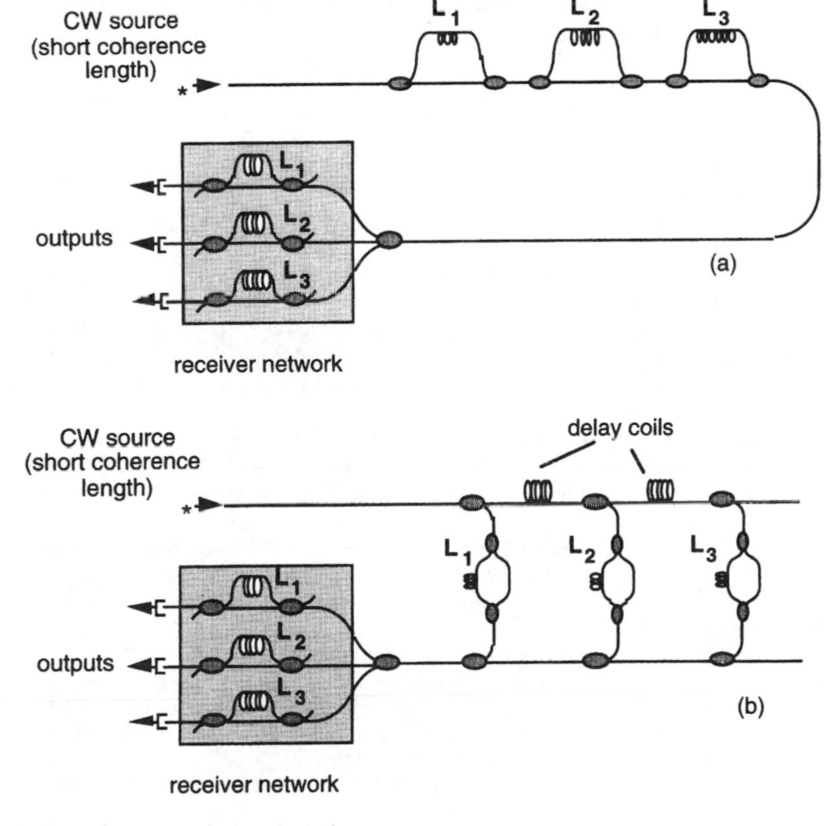

Figure 15.31 Coherence-multiplexed interferometric array.

addressing scheme has been reported. In this case the ultrahigh-phase detection sensitivity normally attainable in interferometric sensor systems was not required, and crosstalk levels of ~ −40 dB were tolerable. The coherence-addressing and multiplexing of polarimetric sensors [113,114] and Fabry-Perot sensors [115] have also been recently reported.

15.5 CONCLUSIONS

This chapter has described the various multiplexing techniques that can be applied to fiber sensors; however, to avoid listing all possible approaches, the coverage is not exhaustive, but has discussed the more successful approaches. Much of the work in this area has been driven by the need to multiplex interferometric sensors for applications in underwater acoustic sensing. More recently, however, there has been a shift in this emphasis to multiplexed systems for sensors such as Bragg gratings, which have inherent wavelength-encoded multiplexing capabilities. Ironically, these very devices also open up new capabilities for multiplexing inline interferometric sensors that may yield very cost-effective and robust multiplexing arrangements in the near future.

References

[1] Barnoski, M. K., and Jensen, S. M., "Fiber Waveguides: A Novel Technique for Investigating Attenuation Characteristics," *Appl. Optics*, 15, 1976, p. 2112.

[2] Dakin, J. P., "Distributed Optical Fiber Sensors," *Proc. SPIE*, 1797, 'Distributed and Multiplexed Fiber-optic Sensors II,' 1992, p. 76.

[3] Kersey, A. D., and Dandridge, A., "Distributed and Multiplexed Fiber-optic Sensor Systems," *J. IERE*, 58, 1988, p. S99.

[4] Kingsley, S. A., "Distributed Fiber-Optic Sensors: an Overview," *Proc. SPIE*, 566, Fiber-optic and Laser Sensors III, San Diego, CA, 1985, p. 234.

[5] Rogers, A. J., "Intrinsic and Extrinsic Distributed Optical-Fiber Sensors," *Proc. SPIE*, 566, Fiber-optic and Laser Sensors III, 1985, p. 234.

[6] Theochorous, E., "Distributed sensors based on differential absorption," *Proc. of the IEEE Colloquium on Distributed Optical Fiber Sensors (Digest 1986/74)*, paper #13, London, 1986.

[7] Farries, M. C., et al., "Distributed Temperature Sensor Using Nd+3-Doped Fiber," *Electron. Lett.*, 22, 1986, p. 418.

[8] Farries, M. C., et al., "Distributed Temperature Sensor Using Holmium+3-Doped Fiber," *Proc. OFC'87*, Reno, NV., 1987, p. 170.

[9] Heinzmann, P., and Hofstetter, R., "Temperature Dependence of PCS Fiber Characteristics," *Proc. SPIE*, 584, Cannes, 1985.

[10] Oscroft, G., "Intrinsic Fiber-optic Sensors," *J. Optical Sensors*, 2, 1987, p. 269.

[11] Dakin, J. P., and Pratt, D. J., "Temperature Distribution Measurement Using Raman Ratio Thermometry," *Proc. SPIE*, 566, 'Fiber-optic and Laser Sensors III', San Diego, 1985, p. 249.

[12] Dakin, J. P., et al., "Distributed Optical Fiber Raman Temperature Sensor using a Semiconductor Light Source and Detector," *Electron. Lett.*, 21, 1985, p. 569.

[13] Hartog, A. H., et al., "Distributed Temperature Sensing in Solid Core Fibers," *Electron. Lett.*, 21, 1985, p. 1061.

[14] Kurashima, T., et al., "Distributed Temperature Sensing Using Stimulated Brillouin Scattering in Optical Silica Fibers," *Optics Lett.*, 15, 1990, p. 1038.
[15] Horiguchi, T., et al., "Development of a Distributed Sensing Technique Using Brillouin Scattering," *IEEE J. Lightwave Technol.*, 13, 1995, p. 1296.
[16] Bao, et al., "Recent Progress in Experiments on a Brillouin Loss Based Distributed Sensor," *Proc. OFS-10*, Glasgow, 1994, p. 506.
[17] Desforges, F. X., et. al.," Progress in OTDR Optical Fiber Sensor Networks," *Proc SPIE*, 718, Fiber-optic and Laser Sensors IV, Cambridge, MA, 1986, p. 225.
[18] Tardy, A., and Jurczyszyn, M., "Multipoint Fiber-optic Refractive Index Sensors," *Proc. SPIE*, 586, 'Fiber-optic Sensors', 1985, p. 65.
[19] Kingsley, S. A., and McGinniss, V. D., "Distributed Fiber-optic Hot-Spot Sensors," *Proc. SPIE*, 718, 'Fiber-optic and Laser Sensors IV', 1986, p. 218.
[20] Meltz, G., Morey, W. W., and Glenn, W. H., "Formation of Bragg Gratings in Optical Fibers by a Transverse Holographic Method," *Optics Lett.*, 14, 1989, p. 823.
[21] Hill, K. O., et al., "Bragg Gratings Fabricated in Monomode photosensitive Optical Fiber by UV Exposure Through a Phase Mask," *Appl. Phy. Lett.*, 62, 1993, p. 1035.
[22] Malo, B., et al, "Point by Point fabrication of Micro-Bragg Gratings in Photosensitive Fiber Using Single Excimer Pulse Refractive Index Modification Techniques," *Electron. Lett.*, 29, 1993, p. 1668.
[23] Archambault, J. L., et al., "100% Reflectivity Bragg Gratings Reflectors Produced in Optical Fibers by Single Excimer Laser Pulses," *Electron. Lett.*, 29, 1993, p. 453.
[24] Askins, C. G., et al., "Considerations for Producing Single-Pulse Fiber Bragg Gratings," *Proc. SPIE*, 2071, 'Distributed and Multiplexed Fiber-optic Sensors III', 1993, p. 12.
[25] Malo, B., et al., "Single Excimer-Pulse Writing of Fiber Gratings by use of a Zero-Order Nulled Phase Mask: Grating Spectral Response and Visualization of Index Perturbations," *Optics Lett.*, 18, 1993, p. 1277.
[26] Askins, C. G., et al., "Stepped-wavelength Optical Fiber Bragg Grating Arrays Fabricated in Line on a Draw Tower," *Optics Lett.*, 19, 1994, p. 147.
[27] Kashyap, R., "Photosensitive Optical Fibers: Devices and Applications," *Optical Fiber Technology*, 1, 1994, p. 17.
[28] Morey, W. W., Meltz, G., and Glenn, W. H., "Fiber Bragg Grating Sensors, *Proc. SPIE*, 1169, 'Fiber-optic & Laser Sensors VII', 1989, p. 98.
[29] Morey, W. W., et al., "Bragg-Grating Temperature and Strain Sensors," *Proc. OFS'89*, Springer Verlag, Paris, 1989, p. 526.
[30] Morey, W. W., Dunphy, J. R., and Meltz, G., "Multiplexing Fiber Bragg Grating Sensors," *Proc. SPIE*, 1586, 'Distributed and Multiplexed Fiber-optic Sensors ', Boston, Sept. 1991, p. 216.
[31] Melle, S. M., Liu, K., and Measures, R. M., "A Passive Wavelength Demodulation System for Guided-Wave Bragg Grating Sensors," *IEEE Photonics Technol. Lett.*, 4, 1992, p. 516.
[32] Davis, M. A., and Kersey, A. D., "All Fiber Bragg Grating Strain Sensor Demodulation Technique Using a Wavelength Division Coupler," Electronics Letters, *Electronics Letters*, 30, 1994, p. 75.
[33] Kersey, A. D., et al. "Multiplexed Fiber Bragg Grating Strain Sensor System with a Fiber Fabry Perot Wavelength Filter," *Optics Lett.*, 18, 1993, p. 1370.
[34] Xu, M. G., et al., "Novel Interrogation System for Fiber Bragg Grating Sensors using an Acousto-Optic Tunable Filter," *Electron. Lett.*, 29, 1993, p. 1510.
[35] Jackson, D. A., et al., "Simple Multiplexing Scheme for a Fiber-optic Grating Sensor Network," *Optics Lett.*, 18, 1993, p. 1193.
[36] Davis, M. A., et al, "High Sensor-Count Fiber Bragg Grating Sensor System for Large Scale Structural Monitoring Applications," *Proc. SPIE*, 2718, paper # 69, 1996.
[37] Davis, M. A., et al., "A 60 Channel Fiber Bragg Grating Demodulation System," to be presented at OFS-11, Sapporo, 1996.

[38] Kersey, A. D., Berkoff, T. A., and Morey, W. W., "High Resolution Fiber Bragg Grating Based Strain Sensor with Interferometric Wavelength Shift Detection," *Electronics Letters*, 28, 1992, p. 236.

[39] Kersey, A. D., Berkoff, T. A., and Morey, W. W., "Fiber-optic Bragg Grating Sensor with Drift - Compensated High Resolution Interferometric Wavelength Shift Detection," *Optics Letters*, 1993, p. 72.

[40] Kersey, A. D., and Berkoff. T. A., "Fiber-optic Bragg Grating Differential Temperature Sensor," *IEEE Photon. Technol. Lett.*, 4, 1993, p. 1183.

[41] Berkoff, T. A., and Kersey, A. D., "Eight Element Time-Division Multiplexed Fiber Grating Sensor Array With Integrated-Optic Wavelength Discriminator," *Proc 10th. Int. Conf. on Optical Fiber Sensors*, Glasgow, 1994, p. 350.

[42] Askins C. G., et al., "Instrumentation for Interrogating Many-Element Fiber Bragg Grating Arrays Embedded in Fiber/Resin Composites," *Proc. SPIE*, 2444, 'Smart Sensing, Processing and Instrumentation', 1995.

[43] Kersey, A. D., et al., "Single Mode Fiber Fourier Transform Spectrometer," *Electron. Lett*, 21, 1985, p. 463.

[44] Davis, M. A., and Kersey, A. D., "A Fiber Fourier Transform Spectrometer for Decoding Fiber Bragg Grating Sensors," *IEEE J. Lightwave Technol.*, 13, 1995, p. 1289.

[45] Berkoff, T. A., et al., "Hybrid Time And Wavelength Division Multiplexed Fiber Bragg Grating Sensor Array," *Proc. SPIE*, 2444, 1995, p. 288.

[46] Proshaka, J. D., et al., "Fiber-optic Bragg Grating Strain Sensor in Large Scale Concrete Structures," *Proc. SPIE*, 1798, 'Fiber-optic Smart Structures and Skins', 1992, p. 286.

[47] Maaskant, R., et al., "Fiber-optic Bragg Grating Sensor Network Installed in a Concrete Road Bridge," *Proc. SPIE*, 2191, 'Smart Sensing, Processing and Instrumentation', 1994, p. 457.

[48] Davis, M. A., et al., "Distributed Fiber Bragg Grating Strain Sensing in Reinforced Concrete Structural Components," *J. Cement and Concrete Composites*, to be published, May 1996.

[49] Bullock, D., et al., "Embedded Bragg Grating Fiber-optic Sensor For Composite Flexbeams," *Proc. SPIE*, 1798, 'Fiber-optic Smart Structures and Skins', 1992, p. 253.

[50] Friebele, E. J., et al., "Distributed Strain Sensing with Fiber Bragg Grating Arrays Embedded in CRTM Composites," *Electron. Lett.*, 30, 1994, p. 1783.

[51] Ball, G. A., et al., "Low Noise Single Frequency Linear Fiber Laser," *Electron. Lett.*, 29, 1993, p. 1623.

[52] Ball, G. A., and Morey, W. W., "Compression Tuned Single Frequency Bragg Grating Fiber Laser," *Optics Lett.*, 19, 1994, p. 1979.

[53] Ball, G., et al., "Single and Multipoint Fiber Laser Sensors," *Photonics Technol. Lett.*, 5, 1993, p. 267.

[54] Melle, S. M., et al., "A Bragg Grating Tuned Fiber Laser Strain Sensor System," *Photonics Technol. Lett.*, 5, 1993, p. 263.

[55] Ball, G., et al., "Polarimetric Heterodyning Bragg Grating Fiber Laser Sensor," *Optics Lett.*, 18, 1993, p. 976.

[56] Kersey, A. D., and Morey, W. W., "Multiplexed Bragg Grating Fiber Laser Strain Sensor System with Mode-Locked Interrogation," *Electronics Letters*, 29, 1993, p.112.

[57] Kersey, A. D., and Morey, W. W., "Multi-element Bragg Grating Based Fiber Laser Strain Sensor," *Electron. Lett.*, 29, 1993, p. 964.

[58] Alavie, A. T., et al., "A Multiplexed Bragg Grating Fiber Laser System," *Photonics Technol. Lett.*, 5, 1993, p. 1112.

[59] Koo, K. P., and Kersey, A. D., "Bragg Grating Based Laser Sensor Systems with Interferometric Interrogation and Wavelength Division Multiplexing," *IEEE J. Lightwave Technol.*, 13, 1995, p. 1243.

[60] Nelson, A. R., et. al., "Passive Multiplexing System for Fiber-Optic Sensors," *Appl. Optics*, 19, 1980, p. 2917.

[61] Spillman, W. B., and Lord, J. R., "Self -Referencing Multiplexing Technique for Fiber-Optic Intensity Sensors," *J. Lightwave Technol.*, LT-5, 1987, p. 865.

[62] Spillman, W. B., and Lord, J. R., "Self -Referencing Multiplexing Technique for Fiber-Optic Intensity Sensors, *SPIE Proc. Fiber-optic and Laser sensors V*, San Diego, CA, 1987.

[63] Brooks, J. L., et al., "Fiber-optic Interferometric Sensor Arrays with Freedom from Source Phase-Induced Noise," *Optics Lett.*, 12, 1987, p. 473.

[64] Kersey, A. D., et al., "Multiplexing of Interferometric Fiber Sensors Using Time Division addressing and Phase Generated Carrier Demodulation," *Optic Lett.*, 12, 1987, p. 775.

[65] Kersey, A. D., Dorsey, K. L., and Dandridge, A., "Demonstration of an eight-element Time-Division Multiplexed Interferometric Fiber Sensor Array, "*Electron. Lett.*, 24, 1988, p. 689.

[66] Kersey, A. D., and Dandridge, A., "Multiplexed Mach-Zehnder Ladder Array with Ten Sensor Elements, *Electron. Lett.*, 25, 1989, p.1298.

[67] Kersey, A. D., Dandridge, A., and Dorsey, K. L., "Transmissive Serial Interferometric Fiber Sensor Array," *J. Lightwave Technol.*, 7, 1989, p. 846.

[68] Moslehi, B., et al., "Efficient Fiber-optic structure with Applications to Sensor Arrays, *IEEE J. Lightwave Technol.*, 7, 1989, p. 236.

[69] Marrone, M. J., et al., "Fiber-optic Michelson Array with Passive Elimination of Polarization Fading and Source Feedback Isolation," *Proc. 8th Int. Conf on Optical Fiber Sensors*, Monterey, Ca., 1992, p. 69.

[70] Stowe, D. W., et al., "Polarization Fading in Fiber Interferometric Sensors," *J. Quant. Electron.*, 18, 1982, p. 1644.

[71] Frigo, N. J., et al., "Technique for the Elimination of Polarization Fading in Fiber Interferometers," *Electron. Lett.*, 20, 1984, p. 319.

[72] Kersey, A. D., and Marrone, M. J., "Input-Polarization Scanning Technique for Overcoming Polarization-Induced Signal Fading In Interferometric Fiber Sensors," *Electron. Lett.*, 24, 1988, p. 931.

[73] Wanser, K. H., and Safar, N. H., "Remote Polarization Control for Fiber-optic Interferometers," *Optics Lett.*, 12, 1987, p. 217.

[74] Kersey A. D., et al, "Dependence of Visibility on Input Polarization in Interferometric Fiber-Optic Sensors," *Opt. Lett.*, 13, 1988, p. 288.

[75] Kersey, A. D., et al, "Optimization and Stabilization of Visibility in Interferometric Fiber-optic Sensors Using Input Polarization Control," *J. Lightwave Technol.*, 6, 1988, p. 1599.

[76] Kersey, A. D., et al, "Observation of Input Polarization Induced Phase Noise in Interferometric Fiber-optic Sensors," *Opt. Lett.*, 13, 1988, p. 847.

[77] Edge, C., and Stewart, W. J., "Measurement of Nonreciprocity in Single Mode Optical Fibers', *Tech. digest IEEE Colloq. on Optical Fiber Measurements*, (1987/55), 1987.

[78] Pistoni, N. C., and Martinelli, M., "Birefringence Effects Suppression in Optical Fiber Sensor Circuits," *Proc. 7th Int. Conf. on Optical Fiber Sensors*, 1990, p. 125.

[79] Kersey, A. D., et al, "Polarization -Insensitive Fiber-optic Michelson Configuration," *Electron. Lett.*, 26, 1991, p. 518.

[80] Marrone, M. J., and Kersey, A. D., "Visibility Limits in Fiber-optic Michelson Interferometer with Birefringence Compensation," *Electron. Lett.*, 27, 1991, p. 1422.

[81] Kersey, A. D., et al., "Demonstration of a 64 channel TDM Fiber Sensor Array," *Proc. OFC 96*, San Jose, 1996, p. 270.

[82] Dakin, J. P., et al., "Novel Optical Fiber Hydrophone Array Using a Single Laser Source and Detector," *Electron Lett.*, 20, 1984, p. 14.

[83] Lee, C. E., and Taylor, H. F., "Interferometric Optical Fiber Sensors Using Internal Mirrors,," *Electron. Lett.*, 24, 1988, p. 193.

[84] Henning, M., et al., "Improvements in Reflectometric Fiber-optic Hydrophones," *Proc. SPIE*, 586, 'Fiber-optic Sensors', 1985, p. 58.

[85] Henning, M. L., and Lamb, C., "At-Sea Deployment of a Multiplexed Fiber-optic Hydrophone

Array, "*Proc. 5th Int. Conf. on Optical Fiber Sensors*, OSA technical digest, New Orleans, Jan. 1988, p. 84.

[86] Kersey, A. D., et al., "Analysis of Intrinsic Crosstalk in Tapped Serial and Fabry Perot Interferometric Fiber Sensor Arrays," *Proc. SPIE*, 985, 'Fiber-optic and Laser Sensors VI', 1988, p. 113.

[87] Dixon, R. C., *Spread Spectrum Systems*, John Wiley & Sons, 1984.

[88] Prucnal, P. R., et al., "Spread Spectrum Fiber-optic Local Area Network using Optical Processing," *J. Lightwave Technol.*, $F3LT-4, 1986, p. 547.

[89] Everard, J.K.A., "Novel Signal Processing Techniques for Enhanced OTDR Sensors," *Proc. Fiber-optic Sensors II, Proc. SPIE*, 798, The Hague, 1987, p. 42.

[90] Al-Raweshidy, H. S., and Uttamchandani, D., "Spread Spectrum Technique for Passive Multiplexing of Interferometric Fiber-optic Sensors," *Proc. Fiber-Optics'90, Proc. SPIE*, 1314, London, 1990, p. 342.

[91] Kersey, A. D., and Dandridge, A., "Low Crosstalk Code division Multiplexed Interferometric Array," *Electron. Lett.*, 28, 1992, p. 351.

[92] Mlodzianowski, J., et. al., "A Simple Frequency Domain Multiplexing System for Optical Point Sensors," *IEEE J. Lightwave Technol.*, LT-5, 1987, p.1002.

[93] K. I., Mallalieu, et. al., "FMCW of Optical Source Envelope Modulation for Passive Multiplexing of Frequency-Based Fiber-Optic Sensors," *Electron. Lett.*, 22, 1986, p. 809.

[94] Giles, I. P., et al., "Coherent Optical-Fiber Sensors with Modulated Laser Sources," *Electron. Lett.*, 19, 1983, p. 14.

[95] Al Chalabi, S., et. al., "Multiplexed Optical Interferometers - An Analysis Based on Radar Systems," *Proc. IEEE Part J.*, 132, 1985, p. 150.

[96] Sakai, I., et. al., "Multiplexing of Optical Fiber Sensors Using a Frequency-Modulated Source and Gated Output," *IEEE J. Lightwave Technol.*, LT-5, 1987, p. 932.

[97] Dandridge, A., Tveten, A. B., and Giallorenzi, T. G., "Homodyne Demodulation Scheme for Fiber-Optic Sensor Using Phase Generated Carrier," *IEEE J. Quantum Electron.*, 18, 1982, p. 147.

[98] Dandridge, A., et al., "Multiplexing of Interferometric Sensors using Phase Generated Carrier Techniques," *IEEE J. Lightwave Technol.*, $F3LT-5, 1987, p. 947.

[99] Dandridge, A., et al., "AOTA Tow test results," *Proc. AFCEA/DoD Conf. on Fiber-optics '90*, McLean, 1990, p. 104.

[100] Kersey, A. D., et al., "Detection of DC and Low Frequency AC Magnetic Fields Using an All Single-Mode Fiber Magnetometer," *Electron. Lett.*, 19, 1983, p. 469.

[101] Vohra, S. T., et al., "Fiber-optic DC and Low Frequency Electric Field Sensor," *Optics Lett.*, 16, 1991, p. 1445.

[102] Kersey, A. D., et al., "New Nonlinear Phase Transduction Method for DC Measurand Interferometric Fiber Sensors," *Electron. Lett.*, 22, 1986, p. 75.

[103] Bucholtz, F., Kersey, A. D., and Dandridge, A., "Multiplexing of Nonlinear Fiber-optic Interferometric Sensors," *IEEE J. Lightwave Technol.*, 7, 1989, p. 514.

[104] Kersey, A. D., et al., "Interferometric Sensors for DC Measurands - A New Class of Fiber Sensors," *Proc. SPIE*, 718, ' Fiber Optic and Laser Sensors IV', Cambridge, MA, 1986, p. 198.

[106] Okamoto, K., and Inoue, Y., "Silica-Based Plannar Lightwave Circuits for WDM Systems," *Proc. OFC'95*, 1995, p. 224.

[107] Kersey, A. D., and Marrone, M. J., "Nested intererometric Sensors Utilizing Fiber Bragg Grating Reflectors," to be presented at OFS-11, Sapporo, Japan, May, 1996.

[108] Kersey, A. D., and Dandridge, A., "Demonstration of a Hybrid Time/Wavelength Division Multiplexed Interferometric Fiber Sensor Array," *Electron. Lett.*, 2, 1991, p.554.

[109] Al -Chalabi, S. A., Culshaw, B., and Davies, D.E.N., "Partially Coherent Sources in Interferometric Sensors," *Proceedings of the First International Conference on Optical Fibre Sensors (IEEE)*, 1983, p. 132.

[110] Brooks, J. L., et al., "Coherence Multiplexing of Fiber-optic Interferometric Sensors," *IEEE J. Lightwave Technol.*, 3, 1985, p. 1062.
[111] Kersey, A. D., and Dandridge, A., "Phase Noise Reduction in Coherence Multiplexed Interferometric Fiber sensors," *Electron. Lett.*, 22, (11), 1986, pp. 616-618.
[112] O'Connell, D., et al., "Coherence Multiplexed Fiber-optic Temperature Sensor using a Wavelength Dithered Source," *Proc. OFC '89*, Houston, Feb. 1989, p. 145.
[113] Kersey, A. D., et al., "Differential polarimetric Fiber-optic Sensor Configuration with Dual Wavelength Operation," *Appl. Optics*, 28, 1989, p. 204.
[114] Gusmeroli, V., et al., "A Coherence Multiplexed Quasi-Distributed Polarimetric Sensor Suitable or Structural Monitoring," *Proc. OFS'89*, Springer Verlag, Paris, 1989, p. 513.
[115] Nellon, Ph. M., et al., "Absolute Strain Measurements with Multiplexed Low Coherence Demodulated Fiber Fabry-Perot Sensors," *Proc. OFS'94*, Glasgow, 1994, p. 518.

Chapter 16

Fiber Optics and Smart Structures

Dr. W. B. Spillman, Jr.
BF Goodrich Aerospace Aircraft, U.S.A.

16.1 INTRODUCTION

In recent years, a new interdisciplinary field called *smart structures* has emerged from a synergistic combination of research in designed materials, advanced sensing, actuation and communications, and artificial intelligence [1–9]. The aim of this field is the creation of active structures that perform better and cost less than their passive counterparts designed for the same purpose. Fiber optics represents one of the most important technologies fueling the growth of this field, both for communications and for sensing [10]. The ability to sense is critical for any active structure, since without this ability, the results of actions cannot be ascertained and the appropriateness of additional action determined. In this chapter, we will provide a definition of smart structures and show the beneficial uses of fiber optics in their development. Two particular application areas, high-performance composite structures and large civil structures, will be discussed. Finally, the challenges that must be met in order for practical smart structures to be created will be covered.

16.2 DEFINITION OF A SMART STRUCTURE

The field of smart structures had its beginnings in attempts in the early 1980s to develop radar antennas that were conformally integrated into the skins of military aircraft [11]. This so-called "smart skins" activity was soon expanded to include health monitoring of the whole aircraft structure and development activity in the area was termed "smart structures" research. Since that time, civil, marine, automotive, and other structures have been the subject of smart structures research and the amount of worldwide activity has been continually increasing.

When the field of smart structures research is first considered, the question immediately arises, "What exactly is a smart structure?" In general, a smart structure will be a structure that is active in some sense and contains the elements shown in Figure 16.1. It will consist of a structural/load-bearing element, sensors, actuators, communications, processing, and some sort of adaptive control algorithm. Although not specifically shown, some source of energy must also be available, either external or integrated within the structural material. As can be seen from Figure 16.1, two control loops are inherent in the structure, one for observing and acting upon the internal state and one for observing and acting upon the environment. The structure can then be seen to become more "smart" by adding functionality in two ways. The first involves enhancing the basic load-bearing function through internal condition monitoring, self-repair, and adaptive load distribution. The second involves integration into the structure of additional functionalities that are not related to the load-bearing function. This could be considered as analogous to integrating the radar antenna into an aircraft skin as envisioned in the first smart skins research. In the final analysis, however, research in the area of smart structures has evolved into the attempt to create structures that perform better and cost less than purely passive structures designed for the same purpose, through emulation of the integrated designs and adaptive natures of biological or living systems. A smart structure will adapt to

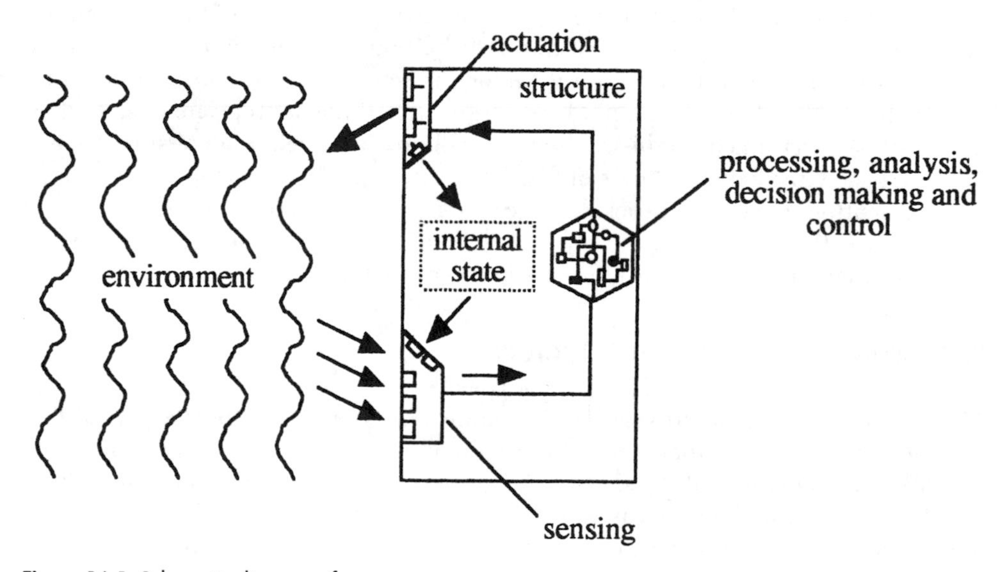

Figure 16.1 Schematic diagram of a smart structure.

changes in its environment and its internal state, so as to optimize itself in order to achieve some specified purpose. A distinction should be made here between behavior that is smart, as in living systems, and intelligent. An intelligent system would develop a world model based on its sensed inputs, predict future conditions, and act now to optimize itself both for present conditions and for the anticipated future states predicted by the model. In this context, a smart structure is emphatically not an intelligent structure.

There have been a number of different attempts to define smart structures. Based upon a survey of workers in the field, however, a general definition has been developed [12] that encompasses most current and planned research. According to this definition, a smart structure is a nonbiological physical structure having the following attributes:

1. A definite purpose;
2. Means and imperative to achieve that purpose;
3. A biological pattern of functioning.

The purpose of the smart structure is designed into it at all levels, based upon some specification. The means and imperative to achieve that purpose are provided by having some ability to sense the external environment and/or the internal structural state, some ability to change or act on the environment and/or the internal state, a method of communications between the structure functions, a data processing and control capability, and, finally, some form of control algorithm, all integrated within the structure itself. The integrated smart structure will act according to its control algorithm in such a way to optimize itself to achieve its purpose in the same way a biological system would (i.e., it will be adaptable and utilize energy as efficiently as possible).

In comparing the functionalities involved in a smart structure, it is easily seen that fiber-optic technology offers the potential to provide either the communications capability, sensing capability, or both. Since the communications aspect of the application of this technology is quite straightforward, this chapter will focus upon the contribution fiber-optics technology can make as the sensing function for smart structures.

16.3 PRACTICAL APPLICATION AREAS FOR SMART STRUCTURES TECHNOLOGY

In the near term, there are two areas in which smart structures technology can have a significant impact. The first of these involves high-performance composite structures fabricated from the new designed materials that are becoming increasingly available while the second involves the design, construction, and maintenance of large civil structures such as buildings, bridges, dams, and roads.

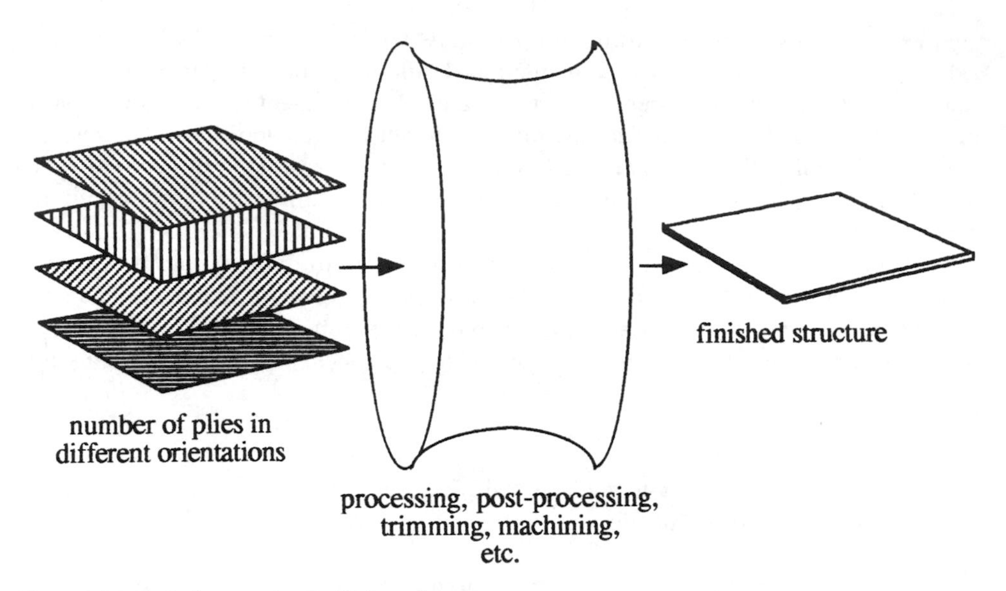

Figure 16.2 Typical processing for high-performance composite structures.

The first near-term application area for smart structures technology encompasses the design, construction, test, and use of structures based upon the wide range of new composite materials now available. Graphite-epoxy composite structures are one example in which the structure can be designed to meet very complicated patterns of anticipated stress and strain. The process of fabrication is shown in Figure 16.2, in which a number of different layers or plies of composite material are assembled together. Each ply will generally consist of a number of (eventually) load-bearing fibers (nonoptical) oriented or arranged in a particular pattern (parallel, interwoven at some angle, etc.) embedded in some epoxy matrix. After the plies are assembled together, they are generally compressed and heated for some period of time until the (now) composite structure has formed. Extraneous material is then removed by trimming and machining, resulting in the finished structure. Such structures are usually stronger and lighter than their metal counterparts. The lack of widespread acceptance and use of such structures is due to two factors. First, there is a very complicated interaction between the uncured structural elements and the fabrication/curing process. It is often very difficult to validate that the final structure realizes the intended design parameters. Unplanned residual strains in the structure can cause it to fail in service even when loading does not exceed specified design levels. The second problem involves the lack of engineering data on the long-term performance of these structures, particularly their failure modes. A chicken and egg situation exists. Structures made of the new designed materials are not widely used, due to the lack of extensive engineering data, while at the same time, the data cannot be obtained because the structures are not in wide use. Finally, when structures are made from these materials, the lack of engineering data results in overdesign to provide sufficient safety margins, so that the benefits of lighter weight and higher strength inherent in the designed materials are largely eliminated. The use of real-time condition-monitoring systems can provide a way to permit structures based on these new materials to finally achieve their potential. Fiber-optic sensors offer the advantage of being very small and unobtrusive. They have been shown not to adversely affect structural strength [13]. They can also be used to both monitor the structure during its cure process and to continue to monitor structural condition throughout the service life of the structure.

The second near-term application area for smart structures technology is concerned with large civil structures. There can be no doubt that the civil infrastructure around the world is in increasing need of renewal [14]. Bridges represent a particularly critical problem, since many have now exceeded their design lifetimes. Every year, numerous tragedies are reported in which bridges have failed and lives have been lost. The question then becomes how to most cost-effectively ascertain the condition of existing bridges and how to create new bridges whose structural condition can be obtained "on demand," to allow potential catastrophes to be identified and steps taken so that they do not occur. Other large civil structures have similar types of problems. What is needed, then, is a method of retrofitting condition-monitoring (sensing) systems on existing structures and a method of integrating structural-monitoring systems into new structures. Fiber-optic sensors appear attractive both for the retrofit of existing structures and for integration into new structures. The key issue for civil structures is their sheer size. Extremely large areas and/or volumes must be monitored and cost must be considered not just in terms of the sensing system acquisition cost, but also its operating costs. The chief attractiveness of fiber optics here is that the technology permits basically lossless transmission of optical signals on the distance scales seen in civil structures. Once into the optical domain, powers are most often measured in milliwatts, not the watts or kilowatts typical of electrical systems. Since the transmission is essentially lossless, sensors and signal-conditioning electronics can be arbitrarily separated in distance without increasing the system energy requirements. Fiber optics also offers the advantages that individual transducer mechanisms exist that can be used as the basis of sensors for all parameters of interest, optical transmission lines are extremely cost-competitive with electrical cables and generally are much lighter and smaller and basically immune to electromagnetic interference. Finally, the fiber-optic transducers can be configured as point sensors, integrating sensors, quasi-distributed sensors and fully distributed sensors that can all be made the basis of large spatially extended systems possessing considerable geometric flexibility due to the low-loss transmission characteristic previously mentioned.

To summarize, development of smart structures in the near future will almost certainly involve the use of fiber-optic sensing. Fiber optics can provide common transducer mechanisms for sensing multiple different parameters, essentially lossless transmission of optical power and data at very high rates, small size and weight, dielectric and relatively chemically inert composition, and, finally, great geometric flexibility. We will next examine the aspects of fiber-optic sensing for smart structures in more detail.

16.4 SENSING OPTIONS, CRITICAL PARAMETERS, AND MEASUREMENT DOMAINS

Depending upon the particular application, and independent of the sensing technology used, there exist a number of options for the creation of sensor systems for smart structures [15]. These can be classified according to how the sensed information is collected spatially. The simplest form of sensor is a point sensor. It measures the value of some parameter at some specific location. One might note in passing that even if a measurement is made at only one point, (depending upon the parameter sensed, an acoustic signal for example), the sensed parameter could represent the integrated and appropriately time delayed contributions from a large spatial region around the point (a fact exploited when carrying out analyses using Green's functions). We will limit our classification of sensing options, however, to a consideration of the spatial arrangement of the sensing systems themselves.

The point sensor generally returns a single piece of information about a single point in space at a single time. An integrating sensor, on the other hand, also provides a single piece of information for a specific time. The information involved, however, represents the integrated value of a parameter or parameters over some spatial domain. This type of sensor, if configured properly, offers the potential to provide simple analog data processing in addition to its basic sensing function.

If one requires information about spatially distributed parameter fields in detail, rather than some integrated or average value, then the spatial domain of interest must be sensed at an appropriately high density of points and data returned individually from each of those points. This can be done in a number of ways using either a number of individual point sensors, a multiplexed point sensor array, a quasi-distributed sensor, or a fully distributed sensor. Each of these types of sensing systems have been demonstrated using fiber-optic sensors. The variety of sensing options is illustrated in Figure 16.3.

Any sensing system for a smart structure must be prepared to answer the following questions. First, has the structure been fabricated correctly? Secondly, is the structure functioning as designed? Is the structure damaged and, if so, what is the extent and location of the damage? Finally, what is the remaining life of the structure? If the integrated sensing system can answer all of these questions, the structure has added a function that is necessary, though not sufficient for the structure to be called smart.

In terms of what needs to be sensed for the two primary applications of smart structures, one must first consider the parameters that define the fabrication process

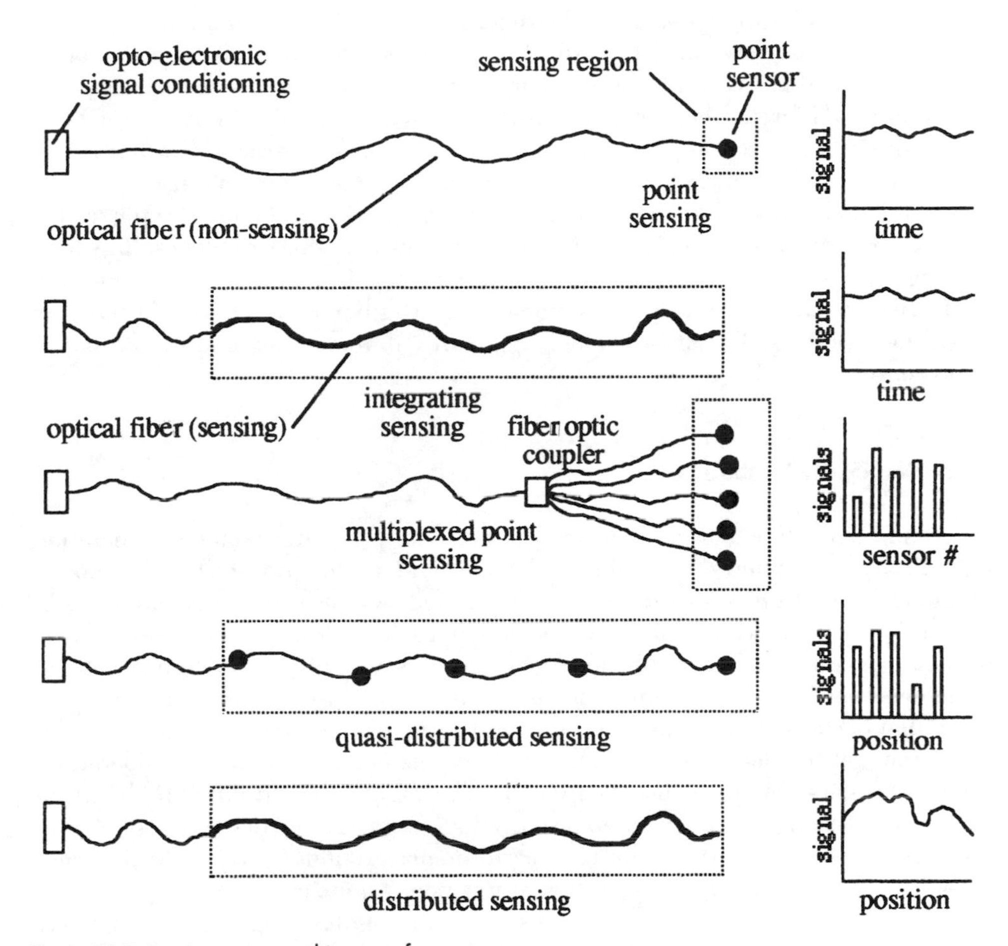

Figure 16.3 Sensing system architectures for smart structures.

for high-performance composite structures. Generally, elevated temperatures are present along with high pressure. The density of the structure changes during the cure process along with its modulus. Finally, chemical processes occur during the cure as well. After fabrication and during service, the primary measures of structure health involve stress and strain distributions under various loading conditions. The size of the structures rarely exceeds a few meters in any dimension. Consideration of these points identifies the need to sense temperature, pressure, stress, strain, and particular chemical species. The size of the structures indicates that in all likelihood, only a limited number of point sensors will be required to carry out the monitoring of the internal structural state and the surrounding environmental conditions.

For the monitoring of large civil structures, the parameters of interest for composite structures are also important. In addition, one would like to obtain information on crack formation, size and location, corrosion, and (in some instances) vibration. The primary difference between the sensing requirements for large civil structures and high-performance composite structures lies in size. Civil structures will generally be hundreds of meters in length, making the use of arrays of point sensors impractical not only because of the large number of points to be sensed and computer processing power required, but also (and most importantly) because of the resulting high sensing system cost. This leads to the conclusion that the sensing systems for large civil structures will probably consist of long gauge length integrating sensors and fully distributed sensors. To a lesser extent, quasi-distributed sensors and multiplexed point sensors might be used.

16.5 INTEGRATION ISSUES

When additional functions are integrated into a smart structure, some consideration must be given as to how this should best be done and, as importantly, how will power and data be transmitted to and from the integrated systems if that is required. As in most things dealing with smart structure design, each application is unique. However, some general guidelines can be identified. There is a significant difference between structures that must operate under loading conditions of torsion and tension versus those that operate in primarily compressive loading regimes. The acceptable ratio of the volume of the defects to total structure volume is also a critical consideration. Finally, the question of whether the functions are going to be retrofitted to the structure or whether they are intended to be part of a new integrated design is also important. A consideration of the use of high-performance graphite/epoxy to create structures for two different applications shows these points quite clearly.

Due to the advantages of higher strength and lighter weight, graphite epoxy structures are being considered for use in both aerospace and undersea applications. For aerospace applications, the structures fabricated tend to be thin-walled and operate under conditions of torsional loading and/or tension. Even a sensing system fabricated from elements as small as optical fibers represents a level of inclusion and potential failure that is unacceptable to the structure designer. For that reason, fiber-optic sensing systems for aerospace composite structure monitoring will probably be attached to the exterior of the load-bearing part of the structure, but still be integrated within the total structure by being placed beneath the nonload-bearing coating layer that is usually present. The system will still provide useful information due to the fact that the structure is thin-walled so that the surface mounted sensor system will be able to detect changes within the structure due to the proximity of the sensing elements to all points in the interior. For undersea applications, the situation is quite different. The structures are thick-walled and the loading is compressive. This has two implications. First, a sensing system mounted on the surface would be too far away from points in the interior of the structure to provide useful data on conditions there. Secondly, a larger size of inclusion is acceptable because the loading, being compressive, tends to hold the structure together in spite of defects; it mitigates their effects. For aerospace structures, the loading involved exacerbates those same effects. In sum then, for two different applications using the same advanced material, the sensing system requirements using fiber-optic technology are quite different. Both systems should be integrated into the structure. One, however, cannot be integrated into the load-bearing elements, while the second must be.

No matter which type of system is required, the problem of ingress/egress to that system remains a problem [16]. It is somewhat surprising that so much time and effort has been spent trying to develop embeddable fiber-optic sensing systems and how little time has been spent trying to develop acceptable methods of interconnecting to them. Figure 16.2 can be used to illustrate the problem. If a fiber-optic sensor system is integrated into the structure shown, then prior to the processing of the structure, some kind of special mold or fixture must be designed and utilized to protect the fibers where they enter and exit the structural material. The sensing system can then be used to monitor the fabrication process. When that process is finished, however, parts would be trimmed and machined in normal fashion. Any fibers would then be sheered off and/or fractured at the edges of the structure. Most of the ways proposed to deal with this problem involve modifying the structure design, and fabrication and machining processes to permit survival of the fibers. Every change involves adding significant cost and must be considered carefully. One other technique has been investigated recently [17] that seeks to address this problem by integrating all of the optoelectronics associated with the fiber-optic sensor into the structure along with the sensor, and powering and interrogating those optoelectronics via inductive coupling. A schematic diagram of an experiment that was conducted is shown in Figure 16.4. It was demonstrated that a laser diode could be powered inductively through ~5-mm thickness of graphite epoxy, used to provide optical power to a modal-domain sensor and transmit the information back the same way. The success of this experiment demonstrates some potential for this technique that would allow the creation of structurally integrated fiber-optic sensing systems without needing significant changes in the structure fabrication/manufacturing process.

16.6 SIMPLE FIBER BREAK DETECTORS

Having now discussed the concept of smart structures and some of the ways that fiber-optic sensor systems can be used in them in general terms, it is now time to investigate some ways in which fiber-optic sensing can be used in more detail. The first and most simple fiber-optic sensor is the binary break detector. The fiber either transmits or does not transmit light. If it transmits light in a nominal fashion, the design failure

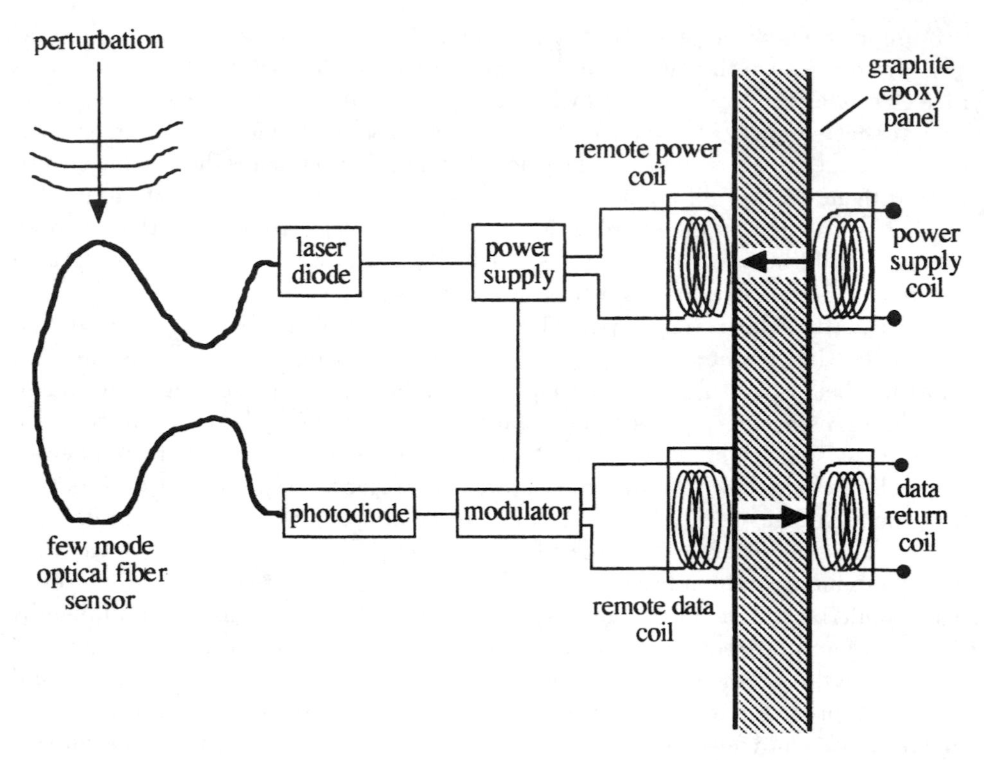

Figure 16.4 Fiber-optic sensor and optoelectronics inductively powered and interrogated through a graphite epoxy panel.

conditions for the fiber have not been exceeded along its length. If it fails to transmit light (and the optical source and detector are shown to be operative), then the design failure loads for the fiber have been exceeded and/or it has been subjected to some sort of catastrophic damage. This damage has then been localized as being somewhere along that fiber length. If a network or grid of fibers are used, the damage can be even further localized. Some early composite damage detection systems were based on this concept [18]. In large civil structures, structural damage location can be carried out using existing fiber-optic local area networks (LANs). An example of how to carry out such damage localization is shown in Figure 16.5, where the fiber-optic LAN system on one floor of a building in an earthquake-prone area such as southern California in the United States is depicted. Figure 16.5(a) shows the building before an earthquake, while Figure 16.5(b) shows it afterward. A self-test of the system after the earthquake indicates that the optical connection between local LAN terminals 6 and 7 is nonfunctional as is the optical connection between master LAN terminal 8 and local LAN terminal 7. An analysis of this data from the communication system indicates that the structure is probably damaged in the upper right-hand corner and that

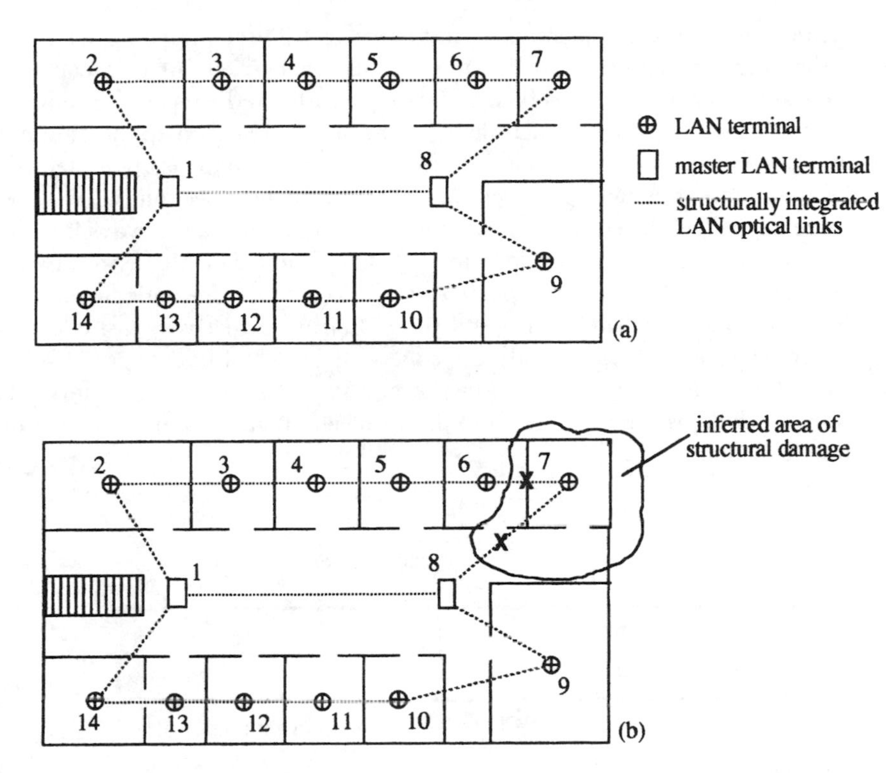

Figure 16.5 Damage location through analysis of LAN status: (a) before earthquake and (b) after earthquake.

area should be subjected to a detailed structural inspection. There is very little overhead required to implement systems such as this with existing LANs other than the addition of software algorithms. In addition, if the optical communications cables are designed so that their failure conditions match those of the structure, enhanced performance can be obtained. Even with systems as simple as this, much useful information can be obtained very cost effectively.

16.7 POINT SENSING

In considering the alternatives for fiber-optic point sensing, two categories of sensors are currently receiving the most interest. These are Fabry-Perot sensors and Bragg grating sensors. A number of other potential fiber-optic point sensings options exist such as those based on microbending, the photoelastic effect, and others, but most current research is in the first two areas mentioned. Fabry-Perot and Bragg grating sensors can be configured as single-point sensors and both types are compatible with various optical multiplexing schemes, so that quasi-distributed systems are also possible. Moderate gauge length strain sensing can also be carried out with each type. The basic transducer mechanisms involved can be configured so as to sense most if not all of the critical parameters identified for the smart structure application areas being considered here. The descriptions of the sensors will then focus upon their basic principles of operation, which are based on strain. In the next section, we will focus on very long gauge length sensors that can be created in a number of ways including intermodal interference in a multimode fiber, Mach-Zehnder or Michelson interferometry with single-mode fibers, and polarization effects in single-mode fibers.

The Fabry-Perot sensors can be fabricated either directly in a single-mode fiber [19] or external to the fiber [20]. Schematic diagrams of both types of sensors are shown in Figure 16.6. In each case, the power returned from the cavity formed by the reflective surfaces is directly related to their reflectivity and separation according to the standard equation [21]

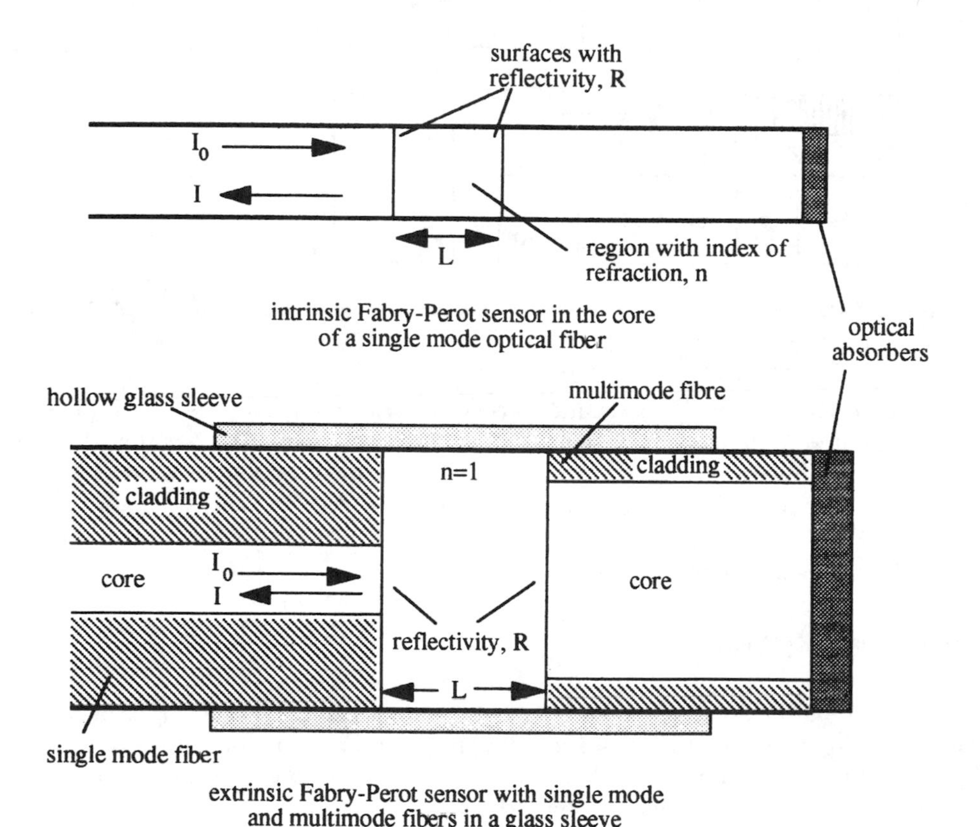

Figure 16.6 Schematic diagram of fiber-optic Fabry-Perot sensor.

$$\frac{I}{I_0} \approx 2R\left[1 + \cos\left(\frac{4\pi nL}{\lambda}\right)\right] \tag{16.1}$$

where I_o is the optical power incident on the cavity, I the power reflected, R the (equal and small) reflectivity of the surfaces, n the index of refraction between the reflective surfaces, λ the free-space optical wavelength, and L the separation between the reflective surfaces.

As can be seen from (16.1), this sensor can be interrogated an intensity-modulating sensor or as a wavelength-modulating sensor. If the extrinsic version of this sensor is considered, and the longitudinal strain ϵ_L is assumed to be zero when the cavity width is L, then with ϵ_L defined as $\Delta L/L$ and $n = 1$, for a narrowband input signal centered at wavelength λ, (16.1) may be rewritten as

$$\frac{I + \Delta I}{I_o} \approx 2R\left\{1 + \cos\left[\frac{4\pi L}{\lambda}(1 + \epsilon_L)\right]\right\} \tag{16.2}$$

so that from a knowledge of I, I_o, R, λ, and L and a measurement of ΔI, the longitudinal strain can be determined. Alternately, an inspection of (16.1) indicates that the returned signal will have maximum reflected powers when

$$\frac{4\pi L}{\lambda_m} = 2m\pi \tag{16.3}$$

where m is an integer. If a relatively broadband wavelength signal is injected into the fiber, a number of return peaks as a function of wavelength will be observed. If the wavelengths of two adjacent peaks, λ_m and λ_{m+1}, are measured under zero-strain conditions, then L can be determined; that is,

$$L = \frac{\lambda_m \lambda_{m+1}}{2(\lambda_m - \lambda_{m+1})} \tag{16.4}$$

Since L and λ_m are now known, m can now be determined from (16.3). Next, define the separation between the peaks as w. From (16.3), the relationship between w and L can be calculated to be

$$L = \frac{m(m+1)w}{2} \tag{16.5}$$

If the cavity width is now changed from L to $L + \Delta L$ due to strain, w must change in a similar manner to $w + \Delta w$. Inserting these values in (16.5) and rearranging terms to solve for the longitudinal strain yields

$$\epsilon_L = \frac{m(m+1)\,(w + \Delta w) - 2L}{2L} \tag{16.6}$$

Depending on the system requirements, numbers of these types of sensors can be multiplexed together through standard time-domain reflectometry (OTDR) techniques or through coherence multiplexing [22]. A key feature of the extrinsic Fabry-Perot sensor is the fact that it only couples to the longitudinal strain component. Sensors of this type have been used in smart structure applications to make measurements in high-performance composite structures [23] and in large civil structures [24].

Bragg grating sensors are also being intensively developed for smart structure applications. These sensors are based upon the creation of periodic index of refraction variations within single-mode optical fibers [25]. The gratings are written in the fiber, as shown in Figure 16.7, by exposing it to high-intensity short-wavelength optical radiation with a periodic intensity distribution that is created by intersecting two beams from the same laser at an angle of 2θ [26]. The period s of the longitudinal grating thus created is given by

$$s = \frac{\lambda_0}{2 \sin \theta} \tag{16.7}$$

where λ_0 is the wavelength of the laser writing the grating. If a relatively broadband optical signal is injected into the fiber, a narrowband portion of that signal centered around a specific wavelength, λ_b, will be reflected by the grating, as shown in

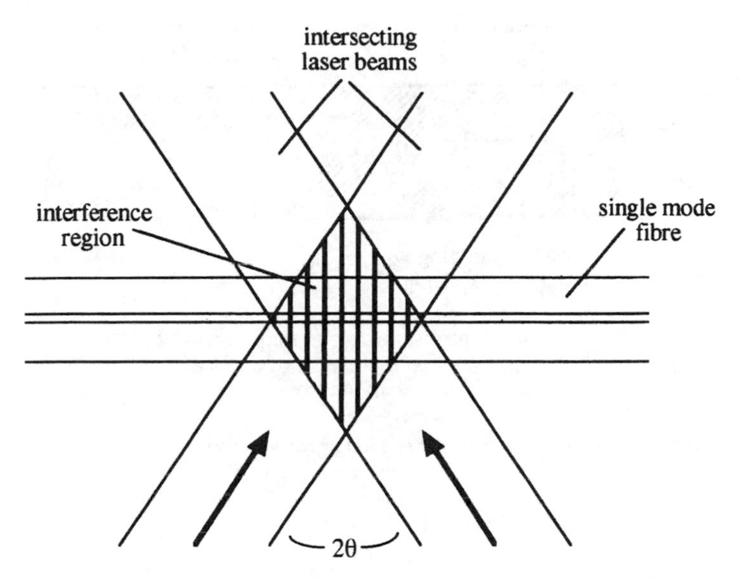

Figure 16.7 Using intersecting laser beams to write a Bragg grating into an optical fiber.

Figure 16.8. The actual effect on the broadband signal spectrum is shown in Figure 16.9. The center wavelength of the narrowband reflected signal is given by

$$\lambda_b = \frac{\lambda_0 n_e}{\sin \theta} \tag{16.8}$$

where n_e is the effective index of refraction of the core of the fiber. For constant temperature, the shift in λ_b due to longitudinal strain in the fiber is given by

$$\frac{\Delta\lambda_b}{\lambda_b} = (1 - \overline{p})\epsilon_L \tag{16.9}$$

where the effective photoelastic constant is defined as

$$\overline{p} \equiv \frac{n^2}{2}\left[p_{12} - \nu(p_{11} + p_{12})\right] \tag{16.10}$$

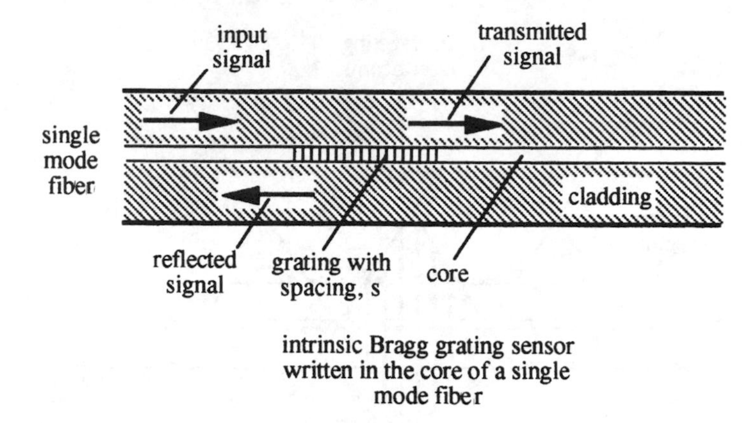

Figure 16.8 Functional diagram of an optical-fiber Bragg grating sensor.

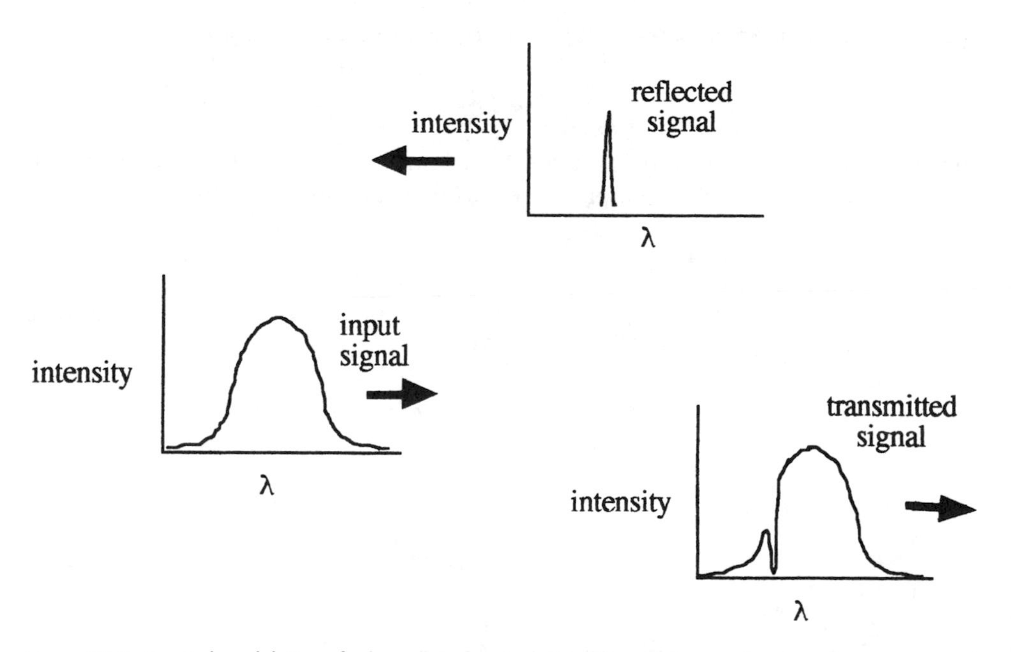

Figure 16.9 Spectral modulation of a broadband optical signal by a Bragg grating sensor.

In this case, n is the index of refraction of the fiber core, p_{12} and p_{11} are components of the strain optic tensor and ν is Poisson's ratio. The longitudinal strain can then be determined from a knowledge of the material constants of the fiber and λ_b combined with a measurement of $\Delta\lambda_b$; that is,

$$\epsilon_L = \frac{\Delta\lambda_b}{\lambda_b(1 - 1\overline{p})} \tag{16.11}$$

The physical properties of Bragg gratings give sensor and system designers numerous options in configuring them for particular applications, and intensive activity is currently taking place. In particular, like the Fabry-Perot sensors, the Bragg grating sensors have also been used in sensing applications involving high-performance composite structures [27] and large civil structures [28]. Numerous options also exist for multiplexing together numbers of Bragg sensors as well [29,30].

16.8 SENSING OVER LARGE SPATIAL DOMAINS

In order to monitor very large spatial domains, a number of options exist, as shown in Figure 16.3. The most straightforward option is to use a large number of point sensors, each with an individual connection back to some signal-processing location. This will always work, but due to the large number of components involved, it can add unacceptable weight and cost to a system. A second option is to use a large number of sensors, but to multiplex them somehow on a single connection back to the signal processor. This quasi-distributed architecture reduces the weight and cost of components, but increases the cost of the signal conditioner. A more sophisticated option is the fully distributed system, in which the single connection back to the signal conditioner also serves as a distributed sensing element with arbitrarily fine resolution. Component cost is again reduced at the expense of much more complex signal conditioning. A final option is the long gauge length integrating sensor, where the single connection back to the signal conditioner serves as a sensing element that integrates the effects of changes along its length on the optical signal. Component costs are minimized. The major cost elements for this option end up in the nonrecurring engineering design required to configure the sensor and make it perform as an appropriate analog preprocessor. This type of sensor offers the promise of being a very cost-effective solution for certain types of smart structure applications and will be examined in detail shortly. First, however, it is important to understand the implications of increasing the size of a structure on the cost/complexity of any required sensing system.

One can envision a structure characterized by some scale parameter, L, so that as L increases, the dimensions of the structure increase proportionally. At this point,

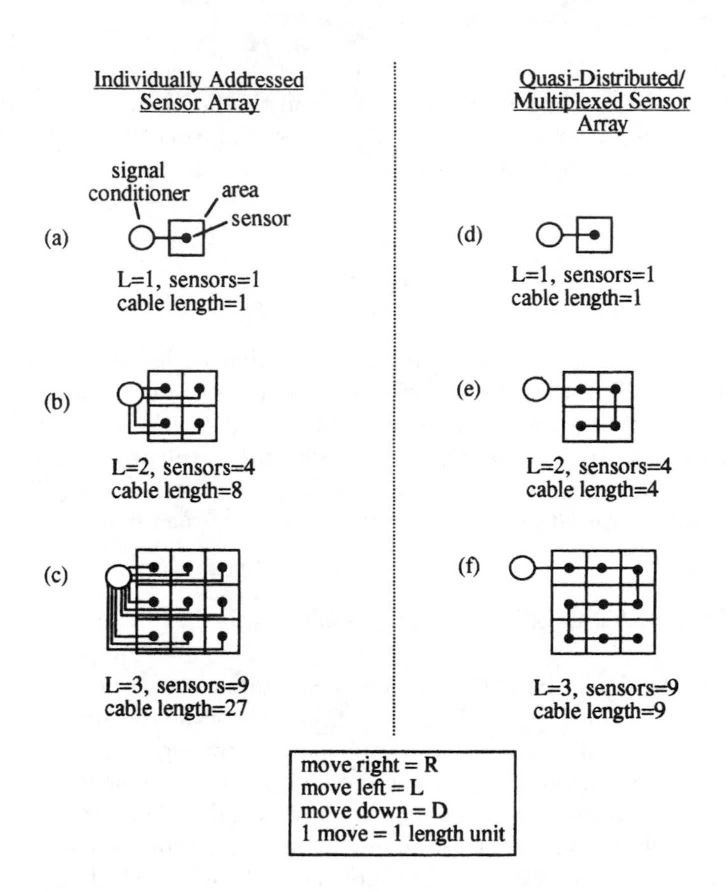

Figure 16.10 The effect of length scaling on sensor system requirements.

we assume that we have an application where it is required to sense the surface of a structure at a density equal to 1 per unit of area. As an example, consider the effects of increasing the scale factor L, as shown in Figure 16.10, for architectures consisting of individually connected point sensors (Figure 16.10(a–c)) and quasi-distributed point sensors (Figure 16.10(d–f)). For the purposes of simplicity, we assume that the sensor cables can only be placed along the vertical or horizontal directions. The number of sensors needed clearly scales as L^2, but the cable lengths needed scale in quite different ways for the two systems. The cable lengths/sensor for each architecture/scale factor are shown in Table 16.1, where each extension of a cable to the right (R), down (D), or to the left (L) is equal to 1 unit of length. The advantage of using multiplexing is clearly shown since the required cable length for the individually connected sensors increases as L^3 while the required cable length for the quasi-distributed system only increases as the number of sensors or L^2. For the same

Table 16.1

Cable Length Calculation Matrix for Structure Size Scaling

(a)	R			(d)	R		
	1 total move				1 total move		
(b)	R	RR		(e)	R	R	
	DR	DRR			L	D	
	4 total moves				8 total moves		
(c)	R	RR	RRR	(f)	R	R	R
	DDR	DDRR	DDRRR		D	R	R
	27 total moves				9 total moves		

required sensing density, one can easily see that increasing a structure's size rapidly makes the use of individually connected sensors impractical from the standpoint of both weight and cost. The situation becomes even worse when the structure must be sensed throughout its volume, rather than just on its surface.

For the different sensing architectures that might be applicable to sensing large spatial domains, a qualitative comparison is given in Table 16.2 for the uniform-density surface sensing case. Depending upon the application and its cost/performance requirements, any of the architectures shown might provide an optimal solution. With the exception of the long gauge length option, these architectures have been discussed in detail elsewhere [31,32]. In general, as a structure increases in size, the optimal sensor architecture will change from individually connected point sensors to a quasi-distributed array of those sensors to the fully distributed sensor case. The long gauge length option appears to offer advantages at all scales if its use is at all appropriate. An examination of Table 16.2, in fact, indicates that the long gauge length architecture will represent the most cost-effective system for a very large structure. For that reason, we will now examine it in some detail.

It has been demonstrated [33] that under certain conditions, a long gauge length sensor can scale its sensitivity in regions where a parameter of interest varies spatially and can be configured so as to produce antenna gain (i.e., since the spatial distribution of a desired parameter field was known in the case cited, the sensor was designed as a matched filter for that distribution). The spatial distribution of the sensor then provided a preprocessing and data reduction function and acted as a simple analog computer. Although the sensor used in the demonstration cited was based on the statistics of the speckle pattern output from a multimode fiber when coherent light is transmitted down it, numerous other options for creating fiber-optic long gauge length sensors exist, some of which are shown schematically in Figure 16.11. In order to understand how to utilize sensing based upon an integrating linear architecture, we

Table 16.2

Qualitative Comparison of Sensing Architectures for Smart Structures

Sensing Architecture	*Number of Sensors*	*Cable Length*	*Number of Sources*	*Number of Detectors*	*Cost/.Complexity of Signal Conditioner*
Individually addtressed sensors array	αL^2	αL^3	αL^2	αL^2	*2.5*
Quasi-distributed multiplexed sensor array	αL^2	αL^2	1	1	2.0
Integrating sensor	1	αL^2	1	1	1.0
Fully distributed sensor		αL^2	1	1	3.0

Note: Cost/complexity of signal conditioning: 1 = low/simple; 2 = moderate/moderate; 3 = high/complex.

must first consider the mathematics relevant for long gauge length sensors. An optical phase-modulating sensor will then be analyzed as an example.

In Figure 16.12, a generalized long linear sensor is depicted. A signal, s_0, is injected at time $t = 0$ into the sensor, which is assumed to consist of N different segments. Each of these segments is centered in space at $x(i)$. The signal that leaves the Nth segment is then **s**N, and the signal that leaves the **i**th segment is s_i. The total time for the signal to pass through the sensor is assumed to be equal to t, with the time at which the signal **s**i leaves the ith segment being equal to **t**i. Each segment is assumed to be immersed in parameter fields $\{P_j\ (i)\}$ and the signal is assumed to have $\{y_k\}$ properties of interest associated with it, so that the signal leaving the ith element would be $s_i(\{y_k(i)\}$. Associated with each discrete sensor segment is an operator, G_i, which takes the signal exiting from segment $I - 1$, s_{i-1}, and acts on it so as to transform it into signal s_i with modified properties $\{y_k(i)\}$ after a time delay Δt_i, with $t_i = t_{i-1} + \Delta t_i$. In terms of this particular formalism, then, sensor operation begins when the input signal s_0 is injected into the first segment so that

$$s_1 = G_1\, s_0 \tag{16.12}$$

and

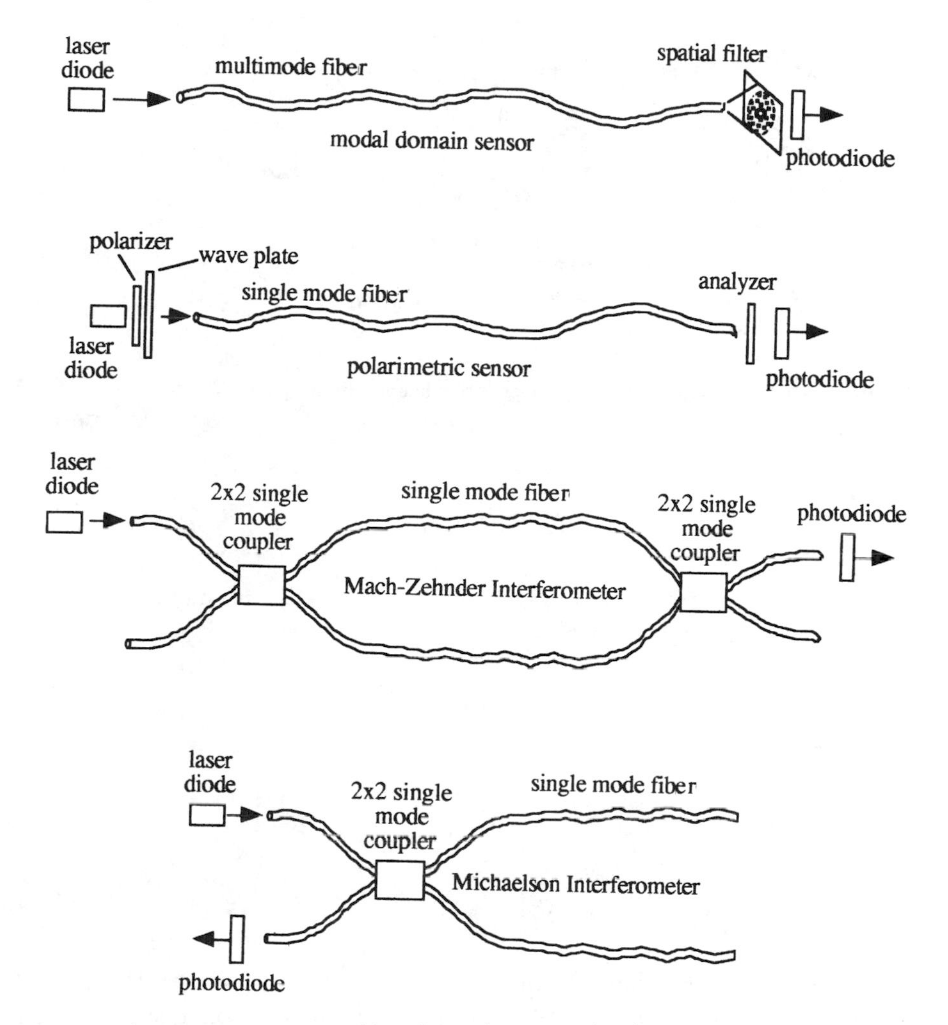

Figure 16.11 Fiber-optic sensors compatible with very long gauge length sensing.

$$s_2 = G_2\, s_1 = G_2\, G_1\, s_0 \tag{16.13}$$

so that continuing in this way, the final result is obtained:

$$s_N = G_N\, s_{N-1} = G_N\, G_{N-1} \cdots G_2\, G_1\, s_0 \tag{16.14}$$

or

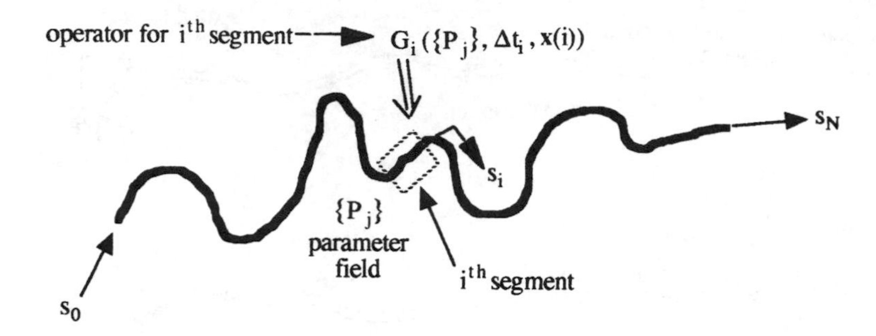

Figure 16.12 Schematic diagram of a long gauge length linear sensor.

$$S_N = \left[\prod_{i=1}^{N} G_i \right] s_0 \tag{16.15}$$

Each operation by an operator changes the time by an amount Δt_i and also changes the signal parameters; that is,

$$t_i \leftarrow t_{i-1}$$

$$s_i \leftarrow s_{i-1}$$

$$\{y_k(i)\} \leftarrow \{y_k(i-1)\} \tag{16.16}$$

In a Mach-Zehnder phase-modulating sensor such as depicted in Figure 16.11, the key parameter of the signal is its phase, φ. The input signal can be simplistically represented as

$$s_0 = e^{i\varphi_0} \tag{16.17}$$

while the output signal would be

$$s_N = e^{i\varphi_N} \tag{16.18}$$

In this case, the operator representing each segment would be given by

$$G_i = e^{i\Delta\varphi_i} \tag{16.19}$$

so that

$$s_N = e^{i\varphi_N} = \left[\prod_{i=1}^{N} G_i\right] e^{i\varphi_0} = \left[\prod_{i=1}^{N} e^{i\Delta\varphi_i}\right] e^{i\varphi_0} \tag{16.20}$$

or

$$s_N = e^{i\varphi_N} = e^{i\left[\sum_{i=1}^{N} \Delta\varphi_i + \varphi_0\right]} \tag{16.21}$$

and since φ is the parameter of interest,

$$\varphi_N = \sum_{i=1}^{N} \Delta\varphi_i + \varphi_0 \tag{16.22}$$

If $\Delta\varphi_i$ is written in terms of a phase modulation density, $\tilde{\varphi}_i$, (16.22) may be written as

$$\varphi_N = \sum_{i=1}^{N} \tilde{\varphi}_i \, \Delta l_i + \varphi_0 \tag{16.23}$$

which may be expressed in integral form as

$$\varphi_N = \int_0^L \tilde{\varphi}(l) \, dl + \varphi_0 \tag{16.24}$$

As can be seen, the parameter of interest, phase, is the directly integrated quantity. The use of phase-modulating sensors is attractive due to the fact that the phase change scales directly with length. One must be careful, however, to ensure that the total phase change does not exceed 2π rad and the sensor is overdriven.

One design process that might be followed to produce a long gauge length sensor that provides an analog preprocessing function for a particular structure may be described as follows:

1. Determine critical parameter regions, V_i and the parameter values that would be present for the structural state one wishes to detect;
2. Map out the shortest linear simply connected path (length L_0) from the signal-conditioning location to, within, and between all of the critical parameter

regions;

3. Determine which segments of the path correspond to which critical parameter regions and values; that is,

$$V_1, P_1: L_{1a} \leftrightarrow L_{1b}$$

$$V_2, P_2: L_{2a} \leftrightarrow L_{2b}$$

$$\vdots \qquad \vdots$$

$$V_N, P_N: L_{Na} \leftrightarrow L_{Nb} \tag{16.25}$$

4. Configure the sensor so that the operators corresponding to the length segments within each critical parameter region will produce a maximum signal modulation when the parameter of interest in that region reaches its critical value; that is,

$$\left.\frac{dG_i}{dP}\right]_{P = P_{\text{critical}}} = 0$$

$$\left.\frac{d^2G_i}{dP^2}\right]_{P = P_{\text{critical}}} < 0 \tag{16.26}$$

(segments between critical parameter regions should be de-sensitized so that they produce no modulation).

The resultant long linear sensing element might be configured as shown in Figure 16.13. If the sensor is a fiber-optic phase modulator, then its output under different state conditions might appear as a function of time, as shown in Figure 16.14. After detecting and demodulating the phase signal, the magnitude of that signal could then be compared with a threshold to determine whether some critical failure condition (for example) might be imminent.

It is seen then that long gauge length sensors, particularly fiber-optic sensors, offer the potential for very effective sensing of large areas and volumes if the spatial distributions of critical parameters have been identified and localized. There are two major difficulties in using this approach at present. First, a large amount of domain knowledge is required about the structural system to which the sensor architecture is to be applied, and secondly, methods of selectively sensitizing fiber segments to produce maximum modulation effects at arbitrary values of parameters of interest have

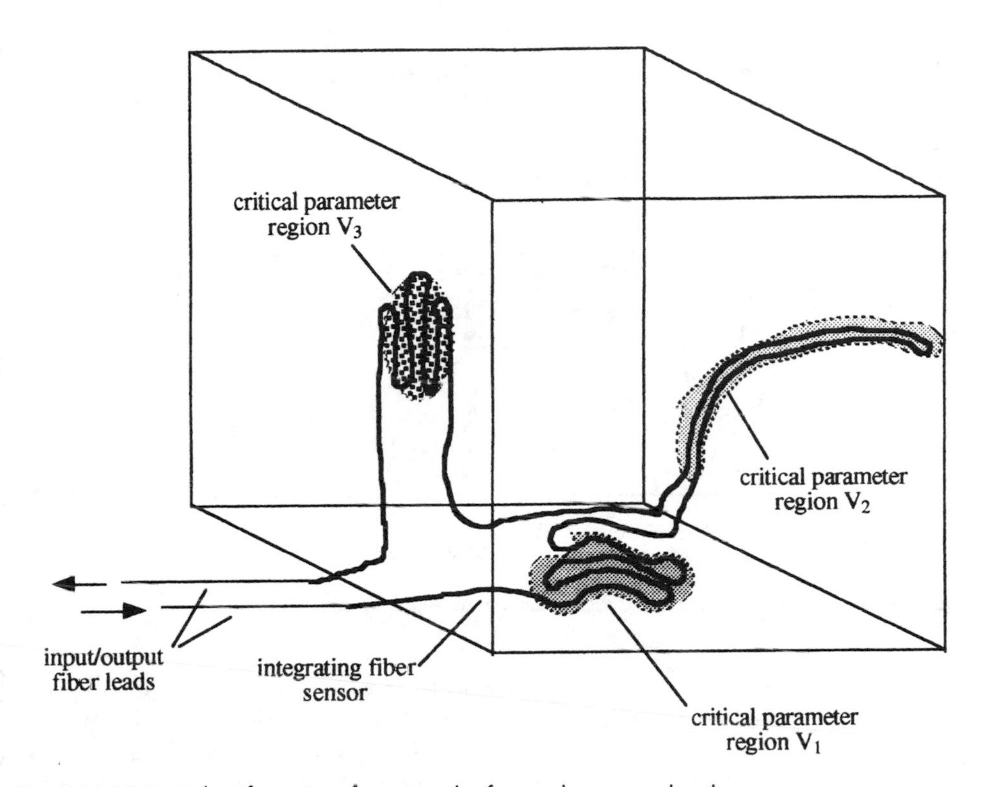

Figure 16.13 Spatial configuration of an example of a very long gauge length sensor.

not significantly advanced beyond some of the excellent early work done as part of the U.S. Navy Fiber-Optic Sensor System (FOSS) program in the 1980s [34].

16.9 TECHNOLOGY CHALLENGES

As a new field, research activity into smart structures has of necessity focused upon its enabling technologies. These are required to make smart structures an economically viable alternative to their existing passive counterparts. Smart structures can be created now in a "bolt-together Frankenstein approach" [35] but they are just too expensive. It is unclear that advances in the enabling technologies alone will be enough to make smart structures attractive enough to become generally used. What is needed, in addition, is much more work in the areas of system design and integration. In particular, the use of adaptive control algorithms in system architectures with hierarchical locally autonomous elements will probably be the key [36].

In terms of the uses of fiber optics for future smart structures, a number of areas for future work suggest themselves. First and foremost is the needed development of a fiber interconnection technology that is compatible with structural fabrication

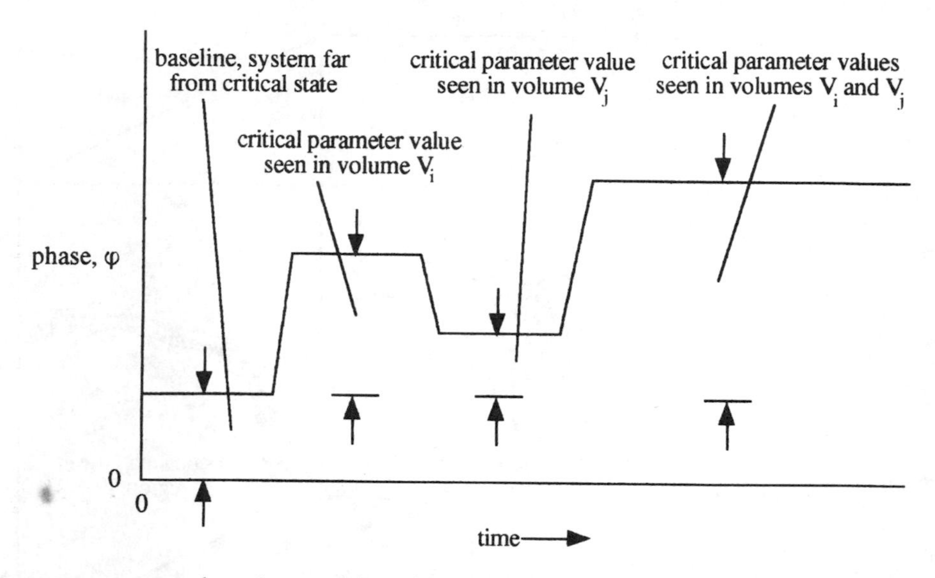

Figure 16.14 χOutput from phase-modulating long gauge length sensor.

processes and is cost-effective. The problem of ingress/egress to embedded optical systems in structures must also be solved or the use of fiber optics for condition monitoring and control will remain problematical. Another problem that needs attention is separation of variables. Virtually every fiber-optic transducer mechanism is sensitive to a number of parameters simultaneously. In particular, these transducer mechanisms are sensitive to temperature. Some fiber-optic temperature sensors have been demonstrated that sense temperature in a very pure fashion [37], but their transducer techniques are so unique that it is difficult to see how they could be effectively integrated into systems. Development of a common fiber-optic transducer mechanism in which sensitivity to a parameter of interest could be enhanced, either at points or along long gauge lengths, while suppressing sensitivity to other parameters would be of great benefit for the creation of practical sensing systems. Optical multiplexing techniques also need considerably more development. Many techniques have been demonstrated, but their cost-effectiveness and practicality remain questionable. Finally, the ability to create optical waveguides within a structure for both communications and sensing during synthesis can be seen as an important long-term research objective.

16.10 SUMMARY

The ultimate goal of smart structure research is the development of active structures that perform better and cost less than passive structures designed for the same purpose. The approach taken to achieve that goal is to create structures that act or are adaptive in the same way as living or biological systems, emulating their integrated hierarchical system design with local autonomy and efficient use of energy. Fiber-optic technology appears to offer considerable promise in contributing to the development of practical smart structures, providing both communications and sensing functions. In particular, the use of very long gauge length fiber-optic sensors, whose sensitivities have been tailored along their lengths and spatially configured to provide antenna gain and selectivity to some spatially distributed state variable of interest, seems to offer considerable potential for reducing sensing system costs. Finally, near-term applications of smart structures are beginning to appear and longer term prospects for the results of the research would seem to be even more favorable.

References

[1] *Proc. SPIE*, 986, 1988.

[2] *Proc. SPIE*, 1170, 1989.

[3] *Proc. SPIE*, 1370, 1990.

[4] *Proc. SPIE*, 1588, 1991.

[5] *Proc. SPIE*, 1798, 1992.

[6] *Proc. 1st European Conference on Smart Structures and Materials*, Glasgow, 1992.

[7] *Proc. 1993 North American Conference on Smart Structures and Materials*, Albuquerque, 1993.

[8] *Proc. 1994 North American Conference on Smart Structures and Materials*, Orlando, 1994.

[9] *Proc. 2nd European Conference on Smart Structures and Materials*, Glasgow, 1994.

[10] Udd, E., (Ed.), *Fiber Optic Smart Structures*, John Wiley & Sons, New York, 1995.

[11] Mazur, C. J., et al., "Air Force smart structures/skins program overview," *Proc. SPIE*, 986, 1988, pp. 19–31.

[12] Spillman, W. B., Jr., et al., "The field of smart structures as seen by those working in it: survey results," *Proc. SPIE*, 2444, 1995, pp. 18–29.

[13] Rudd, R., and Goddard, K., "Composite Integrity Monitoring," Final Report for Program WRDC-TR-89-3031, Flight Dynamics Laboratory, Wright Research and Development Center, Air Force Systems Command, Wright-Patterson Air Force Base, 1989.

[14] Denker, K. F., and Rabbat, B. G., "Why America's Bridges Are Crumbling," *Scientific American*, March, 1993, pp. 66–76.

[15] Spillman, W. B., Jr., "Sensing and processing for smart structures," *Proc. IEEE*, Jan. 1996.

[16] Spillman, W. B., Jr., and Lord, J. R., "Methods of Fiber Optic Ingress/Egress for Smart Structures," in *Fiber Optic Smart Structures*, Udd, E., (Ed.), John Wiley & Sons, New York, 1995, pp. 121–154.

[17] de Vries M., et al., "Completely Embedded Fiber Optic Sensor," Proc. SPIE, 2191, 1994, pp. 373–379.

[18] Dubois, J.M.S., "Research Toward the Development of a Structurally Integrated Optical Fibre Sensor System for Impact Detection in an Aircraft Composite Leading Edge," M.A. Sc. Thesis, University of Toronto, 1989.

[19] Lee, C. E., and Taylor, H. F., "Interferometric Optical Fibre Sensors using Internal Mirrors," *Electron. Lett.*, 24, 1988, pp. 193–196.

[20] Murphy, K., et al, "High Temperature Sensing Applications of Silica and Sapphire Optical Fibers," *Proc. SPIE*, 1370, 1990, pp. 169–175.

[21] Udd, E., (Ed.), "Fiber Optic Sensors," in *Critical Reviews of Optical Science and Technology*, CR44, SPIE Optical Engineering Press, Bellingham, 1993.

[22] Bhatia, V., et al., "Applications of "Absolute" Fiber Optic Sensors to Smart Materials and Structures," *Proc. 10th Optical Fibre Sensors Conference*, Glasgow, 1994, pp. 171–174.
[23] Paul, C. A., et al., "Detection of the onset of damage using an extrinsic Fabry-Perot interferometric strain sensor," *Proc. SPIE*, 2361, 1993, pp. 154–164.
[24] deVries, M. J., et al., "Applications of absolute extrinsic Fabry-Perot interferometer (EFPI) fiber opti sensing system for measurement of strain in pre-tensioned tendons for prestrained concrete," *Proc. SPIE*, 2446, 1995, pp. 9–15.
[25] Meltz, G., et al., "Formation of Bragg Gratings in Optical Fiber by a Transverse Holographic Method," *Optics Lett.*, 14, 1989, pp. 823–825.
[26] Hutley, M. C., *Diffraction Gratings*, Academic Press, New York, 1982.
[27] Frieble, E. J., et al., "Distributed strain sensing with fiber Bragg grating arrays embedded in Continuous Resin Transfer Molding (CRTM) " composites," *Proc. SPIE*, 2361, 1994, pp. 338–341.
[28] Maaskant, R., et al., "Fiber optic Bragg grating sensor network installed in a concrete road bridge," *Proc. SPIE*, 2191, 1994, pp. 13–18.
[29] Morey, W. W., et al., "Multiplexing Fiber Bragg Grating Sensors," *Proc. SPIE*, 1586, 1991, pp. 216–224.
[30] Kersey, A. D., et al., "High Resolution Fiber Grating Based Strain Sensor with Interferometric Wavelength Shift Detection," *Electron. Lett.*, 28, 1992, pp. 236–239.
[31] Kersey, A. D., "Fiber Optic Sensor Multiplexing Techniques," in *Fiber Optic Smart Structures*, Udd, E., (Ed.), John Wiley & Sons, New York, 1995, pp. 409–444.
[32] Dakin, J. P., "Distributed Optical Fiber Sensors," in *Fiber Optic Smart Structures*, Udd, E., (Ed.), John Wiley & Sons, New York, 1995, pp. 373–408.
[33] Spillman, W. B., Jr., and Huston, D. R., "Scaling and antenna gain in integrating fiber optic sensors," J. *Lightwave Technology*, 13, (7), 1995, pp. 1222–1230.
[34] Lagakos, N., et al, "Optimizing fiber coatings of interferometric acoustic sensors," *IEEE J. Quantum Electronics*, 18, 1982, pp. 683–695.
[35] Friend, C., University of Strathclyde Short Course on Smart Structures and Materials, Glasgow, 1995.
[36] Spillman, W. B., Jr., "Instrumentation architecture development for smart structures," *Proc. SPIE*, 1918, 1993, pp. 165–171.
[37] Lieser, B. J., et al, "Design and test of an optical engine inlet temperature sensor," *Proc. SPIE*, 2295, 1994, pp. 151–155.

Chapter 17

Fiber-Optic Sensors: Commercial Presence

Sam Crossley
AOS Technology Ltd., England

17.1 INTRODUCTION

This chapter presents a brief overview of some of the optical-fiber sensors that have matured to the point of commercial availability. In total, this amounts to something over two hundred different devices and systems, of which about half are accounted for by three measurands: temperature, displacement, and pressure.

The sensors may be crudely categorized in terms of the exploitation of the optical fiber. In most cases, this is just a light guide, often simply facilitating the remote operation of a conventional optical sensor. The majority of the displacement sensors are included in this category (remote photoelectric sensors), a third of the temperature sensors (remote pyrometers), and the majority of the gas/chemical sensors that are based on remote absorption spectroscopy. In the main, these are sensors produced by well-established manufacturers via the modification of existing non-FO product, and market acceptance is generally good.

The more technically advanced extrinsic and the few intrinsic (largely distributed temperature and microbend) sensors form a much smaller and less well-developed group of devices, few of which are significant in the industrial market. The obvious exceptions here are the 3M/CDI and Biomedical Sensors blood gas sensors, the Luxtron temperature sensors, and the Herga distributed microbend pressure switch. Arguably more representative of this category is the excellent Babcock & Wilcox microbend technology, which after many man-years of technical development is only now becoming commercially available.

In the following subsections, a number of the more important sensors that have made it to the marketplace are reviewed. Tables of sensor performance data are provided that, while being far from an exhaustive compilation, are intended to illustrate the breadth of device availability.

17.2 TEMPERATURE SENSORS

As might be expected, temperature sensors form the largest class of commercially available fiber-optic sensor. Three main types of sensors are available:

- Remote pyrometers and blackbody sensors;
- Distributed temperature sensors;
- Fluorescence-modulation sensors.

17.2.1 Remote Pyrometers and Blackbody Sensors

Remote pyrometers are in the main a straightforward extension of existing pyrometric devices, the provision of fibers simply permitting greater ease of use. Devices are available from established manufacturers such as ACME Namco, Vanzetti, and Babcock & Wilcox. An interesting newcomer is Celect Electronics, a small UK company that has developed a sensor with—for a pyrometer—a very low temperature response, full accuracy measurement being claimed from 20°C. The sensors from ARi and Luxtron are remote blackbody devices, the latter using a sapphire-tipped probe. This is the Accufiber high-temperature sensor, which was one of the earlier commercially available sensors, Luxtron having taken over the company in 1993.

17.2.2 Distributed Temperature Sensors

Distributed temperature sensors (DTS) would at last appear to have begun to make a significant impact on the commercial scene. The basic operating principle of the DTS is illustrated in Figure 17.1.

A brief, intense pulse of light is launched into a length of optical fiber and the light reflected by Rayleigh scattering is analyzed by what is in effect a wavelength-specific OTDR. The spectrum of the backscattered light contains three elements: the dominant and temperature-insensitive Rayleigh component, and the much weaker Stokes and anti-Stokes components, from which temperature information can be extracted.

DTS systems are now available from a number of suppliers, including York Sensors, who have recently released EMC-compatible versions of the DTS80 (renamed the DTS800) for the European market, and an ultralong range, single-mode fiber instrument capable of operating up to 30 km, single ended. Other manufacturers

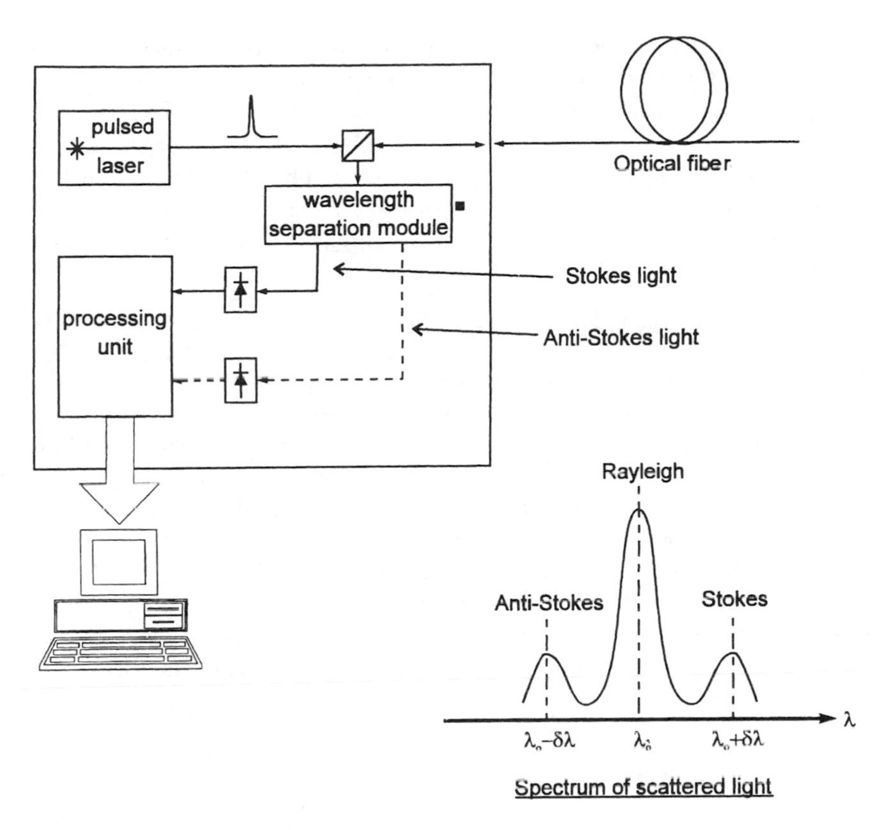

Figure 17.1 Typical distributed optical-fiber temperature sensor.

include Sumitomo, Fujikura, Hitachi, and Ericsson, who market two systems; a recently released, true DTS based on the original CERL/Cossor Thermotrack technology and the Erica sensor, a hot spot locator. This requires the use of a special sensing cable, which is a loose tube construction carrying a single optical fiber, around which is wound a hard filament. The interstitial space is filled with a wax that melts —and hence expands—at a known temperature. The expanding liquid is trapped between the cable walls and cooler, solid-wax-filled sections, and the resulting internal pressure rise exerts a force on the spiral filament, which in turn causes light to be leaked from the fiber via microbending.

Another, unconventional DTS is the cryogenic leak detection system available from Pilkington. This exploits the temperature sensitivity of refractive index, which in PCS fiber reduces the core/cladding refractive index difference to zero at about −50°C . At this temperature, the fiber no longer acts as a light guide, thereby providing a fail-safe detection mechanism for the leakage of cryogenic liquids.

Perhaps the most interesting indicator of the maturity of DTS systems is their increasing take up by intermediary companies under licensing, or similar arrangements. BICC, for example, has recently licensed the Hitachi DTS system for use in power cable applications; and Sensor Highways, a recently formed UK startup company, is exploiting York DTS technology in downhole and topside applications in the oil and gas industry.

17.2.3 Fluorescence-Modulation Sensors

The remaining section of the temperature sensor group contains a number of sensors that exploit the effect of temperature on fluorescence. This can be used in two ways, since temperature affects both the relative amplitudes of components within the fluorescence spectrum and pulsed fluorescence decay time. A typical example of the former is shown in Figure 17.2., which depicts the sensor available from Takaoka, which was originally developed and commercialized by ASEA.

Sensors exploiting this phenomenon are also available from Nortech Fibronics, a Canadian company with a device operating from −40°C to 250°C; while sensors based on fluorescence decay time modulation are available from Luxtron, another of the early players; and Rosemount, which is marketing a sensor intended for aircraft engine applications.

17.3 PRESSURE SENSORS

Fiber-optic pressure sensors are available in two main technologies: 1) microbend devices and 2) those based on the Fotonic principle, although there are a few devices that can loosely be grouped under the heading of wavelength modulation. Foremost in the first category is the sensor from Babcock & Wilcox, a high-precision unit suitable for harsh environments.

17.3.1 Microbend Sensors

In a microbend sensor, a multimode optical fiber is guided between the interlocking sections of two corrugated deformers. As these are brought into close proximity, the fiber is forced into a series of undulations, which causes light to be coupled out of the core into the cladding and from the cladding into the primary coating via microbending. This is depicted schematically in Figure 17.3.

This mechanism forms the basis of a displacement transducer, which can be converted into a pressure transducer by the addition of a compliant member. While in concept this is a very simple device, there are many problems awaiting the designer of a precision instrument based on this principle. The transduction function is

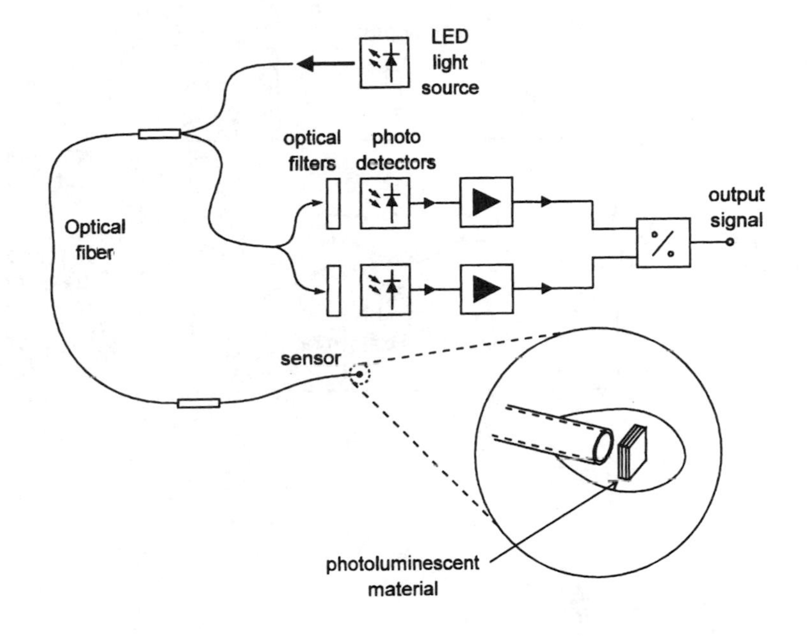

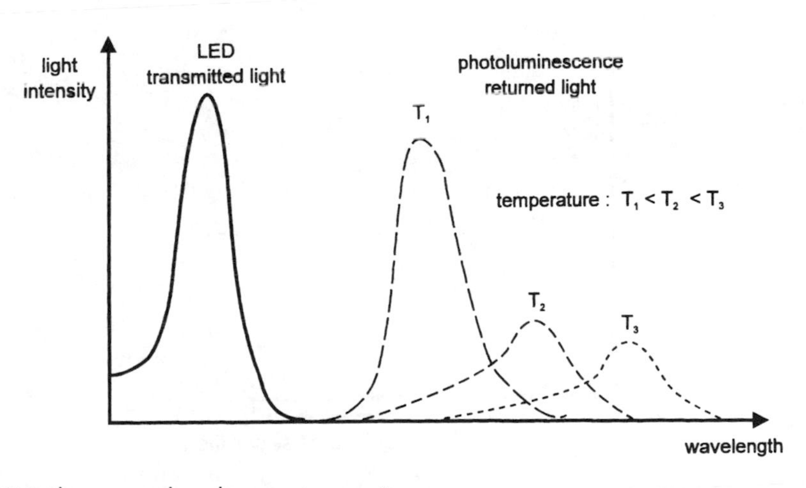

Figure 17.2 Fluorescence-based temperature sensing.

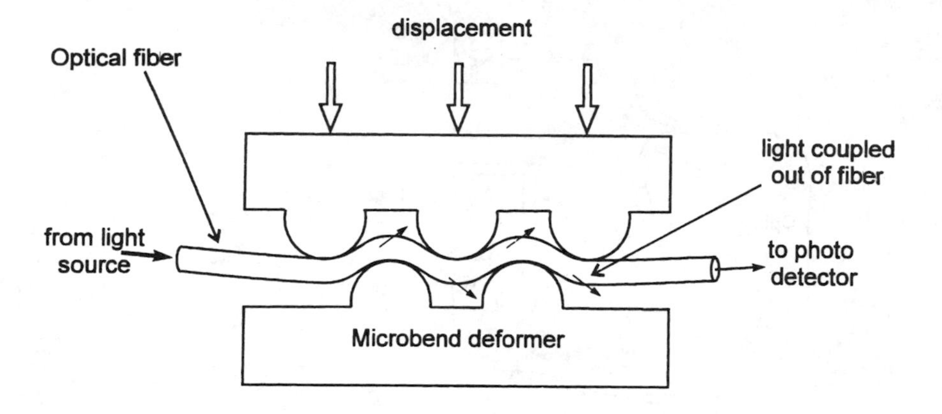

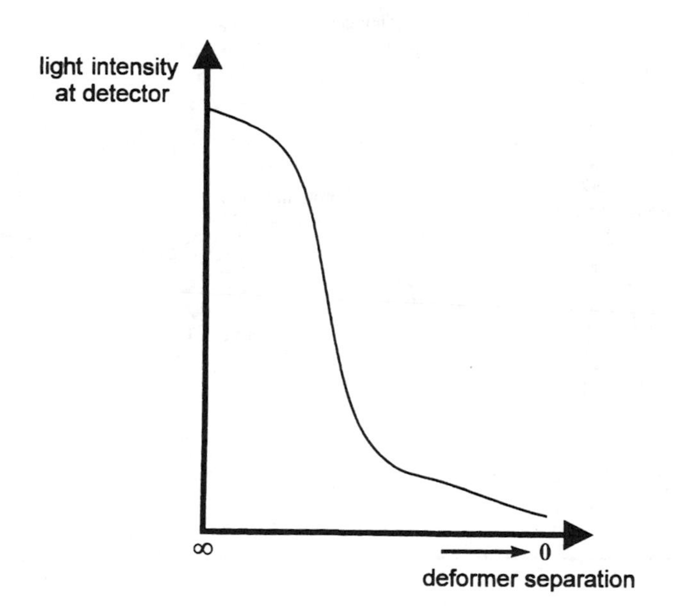

Figure 17.3 Microbend displacement sensor.

highly nonlinear and great care must be exercised in mechanical design, temperature correction, and fiber selection.

The microbend principle is also used in the two distributed pressure sensors that have reached commercial maturity. The Herga sensor was the first of the distributed microbend sensor applications to use a special cable construction comprising a multimode optical fiber and a hard plastic spiral overwinding. This enhances the inherent-and usually unwanted-microbend sensitivity of the fiber by several orders of magnitude, and the cable can therefore be used to detect even very small physical deformation. Originally intended as a distributed switch for machinery guards, the sensor also has applications in bumpers (fenders) for autonomous guided vehicles and in perimeter control. The sensor from C2 Systems Corp. is a more recent device intended for multiparameter (temperature, pressure, and strain) measurement in arduous environments.

17.3.2 Fotonic Sensors

In a Fotonic sensor, the measurand modifies the coupling of light between one fiber and another via a reflective surface. In its simplest realization, this comprises a pair of fibers arranged side by side and a mirror located in a plane orthogonal to the longitudinal axis of the fibers. Light launched into one of the fibers is emitted from the distal end, reflected by the mirror, and coupled into the second fiber. The coupling ratio is a function of the mirror-to-fiber separation (which becomes the primary variable), various sensor-specific parameters such as the fiber NA (which governs the emission and acceptance cone angle), and a number of time-dependent phenomena such as mirror degradation. Various sensor configurations have been proposed, including plane and curved mirrors, lensed arrangements for increased range, and stand-off multifiber systems and the fiber bundle sensors originally developed by MTI.

The Fotonic sensor is essentially a displacement transducer, and its use as a pressure transducer requires that the displacement be controlled by a compliant member.

The commercially available Fotonic sensors are all relatively new to the market. The very elegant sensor from FST is a miniature device aimed at the medical sector, in this case to monitor bladder pressure during diagnosis of urological dysfunction. This is a highly developed package with excellent temperature compensation, self-referencing, and a well-designed electronics and display package. A similar, although physically larger, device intended for the sensing of intercranial pressure is also available from Camino Labs, who have been operating in this market since 1982. Optrand's sensor is part of a temperature/pressure unit aimed at engine control applications and is now at a late stage of development. Dynisco is another relative newcomer to the FOS market, their sensor addressing the needs of high-temperature plastics extrusion, for which the Dynisco group of companies produces a range of equipment. This sensor, which is somewhat atypical of the type, is depicted in Figure 17.4.

17.3.3 Wavelength and Birefringence Modulation Sensors

A number of pressure sensors based on wavelength-modulation effects have become commercially available, and others from Litton Polyscientific and Opticable, for example, have come close. Unfortunately, none would appear to have been particularly successful and few now remain on the market. Perhaps the best known is the sensor now produced by Photonetics SA, which was originally designed by Technology Dynamics Inc. (which subsequently became MetriCor). This sensor is based on a Fabry-Perot cavity design, adapted for the measurement of temperature, pressure, displacement, and refractive index. A more recent Fabry Perot design is available from FPPI, intended for dynamic pressure measurements, the device offering 2,500 psi range at up to 10 kHz. Lucas Control Systems market, to special order, a birefringence pressure sensor using a chromatic readout offering a 1:4,000 resolution. The most technologically advanced device in this category is marketed by Sensor Dynamics U.K. and is intended for downhole monitoring in oil and gas production. This miniature sensor is capable of measuring up to 20,000 psi at temperatures exceeding 250°C.

17.4 DISPLACEMENT SENSORS

The most numerous of the many commercially available displacement sensors is the remote photoelectric device, in which a photocell unit is modified by the addition of light guides, usually random fiber bundles. These are the simplest of the fiber-optic sensors that are included in this review, particularly as their transduction function is usually binary rather than continuous, being intended for presence/absence applications rather than anything more demanding. Three Fotonic devices are available, the originator of the technology, MTI Instruments, still in the market and other units being available from Philtec and Optech.

FO-linked code-plate transducers measuring both linear and angular displacement are available from Computer Optical Products. At one time this was an active area of sensor development, but companies such as Litton Polyscientific, Smiths Industries, and Teledyne Ryan would appear to have ceased their involvement.

Two sensors based on OTDR principles are available from Dr. Johannes Riegl GmbH and from SCICOM. The former is a simple optical radar, offering 500m range and a remote head linked to the optoelectronic processing unit by fiber-optic cables. The SCICOM unit comprises a series of partial reflectors formed within an optical fiber cable and is intended for crack-width monitoring.

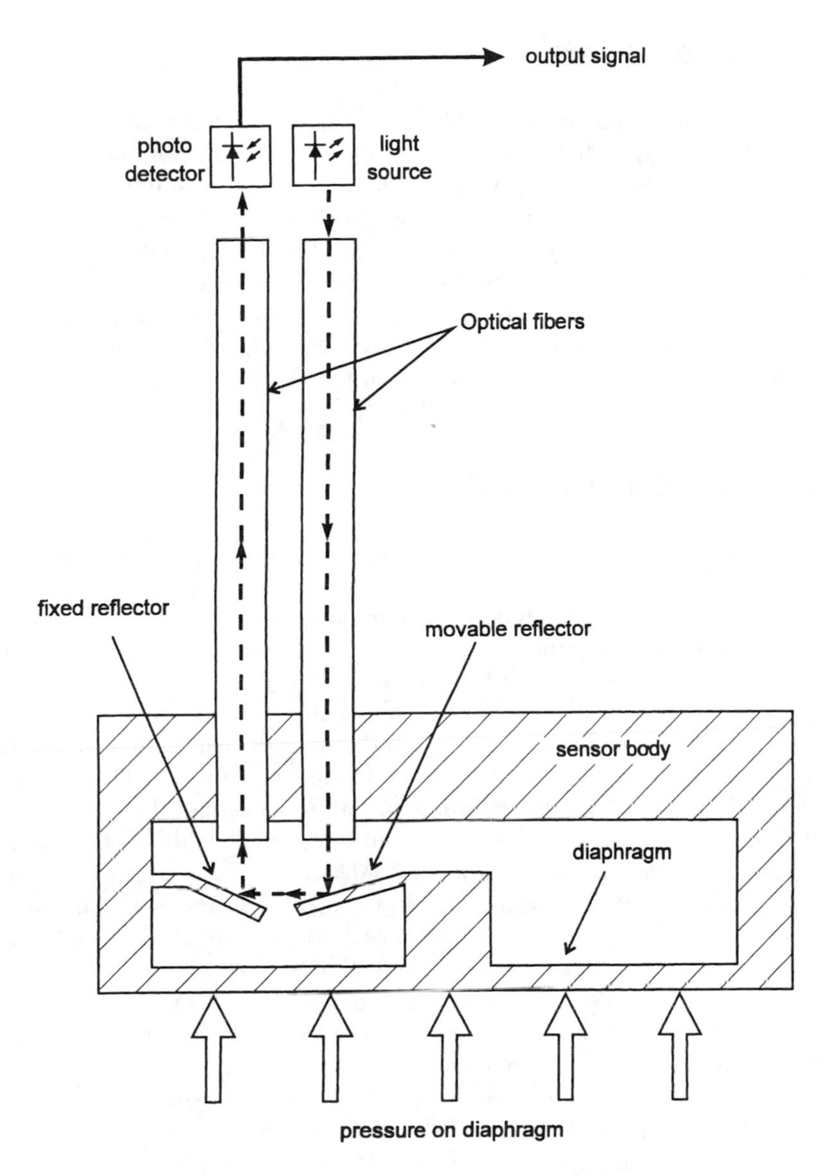

Figure 17.4 Dynisco pressure sensor.

17.5 LIQUID-LEVEL SENSORS

The inclusion of the commonly available optical liquid-level transducers in a review of fiber-optic sensors is perhaps stretching the definition to the limit of credibility, but many do contain a glass or plastic rod to conduct light to and from the sensor tip. The devices, which are available from a number of manufacturers including Honeywell, Keyence Corp, Phoenix Analytec, and Carlo Gavazzi Electromatic, operate on the total internal reflection (TIR) principle, as shown in Figure 17.5, and are thus switches, as opposed to analog, output devices. The only truly analog sensor-specific level sensing (as opposed to the use of pressure gauges, for example) is the Lucas angular rotation sensor, which is available coupled to a float gauge.

17.6 GAS AND CHEMICAL SENSORS

Gas and chemical sensors may be divided into two broad categories: those that are essentially standard optical instruments, fitted with optical fibers in order to permit remote sampling, and those that have been more purposefully designed as optical fiber sensors. Predominant among the former group is the large number of general-purpose spectrometric instruments that could be used to measure a wide variety of species. These are now available from most of the major manufacturers, including Servomex, Rosemount, Perkin Elmer, Bowmem, Seres, and others. PC-card level instrumentation is also now available from, for example, Ocean Optics and American Holographic, and it will be interesting to watch the evolution and improvement of these devices.

Optical-fiber sensors are of particular interest in the healthcare field and several devices are available. The well-established 3M/CDI now has competition from AVL Photronics, Lightsense Corporation, and Biomedical Sensors (which is part of the Pfizer group). All offer the measurement of critical blood gases, CO_2 and O_2, using a broadly similar fluorescence technology, which is illustrated in Figure 17.6.

pH sensing, again using a fluorescence-based technology, is offered by three companies: 3M/CDI and Lightsense in connection with their blood gas monitors, and Optical Systems and Sensors (who are believed to have a more general-purpose unit available). Synectics Medical is commercializing a bile sensor, based on dual-wavelength absorption, developed by CNR IROE in Italy.

Most of the remaining sensors are aimed at the needs of environmental monitoring. This includes the gas correlation spectroscopy instrument offered by Rosemount, intended for flue gas applications; the Seres fiber-linked spectrometer; and the interesting Fibrechem refractive index probe aimed at THC (in groundwater) applications. These are complemented by more general spectrometric instruments from, for example, Seres and Rosemount, intended for the detection of liquid pollutants, and a range of similar instruments configured for gas measurements.

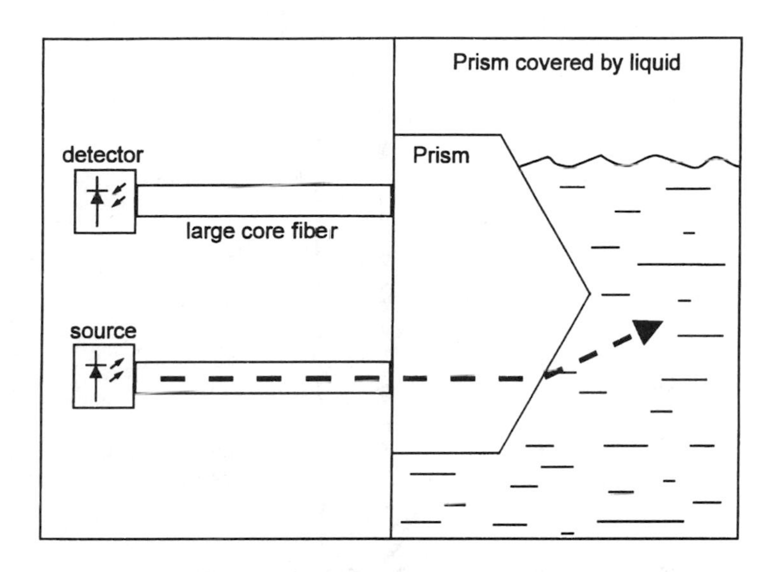

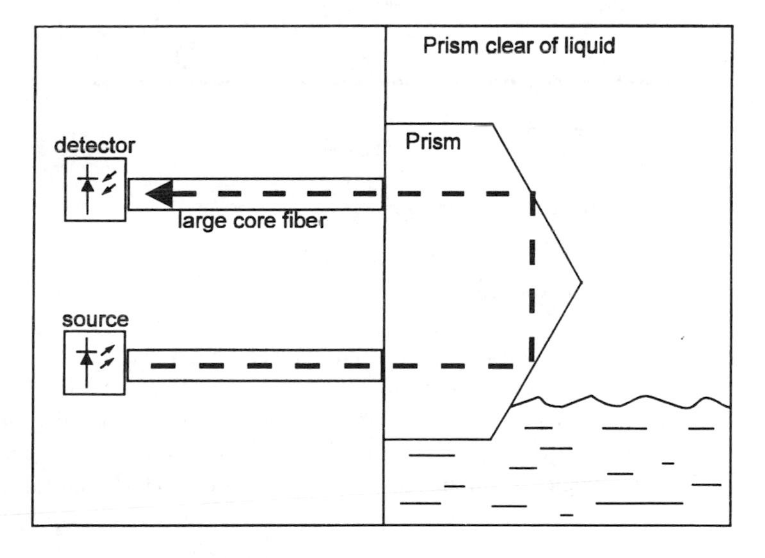

Figure 17.5 Total internal reflection liquid-level sensor.

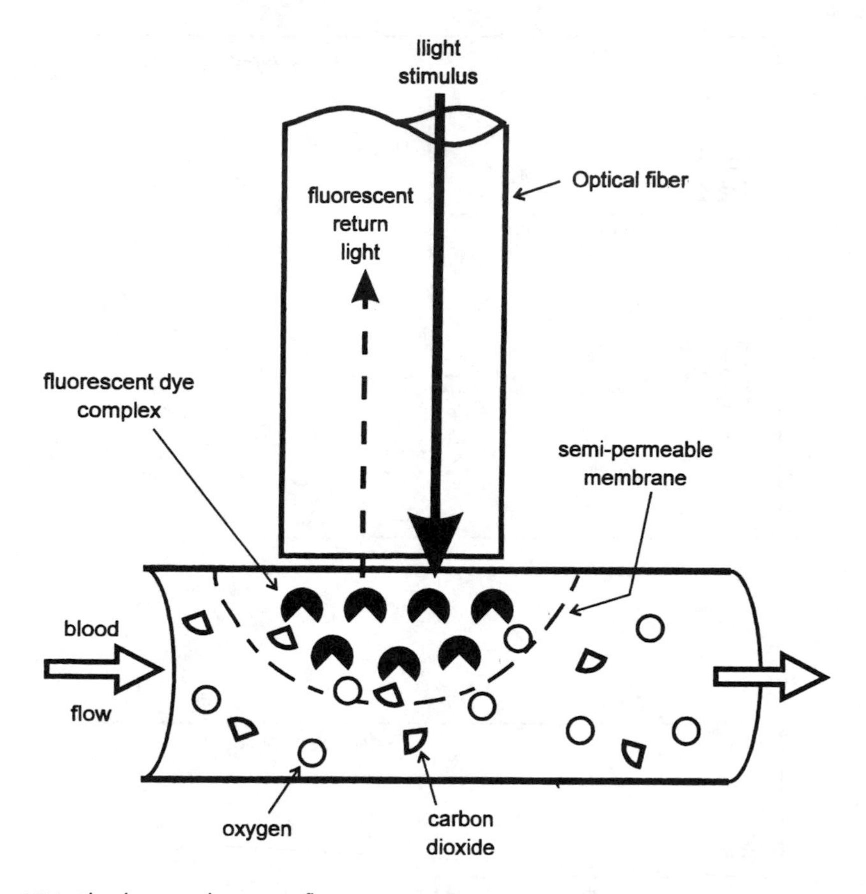

Figure 17.6 Blood gas analysis using fluorescence.

17.7 MISCELLANEOUS SENSORS

In the foregoing sections, sensors have been categorized when reasonable numbers of devices addressing a common measurand are commercially available. There exists however a diverse range of sensors intended for particular niche applications, where the market cannot support the presence of more than a few, and in some cases, more than a single device. Sensors in the miscellaneous category include the following.

- Acceleration;
- Color;
- Current;
- Dust density;

- Flow;
- Hydrophone;
- Intruder detection;
- Oil discharge;
- Particle size;
- Petroleum discharge;
- Petroleum octane;
- Position;
- Refractive index;
- Relative humidity;
- Spark detection;
- Strain;
- Tachometer;
- Thickness;
- Turbidity;
- Velocity;
- Voltage.

Further details may be obtained from the accompanying Tables 17.1–17.6.

17.8 CONCLUSIONS

Optical-fiber sensors began to appear in the commercial marketplace in the late 1970s and early 1980s. Twenty years on, many of the original players are still active. Furthermore, there is some evidence to suggest that the smaller enterprises have fared rather better than their larger competitors, many of whom were experimenting with the technology in the mid to late 1980s but subsequently discovered that fiber-optic sensors offered perhaps a lower turnover than that which could sensibly be sustained. Some of the original manufacturers, for example, 3M/EOTec and Aster no longer produce optical-fiber sensors, but remain in the optical components area, manufacturing specialist fibers and optical couplers. Others, like Luxtron, with annual sales in excess of $10M, Herga, York, and Photonetics are all active and profitable in the high-tech sector of the market. The low-tech sector, which encompasses many of the remote photoelectric-type sensors, also continues to thrive. A rather large number, however, have stayed the course, spent a great deal of time and money in development, and still don't appear to be selling product on a reasonably commercial basis, although there are at last some signs that this is soon to change.

The overall conclusion is then cautiously optimistic; fiber-optic sensors are now available for a wide variety of measurands, and in some cases represent unique measurement solutions. Market acceptance appears to be growing as the technology matures, although this is predominantly within relatively narrow niche applications.

Table 17.1
Temperature Sensors

Name	*Measurement Technique*	*Range*	*Accuracy*	*Resolution*
ACME Namco	Remote IR detector	350, 450, & 800°C		
ARi Industries Inc.	Blackbody	300°C to 2,000°C	1%	N/A
Babcock & Wilcox	Remote pyrometry	120°C to 180°C	±1°C	
Celect Electronics Ltd.	Remote pyrometer	20°c to 800°C		
Ericsson AB	Distributed : microbend	40°C to 90°C		±1°C
Ericsson AB	Distributed : Raman	−200→1,100°C	±1°C	1°C
Fujikura	Distributed : Raman			1°C to 3°C
G2 Systems Corp	Distributed : Microbend	2200°C to 600°C	±3°C	
Hitachi	Distributed : Raman	−200→1,100°C	±1°C typ	±0.1°C typ
Luxtron	Fluorescent decay time	2200°C to 450°C	0.2°C	0.1°C
Luxtron	Blackbody	150°C to 3,000°C	0.2%	0.00002°C
Mikron Instrument Co.	Remote blackbody			
Nortech Fibronics	Fluorescence	240°C to 250°C	±1°C	±0.1°C
Optech Inc.	Microbending	0.3°C/min to 11°C/min	±5% FS	0.3°C/min
Optrand Inc.	Dual wavelength	0°C to 700°C	±1%	
Photonetics SA	FP cavity spectral modulation	240°C to 300°C	±1% FS	N/A
Pilkington Security Ltd.	Distributed : intensity	switch		
Rosemount Inc.	Fluorescence decay time	260°C to 350°C	±0.5°C	
Sumitomo Electric USA	Distributed Raman backscatter	−200→1,100°C	±1°	±3°C
Takaoka Electric Ltd.	GaAs fluorescence	0°C to 200°C	±1°C	0.1°C
Vanzetti Systems Inc.	Remote pyrometry	50°C to 2800°C	±1%	±°1C
York Sensors	Distributed Raman	−200→1,100°C	±0.3°C typ	±0.1→1°C typ

Table 17.2
Pressure Sensors

Name	*Measurement Technique*	*Range*	*Accuracy*	*Resolution*	*Temp Range*
Babcock & Wilcox	Microbend	0 to 3,000 psi	±0.05%		
Dynisco Inc.	Fotonic	0 to 20,000 psi	±0.5% FS	±0.01%FS	400°C max
FiberOptic Sensor Tech	Fotonic	–50 to 250 mm/Hg			
FFPI	Fabry Perot cavity	0 to 2,500 psi	±2%		–50°C to 300°C
Fuji Electric Co. Ltd.	Capacitative sensor E/O convertor	0 to 500 kg/cm2	±0.1%		–30°C to 100°C
G2 Systems Corp	Distributed : microbend	2g to 3 kg			
Herga Electric Ltd.	Microbending	N/A	N/A	N/A	
Lucas Control Systems	Birefringence			0.025%	10°C to 35°C
Optech Inc.	Microbend				0°C to 60°C
Optelecom	Wavelength modulation	Various	0.05% FS	0.003% FS	
Optrand Inc.	Fotonic (Dual diaphragm)	0 to 500 psi	±0.25%		0°C to 650°C
Photonetics SA	FP cavity spectral modulation	0 to 1,500 psi	±1% FS	N/A	5°C to 50°C
Sensor Dynamics	Birefringence	0 to 20,000 psi	1 psi	< 0.1 psi	> 250°C

Table 17.3
Displacement Sensors

Name	*Measurement Technique*	*Range*	*Accuracy*	*Resolution*	*Temp Range*
Carlo Gavazzi Electromatic	Remote photoelectric	0 to 10m max			–25°C to 55°C
Computer Opt Products	Angular : code plate	360°	N/A	to 12 bits	250°C to 150°C
Computer Opt Products	Code plate	25 cm to 50 cm	N/A	8 bits	250°C to 150°C
Dolan Jenner Industries	Remote photoelectric	switch			600°C max
Dr. Johnnes Riegl GmbH	OTDR	1 to 500m	±.0.01m0 002%	0.001m	
Erwin Sick GmbH	LIDAR	130m	±5 mm		225°C to 55°C
Erwin Sick GmbH	Remote photoelectric	to 60m			225°C to 55°C
Herga Electric Ltd.	Distributed microbending	N/A	N/A	N/A	
IDEC GmbH	Remote photoelectric	Analog o/p			
Keyence Corp	Remote photoelectric (laser)	0.1 mm @ 100 mm			
Koden Industry Co. Ltd.	Remote photoelectric	10 to 35 mm	±5μm		5°C to 40°C
Lucas Control Systems	Wavelength modulation - angular rotation	350°	0.5%	1.3°	0°C to 70°C
MTI Instruments	Fotonic	50 μm to few mm	1–2 μm		0°C to 80°C
Optech Inc.	Fotonic	0–400 μm	±1% FS	< 1 μm	
Philtec Inc.	Fotonic	0.05 mm to 250 mm	N/A	0.5μm typ	270°C to 300°C
Pulnix America Inc.	Remote photoelectric	0 to 3m	±0.1 mm	±0.001 mm	250°C to 70°C
SCICOM	OTDR	0.5 mm typ		0.1 mm	
SCICOM	Microbend : spiral overwind	switch		0.001 mm	
Seeka	Remote photoelectric	switch			
Spindler & Hoyer GmbH	Fotonic				
Volpi AG	Remote photoelectric				

Table 17.4
Liquid-Level Sensors

Name	*Measurement Technique*	*Range*	*Accuracy*	*Resolution*	*Temp Range*
Able Instruments & Controls Ltd.	TIR	switch			0°C to 65°C
Carlo Gavazzi Electromatic	TIR	switch			–20°C to 60°C
Fuji Electric Co. Ltd.	Pressure E/O convertor	0 to 32m H2O	±0.25%		230°C to 100°C
Honeywell Inc.	Total internal reflection	±1 mm	1 mm	1 mm	–50°C to 125°C
IMO Industries	TIR	Switch			220°C to 80°C
Keyence Corp	TIR	Switch			
Lucas Control Systems	Wavelength modulation - float gauge	270°	5% FS	2°	–20°C to 40°C
Phoenix Analytec	TIR	Switch	±0.5 mm		230°C to 95°C
Vickers Incorporated	TIR	N/A	N/A	N/A	240°C to 110°C

Table 17.5
Gas and Chemical Sensors

Name	*Measurement Technique*	*Range*	*Accuracy*	*Application Areas*
ADC Ltd.	Absorption - interference filter	to 100%	±0.1% FS	multigas
AVL Photronics	Fluorescence			Blood gas analysis - CO_2 (offline)
Biomedical Sensors	Fluorescence	6.8 to 7.8 pH	± 0.03	Blood monitoring - pH
Biomedical Sensors	Fluorescence	10 to 80 mm/Hg	±3 mm/Hg	Blood gas monitoring - CO_2
Biomedical Sensors	Fluorescence	20 to 500 mm/Hg	±5% to ±10%	Blood gas monitoring - O_2
CDI/3M Healthcare	Fluorescence	7.1 to 7.8 pH	< 0.03	Blood monitoring - pH
CDI/3M Healthcare	Fluorescence	15 torr to 75 torr	±2.5 torr	Blood gas monitoring - CO_2
CDI/3M Healthcare	Fluorescence	40 torr to 220 torr	±8.5 torr	Blood gas monitoring - O_2
Ciencia Inc.	*Fluorescence			Aromatic monitoring
Fiberchem Inc.	Refractive index change in cladding			Hydrocarbon detection
Lightsense Corporation	Fluorescence	N/A	N/A	pH
Lightsense Corporation	Fluorescence	N/A	N/A	Medical & Industrial - PCO2, PO2
Photonetics SA	Remote spectrophotometer	N/A	N/A	
Rosemount Analytical Inc.	Gas correlation spectroscopy	0 to 10,000 ppm	±20 ppm ±6%	Flue gas - CO
Seres	Spectrometer	0 to 30 mg/l	± 3% FS	THC in water
Seres	Spectrometer	0 to 200 mg/l	± 2% FS	Nitrate monitoring in water
Servomex	NIR spectrophotometer		< 1% FS	CO, CO2, CH4, NOx, SOx
Synectics Medical	Dual-wavelength absorption	1 to 100 μmol	4 μmol	Bile detection
UOP / Guided Wave	NIR spectrophotometer			Petroleum octane, total aromatics

Table 17.6
Miscellaneous Sensors

Name	*Type*	*Measurement Technique*	*Range*	*Accuracy*	*Application Areas*
Takaoka Electric MGF Co. Ltd.	Acceleration	Dielectric mirror fluorescence	0.1 m/s/s to 700 m/s/s		
Babcock & Wilcox	Acceleration	Microbend	±100g	±0.02g	
Optech Inc.	Acceleration	Microbend	3 μg/min		Geophysics
Seeka	Color	Wavelength filters	3 colors		
Lucas Control System Products	Current	Faraday effect	0 to 2 kA		Power distribution
Optech Inc.	Current	N/A	mA to > 100 kA	±0.3%	HV installations
Photonetics SA	Current	FP cavity spectral modulation	N/A	N/A	EMC testing
Rosemount Analytical Inc.	Dust density	Absorbance	1% min	±1%	Flue gas
SRICO	Electric field	IO interferometric	0.5 to 150 kV/m		EMC testing
Flowdata Inc.	Flow	Paddle wheel/positive displacement FO encoder	various	±0.1%	
Photonetics SA	Flow	Reflection from flow stream - Intensity based	N/A	N/A	Heat exchangers, refrigerators, chemical reactors
Optech Inc.	Hydrophone	Interferometric	60 dB re μPa min	N/A	Towed/static array
Ispra Product Research Co.	Oil Discharge	Refraction change	N/A	N/A	Marine installations, underground storage
Seres	Oil discharge	Scattering of IR	0 to 1,000 ppm	±2% FS typ	Bilge/Discharge monitoring, Oil in condensates
Sympatec GmbH	Particle size	Laser diffraction	0.1 μm to 3,500 μm		

Table 17.6
(Continued)

Name	*Type*	*Measurement Technique*	*Range*	*Accuracy*	*Application Areas*
Boston Advanced Technology	Petrol discharge	Refractive index change			Petrol pump sumps
Bertin et Cie	Petroleum Octane	Spectrophotometer	380 nm to 750 nm		Online petroleum analysis
Photonetics SA	Refractive index	FP cavity spectral modulation	1.32 to 1.60	±1% FS	Chemical processing
Protimeter Ltd.	Relative Humidity	Remote chilled mirror	20% to 98% RH	<±2%	
Firefly AB	Spark detection	Direct optical detection	400°C/min		Conveyors, ducts, dryers
FIMOD	Strain	OTDR	±10,000 μstrain	±1%	Civil structures, aircraft, composites
Optech Inc.	Strain	Microbending	2μm	±5% FS	Machinery/ structures
G2 Systems Corp	Strain	Distributed : microbend	10 to 1,500 μstrain		
Fiber & Sensor Technologies	Strain	Extrinsic fiber FP cavity			Embedding: smart structures
Optech Inc.	Tachometer	Remote chopper/ reflector	25,000 rpm max	< 1% reading	
Aurora Optics Inc.	Tachometer	Remote shaft encoder		0.001%	
Polytec Gmbh	Thickness	Absorbance	100 μm to 10 cm		on-line coatings
Seres	Turbidity	Absorbance	0 to 10,000 NTU	±1% FS	Water supply, filter monitoring
Mettler Toledo AG	Turbidity	Backscatter	0.1 NTU/FTU to 250 g/l		Filter breakdown, biomass detection typ
Polytec Gmbh	Velocity	LDV	30 mm/s to 30 m/s	±0.1%	
Mindrum Precision Products	Voltage	EO modulator - mm fiber	10 mV/min		Military avionics

About the Authors

Brian Culshaw was born in Ormskirk, Lancashire, England on 24 September 1945 and graduated with a BSc in Physics in 1966 and a Ph.D. in Electrical Engineering in 1970 both from University College London. He joined Strathclyde University as Professor of Electronics in September 1983 with previous appointments as a postdoctoral fellow at Cornell University; technical staff member at Bell Northern Research, Ottawa, Canada; lecturer, later reader, at University College London; and senior research associate in the Applied Physics Laboratory at Stanford University. He worked on microwave and semiconductor devices (their design and technology) until 1975 when his interests evolved into guided wave optics with particular applications in sensing, signal processing and instrumentation. His interests include optical fiber gyroscopes, hydrophones, accelerometers, temperature probes, strain and pressure measurement, sensors, and a host of other measurement systems. He is also venturing into signal processing architectures and high-speed network design. It was from this background that his interests in smart structures evolved through the appreciation that guided wave optics could make a significant contribution to structural instrumentation. Subsequently he has become involved in a number of projects on smart structures, especially in composite materials and civil engineering. He has written extensively on microwave semiconductors, fiber optics and smart structures and materials having authored or co-authored over 300 papers and five textbooks. He has also chaired major international conferences in these areas and currently acts as vice dean of the Engineering Faculty at Strathclyde and is a director of SPIE.

Francesco Baldini was born in 1961. He graduated in physics from the University of Florence magna cum laude in 1986. Since 1986 he has been involved in optical-fiber research and spectrophotometric analysis at IROE-CNR. His research activity is mainly devoted to optical fiber sensors for chemical parameters and optical methods used for the restoration of paintings and frescoes, and he is author of more than 40 publications on the subject in international journals, scientific books, and international conference proceedings in which he participated as an invited speaker. He was guest editor of a special issue of *Sensors & Actuators* on the Europt(r)ode II Conference held in Florence in 1994, and he is an associate member of the International Union Of Pure and Applied Chemistry.

Anna Grazia Mignani was born in Bologna, Italy in 1957. She graduated in physics from the University of Florence, and in 1984 she joined the optical fiber group at IROE-CNR in Florence. Her research work includes fiber optic sensors and related passive components, as well as fiber optic sensor networks, documented by more than 70 journal and conference publications and five international patents. She is a member of the International Steering Committee for the Conference on Optical Fiber Sensors (OFS) and has served on its Technical Program Committee since 1991. She was a member of the Italian Board of the International Union of Radio Science (URSI), and she recently became a member of the Board of the Italian Society for Optics and Photonics (SIOF).

James Barton graduated in physics from Cambridge University and received the Ph.D. degree in 1972 from Liverpool University for research in experimental high energy physics. He joined the Physics Department at Heriot-Watt University, Edinburgh, in 1978 to work on meteorological instrumentation. Now his main research interests are in the application of optical fiber techniques to engineering measurements, in the areas of heat transfer in aerodynamic testing and condition monitoring by acoustic emission measurement, together with studies of propagation in fibers relating to bending loss.

Julian D. C. Jones is Professor of Engineering Optics at Heriot-Watt University. He obtained his B.Sc. degree in physics at the University College of Wales, Aberystwyth, in 1976 and remained there to carry out his Ph.D. and postdoctoral research on the plasma physics of excimer lasers. In 1982 he took up a lectureship at the University of Kent at Canterbury, before moving to Heriot-Watt University in 1988. His research interests are in the physics and technology of optical fiber systems used in sensing, measurement, beam delivery, and communications. He has published over 300 journal articles, conference presentations, and patents on these subjects. Professor Jones is editor of the *Institute of Physics Journal of Measurement Science and Technology* and chairman of the Optical Group of the Institute of Physics.

After graduating from Huddersfield Polytechnic with a degree in electrical and electronic engineering, **Sam Crossley** moved to University College London, where, eventually, he gained his Ph.D. in optical fiber sensing and communications. Since then he has pursued his research and business interests in optical sensing both in academia (Huddersfield Polytechnic and Bradford University) and industry (ERA Technology Ltd and British Gas Research + Technology). He recently left British Gas to become one of the founding directors of AOS Technology Ltd, an SME engaged in the research, development, and manufacture of advanced optical systems. Dr. Crossley is a chartered engineer and a member of the IEE.

Alan D. Kersey received the B.S. degree in physics/electronics with honors from the University of Warwick, United Kingdom, in 1977, and the Ph.D. from the University of Leeds, United Kingdom, in 1985. In his graduate work, he studied magneto-optic spectroscopic effects in atomic vapors. In 1984, he joined the Naval Research Laboratory, where he began work on multiplexing techniques for interferometric sensors. He has worked in the area of fiber optics for the past 15 years and is currently head of the Fiber Optic Smart Structures Section. His research group is involved in a diverse range of sensor development programs but focuses strongly on multiplexed and distributed sensors systems, Bragg grating based sensors, fiber laser sensors, and general aspects of embedded fiber optic sensors for smart structures.

Dr. Kersey's work in fiber optic sensors has led to over 300 journal and conference publications in the area of fiber optics and chapters in two books. He currently holds over 20 patents and has several other applications pending. Dr. Kersey is a fellow of the Optical Society of America and serves on several program committees for international conferences in the field of fiber optics, such as the Optical Fiber Communications (OFC) conference, the Conference on Lasers and Electro Optics (CLEO), the Optical Fiber Sensors Conference (OFS). In addition, he is associate editor of the *Journal of Lightwave Technology*.

Tsuneo Horiguchi was born in Tokyo, Japan, in 1953. He received the B.E. and Dr. Eng. degrees from the University of Tokyo, Japan, in 1976 and 1988, respectively. In 1976, he joined the Ibaraki Electrical Communication Laboratory, NTT, Ibaraki, Japan. He is presently a leader of the lightwave propagation research group of NTT Access Network Systems Laboratories. His current research interests include photonic network systems and related measurement technology.

Alan J. Rogers graduated with a double first in the Natural Sciences Tripos at Cambridge University. He took a Ph.D. in radio astronomy and space physics and then spent three years as a lecturer in physics in the University of London, where he researched in ionospheric physics and solar terrestrial interactions. In 1969 he joined the Central Electricity Research Laboratories (CERL) of the CEGB where he led a research group active, at various times, in the areas of microwave communications, mobile radio communications, optical communications, signal processing, optical fiber sensors, laser diagnostics, and on-linear optics. In 1985 he became Professor of Electronics at King's College, London, where he now leads a group of researchers active in optical sensors, nonlinear optics, and optical signal processing. In 1991 he became head of the department of Electronic and Electrical Engineering at King's College London. He has published well over 150 papers in learned journals and has initiated 8 patents. He is a fellow of the Institute of Physics and of the Institution of Electrical Engineering and is a senior member of the Institute of Electronic and Electrical Engineering.

Otto S. Wolfbeis received a Ph.D. in chemistry at the Karl-Franzens University in Graz, Austria in 1972. After several years at the Max-Planck Institute of Radiation Chemistry in Mülhim (FRG) and at the Technical University of Berlin, he was a Professor of Chemistry at the Institute of Organic Chemistry of the Karl-Franzens University in Graz until 1995. From 1990 to 1992 he also acted as the first director of the Institute for Optical and Chemical Sensors at Joanneium Research in Graz. Since 1995 he has held a chair on Analytical Chemistry at the Institute of Analytical Chemistry, Chemo- and Biosensors at the University of Regensnburg, Germany. Dr. Wolfbeis has authored more than 260 papers, reviews, and books on optical sensors, fluorescence spectroscopy, optical sensors and fluorescent dyes and molecular robes. The main focus is on optical chemical sensors and biosensors. He has given numerous invited lectures at international meetings and guest lectures at universities and institutes. He has served as a symposium chair or co-chair for many conferences and has (co)organized a number of meetings related to fluorescence technology and optical chemical (bio)sensors.

John P. Dakin is a principal research fellow at Southampton University, supervising research and development in optical instrumentation, having particular interests in optical fiber sensors and other optical measurement instruments. He was formerly a student for his B.Sc. and Ph.D. degrees at Southampton University and remained there as a research fellow until 1973. He has since spent two years in Germany at AEG Telefunken, 12 years at Plessey UK, and two years with York Lt before returning to the university. He is the author of over 130 technical and scientific papers and over 120 patent applications and has edited or coedited four books on optical fiber sensors. He was previously a visiting professor at Strathclyde University and was the technical program committee chairman of the recent OFS'8 conference in Paris. He is frequently an invited speaker and chair at major international conferences and serves on the editorial board of a number of optoelectronics journals.

Dr. Dakin has won a number of awards, including "Inventor of the Year" for Plessey Electronic Systems Ltd, the Electronics Divisional Board Premium of the IEE, and open scholarships to both Southampton and Manchester Universities. He has also been responsible for a number of key electro-optic developments. These include the sphere optical fiber connector, the first WDM optical shaft encoder, the Raman optical fiber distributed temperature sensor, and the first realization of a fiber optic passive hydrophone array sensor.

Kazuo Hotate was born in Tokyo, Japan on 20 June 1951. He received the B.E., M.E. and Dr. Eng. degrees in electronics engineering from the University of Tokyo in 1974,

1976, and 1979, respectively. In 1979 he joined the University of Tokyo as a lecturer. In 1987 he became an associate professor, and since June 1993 he has been a professor in the Research Center for Advanced Science and Technology (RCAST) at the University of Tokyo. He has been engaged in research of projection type holography and measurement and analysis of optical fiber characteristics. At present, he is working on photonic sensing and photonic computing. He is a co-author of the books *Optical Fibers and Optical Fiber Sensors*. Dr. Hotate received the Achievement Award in 1979 and the Book Award in 1984 both from the Institute of Electronics, Information and Communication Engineers of Japan; the Paper Award on Optical Fiber Gyro in 1984 from the Society of Instrument and Control Engineers of Japan; and the Best Papers Award of the fourth Optoelectronics Conference in 1992.

He is a member of IEEE, IEICE, SICE, the Institute of Electrical Engineers of Japan, the Institute of Television Engineers of Japan, the Optical Society of Japan, and the Japan Society for Aeronautical and Space Sciences.

Ivan Andonovic, B.Sc., Ph.D., C.Eng., MIEE, MIEEE, MOSA, reader, graduated with honors in electronic and electrical engineering from the University of Strathclyde in 1978. He obtained a doctorate in optical waveguide modulator devices on LiNbO3 in collaboration with nearby Glasgow University. Following a three year period as research scientist responsible for the design, manufacture, and test of guided wave devices for a variety of applications, he joined the Electronic and Electrical Engineering Department at Strathclyde University in 1985. His main interests center on the development of guided wave architectures for implementing optical signal processing and switching functions for optical networks. He has recently returned from a two-year Royal Society industrial Fellowship in collaboration with BT Labs during which time he was tasked with investigating novel approaches to networking. He has edited two books and authored/co-authored four chapters in books and over 90 journal and conference papers. He has been chairman of the IEE professional group E13, has held a BT Short Term Fellowship, and is editor of the *International Journal of Optoelectronics*.

Chris Lloyd graduated with a B.Ed. (Hons) in physics at Sussex University after working as an engineer at Rolls Royce, Glasgow from 1984 to 1988. In 1991 he graduated with an M.Sc. in optoelectronics at Newcastle Polytechnic and then began work at CAMR studying optical analysis of dense suspensions. He is currently working under Professor Clarke with the Drug Delivery Group at Manchester University where they are looking into novel methods of drug targeting, detection, and encapsulation.

George Stewart is currently senior lecturer in the Department of Electronic and Electrical Engineering, University of Strathclyde, Glasgow. He was awarded the B.Eng. Degree with first class honors in 1974 and the Ph.D. degree in 1979, both from the University of Glasgow. He worked for a number of years in the Department of

Electronics and Electrical Engineering at Glasgow University as a postdoctoral research fellow in integrated and fiber optics. His research work involved the development of optical waveguides by ion-exchange methods and the investigation of techniques for hybrid integration to combine optical devices in glass and lithium niobate substrates. In 1985 he joined the Optoelectronics Group at Strathclyde University to work on evanescent wave fiber optic components and chemical sensors. Currently he is actively involved in the development of fiber optic methane sensors and evanescent field sensors using doped sol-gel films. Dr Stewart is author or co-author of over 60 papers in the field of fiber and integrated optics and their application in communication and sensor systems.

Dr. John W Berthold III has been employed at the Babcock & Wilcox Research and Development Division in Alliance, Ohio, since 1979 and previously performed research for the U.S. Department of Defense and Bell Laboratories. He is presently responsible for oversight, guidance, and project management of company R&D efforts in measurement technology. He has experience in the design, installation, and application of standard commercial sensors and instrumentation, including strain gauges, accelerometers, pressure transducers, and flow meters. He leads selected projects and directs R&D contract work for development of advanced fiber optic sensors for measurement of pressure, acceleration, temperature, strain flow rate, liquid level, gas concentration, and process chemistry. He has given invited lectures, made numerous presentations, and published many refereed journal articles. He holds 25 U.S. patents in various areas of applied optics and measurement instrumentation. Dr. Berthold received the B.A. degree in physics from Gettysburg College and the M.S. and Ph.D. degrees from the University of Arizona Optical Sciences Center. He is a fellow of SPIE, a senior member of ISA, a member of OSA, and an IEEE affiliate.

Dr. Craig Michie is senior research fellow within the Electronic and Electrical Engineering Faculty of the University of Strathclyde. He obtained a B.Sc. in electronics and electrical engineering in 1983 and a Ph.D. in coherent optical communications in 1988, both from the University of Glasgow. In between earning the two degrees, he worked as an engineer with the BBC at Broadcasting House in London.

Since joining the University of Strathclyde in 1988 Dr. Michie has worked in a diverse range of optical fiber sensing techniques and has published over 50 papers in this area. His current interests lie in distributed measurements of physical and chemical parameters with a particular bias to structural monitoring and smart structures applications. Dr. Michie currently manages the research activity within the Smart Structures Research Institute of the university.

William B. Spillman, Jr., was born in January 1946. He undertook a basic technical training as a physicist and has developed a background in fiber optic sensors, fiber optic switching devices, integrated optics, ultrasonic sensing, and smart structures/

materials. Since 1991 has served as chief scientist at the BFGoodrich Aerospace Aircraft Integrated Systems Division. There his responsibilities include the development of new devices and systems suitable for incorporation into future products, planning of business unit R&D activities, and presentation of R&D capabilities to present and potential business unit customers as a support to marketing. His present specific research activities include fiber optic sensor systems, very long gauge length sensors with antenna gain, ultrasonic sensors, noncontact sensor powering and interrogation, fly-by-light control systems, automated evolutionary system design, artificial neural networks, genetic algorithms, complex adaptive systems, and smart structures and materials. He served as the executive chairman of the 1996 SPIE International Symposium on Smart Structures and Materials, which included nine separate technical conferences, and is executive chairman of the 1997 symposium. Currently he holds 31 patents with a number of patent applications in process and is the author or coauthor of more than 100 technical publications.

Index

The Artech House Optoelectronics Library

Brian Culshaw, Alan Rogers, and Henry Taylor, *Series Editors*

Acousto-Optic Signal Processing: Fundamentals and Applications, Pankaj Das

Amorphous and Microcrystalline Semiconductor Devices, Volume II: Materials and Device Physics, Jerzy Kanicki, editor

Bistabilities and Nonlinearities in Laser Diodes, Hitoshi Kawaguchi

Chemical and Biochemical Sensing With Optical Fibers and Waveguides, Gilbert Boisdé and Alan Harmer

Coherent and Nonlinear Lightwave Communications, Milorad Cvijetic

Coherent Lightwave Communication Systems, Shiro Ryu

Elliptical Fiber Waveguides, R. B. Dyott

Field Theory of Acousto-Optic Signal Processing Devices, Craig Scott

Frequency Stabilization of Semiconductor Laser Diodes, Tetsuhiko Ikegami, Shoichi Sudo, Yoshihisa Sakai

Fundamentals of Multiaccess Optical Fiber Networks, Denis J. G. Mestdagh

Germanate Glasses: Structure, Spectroscopy, and Properties, Alfred Margaryan and Michael A. Piliavin

High-Power Optically Activated Solid-State Switches, Arye Rosen and Fred Zutavern, editors

Highly Coherent Semiconductor Lasers, Motoichi Ohtsu

Iddq Testing for CMOS VLSI, Rochit Rajsuman

Integrated Optics: Design and Modeling, Reinhard März

Introduction to Lightwave Communication Systems, Rajappa Papannareddy

Introduction to Glass Integrated Optics, S. Iraj Najafi

Introduction to Radiometry and Photometry, William Ross McCluney

Introduction to Semiconductor Integrated Optics, Hans P. Zappe

Laser Communications in Space, Stephen G. Lambert and William L. Casey

Optical Fiber Amplifiers: Design and System Applications, Anders Bjarklev

Optical Fiber Communication Systems, Leonid Kazovsky, Sergio Benedetto, Alan Willner.

Optical Fiber Sensors, Volume Two: Systems and Applicatons, John Dakin and Brian Culshaw, editors

Optical Fiber Sensors, Volume Three: Components and Subsystems, John Dakin and Brian Culshaw, editors

Optical Fiber Sensors, Volume Four: Applications, Analysis, and Future Trends, John Dakin and Brian Culshaw, editors

Optical Interconnection: Foundations and Applications, Christopher Tocci and H. John Caulfield

Optical Network Theory, Yitzhak Weissman

Optoelectronic Techniques for Microwave and Millimeter-Wave Engineering, William M. Robertson

Reliability and Degradation of LEDs and Semiconductor Lasers, Mitsuo Fukuda

Reliability and Degradation of III-V Optical Devices, Osamu Ueda

Semiconductor Raman Laser, Ken Suto and Jun-ichi Nishizawa

Semiconductors for Solar Cells, Hans Joachim Möller

Smart Structures and Materials, Brian Culshaw

Ultrafast Diode Lasers: Fundamentals and Applications, Peter Vasil'ev

For further information on these and other Artech House titles, contact:

Artech House
685 Canton Street
Norwood, MA 02062
617-769-9750
Fax: 617-769-6334
Telex: 951-659
email: artech@artech-house.com

Artech House
Portland House, Stag Place
London SW1E 5XA England
+44 (0) 171-973-8077
Fax: +44 (0) 171-630-0166
Telex: 951-659
email: artech-uk@artech-house.com

WWW: http://www.artech-house.com